BIOCONVERSION OF WASTE MATERIALS TO INDUSTRIAL PRODUCTS

Other titles in the Elsevier Applied Biotechnology Series:

M. Y. Chisti. *Airlift Bioreactors*
T. J. R. Harris (ed.). *Protein Production by Biotechnology*
E. J. Vandamme (ed.). *Biotechnology of Vitamins, Pigments and Growth Factors*
W. M. Fogarty & C. T. Kelly (eds). *Microbial Enzymes and Biotechnology. 2nd Edition*

BIOCONVERSION OF WASTE MATERIALS TO INDUSTRIAL PRODUCTS

Edited by

A. M. MARTIN

Department of Biochemistry, Memorial University of Newfoundland, St John's, Newfoundland, Canada A1B 3X9

ELSEVIER APPLIED SCIENCE

LONDON AND NEW YORK

ELSEVIER SCIENCE PUBLISHERS LTD
Crown House, Linton Road, Barking, Essex IG11 8JU, England

Sole distributor in the USA and Canada
ELSEVIER SCIENCE PUBLISHING CO., INC.
655 Avenue of the Americas, New York, NY 10010, USA

WITH 145 TABLES AND 134 ILLUSTRATIONS

British Library Cataloguing in Publication Data

Bioconversion of waste materials to
industrial products.
I. Martin, A. M.
628.4458

ISBN 1851665714

Library of Congress Cataloging-in-Publication Data

Bioconversion of waste materials to industrial products/edited by A. M. Martin.
p. cm.—(Elsevier applied biotechnology series)
Includes bibliographical references and index.
ISBN 1-85166-571-4
1. Refuse and refuse disposal—Biodegradation. 2. Materials—Recycling. I. Martin, A. M. (Antonio M.) II. Series.
TD796.5.B55 1991
628.4′458—dc20

90-20797
CIP

Printed by The Universities Press (Belfast) Ltd.

PREFACE

When focusing on the latest stages of human development, many factors have been long identified as representative of both mankind's successes and failures. Industrialization, rising standards of living and the exploitation of new materials and energy sources, while characteristics of progress, are also a source of new problems such as overpopulation, increased urbanization, the energy crisis and pollution, to mention only a few. More recently, the problem of wastes from processing operations and their disposal has gained full-fledged public recognition.

In the past, problems associated with wastes were not given special treatment by society, and generally they were recognized as specific problems of the institutions which generated them: cities, industries, and agriculture. Indeed, before the advent of the modern chemical and processing industry of the present century, most of the wastes were recycled, and before the population explosion of the last decades, it appeared that there was enough space on earth to simply dump wastes and allow nature to dispose of those which were biodegradable.

This situation is no longer sustainable. There is increasing recognition of the pollution-associated problems created by wastes, the scarcity of places to dispose of them, and the need to save the valuable resources which are part of the refuse of our present 'throwaway' society. The recycling of resources is becoming a valid and viable economic activity and is increasingly mentioned as a solution to some of the most pressing problems which will affect mankind's future economic performance.

Bioconversion of wastes has been the natural way to return to the environment the resources previously extracted from it. It is expected that the development of biotechnology will facilitate the acceleration of this natural recycling process, which is being made necessary by our present and future levels of population densities and their increasing demands.

This book is designed as a study of the biotechnological methods for the recovery of wastes, emphasizing the recycling objective of the bioconversion, i.e. the production of industrial products from wastes. In conducting this study, it is this book's objective that its various chapters deal with the scientific and technological bases of the bioconversion processes involved, the problems and advantages associated with each, the products arising from the operations, and future trends and possibilities. If relevant to their content, individual chapters also deal with processing methods required to concentrate and purify the complex mixture of waste materials, applied before the biological step ('upstream' operations), and present an overview of the economic basis for the bioconversion process discussed.

This book is not intended to be a specialized study of the biodegradation processes involved, which will be presented in a second volume currently being prepared. Although the bioconversion processes have been traditionally applied to products of biological origin, such as agricultural, fisheries, forestry and food processing wastes, the present volume also deals with areas where novel bioconversion processes are also applicable, such as some mineral- and hydrocarbon-based industrial operations.

The contents of the book have been organized into two sections. Chapters in Section 1 are oriented more toward the principles or fundamentals of the bioconversion operations, while Chapters in Section 2 consider the characteristics of the main groups of waste materials and the specific technologies for bioprocessing and recycling them.

This book cannot cover all of the areas of a presently increasing, expanding field of research. However, it is expected that this book will provide useful and updated information to the academic, industrial and scientific communities, including ecologists and environmentalists. By including technological examples which will allow and encourage the use of bioconversion methods for the solution of waste disposal problems, this book could act as a guide to administrators, consultants, and governments.

ANTONIO M. MARTIN

CONTENTS

LIST OF CONTRIBUTORS

A. Alpuche-Solís
Lab. de Biotechnologia de Alimentos, Unidad Irapuato, Centro de Investigación y de Estudios Avanzados del IPN, Apdo Postal 629, 36500 Irapuato, Gto., Mexico

V. Bisaria
Biochemical Engineering Research Centre, Indian Institute of Technology, Hauz Khas, New Delhi-110 016, India

D. S. Chahal
Institut Armand-Frappier, 531, boulevard des Prairies, C.P. 100, Succ. L-Q-R, Ville de Laval, Québec, Canada H7V 1B7

D. Couillard
Institut National de la Recherche Scientifique, Université du Québec, INRS-Eau, 2700 rue Einstein, C.P. 7500, Sainte-Foy, Québec, Canada G1V 4C7

Severian Dumitriu
Polytechnic Institute of Jassy, Department of Organic and Macromolecular Chemistry and Technology, 23 August St. 11, R-6600 Jassy, Romania

W. R. Finnerty
Finnerty Enterprises Inc, 160 Chinquapin Place, Athens, Georgia, 30605, USA.

Peter Gacesa
Department of Biochemistry, University of Wales College of Cardiff, PO Box 903, Cardiff CF1 1ST, UK

David S. Holmes
Department of Biology, Clarkson University, Potsdam, New York 13699, USA

John Hubble
School of Chemical Engineering, University of Bath, Claverton Down, Bath BA2 7AY, UK

D. Kluepfel
Institut Armand-Frappier, 531, boulevard des Prairies, C.P. 100, Succ. L-Q-R, Ville de Laval, Québec, Canada H7N 4Z3

N. Kosaric
Department of Chemical and Biochemical Engineering, Faculty of Engineering Science, The University of Western Ontario, London, Ontario, Canada N6A 5B9

G. Th. Kroyer
Institute of Food Chemistry and Technology, Technical University Vienna, Getreidemarkt 9, A-1060 Vienna, Austria

Byong H. Lee
Agriculture Canada, Food Research Development Centre, 3600 Casavant Blvd West, St Hyacinthe, Québec, Canada J2S 8E3 and Department of Food Science, Macdonald College, McGill University, Ste-Anne-de-Bellevue, Québec, Canada H9X 1C0

Ewa Lie
Applied Microbiology, Chemical Center, University of Lund, PO Box 124, S-221 00 Lund, Sweden

A. M. Martin
Department of Biochemistry, Memorial University of Newfoundland, St John's, Newfoundland, Canada A1B 3X9

S. P. Mathur
Land Resource Research Centre, Central Experimental Farm, Agriculture Canada, Ottawa, Ontario, Canada K1A 0C6
Present address: Environmental Issues Working Group, Office of the Associate Deputy Minister, Agriculture Canada, Ottawa

Göran Molin
Applied Microbiology, Chemical Center, University of Lund, PO Box 124, S-221 00 Lund, Sweden

Shiro Nagai
Department of Fermentation Technology, Faculty of Engineering, Hiroshima University, Kagami-yama, Higashi-Hiroshima 724, Japan

NAPAVARN NOPARATNARAPORN
Department of Microbiology, Faculty of Science, Kasetsart Univeristy, Bangkok 1900, Thailand

O. PAREDES-LÓPEZ
Lab. de Biotechnología de Alimentos, Unidad Irapuato, Centro de Investigación y de Estudios Avanzados del IPN, Apdo Postal 529, 36500 Irapuato, Gto., Mexico.

T. R. PATEL
Department of Biochemistry, Memorial University of Newfoundland, St John's, Newfoundland, Canada A1B 3X9

KEN SASAKI
Hiroshima-Denki Institute of Technology, Nakano, Akiku, Hiroshima 739–03, *Japan*

TATSUROKURO TOCHIKURA
Department of Food Science and Technology, Faculty of Agriculture, Kyoto University, Sakyo-ku, Kyoto 606, Japan

R. D. TYAGI
Institut National de la Recherche Scientifique, Université de Québec, INRS-Eau, 2700 rue Einstein, C.P. 7500, Sainte-Foy, Québec, Canada G1V 4C7

R. VELAYUDHAN
Department of Chemical and Biochemical Engineering, Faculty of Engineering Science, The University of Western Ontario, London, Ontario, Canada N6A 5B9

FUNDAMENTALS OF BIOCONVERSION OPERATIONS

Chapter 1

BIORECOVERY PROCESSES: FUNDAMENTAL AND ECONOMIC CONSIDERATIONS

N. Kosaric & R. Velayudhan

Department of Chemical and Biochemical Engineering, Faculty of Engineering Science, The University of Western Ontario, London, Ontario, Canada N6A 5B9

CONTENTS

1 INTRODUCTION

With respect to humans, the degradation and reconstitution of wastes by microorganisms could be either harmful (biodeterioration) or beneficial (biodisintegration). This can be summarized in a cycle as shown in Fig. 1 (Barton, 1979).

The bioconversion of wastes to useful industrial products (biosynthesis) as per the above scheme has tremendous potential in that it can help meet the increasing world demand for food and energy. There is no doubt that organic waste materials like carbohydrates and cellulose present in agricultural, industrial and municipal wastes, are very good microbiological nutrients and this fact should be judiciously exploited.

Since the world's recoverable reserves of minerals and fuels are limited, as shown in Fig. 2, it is a good idea to look for alternative sources of energy from wastes (Barton, 1979).

A look at a simplified model of the terrestial ecosystem (Fig. 3) shows why recycling of wastes is so essential to maintain this delicate balance (Priestley, 1976).

2 SOURCES OF WASTE MATERIALS, QUANTITIES AVAILABLE AND COMPOSITION

2.1 Agricultural Wastes

2.1.1 Animal Wastes

Table 1 (Runov, 1977) presents typical input–output figures for cattle, swine, poultry, sheep and horse feedlots.

Waste production may be assumed to be directly proportional to the total live weight (TLW) of animals. The daily excretion of wastes as percentages of the TLW ranges from 3·6% for sheep to 9·4% for dairy cows (Taiganides, 1977).

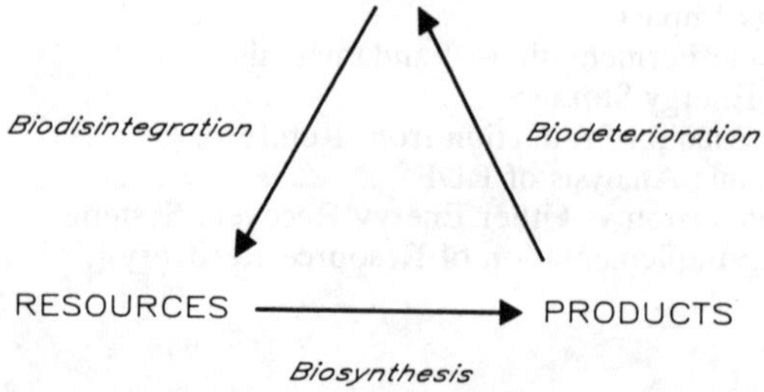

Fig. 1. Microbiological recycling (from Barton, 1979).

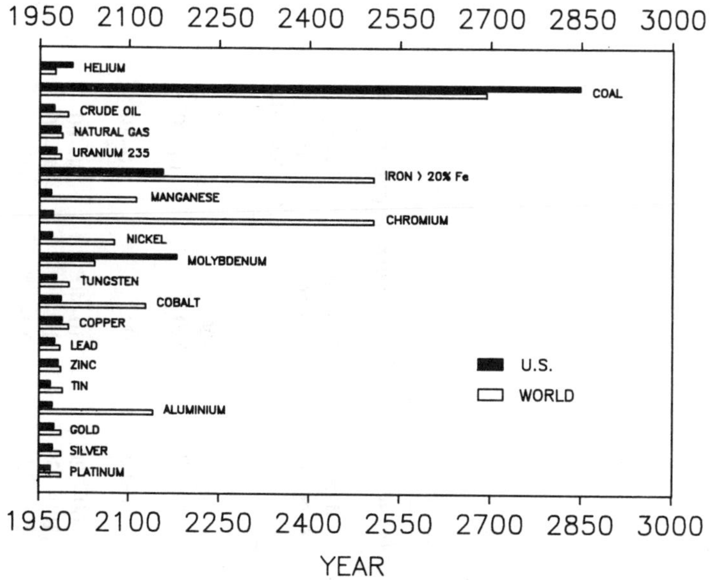

Fig. 2. Estimated lifetimes of known recoverable reserves of mineral commodities at currently minable grades and at rates of consumption based on a stable world population of $3{\cdot}5 \times 10^9$ and current rates of use (from Barton, 1979).

Unlike municipal and other industrial wastes, feedlot wastes do not contain metals or biocidal chemicals in concentrations high enough to be toxic.

Production of wastewaters from seven modern large swine feedlots is given in Table 2 (Jelinek, 1977).

Table 3 shows the typical composition of cattle manure collected from the farmyard and containing straw, or collected as liquid manure (Skarda, 1977).

2.1.2 Farm and Forestry Wastes

Estimates of world waste products are given in Table 4 (Hungate, 1977).

Logging residues are a valuable source of energy and significant amounts are produced in the US as seen in Table 5 (Grantham & Howard, 1980).

World lignocellulosic residues for major agricultural harvests are presented in Table 6 (Chornet & Overend, 1986) while fruit processing and vegetable

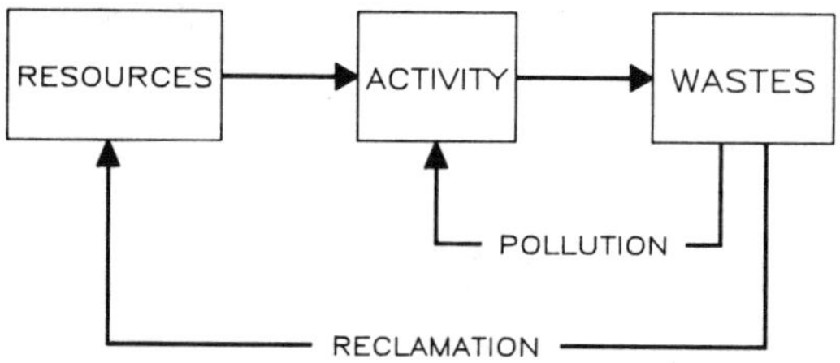

Fig. 3. Simple representation of terrestial ecosystem (from Priestley, 1976).

Table 1
Typical Input–output Data in Modern Animal Feedlots

Animal type	Feedlot type	Animal size range (kg/head)	Time in feedlot (days)	Materials input		Materials output	
				Feed (kg/head/day)	Water (litre/head/day)	Product (kg/head)	Waste (manure) (kg/head/day)
Cattle							
Beef	Open	250–500	100–180	8–16	40–120	500	2–20
	Housed (slotted floor)	250–500	100–180	8–16	40–120	500	10–30
	Housed (solid floor)	250–500	100–180	8–16	40–120	500	10–30
Dairy	Stalls	500–650	—	15–25	60–320	10–40[a]	40–60
	Free stalls	500–650	—	15–25	100–130	10–40[a]	40–60
	Cow yard	500–650	—	15–25	120–320	10–40[a]	40–60
Swine							
Pork pigs	Open (dirt)	20–100	150–180	1–3	4–20	100	1–3
	Housed (slotted floor)	20–100	150–180	1–3	4–20	100	1–5

	Open (solid floor)	20–100	150–180	1–3	4–20	100	1–5
Poultry							
Broilers	Housed (litter)	0–2	40–60	0·05–0·1	0·1–0·2	1·5–2	0·05–0·06
Layers	Housed (cages)	1·5–2	400	0·1–0·12	0·15–0·2	0·6–0·9[b]	0·1–0·2
	Housed (litter)	1·5–2	400	0·1–0·12	0·15–0·2	0·6–0·9[b]	0·1–0·2
Turkeys	Open	2–14	120–170	0·2–0·3	0·3–0·5	8–14	0·3–0·6
Ducks	Open, wet	0·5–4	40–60	0·2–0·3	40–130	3–4	—
Sheep							
Lambs	Housed	30–60	40–150	2–3	4–7	40–60	1·5–3
Sheep	Housed	50–100	40–150	2–4	7–13	Wool and mutton	2–4
	Open	50–100	40–150	2–4	7–13	Wool and mutton	2–4
Horses	Stable	300–600	—	9–14	30–40	Recreation, work	20–60

[a] Milk in kg/cow/day.
[b] Eggs/hen/day.
Source: Runov, 1977.

Table 2
Wastewater Quantities from Larger Confinement Swine Feedlots

Feedlot no.	*Pig–Days*	*Wastewater (kg/100 kg live weight)*	*Dilution water*[a]
1	381 294	16·3	61
2	1 406 422	10·2	31
3	657 730	12·3	49
4	331 405	10·8	37
5	505 817	11·8	38
6	599 188	15·2	60
7	1 289 180	10·9	39

[a] Added water is calculated as the difference between the measured wastewater quantity and the quantity of excrements.
Source: Jelinek, 1977.

processing wastes generated in the US are shown in Table 7 (Kosaric *et al.*, 1984).

The projected biomass energy in the US is shown in Table 8 (Kulp, 1985).

2.2 Municipal Wastes

Table 9 shows the estimated and projected compositions and the annual tonnage of municipal refuse in the US (Pfeffer, 1988), while Table 10 shows

Table 3
Composition of Cattle Farmyard and Liquid Manure and Daily Production per Animal Equivalent (AE)[a]

	Faeces	*Urine*	*Added*		*Total (% of total)*
			Straw	*Water*	
	(kg/AE/day)				
Farmyard manure	26	4	3	5	38
	(68)	(11)	(8)	(13)	(100)
Urine	—	11	—	5	16
	—	(69)	—	(31)	(100)
Total	26	15	3	10	54
	(48)	(28)	(6)	(18)	(100)
Liquid manure	26	15	—	14	55
	(48)	(27)	—	(25)	(100)

[a] 1 AE is an animal of 450 kg live weight.
Source: Skarda, 1977.

Table 4
Accumulation of World Wastes

	Wet ($\times 10^{15}$ *kcal/year*)	*Dry* ($\times 10^{15}$ *kcal/year*)
Urban:		
Municipal	0·66	
Sewage		0·25
Agriculture:		
Cereal straws	0·54	
Other residues	0·9	
Manure		0·86
Forest:		
Logging residues	0·2	
Manufacturing residues	0·06	
Total	2·36	1·11

Source: Hungate, 1977.

typical compositions of Municipal Solid Waste (MSW) in Europe (White & Plaskett, 1981).

Incinerator residues represent a sizable input of municipal waste. In the Federal Republic of Germany (Reimann, 1989), the MSW typically contains significant amounts of chlorine (8000 g/Mg), zinc (5000 g/Mg) and lead (1200 g/Mg).

Household wastes are composed of many heterogeneous objects and materials. These have been divided into nine categories: fine (≤20 mm);

Table 5
Quantity of Logging Residue (10·2 cm and Larger in Diameter) in the United States

	Volume (*thousand* m^3)	*Weight* (*oven dry tons*)
North:		
Northeast	14 967	8 400 000
North Central	6 913	3 900 000
South:		
Southeast	19 235	10 360 000
South Central	16 380	8 820 000
Intermountain	7 026	3 400 000
Pacific Coast	29 399	14 430 000
Total USA	93 920	49 310 000

Source: Grantham & Howard, 1980.

Table 6
Lignocellulosic World Residues from the Main Harvests, According to the FAO Report, Vol. 36 (in 1000t, 1983)

Product	*Africa*	*Asia (without USSR)*	*Latin America*	*North America*	*Europe*	*USSR*	*Oceania*
Wheat	12 047	229 085	27 195	124 970	167 970	98 000	24 538
Rice	12 015	606 476	23 586	11 929	2 743	10	921
Barley	4 398	19 182	1 482	28 651	95 342	5 460	7 271
Corn	22 201	100 157	51 742	218 437	60 264	13 000	549
Rye	15	2 720	210	13	23 835	14 700	13
Oats	354	1 562	1 452	13 460	20 864	2 100	2 135
Millet	12 470	27 392	190	—	42	3 080	42
Sorghum	12 603	30 502	22 793	41 083	770	280	2 639
Peanuts	3 832	13 840	846	2 196	24	2	60
Tubers	54 989	146 142	25 775	14 017	64 353	51 180	1 672
Grains with a high oil content (i.e. lin)	8 910	80 196	15 968	30 690	22 046	26 802	1 180
Vegetables	8 314	50 700	41 100	78 075	7 040	10 867	938
Cane sugar	9 592	48 835	64 746	25 520	49	—	4 226
Cocoa bean	624	10 104	787	498	—	—	902

Source: Chornet & Overend, 1986.

Table 7
Representative Fruit and Vegetable Processing Wastes in the United States

Product and process employed	*Major season*	*Residual wet weight (tons/year)*	*Residual saccharide content (tons dry weight)*	*Percent disposed (not used as by-product)*
Apple	Aug.–Jan.			
Peel and core (69%)		284 400	38 376	31
Crush (31%)		42 600	5 924	
Citrus crush (100%)	All year	3 555 640	944 378	3
Grape crush (100%)	Aug.–Nov.	438 000	54 750	20
Peach peel and pit (100%)	July–Oct.	192 500	86 625	68
Tomato	July–Nov.	70 200	1 543	78
Peel (10%)				
Crush (90%)				

Source: Kosaric *et al.*, 1984.

Table 8
Biomass Energy in the United States

Source	*Biomass energy (quads (10^9 GJ/year)*	
	1980	*2000*
Wood	2·2	4
Agricultural wastes	0·1	0·5
Municipal solid waste	0·2	1
US Total energy use	75	95

Source: Kulp, 1985.

Table 9
Composition of Muncipal Refuse in the US

Constituent	*As received composition (% wet wt)*		
	1970	*1984*	*2000*
Paper and paper products	33·1	37·1	41·0
Garbage (food wastes)	11·5	8·1	6·8
Yard and garden waste	19·0	17·9	15·3
Metals	12·2	9·6	9·0
Glass	12·5	9·7	7·6
Plastic	2·7	7·2	9·8
Miscellaneous organic solids	8·3	8·4	8·4
Miscellaneous inorganic solids	1·7	1·9	2·1
Total (million tons)	110·2	133·0	158·8

Source: Pfeffer, 1988.

Table 10
Typical Composition of Municipal Solid Waste in Europe

	Percent by weight
Combustibles:	
Paper	35–60
Garden wastes	2–35
Food	2–8
Cloth	1–3
Plastics	1–2
Non-combustibles:	
Metal	6–9
Glass	5–13
Dirt	1–5
Moisture	20–40

Source: White & Plaskett, 1981.

Table 11
Biomass Generated in the US

Source	*Amount generated (10^6 tons/year)*
Municipal solid waste	140
Agricultural residue	300
Forestry residue (total):	168
Forest residues	145
Wood processing residues	20
Pulp and paper mill waste	3

Source: McGinnis *et al.*, 1984.

organic matter (OM), bone, wood; paper and cardboard; textiles, leather and rubber; plastics; glass; metals; and special wastes (Rosseaux *et al.*, 1989).

There are three major sources of residual biomass in the US: agricultural waste, MSW and forestry residues. Table 11 gives their relative amounts (McGinnis *et al.*, 1984).

Table 12
Total Quantity of Industrial Waste Ordered by Manufacturing Industries that Produce Hazardous Residual Waste

Industries	*Percent of total quantity*[a]
Chemicals	37·6
Primary metals	29·1
Fabricated metals	7·7
Machinery	6·5
Paper	4·6
Transportation	4·0
Food	2·7
Petroleum	2·4
Stone	2·0
Electrical	0·7
Rubber	0·7
Leather	0·5
Lumber	0·4
Instruments	0·3
Miscellaneous manufacturing	0·2
Furniture	0·2
Textiles	0·2
Printing	0·1
Tobacco	<0·1
Apparel	<0·1

[a] The total quantity comprises 100% of 27·8 million tons of wastes reported from 21 US States.
Source: Noll *et al.*, 1985.

2.3 Industrial Wastes

Table 12 gives the rank order of US industries generating residues. Industrial residuals are those considered 'hazardous' under the Resource Recovery Act and Toxic Substances Control Act which make up rougly half the total industrial discharge (Noll *et al.*, 1985).

3 BIOLOGICAL WASTE RECYCLING

3.1 Biohydrometallurgy

For many years, bacterial leaching processes have been commercially used for the recovery of copper and uranium from rejects, waste and low-grade materials. This method is governed by the action of chemolithoautotrophic bacteria such as *Thiobacillus ferrooxidans* and *T. Thiooxidans* which use organic substrate as a source of energy (Bosecker, 1988). Sulphur, phosphorus and metals such as calcium, iron and copper can also be leached out of waste materials (Barton, 1979).

3.1.1 Economics

Biohydrometallurgy is considered to be of prime importance for the future because of favourable economics like low capital and energy costs, absence of environmental pollution, and the ease of operation (Bosecker, 1988).

3.2 Aerobic Degradation of Wastes (Composting)

3.2.1 Process

Composting is a biological process in which organic biodegradable wastes are stabilized and converted by the action of bacteria, actinomycetes and fungi into a product (humus or compost) to be used as a soil conditioner and organic fertilizer (Barton, 1979; Domino, 1979; Golueke, 1989; Willson, 1989). Waste disposal is combined with reclamation in a 'natural' manner.

When an organic waste pile (windrow) is well aerated and has a moisture content of about 50–60%, the existing mixed microbial populations metabolize rapidly, resulting in the formation of compost as per the scheme in Fig. 4 (Barton, 1979).

Several industrial and agricultural wastes are good compost materials, as seen in Table 13 (Toth, 1973).

3.2.2 Propularity of the Process

Composting is gaining wide acceptance as a good biological method of recycling wastes to fertilizers. Recently, Oregon's Metropolitan Service Distr-ict unanimously approved a $155 million, 20-year contract with Riedel Environmental Technologies to build a MSW composting facility in Portland

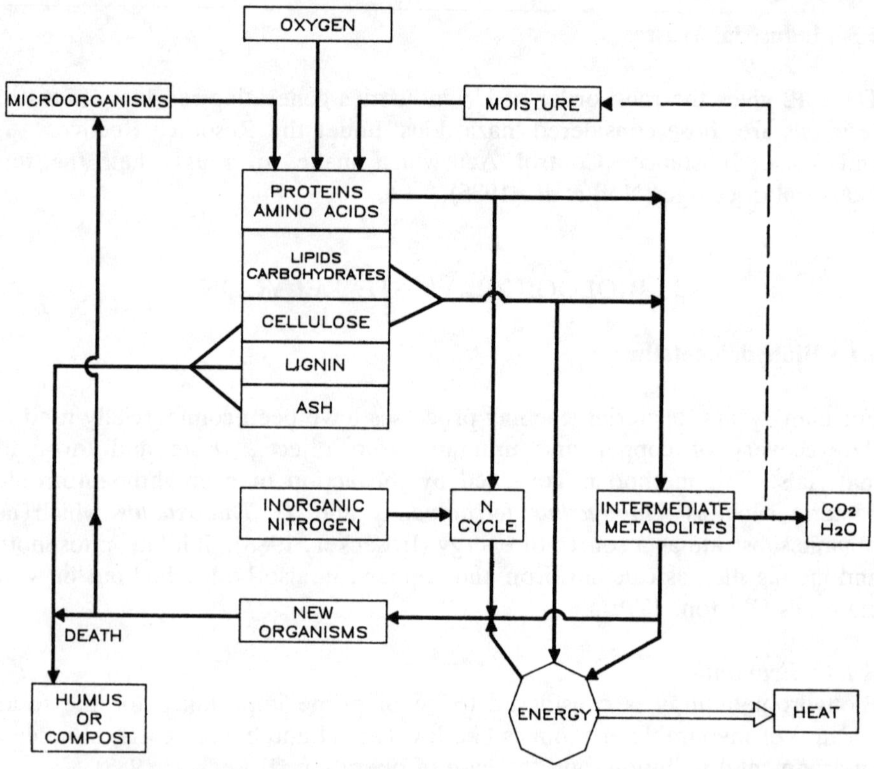

Fig. 4. Overall flow sheet for the composting process (from Barton, 1979).

(Anon, 1989*a*). Also, Maine's potato industry produces a yearly average of 100 000 tons of cull potatoes. These wastes combined with other potato processing wastes, sawdust, small amounts of wood ash, and paper mill sludge have been turned into compost (Anon., 1989*b*). The Region of Durham (near Toronto, Ontario) recently made corn starch-based biodegradable plastic bags for collection of yard waste as part of a testing program of a composting project (Glenn, 1989).

The inherent advantage of composting is the wide range of wastes that can be used which include vegetable trimmings discarded by grocery stores, dry bones, straw, sawdust and waste paper. In the US, several states are actively involved in composting, e.g. Wisconsin, Colorado, Delaware, New York, Minnesota, Florida, Massachusetts, Maine and California (Goldstein, 1989).

3.2.3 Marketing

Marketing plays the key role in the ultimate effectiveness of the process. Securing a market for the compost product accomplishes two primary objectives (Eggerth *et al.*, 1989):

(a) to provide a use for the end product and thus lessen the quantity of waste requiring landfilling;
(b) to provide revenue from the sale of the compost.

Table 13
Composition of Various Industrial and Agricultural Organic Wastes

	Moisture (%)	*Ash (%)*	*N (%)*	P_2O_5 *(%)*	K_2O *(%)*	*N Availability*[a]
Industrial						
Antibiotic wastes:						
Penicillin	75·3	29·5	3·85	4·13	1·08	0·4
Streptomycin	62·6	71·9	2·20	0·52	0·06	0·7
Botanical drug wastes:						
Cascara bark	75·8	9·7	0·54	0·03	0·01	0·0
Licorice roots	—	—	2·20	0·11	1·12	0·1
Pyrethrum flowers	25·0	12·1	1·07	0·58	2·58	0·0
Scammony roots	36·4	13·9	1·12	0·47	2·22	0·0
Cannery wastes:						
Asparagus	93·8	12·5	3·96	0·91	3·54	0·6
Beet (red)	94·2	7·0	2·57	0·52	1·92	0·5
Spinach	85·7	58·4	3·21	1·14	1·07	0·9
Sweet potatoes	90·0	10·2	1·84	0·46	0·72	0·2
Spent spice mare:						
Carolina chili	16·2	8·3	3·41	1·03	3·84	0·5
Ginger	11·8	5·5	1·74	0·35	1·80	0·1
Nutmeg	37·3	4·3	1·25	0·40	1·26	0·1
Patchouli leaves	11·0	19·5	3·34	0·60	3·96	0·7
Coffee wastes:						
Chaff	3·0	5·5	2·58	0·19	2·10	0·0
Grounds, fresh	62·9	0·5	1·84	0·03	0·12	0·0
Grounds, composted	58·0	—	1·65	0·22	3·00	0·2
Snuff wastes:						
Stem sand	7·5	38·8	2·81	0·77	4·54	0·2
Leaf sand	3·5	74·4	0·94	0·21	1·62	0·0
Sweepings	16·9	25·1	2·91	0·49	6·90	0·4
Cocoa shells	—	—	2·71	1·17	3·06	0·2
Seaweed	—	—	3·17	—	—	0·8
Spent hops	79·1	4·4	2·13	0·66	0·42	0·3
Tea leaves	86·0	3·2	4·41	0·29	0·24	0·2
Agricultural						
Apple pomace	—	—	1·70	—	—	0·0
Cow manure	—	—	2·50	0·50	1·40	0·8
Duck manure	—	—	3·55	—	—	0·9
Horse manure	—	—	2·85	—	—	0·5
Poultry manure and bedding	22·6	25·6	3·58	3·02	1·62	0·8
Sawdust:						
Oak	45·6	2·1	0·12	0·002	0·12	0·0
Poplar	43·3	2·5	0·13	0·001	0·15	0·0
Soybean mash	84·0	15·1	4·81	1·49	0·78	0·8
Tobacco stems	20·0	21·5	1·96	0·63	10·80	0·4
Garbage	49·3	28·5	1·07	1·16	0·12	0·0
Garbage compost	10·0	55·0	1·00	2·7	0·69	0·0
Sewage sludge	38·0	15·0	2·00	1·7	0·10	0·0

[a] Based on $(NH_4)_2SO_4$ nitrification value of 1·0.
Source: Toth, 1973.

The primary markets for yard waste compost depend on (Cox, 1989):

(a) whether the composting operation is publicly or privately run:
(b) the existence of a profit-making incentive;
(c) the quality of the compost;
(d) the quantity of the available compost;
(e) the local industries.

The means of expanding the market for compost are several:

(a) focus on increasing the use of compost in general (home delivery accounts for 60–70% of all compost sold);
(b) preferential procurement of recycled products by government agencies;
(c) increasing the marketability of a facility's specific compost;
(d) market analyses and education.

Yard waste composting is currently being called 'the most logical first step in managing solid wastes' (Cox, 1989).

3.2.4 Economics

The pros and cons of composting have been evaluated (Kashmanian & Taylor, 1989) as given below:

Costs involved	*Benefits received*
(a) Collection (b) Transportation (c) Processing (d) Compost storage and marketing (e) Program administration (f) Public education (g) Technical assistance	(a) Revenues received from selling the finished compost (b) Avoided costs (rather than earned revenues) from using the finished compost instead of purchasing a substitute product (c) Avoided tip fees from not landfilling (or incinerating) the yard wastes (d) Reduction of potential environmental problems by not landfilling (or incinerating) yard wastes

A study funded by the US Environmental Protection Agency characterized eight yard waste composting programs in the US as shown in Table 14 (Kashmanian & Taylor, 1989).

Composting tends to 'even out' the peaks in MSW generation and dampen the impact on the household garbage collection cost. Table 15 examines the costs and revenues of yard waste composting for the above study. The total costs include yard waste collection, transport and processing (including reject disposal) costs. The net total cost is the difference between compost revenue (or avoided cost) and the total cost. The full cost is the net total cost less the landfill tip fee (the alternative yard waste management method).

As is evident from Table 15, as landfill tip fees continue to soar, reflecting

Table 14

1. Background Information on City/County Selected

City/county	*Total population*	*Total households*	*Total yard waste stream (tons/year)*	*Year of Data*
Davis, CA	44 000	10 000	5 475[a]	1987
East Tawas, MI	2 600	1 350	350	1987
Montgomery Co.,[b] MD	633 000	244 000	110 000	1987
Omaha, NE	350 000	100 000	48 000	1988
Seattle, WA	500 000	229 000	92 000[a]	1988
Wellesley, MA	27 000	8 500	8 000	1987/1988
Westfield, NJ	30 000	10 400	n/a	1987/1988
Woodbury, MN	13 520	4 790	1 092[c]	1987

[a] Estimate of total yard waste stream does not include amount generated and collected by lawn service companies and public work crews.
[b] Population and household estimates based on 1986 data.
[c] Yard waste estimate does not include brush.
n/a Not available.

2. Composting Program Effectiveness

City/county	*Total pop. served by composting*	*Total households served by composting*	*Total yard waste stream composted (tons/year)*	*% of total yard waste stream composted*
Davis	44 000	10 000	500	9
East Tawas	2 600	1 350	138[a]	39[a]
Montgomery Co.	282 000	75 000	15 600	14
Omaha	3 735	830	500	1
Seattle	500 000	229 000	3 600	4
Wellesley	27 000	8 500	6 500	81
Westfield	30 000	10 400	4 687[a]	n/a
Woodbury	2 329	825	~116	11[a]

[a] Does not include amount of brush chipped or shredded.
n/a Not available.
Source: Kashmanian & Taylor, 1989.

their own full costs, yard waste composting will continue to become a more preferred solid waste management alternative for communities on economic grounds alone.

Putrescible material unsuitable for Refuse Derived Fuel (RDF) production can be diverted to compost production which in turn benefits the economics of the RDF process. This ecological concept is being promoted as an aid to the RDF process (Porteous, 1981). Almost any organic waste material can be made into compost, if a market can be found that warrants the expense (Willson, 1989).

Table 15
Costs and Revenues of Yard Waste Composting[a]

City/county	*Collection & trans. cost for yard wastes ($/ton rec'd)*	*Process cost for yard wastes[b] ($/ton rec'd) (+)*	*Total compost. cost ($/ton rec'd) (=)*	*Revenues for compost. cost[c] ($/ton rec'd) (−)*	*Net total compost. cost[c] ($/ton rec'd) (=)*	*Local landfill tip fee ($/ton rec'd) (−)*	*Full compost. cost[d] ($/ton rec'd) (=)*
Davis[e]	n/a	n/a	n/a	$0.00	n/a	$8.00	n/a
East Tawas[f]	$10.00	<$10.00	<$20.00	$0.00	<$20.00	$5.25	<$14.75
Montgomery Co.[g]	$83.33	$18.60	$101.93	$4.31	$97.62	$46.00	$51.62
Omaha	$40.16	$3.60	$43.76	$6.30[h]	$37.46	$6.40	$31.06
Seattle	$12.00[i]	$22.50	$34.50	$10.00	$34.50	$31.50	$3.00
Wellesley[j]	$0.00	$16.67	$16.67[j]	$4.05[k]	$12.62[j]	$52.00	−$39.38[j]
Westfield[lm]	$70·69	$31.58	$102.27	$6.54	$102.27	$137.00[n]	−$34.73
Westfield[lo]	—[p]	$33.33	$33.33[p]	AC[q]	$33.33[p]	$137.00[n]	−$103.67[p]
Woodbury	$43.00	$15.00	$58.00	AC[r]	$58.00	$30.00	$28.00

[a] Costs for Davis, East Tawas, Montgomery County, and Wellesley programs are based on 1987 $; cost for Omaha, Seattle, and Westfield programs are based on 1988 $; costs for Woodbury program are based on 1987–1988 $.

[b] Processing costs for Seattle, Westfield and Woodbury programs represent payments made to privately operated facilities (Montgomery County's facility is public/private); cost for disposal of reject material is not included for private facilities; Montgomery County's and Westfield's facilities use windrow turning machines, the other facilities use front end loaders.

[c] Net total cost of composting represents summation of costs for collection, transport, and processing, including cost for disposal of reject material (i.e., the total cost of composting), minus revenues for compost; revenue for private facilities is not included in calculating net total cost.

[d] Full cost of composting represents net total cost minus landfill tip fee (other avoided costs, e.g. avoided garbage collection cost, are not included); negative sign implies a benefit.

[e] Cost charged to Davis residents for collection, transport and composting of yard wastes is $3.46/month/household.

[f] Cost for equipment shared with Department of Public Works not included in composting costs; no information available on avoided cost from using finished compost.

[g] Processing cost does not include cost for land nor amortized capital cost.

[h] Derived from an avoided cost of $8–$10/ton of landfill cover and soil amendment not purchased.

[i] Collection cost not included in 1988 estimate, $56/ton in 1989.

[j] Yard wastes are dropped off at two composting facilities; therefore, collection and transport costs are $0/ton; processing cost does not include cost of land; $52/ton tip fee includes transport cost to landfill.

[k] Derived from selling price of $0.00–$0.50/ton sold and avoided cost of $15/yd^3 of loam not purchased.

[l] Westfield pays $10/yd^3 (3·3 yd^3/ton) to Alternate Disposal Systems Inc. (ADSI) to receive and shred its brush and other woody materials (not included in the cost estimates in the table; 1423 tons, 1988); ADSI sells fine mulch ($12/yd^3), coarse mulch ($10/yd^3), topsoil ($10/yd^3) and boiler fuel ($3–$4/yd^3). ($/outgoing yd^3 sold).

[m] Middlebush Compost Inc. composts leaves from Westfield (3287 tons, 1988); conversion factor used is 475 lb/yd^3 for leaves received.

[n] $137.00/ton is tip fee at the transfer station.

[o] Woodhue Ltd composts grass clippings from Westfield (1400 tons, 1988); conversion factor used is 600 lb/yd^3.

[p] Collection cost of grass clippings (and brush) for Westfield is $0/ton with resident drop-off; cost for transport to Woodhue (and ADSI) was not estimated.

[q] Avoided cost of $35–$65/acre for on-site use on farm as fertilizer supplement.

[r] Avoided cost of use of nursery's land by exchanging compost (no estimate available).

AC Avoided Cost.
n/a Not available.
Source: Kashmanian & Taylor, 1989.

3.2.5 State of the Art

In Japan, considerable amounts of organic wastes (from the food industry and sludge from water treatment facilities) are produced each year. Conversion of food wastes to livestock feed has become uneconomical due to the availability of cheap imported feed. Hence composting is the only economical alternative. The composting methods used and the process economics are outlined in Tables 16–19 (Kubota & Nakasaki, 1989).

In Japan, the amount of chemical fertilizers applied per unit of farm land area is almost four times that in the US. As per estimates, 24% of the N_2, 63% of the P_2O_5 and 25% of the K_2O in fertilizers could be replaced by applying compost.

3.3 Food from Waste

The world food problem, especially in the Third World countries, is clearly evident. The steep rise in grain prices has driven the prices of other foodstuffs beyond the reach of the common man. The production of food from waste certainly looks like an attractive solution to these problems (Kapsiotis, 1976). Recycling of waste nutrients from agricultural wastes can be achieved both naturally and artificially by microorganisms (Seal & Eggins, 1976).

Table 16

Capital Costs and Operating Costs in compost Plants

Raw material	*Garbage*	*Sewage sludge*	*Manure*	*Bark*	*Others*
Capital cost, 10^3 \$/wet ton raw material/day:					
Range	24·6–2 990	13–12 200	12–1 030	—	—
Average	390	1 520	140	46	11
Operating cost, \$/wet ton compost:					
Range	61–530	62–1 300	3·1–115	56–74	—
Average	320	280	110	65	—
Average breakdown of operating cost, %:[a]					
Raw material	0	5	0	27	—
Additives	13	26	28	15	—
Power and fuel	24	16	23	8	
Labour	32	29	24	31	—
Maintenance & others	31	24	25	19	—
Plant capacity, wet ton raw material/day:					
Range	2·0–53·7	0·9–29·8	7·8–32	20–40	6–25
Average	14·9	12·9	21·3	30	17

[a] Values are taken only in plants where additives are used.

Source: Kubota & Nakasaki, 1989.

Table 17
Average Nutrient Contents and Sale Prices of Compost Products

Raw material	*Garbage*	*Sewage sludge*	*Manure*	*Bark*
Nutrient content, % dry wt:				
N	2·1	2·3	1·7	0·8
P_2O_5	2·0	4·0	3·2	1·7
K_2O	1·4	0·4	1·9	0·5
Price of product, $/wet ton:				
Sale price in bulk:				
Range	8–85	12–85	12–138	—
Average	35	35	52	77
Sale price of bagged products:				
Range	103–192	58–135	15–256	—
Average	149	96	146	128
Price converted from nutrient value[a]	60	65	77	22

[a] For the conversion, the following are used: $1.38/kg N, $1.84/kg P_2O_5, $0.975/kg K_2O.
Source: Kubota & Nakasaki, 1989.

3.3.1 Single Cell Protein (SCP) Production

A general scheme for SCP production from wastes is shown in Fig. 5 (Tannenbaum & Pace, 1976).

Cheese whey is a good example of a food industry waste (by-product) material that can be converted to various higher value products. Whey, a by-product of cheese manufacture, has high quality protein (0·6%) and a readily fermentable sugar, lactose (5%). Annual production of whey is estimated at 72 million tons which contains about 200 000 tons of protein and 1·2 million tons of lactose (Kosaric & Asher, 1982). On fermentation, whey

Table 18
Estimated Amounts of Organic Wastes and Their Composted Products in Japan

Waste type	*Generated (10^3 dry ton/year)*	*Land-reclaimed (10^3 dry ton/year (%))*	*Composted (10^3 dry ton/year (%))*	*Number of compost facilities*
Garbage	2 560	—	12(0·5)	22
Sewage sludge	520	114(22)	47(9)	31
Sludge from night soil treatment	260	66(25)	unknown	9
Manure	13 960	10 890(78)	unknown (<5)	2 000
Bark	782	—	104(13)	70

Note: There is a considerable amount of industrial organic wastes, which is partly composted, but the detail is unknown.
Source: Kubota & Nakasaki, 1989.

Table 19
Types of High-Rate Fermenters Used for Garbage and Dewatered Sludge Cake Composting

Raw material	*Nonreactor type*		*Reactor or enclosed type*		
	Windrow	*Aerated pile*	*Rotary kiln*	*Vertical*	*Horizontal trough*
Garbage or separated refuse			2(3)[a]	5	13
Dewatered sludge cake[b]	4	6		17	10

[a] In three plants, rotary kilns are used with multi-stage vertical type fermenters.
[b] Sludge cake from both municipal sewage treatment and night soild treatment plants.
Source: Kubota & Nakasaki, 1989.

yields several useful products which have important industrial applications as shown in Fig. 6 (Kosaric & Asher, 1982).

Agricultural residues represent an inexpensive source of substrates for SCP production. These wastes contain lactose (easily metabolized by *Kluyveromyces fragilis* and *K. lactis*), cellulose (readily metabolized by

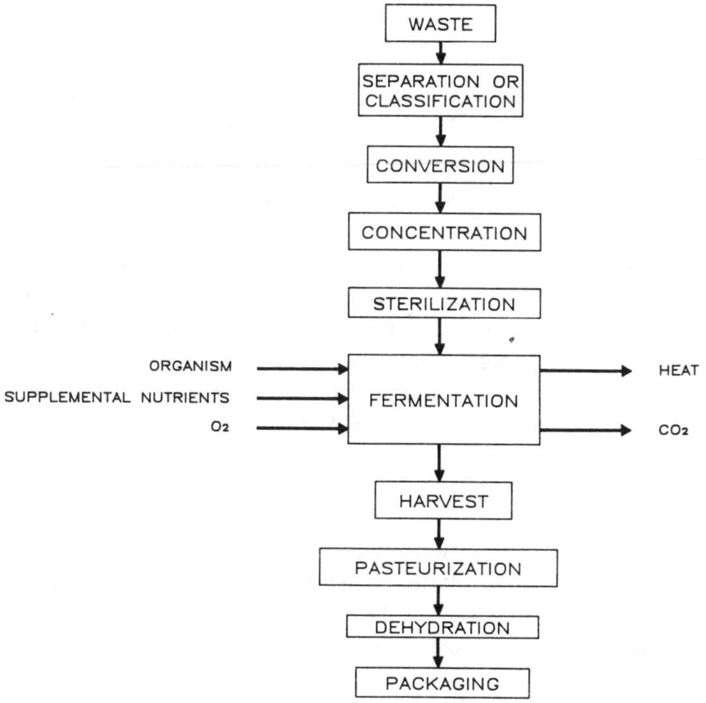

Fig. 5. General scheme for SCP production from waste (from Tannenbaum & Pace, 1976).

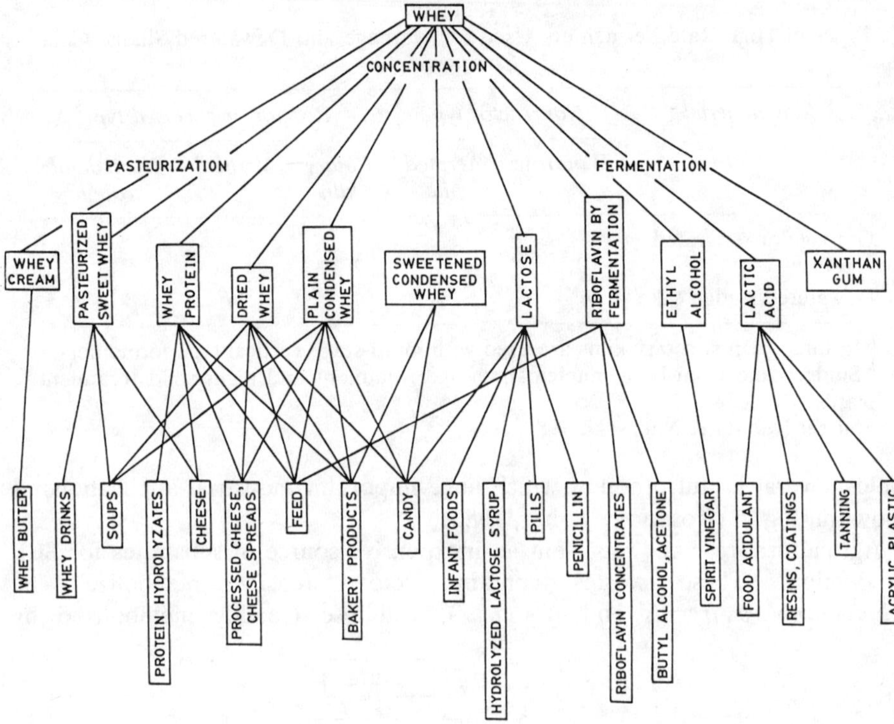

Fig. 6. Some proposed processes and products associated with the utilization of whey (from Kosaric & Asher, 1982).

Candida utilis and *Saccharomyces cerevisiae*), hermicellulose and starch (Batt & Sinskey, 1987).

Agricultural and food processing wastes are usually too fibrous to be digested by animals and the process to recover organic matter is usually not economically feasible. Several microorganisms (mainly fungi) utilize this organic matter as an energy source to synthesize SCP (Worgan, 1976). Looking at the annual production of some of these wastes (Table 20), one can see why it is imperative that we recycle them to food as shown in Table 21 (Worgan, 1976).

ECONOMICS

Operating costs in the fermentation include costs for nutrients, oxygen transfer and heat removal. The costs involved include cell recycle from the effluents and also the pretreatment of wastes. The market value of the SCP will depend on whether (Tannenbaum & Pace, 1976):

(a) whole whey is fermented;
(b) whey protein is recovered and lactose is discharged;
(c) a combined process is used.

Table 20
Annual World Production of Carbohydrates in Wastes from Crop Plants

	Carbohydrates from agricultural waste (× 1000 ton)	*Carbohydrates from processing waste (× 1000 ton)*
Wheat straw	286 600	
Wheat bran		57 300
Maize stover	120 000	
Maize cobs		30 100
Sugar cane bagasse		83 000
Molasses		9 300

Source: Worgan, 1976.

There are four difficult but critical issues that complicate economic analyses designed to investigate the viability of SCP manufacture (Kosaric & Bell, 1986):

(a) how to determine the product price;
(b) how to handle the capital cost of building the plant;
(c) how to assess the impact of inflation;
(d) how to assess the impact of the tax system.

The revenue from sale is the primary source of income but the evaluation of a reasonable market price is very difficult. The capital cost of equipment for SCP from solid agricultural wastes is low in a small-scale operation but usually the operating costs per ton of product are high assuming a 5-year amortization period (Imrie & Righelato, 1976). The process could be more economical on the 100 tons/year scale only in countries with cheap labour, since it is clear that labour forms the major portion of the operating costs. As the scale of operation is increased, then the unit price of SCP decreases.

Table 21
Yields of Mycelial Protein (*Fusarium semitectum*) from Various Lignocellulosic Wastes

Waste	*g/100 g Waste*	*g/100 g C source*
Oat husk	11·0	16·0
Sugar cane bagasse	5·6	9·9
Cocoa shell	5·9	9·1
Spent hops	5·4	9·5
Glucose as C source		10·0

Source: Worgan, 1976.

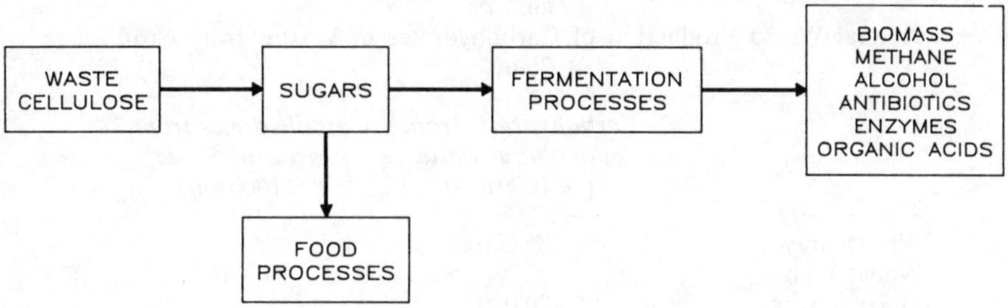

Fig. 7. Potential products from waste cellulose (from Brown & Fitzpatrick, 1976).

3.3.2 Food from Paper Wastes

Cellulose is the main ingredient of most paper and pulp industry wastes and this can be recycled biologically (Fig. 7) to a variety of useful industrial products (Brown & Fitzpatrick, 1976).

3.3.3 Socio-economic Implications of Producing Food from Wastes

In the coming years, the mass production of food from waste will become inevitable (Plaskett, 1976). There are several questions that provide the clues to this:

(a) Will future agricultural output go up exponentially as predicted?
(b) Will synthetic food production lead to a rapid depletion of resources?
(c) How will the public respond to the ever-growing pollution problem?
(d) How long can fossil carbon satisfy all our food needs?
(e) Is it wise to waste the large reserves of unused organic matter in wastes?

The foods produced from wastes can be similar to the foods being consumed today (Plaskett, 1976).

3.4 Energy Generation from Refuse

The concept of producing synfuels and energy products from biomass and wastes is very simple, as shown in Fig. 8. Combining waste disposal and energy recovery processes offers recycling opportunities as well as improved disposal technology, often at lower cost (Klass, 1981). Also, a synfuels industry based on biomass and wastes is independent of foreign price controls and regulations, and technological breakthroughs are not required to develop commercial systems and processes.

3.4.1 Energy Impact

Recent US projections of the potential of energy from biomass and wastes span the entire range of small to large impact. One of the forecasts made by the Office of Technology Assessment (OTA) indicates that in the year 2000, as few as 4–6 quads (1 quad = $1{\cdot}05 \times 10^{18}$ J) or as many as 12–17 quads could be derived from biomass and selected wastes (Jerger *et al.*, 1981; Klass, 1981).

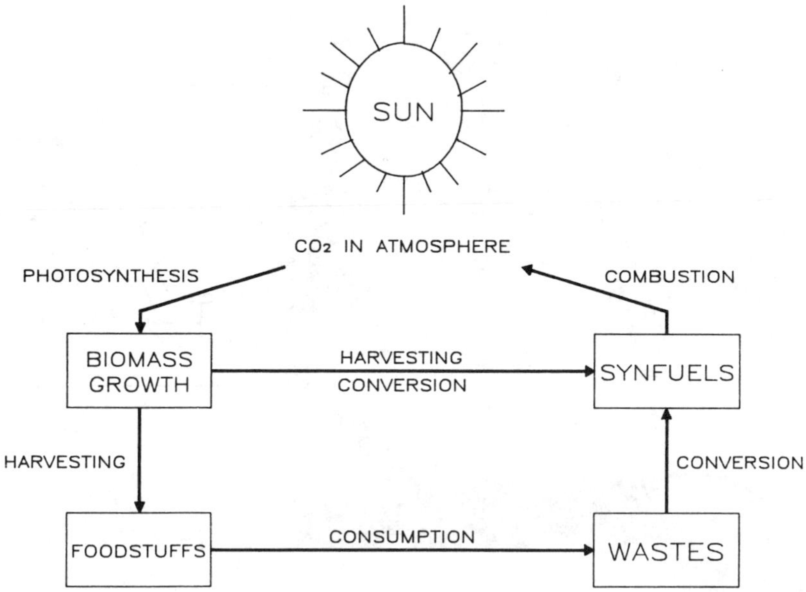

Fig. 8. Schematic representation of synfuel production from biomass and wastes (from Klass, 1981).

The US Department of Energy (DOE) funding for bioenergy R & D for the fiscal year 1984 (Fig. 9) clearly shows that renewable energy is of the utmost importance (Gutstein & Richards, 1985).

At the 1986 production rates, natural gas reserves should last only about 59 years and oil reserves only about 33 years (Anon., 1989*c*). One can see how RDF can assist in this situation, e.g. the energy content of the MSW produced

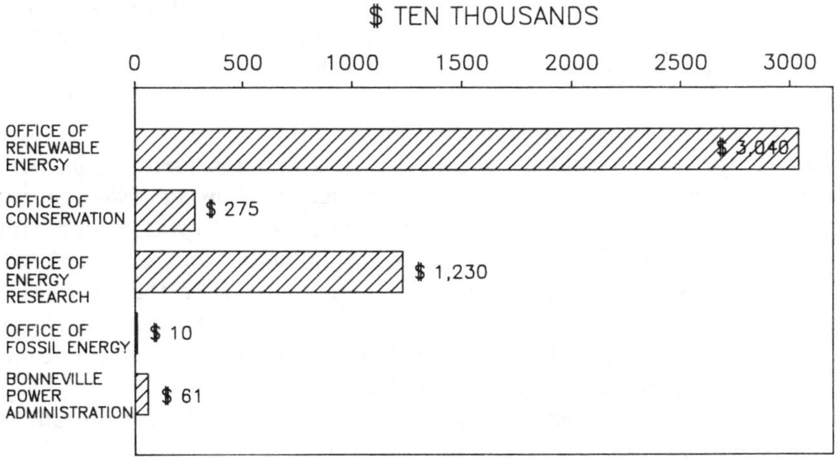

Fig. 9. Fiscal year 1984 funding for US DOE bioenergy R & D (from Gutstein & Richards, 1985).

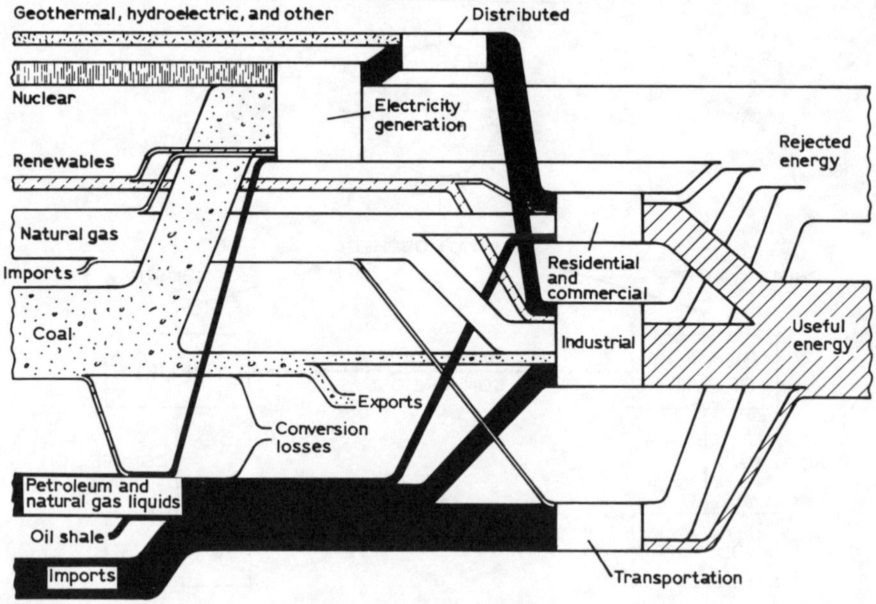

Fig. 10. Projections for energy flow in the US for the year 2000 (from Calvin, 1985).

annually in the US is equivalent to that produced from 50 million tons of coal (Anon., 1989*c*). Considering the fact that 160 million tons of solid waste are produced in the US every year, it is easy to realize how this can be economically converted biologically to fuel. The projection of energy use in the US for the year 2000 (Fig. 10) for the first time includes biomass as a significant component of the total (6% of the total energy requirement) in the form of renewables (Calvin, 1985).

By the year 2000, it is projected that 30·3% of the 182·2 million tons of MSW generated will be recycled. The energy content of all this waste is about $1{\cdot}58 \times 10^{18}$ J/year. The organic matter (paper, food waste and yard debris) is amenable to biological conversion to methane (Pfeffer, 1988).

3.4.2 Methane Fermentation—Fundamentals

Organic compounds present in industrial wastes undergo methane fermentation as shown schematically in Fig. 11 (Albagnac & Roustan, 1988). Fortunately natural methanogenic mixed cultures can be grown on wastewaters either as aggregates with good settling properties or as biofilm on carrier materials. Up until now, methane fermentation was mainly applied to the treatment of wastes containing complex organic compounds and in most cases nutritional requirements of the bacteria are satisfied. However, more recent applications deal with wastewaters from chemical industries.

In the 1960s, manure from chickens was slurried, digested anaerobically and introduced into an algae pond. The algae were then harvested and used as part of the diet for the chickens (Golueke, 1989).

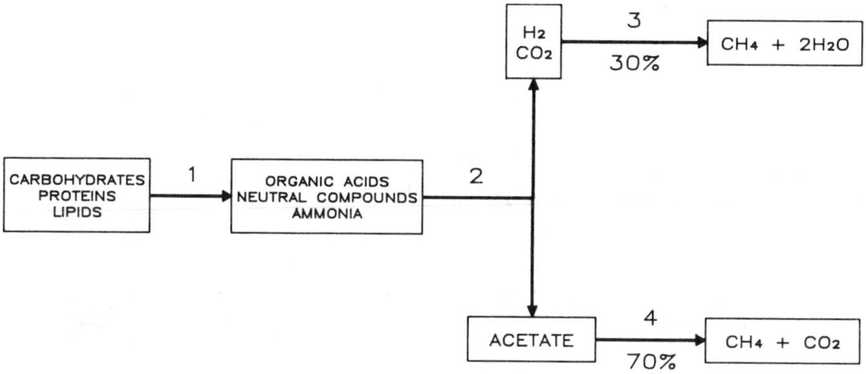

1) HYDROLYTIC AND FERMENTATIVE BACTERIA
2) OBLIGATE HYDROGEN PRODUCING ACETOGENIC BACTERIA
3) HYDROGENOPHILIC METHANOGENIC BACTERIA
4) ACETOCLASTIC METHANOGENIC BACTERIA

Fig. 11. Methane fermentation of organic compounds (from Albagnac & Roustan, 1988).

For several reasons biogasification is more beneficial as it applies to agricultural wastes. It is economically feasible for the farmer, especially with the introduction of low-cost reactors. Also biogasification serves a dual purpose (Golueke, 1981):

(a) production of combustible gas;
(b) simultaneous stabilization of the waste.

SYSTEM CONFIGURATION AND PERFORMANCE

For the Refuse Conversion to Methane (RefCoM) Process (Fig. 12), there are four basic subsystems, i.e. refuse processing and separation, fermentation, residue dewatering and disposal, and gas purification and utilization (Porteous, 1981; Pfeffer, 1988).

ECONOMIC CONSIDERATIONS

The costs associated with the RefCoM Process are presented in Table 22 (Pfeffer, 1988). Careful technical and economic analysis of this system clearly demonstrates the technical feasibility and cost effectiveness of the production of a fuel gas from municipal refuse.

The annual cost of processing and the annual savings estimate are shown in Tables 23 and 24 (Porteous, 1981).

The waste treatment and energy recovery system for the Bacardi Corporation's rum distillery in San Juan, Puerto Rico uses anaerobic digestion to treat distillery wastes and to produce methane gas for use as boiler fuel. The system currently supplies the equivalent of $8{\cdot}44 \times 10^{11}$ J daily, providing an annual estimated fuel saving of $1 million. Operating costs for the above process are very low (Anon., 1984).

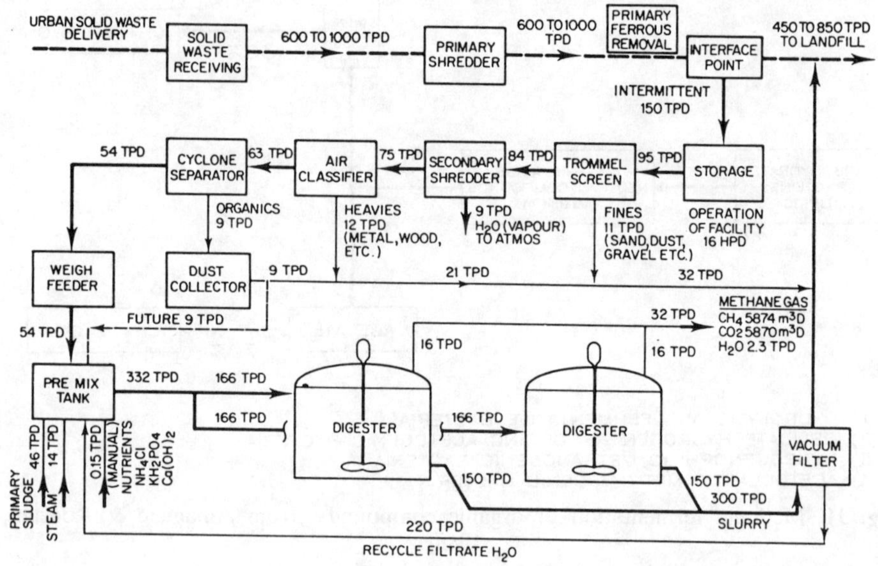

Fig. 12. Refuse Conversion to Methane (RefCoM) flowsheet (from Porteous, 1981).

Table 22
RefCoM Cost Summary

	Construction ($)	*Maintenance ($/year)*	*Power (kWh/year)*
Refuse processing	1 453 500	141 775	611 500
Fermentation	4 842 400	12 414	5 542 400
Dewatering	855 100	8 252	56 214
Incineration	8 904 200	124 780	830 800
Gas processing	2 108 100	29 515	6 090 400
Buildings	1 126 000	—	—
Miscellaneous	1 046 000	6 235	517 000
Total	20 335 300	322 971	13 648 300
First year personnel costs ($/year)	1 082 000		

Source: Pfeffer, 1988.

Table 23
Annual Cost Estimate of Processing Putrescible Residues by Anaerobic Digestion, from a 67 000 t/year Domestic Refuse Plant

Capital charges on £1.25m at 13·8% p.a.	£172 000
Operating cost	£110 000
Total annual cost	£282 000

Source: Porteous, 1981.

Table 24
Annual Savings Estimate for the Plant Costed in Table 23

Savings in cost of electricity used in sorting plant	£60 000
Savings in heating costs in sorting plant	£5 000
Savings in refuse disposal costs, say 10 000 t at £2.50/t transport and tipping costs	£25 000
Total	£90 000

Source: Porteous, 1981.

3.4.3 Solar Energy Storage

Bacteria and fungi consume fuels and are thus called 'reducers', whereas algae and plants are called 'producers' since they convert light energy to chemical energy. Figure 13 shows such a 'producer' in action. Here chemical energy is produced by algae, with chicken manure or sewage as the raw material, using the sludge–algae–methane cycle and closed nutritional and waste streams (Barton, 1979). The aerobic bacteria present in the sewage break down the complex substrates to forms that can be nutritionally beneficial to algae which are then harvested and anaerobically digested.

Another application is in the area of hydroponics (the technique of growing plants directly in nutrient solutions rather than in soils). Organic wastes and effluents from the anaerobic digestion could be the basis for these nutrients. Such a scheme is shown in Fig. 14 (Barton, 1979). This waste-grown algae is a cheap and abundant source of SCP used as animal feed.

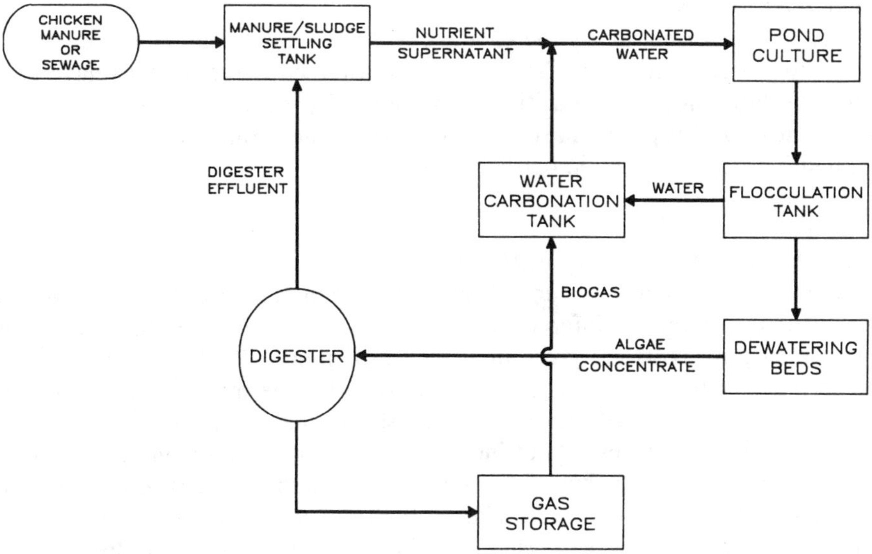

Fig. 13. Flow diagram of sludge–algae–methane system (from Barton, 1979).

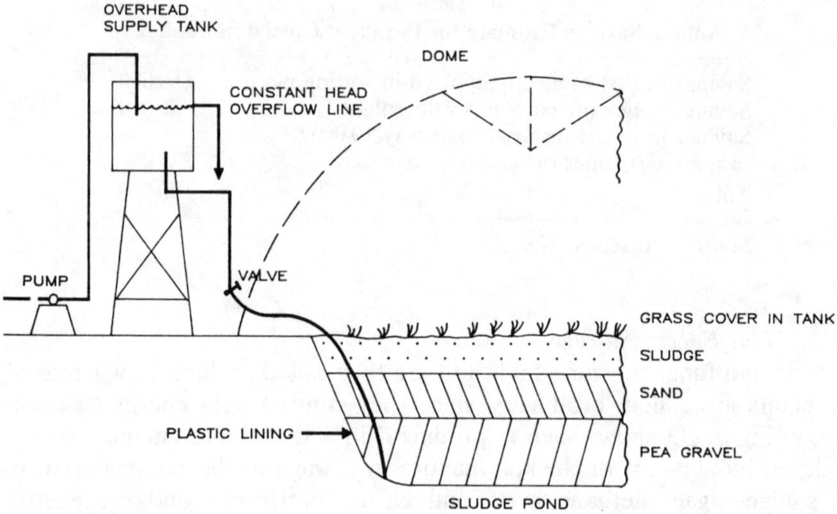

Fig. 14. Hydroponic sludge culture of pasture grasses (from Barton, 1979).

3.4.4 Ethyl Alcohol Production from Refuse

Cellulosic components from refuse (paper and cardboard) can be enzymatically hydrolyzed to yield fermentable sugars which are then fermented to alcohol (Porteous, 1981). Organic acids and SCP can also be recovered in this process.

PROCESS ECONOMICS

This is decided primarily by the capital cost of the plant, the energy input and the selling price of ethanol. Capital and operating cost comparisons have been made for different plant capacities as in Table 25 (Porteous, 1981).

If plant refuse disposal charges can be levied, the economics (Table 26) can be slightly improved (Domino, 1979).

3.4.5 Economic Analysis of RDF

Different operational and environmental parameters make it difficult to effectively compare the different systems recovering energy by bioconversion. The comparative costs are given in Table 27 (Domino, 1979).

The cost of the base unit system is considered to be the gross cuts involved from the point of entry of fuel into the digester to the production of total gas at the back-end. The figures do not include revenues from methane, or credit for fertilizer, or feedstock value of the residues of methane digestion. Operating costs and capital costs together make up the total costs (Domino, 1979).

For any waste-to-energy system to be beneficial, one should evaluate elements of cost *vis-à-vis* those of revenue (Table 28) (Alter & Dunn, 1980).

Table 25
Capital and Operating Costs as Related to Plant Size for the Acid Hydrolysis of Newsprint and Subsequent Production of Fuel Alcohol

Plant capacity (dry t feed/year)	50 000	100 000	250 000	495 000	1 250 000
Ethanol produced (t/year)	8 937	17 873	44 987	89 367	223 418
Capital Costs × £10^3					
Equipment	1 169	1 856	3 434	5 427	9 996
Total fixed investment (TFI)	5 472	8 686	16 071	25 397	46 781
Start-up (8·5% of TFI)	465	738	1 366	2 159	3 976
Working capital	631	799	1 234	1 856	3 517
Operating costs × £10^3/*year*					
Feedstock	0	0	0	0	0
Chemicals, utilities, effluent	235	470	1 183	2 350	5 875
Labour	590	590	590	590	590
Other direct costs	846	1 030	1 455	1 991	3 221
Overheads	623	816	1 259	1 819	3 102
Subtotal(1)	2 294	2 906	4 487	6 750	12 788
10% Contingency	229	291	449	675	1 279
Subtotal(2)	2 523	3 197	4 936	7 425	14 067
Financial Charges:					
Depreciation (£)	547	869	1 607	2 540	4 678
Interest (£)	999	1 563	2 865	4 517	8 330
Total costs £10^3/year	4 069	5 629	9 408	14 482	27 075
Costs of ethanol (99·5%):					
£/tonne	455·3	314·9	209·1	162·1	121·2
£/MJ	0·015	0·011	0·007	0·005	0·004
£/gallon (95%)	1·47	1·01	0·67	0·52	0·39

Source: Porteous, 1981.

3.4.6 *Bioconversion vs Other Energy Recovery Systems*

The comparison above shows that anaerobic digestion is in a cost range that is not uncompetitive with the other technologies. Anaerobic digestion produces both energy in the form of methane and residues with a potential market as a fertilizer or feed, thus eliminating a charge for disposal of the residues of digestion. The major economic output for anaerobic digestion of manures is in the upgrading of the gas produced to pipeline-quality methane. As every effort is being made to optimize the process, it is not inconceivable that the costs will be reduced as a percentage of the total outlay of capital. The digestion of manures can be successfully completed at mesophilic temperatures (30–40°C), thus decreasing the need for external energy to maintain the reaction.

Anaerobic digestion of MSW is a little less attractive than that of manures. The front-end recovery system is expensive to install; however, it does provide additional credits to the system. The residues still require disposal, adding costs to the system. The going market price of methane decides whether digestion of MSW is going to be economically feasible (Domino, 1979).

The hydrolysis and fermentation of MSW or agricultural wastes remain as a

Table 26
Advantages and Disadvantages of Hydrolysis and Fermentation of Cellulosic Wastes

Process	*Advantages*	*Disadvantages*
Enzyme hydrolysis and fermentation	1. Up to 90% conversion of cellulose	1. Initial pretreatment of refuse still to be optimized.
	2. Glucose is easy to store	2. Large scale demonstration still required
	3. Careful operation not needed	3. Limit to possible markets for products(?)
	4. Not an energy intensive process	
	5. Ethanol has a possible use as fuel	
	6. Alternative route to chemicals now made from fossil fuels, therefore conserving resources	
	7. Complete utilization of wastes, lignin in sludge has possible use as fuel	
	8. Yields of scale-up predictably produced	
	9. Products can be cheaply produced	
Acid hydrolysis and fermentation	1. Glucose is easy to store	1. Corrosion resistant equipment needed
	2. Ethanol has possible use as a fuel	2. Careful regulation of temperature and pH required
	3. Alternative route to chemicals now made from fossil fuel	3. Only 50% conversion
	4. Costs of sugar and ethanol highly competitive	4. Treatment of sludge
		5. Energy requirements
		6. Large scale demonstration still to be proven
		7. Limit to possible markets for products(?)

Source: Domino, 1979.

possible alternative attractive means of energy production, now that ethanol is currently being tried out as a motor fuel (Henrich, 1981).

The economics of the bioconversion processes discussed here would not disqualify any of these processes from future consideration as a means of solid waste recycling (Domino, 1979).

Table 27
Comparative Costs of Alternative Systems to Reclaim Energy from Wastes, 1975 Dollars

System	*Gross operational cost per $1{\cdot}055 \times 10^9 J$ net output*	*Gross capital cost per $1{\cdot}055 \times 10^9 J$ net output*	*Gross total cost per $1{\cdot}055 \times 10^9 J$ net output*	*Capacity (J/day)*
1. Anaerobic digestion of manures:				
(a) Jacobs Engineering (200 tpd) 15 year deprec., mesophilic temperature			\$1.39 (low pressure, low purity) \$2.82 (low pressure, high purity) \$3.19 (high pressure, high purity)	$1{\cdot}7 \times 10^{12}$
(b) USDA (9 120 tpd) 10 year deprec., mesophilic temperature	\$1.61	\$0.53	\$2.14	$3{\cdot}599 \times 10^{11}$
2. Anaerobic digestion of MSW:				
(a) Dynatech (1 000 tpd), thermophilic	\$2.45	\$2.77	\$5.22	$3{\cdot}429 \times 10^{12}$
(b) Pfeffer–Biogas Case 1, thermophilic (1 000 tpd)	\$1.99	\$1.68	\$3.67	$3{\cdot}323 \times 10^{12}$
(c) Institute of Gas Tech. (571 tpd) 25 year deprec., mesophilic	\$0.599	\$0.18	\$0.78	$5{\cdot}486 \times 10^{12}$
3. Acid hydrolysis of MSW, fermentation of ethanol (400 tpd)			\$6.46	$1{\cdot}052 \times 10^{12}$
4. Combustion of MSW to produce steam (1000 tpd)	\$0.55	\$0.48	\$1.03	$1{\cdot}055 \times 10^{13}$
5. Combustion of MSW to produce electricity from steam (1 000 tpd)	\$1.53	\$1.65	\$3.18	$4{\cdot}22 \times 10^{12}$
6. Cofiring MSW with coal (980 tpd)	\$1.21	\$0.64	\$1.85	$4{\cdot}136 \times 10^{12}$
7. Pyrolysis:				
(a) To produce fuel oil and char (1 000 tpd)	\$0.79	\$0.42	\$1.21	$6{\cdot}858 \times 10^{12}$
(b) To produce methanol (300 tpd) 15 year deprec.	\$7.93	\$2.16	\$10.09	$5{\cdot}212 \times 10^{12}$

Source: Domino, 1979.

Table 28
Waste-to-Energy Systems: Costs and Revenues

Elements of cost	*Elements of revenue*
Amortized capital cost	Tip fees
Operating costs:	Sale of received materials
Salaries and wages	Sale of energy product
Fringe benefits	
Insurance	
Taxes	
Fuel and other utilities	
Transportation	
Maintenance and supplies	
Disposal of unrecovered fraction and residue	

Source: Alter & Dunn, 1980.

4 FACTORS GOVERNING IMPLEMENTATION OF RESOURCE RECOVERY

In order to successfully plan a recycling plant, be it biological or not, one must consider the following factors (Alter & Dunn, 1980):

(a) Quantity and composition of the waste materials
(b) Markets for recycled goods and specifications
(c) Determining the appropriate technology
(d) Determining the cost of operation
(e) Determining the cost of financing
(f) Determining the method of management
(g) Site location
(h) The procurement of wastes

Economic feasibility of waste recycling processes depends primarily on (Manchester & Vertrees, 1973):

(a) market potential for the products derived from the process;
(b) cost and economics of scale of the process;
(c) cost of waste management alternatives.

The rate of technical development of new uses for wastes is greater than the rate at which new uses are adopted. Also, there is often a gap between technical and economic feasibility, but technical feasibility alone will not place the product in the market. Social acceptance, among other interacting forces of the market, is also of prime importance.

5 CONCLUSIONS

The various aspects of waste recycling through biological means discussed in this chapter lead us to understand the tremendous global impact this can have, particularly in light of the fact that since the Industrial Revolution, the global economy has been characterized by ever-increasing demands for food and energy (Prentis, 1984). Wastes from whichever source, whether agricultural, industrial or municipal, are a complex mixture of nutrients that can be utilized by several microorganisms. This explains why biological methods of recycling can be extremely economical and beneficial in the long run. The main opposition seems to be the lack of awareness and possibly skepticism among the public as to its benefits, and also the competition from well-established nonbiological methods like waste incineration for heating purposes. The industries generating waste should not only attempt to reduce waste generation but also try to recycle these wastes. But for this to happen, they should be provided with the proper incentives so they can harness waste materials efficiently and optimally. So proper government guidance is imperative. The MISA program in Ontario, Canada is one such example. This ensures efficient scavenging of wastes in a private enterprise system. Such systems work well in developing countries where the dependence on biomass energy is anything from 50% to 95% (Hall *et al.*, 1982). The wealth of national economies may well come to depend largely on their ability to biologically recycle organic matter (Plaskett, 1976).

REFERENCES

Albagnac, G. & Roustan, J. L. (1988). In *Proc. 8th Int. Biotechnology Symp.*, Vol. II, eds G. Durand, L. Bobichon & J. Florent, Société Française de Microbiologie, Paris, p. 1169.

Alter, H. & Dunn, J. J. (1980). *Solid Waste Conversion to Energy: Current European and US Practice.* Marcel Dekker, New York.

Anon. (1984). *Civil Engineering/ASCE,* **54,** 69.

Anon. (1989*a*). *BioCycle,* **30**(10), 14.

Anon. (1989*b*). BioCycle, **30**(10), 20.

Anon. (1989*c*). *Environ. Sci. Tech.,* **23**(2), 150.

Barton, A. F. M. (1979). *Resource Recovery and Recycling.* John Wiley, New York.

Batt, C. A. & Sinskey, A. J. (1987). In *Food Biotechnology,* ed. D. Knorr. Marcel Dekker, New York, p. 347.

Bosecker, K. (1988). In *Proc. 8th Int. Biotechnology Symp.*, Vol. II, eds G. Durand, L. Bobichon & J. Florent. Société Française de Microbiologie, Paris, p. 1106.

Brown, D. E. & Fitzpatrick, S. W. (1976). In *Food from Waste,* eds G. G. Birch, K. J. Parker & J. T. Worgan. Applied Science, London, p. 139.

Calvin, M. (1985). In *Energy Applications of Biomass,* ed. M. Z. Lowenstein. Elsevier Applied Science, London, p. 3.

Chornet, E. & Overend, R. P. (1986). *Fractionation of Lignocellulosics.* Centre Québecois de Valorisation de la Biomasse, Sainte-Foy, Québec, Canada.

Cox, K. A. (1989). *BioCycle,* **30**(10), 64.

Domino, F. A. (ed.) (1979). *Energy from Solid Waste—Recent Developments.* Noyes Data Corporation, New Jersey.

Eggerth, L. L., Diaz, L. F. & Gurkewitz, S. (1989). *BioCycle,* **30**(5), 29.

Glenn, J. (1989). *BioCycle,* **30**(10), 28.

Goldstein, N. (1989). *BioCycle,* **30**(10), 50.

Golueke, C. G. (1981). In *Biogas and Alcohol Fuels Production,* Vol. II, Proceedings of the Second Seminar on Biomass Energy for City, Farm and Industry, eds Staff of *BioCycle.* J. G. Press, Emmaus, Pennsylvania, p. 9.

Golueke, C. G. (1989). *BioCycle*, **30**(10), 44.
Grantham, J. B. & Howard, J. O. (1980). In *Progress in Biomass Conversion*, Vol. 2, eds K. V. Sarkanen & D. A. Tillman. Academic Press, New York, p. 1.
Gutstein, M. & Richards, D. (1985). In *Energy Applications of Biomass*, ed. M. Z. Lowenstein. Elsevier Applied Science, London, p. 55.
Hall, D. O., Barnard, G. W. & Moss, P. A. (1982). *Biomass for Energy in the Developing Countries: Current Role, Potential, Problems, Prospects*. Pergamon Press, Oxford, UK.
Henrich, R. A. (1981). In *Biogas and Alcohol Fuels Production*, Vol. II, Proceedings of the Second Seminar on Biomass Energy for City, Farm and Industry, eds Staff of *BioCycle*. J. G. Press, Emmaus, Pennsylvania, p. 25.
Hungate, R. E. (1977). In *Microbial Energy Conversion*, eds H. G. Schlegel & J. Barnea. Pergamon Press, Oxford, p. 339.
Imrie, F. K. E. & Righelato, R. C. (1976). In *Food from Waste*, eds G. G. Birch, K. J. Parker & J. T. Worgan. Applied Science, London, p. 79.
Jelinek, T. (1977). In *Animal Wastes*, ed. E. P. Taiganides. Applied Science, London, p. 165.
Jerger, D. E., Srivastava, V. J., Daniels, E., Novil, M., Tarman, P. B. & Chynoweth, D. P. (1981). In *Biogas and Alcohol Fuels Production*, Vol. II, Proceedings of the Second Seminar on Biomass Energy for City, Farm and Industry, eds Staff of *Biocycle*. J. G. Press, Emmaus, Pennsylvania, p. 135.
Kapsiotis, G. D. (1976). In *Food from Waste*, eds. G. G. Birch, K. J. Parker & J. T. Worgan. Applied Science, London, p. 232.
Kashmanian, R. M. & Taylor, A. C. (1989). *BioCycle*, **30**(10), 60.
Klass, D. L. (1981). In *Fuels from Biomass and Wastes*, eds D. L. Klass & G. H. Emert. Ann Arbor Science, Michigan.
Kosaric, N. & Asher, Y. (1982). *Conservation and Recycling*, **5**, 23.
Kosaric, N. & Bell, P. C. (1986). In *Microbial Biomass Proteins*, eds M. Moo-Young & K. F. Gregory. Elsevier Applied Science, London, p. 61.
Kosaric, N., Bell, P. C., Cosentino, G., Magee, R., Turcotte, G. & Purcell, A. (1984). In *Technological Innovation, Studies Program Research Report*, Office of Industrial Innovation, Regional Industrial Expansion, Government of Canada.
Kubota, H. & Nakasaki, K. (1989). *BioCycle*, **30**(8), 67.
Kulp, J. L. (1985). In *Energy Applications of Biomass*, ed. M. Z. Lowenstein. Elsevier Applied Science, London, p. 43.
Manchester, A. C. & Vertrees, J. G. (1973). In *Symposium: Processing Agricultural and Municipal Wastes*, ed. G. E. Inglett. AVI Publishing Co., Connecticut, p. 6.
McGinnis, G. D., Wilson, W. W. & Biermann, C. J. (1984). In *Bioconversion Systems*, ed. D. L. Wise. CRC Boca Raton, Florida, USA, p. 89.
Noll, K. E., Haas, C. N., Schmidt, C. & Kodukula, P. (1985). *Recovery, Recycle and Reuse of Industrial Wastes* (*Industrial Waste Management Series*), ed. J. W. Patterson. Lewis Publishers, Michigan.
Pfeffer, J. T. (1988). In *Proc. 8th Int. Biotechnology Symp.*, Vol. II, eds G. Durand, L. Bobichon & J. Florent. Société Française de Microbiologie, Paris, p. 1198.
Plaskett, L. G. (1976). In *Food from Waste*, eds G. G. Birch, K. J. Parker & J. T. Worgan. Applied Science, London, p. 264.
Porteous, A. (1981). *Refuse Derived Fuels*. Applied Science, London.
Prentis, S. (1984). *Biotechnology: A New Industrial Revolution*. Orbis Publishing London.
Priestley, G. (1976). In *Food from Waste*, eds G. G. Birch, K. J. Parker & J. T. Worgan. Applied Science, London, p. 114.
Reimann, D. O. (1989). *Waste Mgt Res.*, **7**, 57.
Rosseaux, P., Navarro, A. & Vermande, P. (1989). *BioCycle*, **30**(9), 81.

Runov, B. A. (1977). In *Animal Wastes,* ed. E. P. Taiganides. Applied Science, London, p. 11.
Seal, K. J. & Eggins, H. O. W. (1976). In *Food from Waste,* eds G. G. Birch, K. J. Parker & J. T. Worgan. Applied Science, London, p. 58.
Skarda, M. (1977). In *Animal Wastes,* ed. E. P. Taiganides. Applied Science, London, p. 315.
Taiganides, E. P. (1977). In *Animal Wastes,* ed. E. P. Taiganides. Applied Science, London, p. 131.
Tannenbaum, S. R. & Pace, G. W. (1976). In *Food from Waste,* eds. G. G. Birch, K. J. Parker, & J. T. Worgan. Applied Science, London, p. 8.
Toth, S. J. (1973). In *Symposium: Processing Agricultural and Municipal Wastes,* ed. G. E. Inglett. AVI Publishing Co., Connecticut, p. 172.
White, L. P. & Plaskett, L. G. (1981). *Biomass as Fuel.* Academic Press, London.
Willson, G. B. (1989). *BioCycle,* **30**(8), 82.
Worgan, J. T. (1976). In *Food from Waste,* eds G. G. Birch, K. J. Parker & J. T. Worgan. Applied Science, London, p. 23.

Chapter 2

THE ENZYMIC TREATMENT OF WASTE MATERIALS

PETER GACESA

Department of Biochemistry, University of Wales College of Cardiff, PO Box 903, Cardiff CF1 1ST, UK

& JOHN HUBBLE

School of Chemical Engineering, University of Bath, Claverton Down, Bath BA2 7AY, UK

CONTENTS

1 INTRODUCTION

The use of enzymes in biotechnological processes is part of a long and established tradition. In some cases this has involved the specific extraction of an enzyme, e.g. calf chymosin for cheese making, whereas in other applications the activity of endogenous enzymes has been utilized *in situ*, e.g. the malting of barley. The utilization of enzymes for waste processing is a relatively recent development and has grown out of the increasing demands, both economic and environmental, for acceptable methods for the disposal or recycling of these materials.

2 FACTORS INFLUENCING ENZYME USE

2.1 Sources of Enzymes

There are some 2500 different enzyme-catalysed reactions listed in the current edition of the International Union Handbook of Enzyme Nomenclature (Webb, 1984). However, this is a significant underestimate of the total number of enzymes that have been discovered because many of the reactions may be catalysed by a multiplicity of proteins with different properties. Of this number only 20–30 enzymes are produced on the industrial scale (i.e. kilogram to tonne quantities per annum).

Most enzymes are produced from approximately 20 microbial sources, either bacterial or fungal. This has resulted largely because many of the enzyme manufacturers are significant producers for the food industry where legislation proscribes the range of acceptable microorganisms (see Denner, 1983 for a review). Therefore, the enzyme producers have tended to use these microorganisms in preference to others with the result that, in many cases, more than one enzyme is extracted from each organism (Table 1).

Enzymes are also obtained from several traditional non-microbial sources. A

Table 1
The Major Microbial Sources of Bulk Enzymes

Organism	*Enzyme*	*Applications*
Aspergillus sp.	Amyloglucosidase	Sugar industry
	Cellulase	Cellulose processing
	Hemicellulase	Food processing
	Lactase	Whey processing
	Lipase/esterase	Fat processing
	Pectinase	Pectin degradation
Bacillus sp.	α-Amylase	Saccharification
	β-Glucanase	Glucan degradation
	Glucose isomerase	Sugar industry
	Alkaline proteinase	Detergents
	Neutral proteinase	Protein hydrolysis
	Subtilisin	Detergents
Klebsiella aerogenes	Pullulanase	Debranching of starch
Kluyveromyces lactis	Lactase	Whey processing
Mucor miehei	Microbial rennet	Dairy products
Saccharomyces spp.	Invertase	Confectionery industry
Streptomyces sp.	Glucose isomerase	Sugar industry
Trichoderma sp.	Cellulase	Cellulose processing

group of structurally and catalytically similar sulphydryl proteinases are obtained from a variety of tropical plants including papaya (papain), pineapple (bromelain) and fig (ficin). These are rich sources of proteolytic enzymes which may be obtained readily using unskilled labour. For example, ficin comprises approximately 90% of the protein in the latex tapped from *Ficus glabrata* (Gacesa & Hubble, 1987). As the enzymes are relatively concentrated in these sources they are rarely purified to any degree.

A number of enzymes have been obtained from animals or animal-derived sources. For example, the proteolytic enzyme chymosin, which is extracted from the fourth stomach of unweaned calves, has been used for cheese making for as long as records go back. Canine excrement has proved to be an invaluable (if somewhat unpleasant!) source of pancreatic proteinases for the bating of hides. Although the use of chymosin continues and may be enhanced by the cloning and expression of the enzyme in yeast cells (Beppu, 1983), pancreas-derived enzymes are being replaced by similar products obtained from microorganisms. This is largely because the availability of offal is restricted as the production of enzymes is merely a sideline of meat processing and manufacture.

It has been estimated (Godfrey & Reichelt, 1983) that approximately 80% of enzymes are produced by fermentation. Most of these enzymes are destined for use in detergents and for the processing of starch and dairy-related products (Frost & Moss, 1987) (Fig. 1(a)). The major categories of industrial enzymes are also shown (Fig. 1(b)). There is no simple analysis of the quantities of enzymes used for the recycling of waste *per se*; however, it could be argued that, as a minimum, the proportion attributable to starch processing would fall into this category. Therefore, it is probably safe to estimate that at least 30% of total enzyme production is concerned with the recycling of waste into usable products.

The ability of an organism to produce extracellular enzyme is an important consideration as it simplifies subsequent processing (Fig. 2). Consequently, only one enzyme is obtained on the large scale from a Gram-negative bacterium as the outer membrane of these organisms is an additional barrier to export of protein from the cell. The pullulanase (an essential enzyme in starch processing—see later) of the Gram-negative bacterium *Klebsiella aerogenes/pneumoniae* is a complex lipoprotein which is excreted from the cell but accumulates in the periplasm of this organism where it associates with various membrane structures (d'Enfert *et al.*, 1987). However, the advent (Jensen & Norman, 1984) of a more industrially-applicable form of the enzyme from a Gram-positive bacterium, *Bacillus acidopullulyticus,* may well replace the pullulanase from the *K. pneumoniae*.

Most enzymes that are used for industrial processes require only a minimum of purification and in some cases it is a crude disrupted-cell extract that is used as the catalyst. Numerous combinations of methods have been used to purify enzymes on a laboratory scale (Scopes, 1982; Brummer & Gunzer, 1987) and although the same principles are applicable to large-scale manufacture due

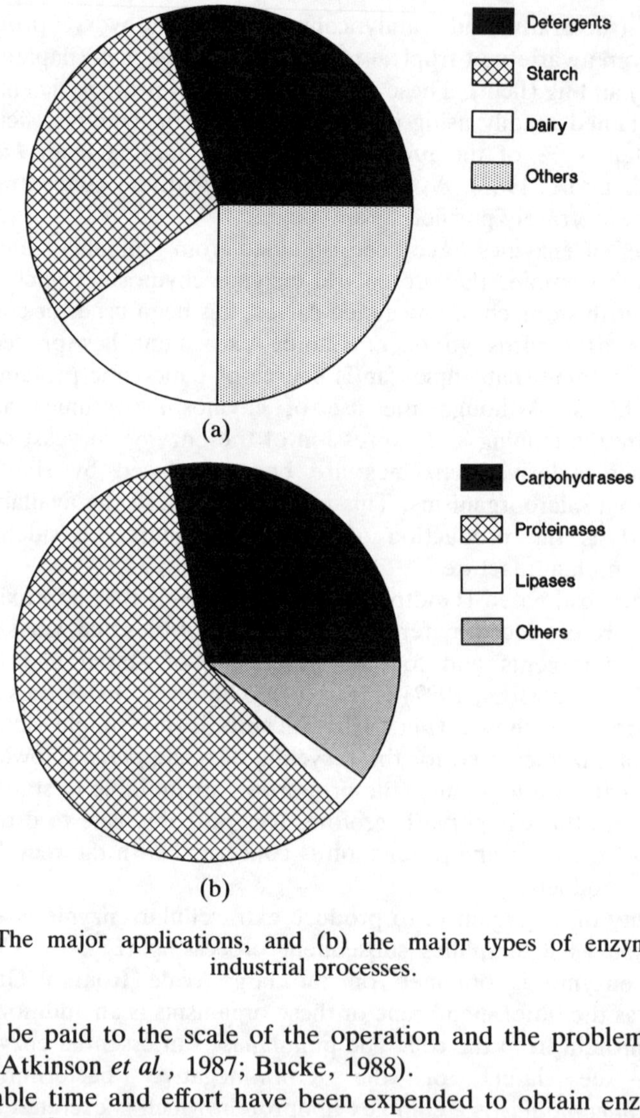

Fig. 1. (a) The major applications, and (b) the major types of enzymes used in industrial processes.

note has to be paid to the scale of the operation and the problems that this introduces (Atkinson *et al.*, 1987; Bucke, 1988).

Considerable time and effort have been expended to obtain enzymes from extremophiles. Some organisms that grow in abnormal environments, e.g. thermophiles, halophiles or alkylophiles, produce extracellular enzymes that are stable and active in these environments and therefore may be useful in specific applications. In particular, numerous enzymes have been isolated from thermophiles, including α-amylases (Aiba *et al.*, 1983; Joyet *et al.*, 1984), cellulases (Durand *et al.*, 1984), proteinases (Fujii *et al.*, 1983), amyloglucosidases (Oren, 1983) and lipases (Ammar & McDaniel, 1984).

The quantities of enzymes and the conditions under which they can be produced may be altered radically by mutagenesis and recombinant DNA techniques. The biosynthesis of most proteins is controlled by induction, e.g.

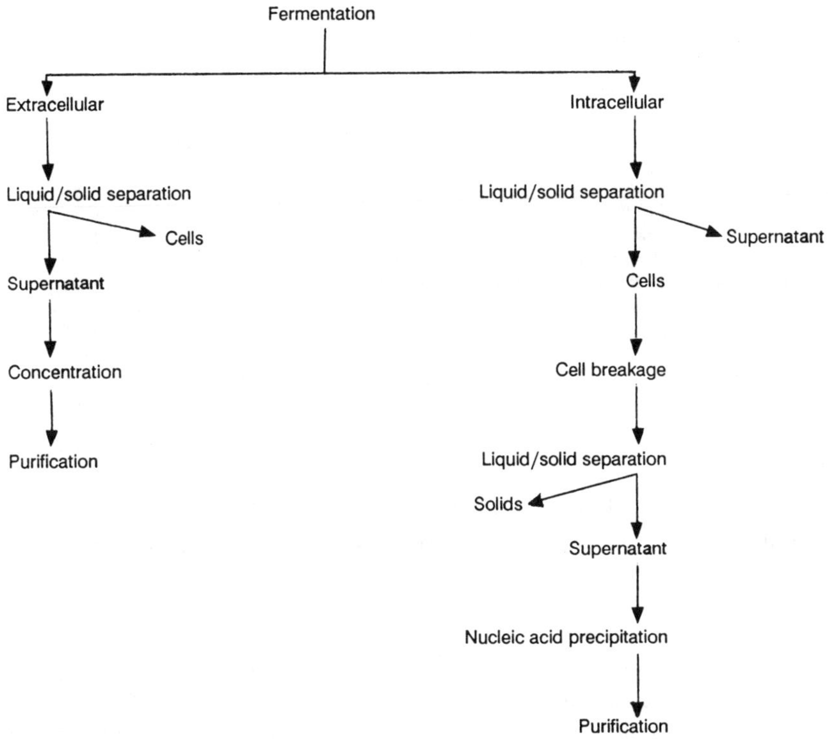

Fig. 2. Comparison of the processing steps involved in the extraction of intra- and extracellular enzymes (adapted from Hacking, 1986).

induction of cellulase by cellulose (Merivuori *et al.*, 1984), and/or by catabolite repression (Demain, 1983), thus limiting the quantities of enzyme that may be produced on non-defined media. These problems may be overcome by using random mutagenesis procedures (Dale, 1988) to obtain constitutive mutants and overproducers.

Recombinant DNA techniques have been widely applied to the cloning and overexpression of industrially significant enzymes (Table 2). In most cloning experiments *Escherichia coli* has been utilized as the host organism because it is the best genetically characterized microorganism, but *Bacillus subtilis* is a more suitable bacterium for industrial processes. Overproduction of enzyme can be achieved by insertion of the DNA fragment into a high copy number plasmid, i.e. a gene-dosage effect, preferably downstream of a high-efficiency promoter sequence. Also, DNA encoding an N-terminal signal peptide is needed if the protein is to be exported from the cell (see Peberdy, 1987). However, in *E. coli* many cloned enzymes are only exported into the periplasm (Gilkes *et al.*, 1984*a*; Keen *et al.*, 1984), although the use of leaky mutants may allow release of some of the enzyme into the medium (Gilkes *et al.*, 1984*b*). Other bacteria, particularly *Bacillus* spp., are better suited for the secretion of

Table 2
Some Examples of Industrially Important Enzymes which have been Cloned

Enzyme	*Source strain*	*Host strain*	*Reference*
α-Amylase	*Bacillus licheniformis*	*B. subtilis*	Piggott *et al.* (1984)
α-Amylase	*B. amyloliquefaciens*	*Escherichia coli*	Palva (1982)
Pullulanase	*Klebsiella aerogenes*	*E. coli*	Takizawa & Murooka (1985)
Cellulases	*Cellulomonas fimi*	*E. coli*	Gilkes *et al.* (1984*a,b*)
Pectate lyases	*Erwinia crysanthemi*	*E. coli*	Keen *et al.* (1984)
Lactase	*B. stereathermophilus*	*B. subtilis*	Hirata *et al.* (1985)
Neutral proteinase	*B. stereathermophilus*	*B. subtilis*	Fujii *et al.* (1983)
Alkaline proteinase	*B. stereathermophilus*	*B. subtilis*	Vasantha *et al.* (1984)
Subtilisin	*B. amyloliquefaciens*	*B. subtilis*	Wells *et al.* (1983)
Chymosin	Bovine	*Saccharomyces cerevisiae*	Mellor *et al.* (1983)
Xylose isomerase	*E. coli*	*Schizosaccharomyces pombe*	Ueng *et al.* (1985)
Lignin peroxidase	*Phanerochaete crysosporium*	*E. coli*	Tien & Kirk (1983)

proteins into the medium, as the absence of an outer membrane precludes the possibility of periplasmic accumulation of the cloned gene product.

2.2. Enzyme Stability

The economics of any process which utilizes enzymes will be considerably influenced by the cost and stability of the enzyme preparation used. Stability factors would usually be expected to have a more significant effect on continuous rather than batch operation, although in situations like waste processing where the feed stream may contain many contaminants, enzyme stability may be severely compromised even in the context of a batch reaction.

The limited stability of enzyme preparations results from their critical dependence on a specific molecular conformation. Given that the conformation of a protein results from the collective effects of a number of weak, usually non-covalent, interactions, the free energy difference between active and inactive states of an enzyme can be as little as 40 kJ/mol. Enzyme inactivation is usually attributed to thermal effects but the temperature dependence of the rate constant for deactivation will be strongly influenced by changes in the local chemical environment. As protein conformation depends on a range of weak interactions (Bailey & Ollis, 1986) whose individual contributions will vary for each case, it is clear that prediction of the resultant changes in enzyme activity is unlikely to be feasible except over a limited range of conditions.

2.2.1 Quantitation of Stability

Although not universally applicable, first-order deactivation kinetics have been widely used to predict the effects of enzyme inactivation. This is described by the equation

$$\frac{\mathrm{d}E}{\mathrm{d}t} = -kE \tag{1}$$

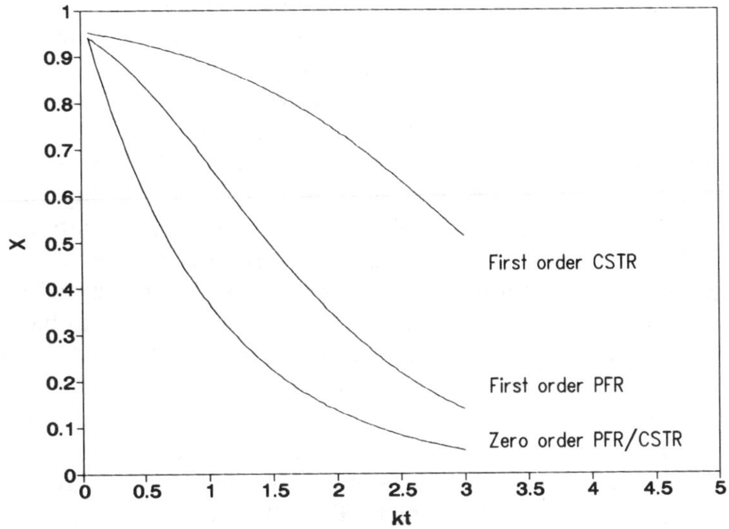

Fig. 3. Effects of inactivation on the performance of immobilized enzyme reactors, showing the variation of fractional conversion (X) as a function of dimensionless time (kt) for a continuous flow stirred tank reactor (CSTR) and plug flow reactor (PFR) under conditions of first and zero order kinetics (Gacesa & Hubble, 1987).

where E represents enzyme and t is time. Integration of eqn (1) gives

$$E_t = E_0 \exp(-kt) \tag{2}$$

which can be used to calculate the decay constant of an enzyme under defined conditions. Conversely, if the decay constant is known eqn (2) can be used to predict the fraction of activity remaining as a function of time. As this simple analytical approach is a gross simplification of the range of factors affecting a real process (for example, it ignores changes related to throughput rather than time) it must be used with caution. The effects of enzyme inactivation on reactor performance, as predicted by eqn (2), have been considered elsewhere (Gacesa & Hubble, 1987) but in the case of continuous reactors are summarized in Fig. 3.

In practice it would be usual to monitor reactor performance throughout the life of the catalyst and to adjust the feed flow rate to maintain fractional conversion. For economic reasons it is usually only practical to operate for three half-life decay periods before changes in feed rate become unacceptable. Within this constraint the use of reactors in parallel allows the overall process throughput to be maintained at a constant rate.

2.2.2. Stabilization of Enzymes

For many applications, particularly where continuous operation is required, it is advantageous to immobilize the enzyme on an insoluble support. Ideally, this improves the handling properties of the enzyme, improves the ease of reuse, and often confers some stability advantages. The basic approaches for enzyme immobilization have been established for many years and are widely reported in the literature (see, for example, Zaborsky, 1973; Messing, 1985). However, the apparent kinetic properties of the enzyme often change as a consequence of enzyme immobilization. This can result from one of a number of effects including conformational changes, microenvironmental effects, and mass transfer limitations (Bailey & Ollis, 1986; Gacesa & Hubble, 1987).

The immobilization process may also affect enzyme stability (Monsan & Combes, 1988), but although widely quoted as improving stability this cannot be assumed as a potential benefit in all cases. Both the thermal stability and the ability to reverse conformational changes can be dependent on the number of bonds between enzyme and support (Koch-Schmidt & Mosbach, 1977). These effects demonstrate that there is a compromise to be made between the benefits of locking the enzyme into an active conformation and the disadvantages of the molecule losing the ability to 'renature' once a conformational change has been induced. Hence the effects of immobilization on protein conformation are complex and in some ways analogous to the effects of intramolecular crosslinking (see later). Given the heterogeneous nature of the bonds formed between enzyme and support, the benefits on stability must again be assessed on a case-by-case basis.

In the context of waste treatment there are probably two significant stability benefits which may be gained from immobilization. The first applies specifically where proteinases are used to modify the structure of waste protein. A major problem with proteinases is autolysis of the catalyst preparation. Immobilization of the enzyme will tend to eliminate this effect by preventing contact between individual proteinase molecules. The second is the use of encapsulation of the enzyme using a semi-permeable membrane to protect against potential denaturants in the process feed.

In addition to those factors described above which affect the intrinsic stability of an enzyme preparation, there are a number of other effects which can be introduced as a consequence of immobilization and lead to deviations from first-order deactivation kinetics. Examples include the formation of sub-populations of immobilized enzymes with different stabilities, and also mass-transfer limitations which can mask the enzyme decay. In this case, while mass transfer is limiting the reactor may appear to give a constant performance over an extended period; however, once a critical enzyme level is reached the effects of inactivation become apparent. Therefore mass-transfer effects can lead to erroneously optimistic assessments of the benefits of immobilization on enzyme stability (Cheetham, 1983).

2.2.3 Stabilization of Soluble Enzymes

Although there has been much interest in the development and use of immobilized enzyme preparations (Gestrelius & Mosbach, 1987), many enzymes are still used in a soluble form. For this reason there has been significant research on the stabilization of enzymes using additives, and by direct chemical modification (Torchilin & Martinek, 1979). Clearly, the cost

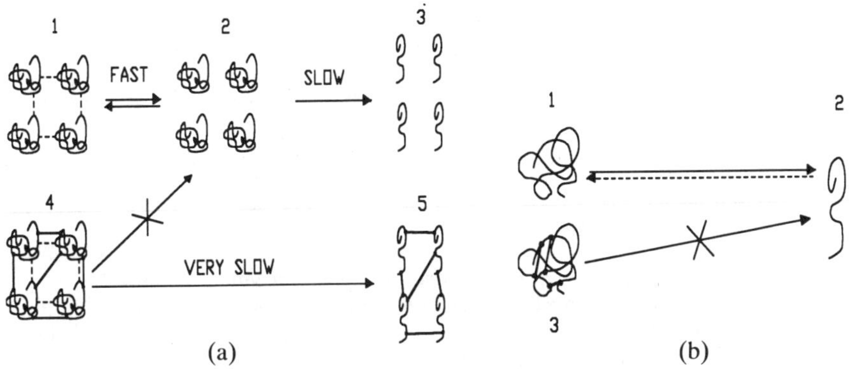

Fig. 4. General scheme of enzyme stabilization effected by intramolecular crosslinking. (a) 1, Native oligomeric enzyme; 2, reversibly dissociated subunits; 3, irreversibly denatured subunits; 4, crosslinked enzyme; 5, irreversibly denatured crosslinked enzyme. (b) 1, Native monomeric enzyme, 2, denatured enzyme; 3, crosslinked enzyme (Martinek & Torchilin, 1988).

and benefits of any stabilization protocol must be carefully balanced. In the case of additives (e.g. chelating agents and microbial growth inhibitors) the costs of subsequent removal must be added to the raw materials cost.

A more direct approach to stabilization is to modify the native structure of the enzyme using chemical agents, for example, bifunctional agents have been used successfully to stabilize chemically enzyme molecules by the introduction of intramolecular crosslinks (Martinek & Torchilin, 1988). The potential advantages to be gained from internal crosslinking for both monomeric and multimeric proteins are shown in Fig. 4. Increases in stability have been demonstrated experimentally for both chymotrypsin (Torchilin *et al.*, 1978) and glyceraldehyde 3-phosphate dehydrogenase (Torchilin *et al.*, 1983). An alternative approach to chemical modification is to use site-directed mutagenesis to introduce additional disulphide bridges into an enzyme molecule. This approach has been demonstrated using the enzyme lysozyme (Perry & Wetzel, 1984).

2.3 Economics of Enzyme-catalysed Reactions

The decision to use enzymes for the transformation of raw materials will depend on many factors. Unless the product is of high added value, which is unlikely for waste processing, the enzymes used must be produced in bulk and at minimal cost. Typically, large tonnage enzymes can be produced for \$10 000–\$100 000 per tonne. It is difficult to translate this into processing costs but it has been estimated that the use of amylases contributes \$5–\$15 to the cost of hydrolysing one tonne of starch (Hacking, 1986). Immobilization and the ability to reuse an enzyme can make processes more economic (Chibata & Tosa, 1977) but this is not always the case; for example, the hydrolysis of starch by α-amylase is invariably catalysed using native enzyme. An interesting

Table 3
Costs of Hydrolysis of the Lactose Component of Whey

Method	*Cost($)/te dry solids*[a]
Batch acid	40–60
Acid resin	90–110[b]
Batch enzyme	50–100
Immobilized enzyme	20–100

[a] These are estimates of total hydrolysis costs (enzyme + reactor costs + labour) but exclude the costs of concentration, transport and other charges.
[b] Includes the cost of pretreatment.
Adapted from Hacking (1986).

comparison has been made for the overall costs of whey processing using either acid or enzymic hydrolysis of lactose (Table 3). The economics of each process depend on a number of factors including the scale of operation. However, it is clear that no one process is paramount and that the optimum operational procedure will have to be determined for each plant.

Clearly, the decision to use enzymes in the processing of wastes has to be made on the basis of a sound economic assessment. Enzymes are inherently unstable, relatively expensive to produce and subject to inhibition, e.g. by products, but the advantages of specificity and the lack of side reactions can often outweigh these disadvantages. Nonetheless in any application the merits of using enzymes should be judged in comparison with the use of whole cells or chemical processing.

3 APPLICATION OF ENZYMES

3.1 Hydrolases

3.1.1 Carbohydrases

Polysaccharides are probably the most adundant waste materials available for recycling into useful products. Cellulose and starch are major by-products of a variety of commercial processes and both are amenable to enzyme treatment. Similarly, lactose, the major waste product from the dairy industry, will also be a major target for enzyme treatment.

STARCH UTILIZATION

The processing of waste starch into a variety of products has been one of the most useful applications of enzyme technology (Fig. 5). Plentiful supplies of inexpensive amylases are available and there has been no serious attempt to produce an immobilized enzyme. Traditionally, α-amylase has been obtained from *Aspergillus oryzae* but the enzyme obtained from various thermophilic

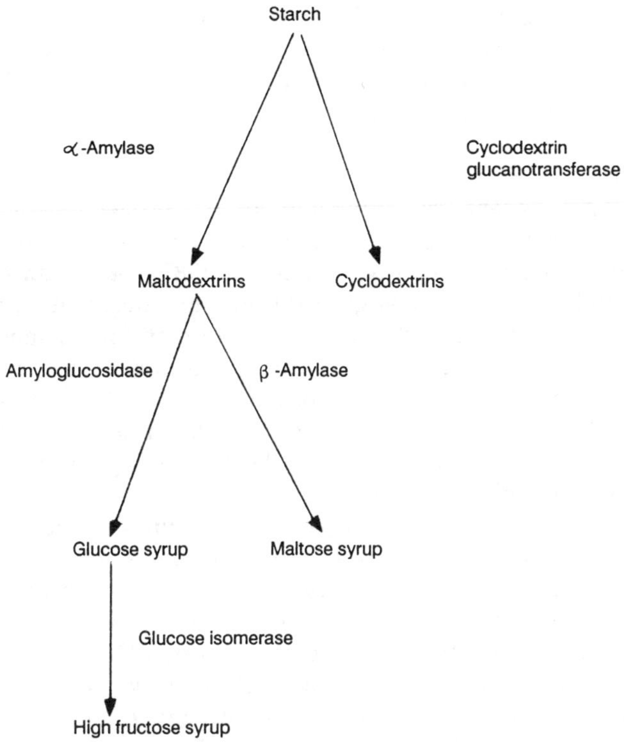

Fig. 5. The use of enzymes for the modification of starch waste into useful products (after Mercier & Colonna, 1988).

Bacillus spp. have the advantage of temperature stability. Although there is considerable sequence homology between the *Bacillus* α-amylases (Mercier & Colonna, 1988) some are much more thermostable than others. The main cause of thermal denaturation in these enzymes is the deamination of asparagine and glutamine residues (Tomazic & Klibanov, 1988). The advent of *Bacillus licheniformis* α-amylase, which operates at temperatures of 95–105°C, has allowed the heat disruption of the starch grain and the enzymic hydrolysis to occur simultaneously, thus simplifying the whole process (Reichelt, 1983). There have been numerous reports describing the cloning and expression of α-amylase genes (see, for example, Aiba *et al.*, 1983; Joyet *et al.*, 1984; Piggott *et al.*, 1984).

The dextrins produced by α-amylase action are further processed by a variety of enzymes including β-amylases for the production of maltose syrups and amyloglucosidase for the production of glucose syrups. Pullulanase, usually derived from *Klebsiella pneumoniae,* is required to hydrolyse the α1-6 branch points in the dextrin structure. Amyloglucosidases are produced by a variety of organisms, although the commercial product is obtained principally from *Aspergillus* or *Rhizopus* (Frost & Moss, 1987). Glucose syrups produced using amyloglucosidase are a useful starting material for a wide variety of processes, including fermentation to products such as alcohol and organic acids (Lehtonen, 1988), or for isomerization into high fructose syrups by the use of glucose (xylose) isomerases (see later).

CELLULOSE UTILIZATION

Cellulose is an abundant waste material for the production of glucose and hence other products; it has been estimated that $1{\cdot}8 \times 10^{12}$ tons of cellulose are biosynthesized annually (Alfani & Cantarella, 1987). Although cellulose is a simple polysaccharide (poly β1-4-glucose), enzymic degradation is hampered by its crystalline structure and the complex mixture of other components which occurs in plant cell-wall material. Typical agricultural wastes, such as straw, contain 38–42% cellulose, 31–34% hemicellulose, 7–11% lignin and 10–13% of other components (Ladisch *et al.*, 1983). Therefore, cellulose-containing material is normally pretreated to disrupt the fibres using a variety of chemical or physical methods prior to enzyme degradation.

Native cellulose is degraded by a complex mixture of enzymes, including endoglucanases, cellobiohydrolases and β-glucosidases, which act in concert (Eriksson & Pettersson, 1975). *Trichoderma reesei* has been one of the most useful organisms for the production of cellulases and overproducing variants have been obtained by mutagenesis (Montenecourt, 1983). A major disadvantage of *T. reesei* is the low specific activity of the enzymes obtained from this organism (Beguin *et al.*, 1988), so consequently other cellulolytic organisms, e.g. *Cellulomonas,* have been investigated and many of the genes for the cellulose-degrading enzymes have been cloned (Gilkes *et al.*, 1984*a,b*).

Attempts have been made to use cellulases to process materials contained in mixed and defined wastes. For example, municipal solid waste has been separated into its various components and the cellulose-containing fraction has been saccharified with bacterial and fungal cellulases (Clanet *et al.*, 1988). Similarly, sawdust (Kumakura & Kaetsu, 1988) and wheat straw (Gonzalez *et al.*, 1988) have been processed using cellulases. It is these types of applications that will result in the large-scale use of cellulases for waste processing.

LACTOSE UTILIZATION

The UK dairy industry produces some 16 Mt/year of combined waste (Wheatley, 1987) of which a significant proportion contains lactose. The lactose in dairy waste is a considerable problem and is the major concern in recycling the material into new products. It has been estimated that up to one-half of the whey produced during cheese making is dumped directly into the drains (Marwaha & Kennedy, 1984), which causes severe pollution problems because of the high biological oxygen demand which averages 35 000–40 000 ppm (Maiorella & Castillo, 1984).

Attempts to use β-galactosidases to hydrolyse lactose into its constituent sugars, glucose and galactose, have met with some success, and a plant in Italy currently processes 10 000 litres of milk per day for human consumption (Harrison, 1987). Although this and other uses of β-galactosidase have been developed primarily to increase the digestibility of milk rather than to process waste material derived from the dairy industry, these processes are applicable to the recycling of wastes, e.g. in the use of lactose-containing whey as an animal-feed supplement.

The use of enzymes for the degradation of lactose has been compared to other methods such as acid hydrolysis (Burgess & Shaw, 1983) and also to microbial fermentation (Maiorella & Castillo, 1984). The syrup derived from the enzymic hydrolysis of lactose in whey has been used commercially for the production of bread (Matveeva *et al.*, 1989). At present enzyme degradation of lactose cannot be considered as the major route for whey processing. A number of appropriate enzymes have been obtained from a wide range of organisms (Frost & Moss, 1987) although many of these are of limited commercial value, often because of susceptibility to product inhibition (van Griethuysen-Dilber *et al.*, 1988). However, the economics of enzymic hydrolysis of lactose in whey can be favourable (Table 3) and there is no doubt that this will be an expanding area of waste recycling.

3.1.2 Proteinases

Proteinases have been widely used to modify the structure, and hence texture and flavour, of proteins in a range of food industry applications (Alder-Nissen, 1986). The ability to change structure and hence modify the digestibility of protein materials offers the potential for upgrading protein wastes for use as animal or human feed supplements. A number of raw materials have been targets for this approach, including hide hair and trimmings (Snitsar *et al.*, 1987) collagen (Bronowski *et al.*, 1979) and blood (Quaglia & Massacci, 1982). Although the reports cited appear to be preliminary studies, the results described for the work with hide trimmings show a significant reduction of protein molecular weight attributable to a two-stage alkali/enzymic treatment, with the product material showing 86% digestibility *in vitro*. Given the scale of the animal processing industry this offers obvious process opportunities. Proteolytic enzymes have also been used successfully to facilitate the extraction of high-grade fat from fleshing-machine waste (Taylor *et al.*, 1989).

Although not leading to a biological product, there have been a number of reports on the use of proteinases for the recovery of silver from photographic emulsions (Phillips, 1977; Krusteva *et al.*, 1987). Recently, a pilot plant capable of processing 900–1000 kg of X-ray film per day has been commissioned for the continuous recovery of silver (Fujiwara *et al.*, 1989). However, it is difficult to assess the viability of enzymic treatment compared with acid digestion, although it has been claimed that processing odours are eliminated by the use of enzymes (Fujiwara *et al.*, 1989).

3.2 Non-hydrolytic Enzymes

3.2.1 Oxidoreductases

The oxidoreductases are a group of enzymes that are not used currently for the processing of waste materials into industrially useful compounds. However, the degradation of lignin, which is essentially an oxidative process, represents the most significant potential application of an oxidoreductase to the recycling of waste materials. Lignin is a complex heteropolymer comprised of a variety of phenolic compounds linked by ether and carbon–carbon bonds. Furthermore, the lignins obtained from softwoods, hardwoods and grasses differ significantly in structure (Sarkanen & Ludwig, 1971; Adler, 1977). The potential of lignin as a raw material has been recognized (Haemmerli *et al.*, 1988), although the difficulties inherent in selectively degrading the polymer present many problems.

It has been recognized for some time that white rot fungi, in particular, are able to degrade lignin (see Kirk & Farrell, 1987, for a review); however, it is only relatively recently that a lignin-degrading enzyme has been isolated and characterized (Tien & Kirk, 1983). A gene for lignin peroxidase has subsequently been cloned and sequenced (Tien & Tu, 1987) and overexpression in a suitable host would make industrial use of the enzyme a more feasible proposition.

The lignin peroxidase enzyme catalyses a variety of reactions, including the oxidation of phenolics, hydroxylation of alkene bonds, cleavage of ether bonds, and aromatic ring opening (Palmer *et al.*, 1987). The enzyme, which exists in multiple isoenzyme forms, contains a single iron protoporphorin IX prosthetic group and catalyses a one-electron oxidation of the substrate to yield radical cations (Schoemaker *et al.*, 1985). It has been argued that lignin peroxidase produces the initial radical cation and that subsequent degradation of the substrate occurs by free-radical attack. However, some workers (Alfani & Cantarella, 1987) believe that free-radical attack might be less important than the combined action of several different extracellular peroxidase-type enzymes. The lignin peroxidase is relatively non-specific for substrate which would be an advantage in commercial applications as the structures of lignins are so heterogeneous.

A perhaps unexpected finding was the discovery that other peroxidases, e.g. horse radish peroxidase, were able to degrade lignin in non-aqueous systems (Van Brunt, 1986). At the time when lignin peroxidases were difficult to purify the use of other peroxidase enzymes was a particularly attractive idea. However, now that the lignin peroxidase has been cloned, adequate supplies of the enzyme are no longer an insurmountable problem.

The potential of using ligninase for the degradation of waste lignin is vast, and the generation of low molecular weight compounds would provide precursors for chemical processing, possibly substituting for raw materials derived from the petrochemical industry (Buswell & Odier, 1987). As lignocelluloses are a major waste-disposal problem, e.g. from paper mills, a process that would enable the effluent water to be treated and also produce new materials would be economically attractive.

Although the degradation of lignin represents a major potential use for peroxidase-type enzymes, other applications have been proposed. For example, it has been suggested that horse radish peroxidase might be used for the removal of carcinogenic aromatic amines from water (Klibanov & Morris, 1981). In fact, it is likely that peroxidases will be useful in a wide range of applications.

3.2.2 Isomerase Enzymes

The enzymic treatment of by-products from the corn industry currently represents the largest-scale application of enzyme technology in the waste treatment area. The success of the process is critically dependent on the ability to meet very stringent glucose/fructose ratios in the final product (Daniels, 1986), and it was the successful solution of the problems associated with the development of an enzyme-catalysed glucose isomerization that was crucial to the formation of this large-scale industry. The history of the development of the isomerization process has been documented (Jensen & Rugh, 1987) and provides an interesting insight into the factors affecting the growth of an industry based on the production of a valuable product from a waste material.

It was shortcomings in the chemical isomerization step that first provided the incentive to investigate an enzymic alternative. Unlike the hydrolytic enzymes that had been used previously on a large scale, the isomerase enzyme was intracellular, showed significantly lower stability, and required close control of reaction conditions. Notwithstanding these problems, the advantages of an enzyme-catalysed isomerization were such that development of the batch process was almost inevitable. However, the batch enzyme method was still labour-intensive and difficult to operate, although solving some of the problems of by-product formation associated with the chemical process (Coker & Venkatasubramanian, 1985). These shortcomings led to the introduction of a continuous process, hence in this case it was 'market pull' rather than 'technology push' which provided the impetus for development.

There are a number of advantages in the use of continuous plug flow reactors compared with the batch process (Table 4). The pertinent factors are the reduction in residence time with concomitant reduction in reactor size, and

Table 4
Glucose Isomerase: Comparison between Batch and Plug Flow Processes

Parameter	*Batch process (1974–1976)*	*Plug flow processes (1976)*	*(1984)*
Syrup conc.	40–45% w/w	40–45% w/w	43–47% w/w
pH	6·5–7·0	8·4 (inlet)	7·5–7·8 (inlet)
Temperature	65°	65°	60°
Co^{2+}	$3{\cdot}5 \times 10^{-4}$M	—	0
Mg^{2+}	$20 \times (Ca^{2+})$	$20 \times (Ca^{2+})$	$20 \times (Ca^{2+})$
Half-life (Enzyme)	300 h	500 h	1 200–1 500 h
Productivity (kg syrup/kg enz.)	900 kg	1 500 kg	3 000–4 000 kg

Source: Jensen & Rugh, 1987.

hence plant installation costs, and the significant increase (>50%) in enzyme productivity. Given the relatively small 'value added' in this process, enzyme costs have been critical in determining economic viability (Daniels, 1988).

3.2.3 Lyases

The only lyase of commercial significance is pectin transeliminase which is one of a series of enzymes involved in pectin degradation. This and other enzymes have been used extensively for the extraction of fruit juice (Fogarty & Kelly, 1983) but not for the recycling of waste materials. However, low-grade agricultural waste materials have been used in solid substrate fermentations for the production of these enzymes (Miura & Endo, 1962).

4 ENZYMES WITH MODIFIED ACTIVITIES

4.1 Applications of Molecular Techniques

The application of enzymes to waste processing has often been hampered by a number of factors, including availability, stability and substrate specificity. However, the advent of molecular biological and other techniques provides opportunities for overcoming these problems. Many examples have already been cited in this chapter of enzymes that have been cloned and overexpressed in appropriate host organisms (Table 2), and there have been many recent reviews on the topic (see, for example, Berry & Senior, 1987; Peberdy, 1987).

The techniques of molecular biology have been further extended by the advent of site-directed mutagenesis. It is now possible to alter the properties of an enzyme in a pre-determined manner (Knowles, 1987), although not all of the modifications produce the expected results! In the field of waste treatment it is often the stability of an enzyme that is of major concern. Several attempts have been made to increase the stability of proteins by the introduction of additional sulphydryl bridges (Katz & Kossiakov, 1986; Perry & Wetzel, 1986; Wells & Powers, 1986) or by the replacement of particular amino acids in the primary structure. For example, subtilisin is unstable in the presence of oxidizing agents but this problem has been overcome by the replacement of a single methionine residue whilst still retaining activity at 80% of the original k_{cat} (Estell *et al.*, 1985). It is clear that site-directed mutagenesis and related techniques will be of paramount importance for the development of novel enzymes; however, a fuller description of these methods is outside the scope of this review.

4.2 Non-aqueous/Low Water Systems

This is an area of research attracting a great deal of interest (Klibanov *et al.*, 1977; Lilly *et al.*, 1987), with particular importance for the treatment and conversion of wastes containing hydrophobic compounds. The problems encountered in the treatment of non-polar materials stems from the fact that most enzyme technology applications to date have been based on aqueous systems. The traditional view was that most enzymes were inherently unstable in organic solvents and therefore could only be used in an aqueous environment. However, it is now appreciated that polar and non-polar solvents have completely different effects on enzyme stability. Polar solvents (e.g. ethanol and acetone) compete for water molecules which are essential for maintaining enzyme structure, resulting in inactivation. Conversely, non-polar solvents can support highly active enzyme conversions providing that a critical level of 'essential' water is present in the system (Halling, 1987*a*).

Although it has been possible to demonstrate the use of enzymes suspended in organic solvents, the possibility also exists of various intermediate approaches for the modification of non-polar molecules. These are mainly based on the use of surface-active agents to create micelles in a two-phase system, i.e. where droplets of one phase are dispersed throughout the second phase. Some experimental techniques for the production of enzyme-containing micelles have been described (Martinek *et al.*, 1986) but in essence these simplify to two basic systems:

(a) Normal micelles—aqueous bulk phase—dispersed organic phase
(b) Reverse micelles—organic bulk phase—dispersed aqueous phase

A refinement of the use of micelles is the liquid membrane where there is a double micelle arrangement (Fig. 6) which can allow additional selectivity to be achieved.

The advantages of micelles is that the enzyme remains in a localized aqueous environment but there is a large interfacial surface area which facilitates mass transfer of substrate and product. An additional advantage is that the choice of organic solvent can be optimized to control the partitioning of reactants and products between the two phases. This has been demonstrated in experimental systems where inhibitory products have been removed by partition from the

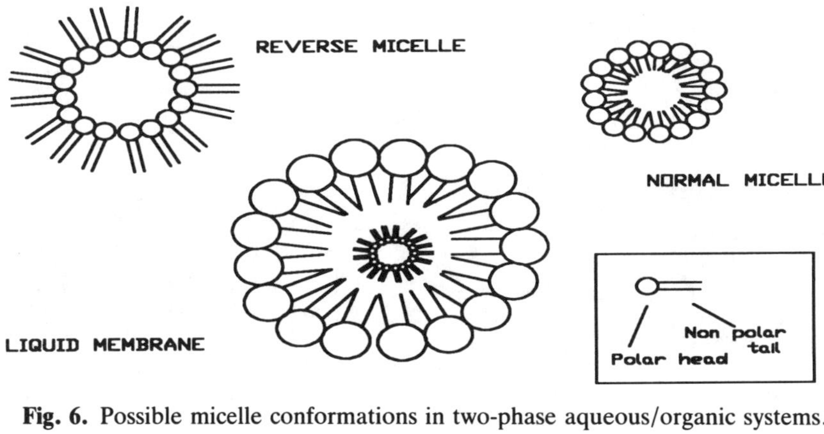

Fig. 6. Possible micelle conformations in two-phase aqueous/organic systems.

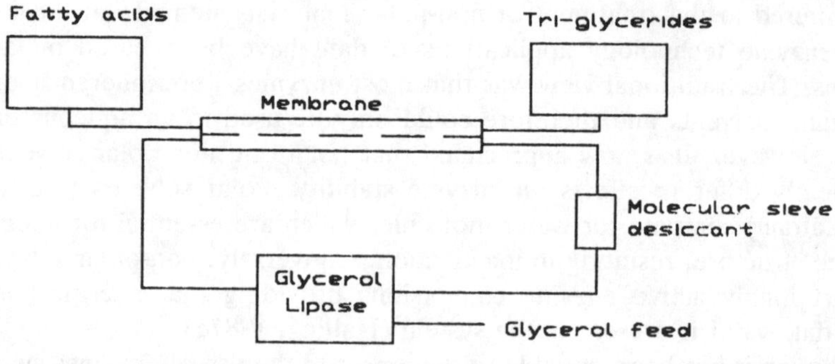

Fig. 7. Two-phase membrane reactor for triglyceride synthesis (adapted from Hoq *et al.*, 1985*b*).

aqueous phase (Schwartz & McCoy, 1977). In the design of a two-phase reactor based on micelles two aspects are critical:

(a) optimization and prediction of mass-transfer effects (Lilly *et al.*, 1988);
(b) separation and recovery of the two phases after completion of the reaction (Lilly *et al.*, 1987).

It was difficulties encountered in the separation of the two phases which provided the incentive for development of the two-phase membrane bioreactor (Hoq *et al.*, 1985*a*). In this case the two phases are physically separated by a micro- or ultrafiltration membrane. This reactor geometry (Fig. 7) has been used for both the synthesis and hydrolysis of triglycerides using lipases (Hoq *et al.*, 1985*a*,*b*; Katsunori *et al.*, 1986) and it would appear to have significant potential applications for conversions based on lipid or fatty-acid-containing waste streams.

The five possible approaches to the use of enzymes in non-aqueous/low water systems are summarized schematically in Fig. 8. Of these the use of a

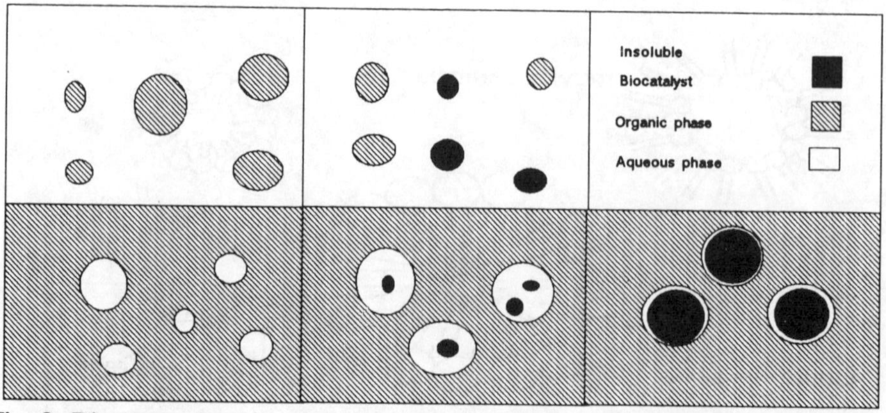

Fig. 8. Diagrammatic representation of reaction systems involving an organic liquid phase (Lilly *et al.*, 1987).

homogeneous organic phase containing enzyme with the quantity of water restricted only to that essential for enzyme activity represents the most radical approach. The exclusion of non-essential water (>99%) can lead to significant changes in enzyme specificity in addition to changes in the equilibrium position of the reaction. One of the most interesting possibilities lies in the use of hydrolytic enzymes to drive synthetic rather than degradative reactions. By excluding water from the reaction mixture, the equilibrium of the hydrolytic reaction is shifted in favour of polymer formation rather than degradation. In this way it is possible to synthesize lipids and peptides from simple precursors (Halling, 1987*b*).

The influence of organic solvents on the enzyme molecule stems from the significance of water molecules in maintaining and stabilizing the non-covalent interactions which determine the three-dimensional conformation of the protein molecule. By using a reaction medium based on an organic solvent, it is possible to control closely the amount of water present in the system and hence vary the hydration state of the enzyme preparation. This allows stability, specificity and catalytic efficiency to be modified systematically to match process requirements (Klibanov, 1989).

Data obtained for chymotrypsin show the effect of solvent hydrophobicity on catalyst activity (Zaks & Klibanov, 1988). In this case there is a 10 000-fold increase in the activity observed in octane compared with pyridine. This difference is attributed to the polar solvent (pyridine) stripping 'essential' water from the enzyme and reducing catalytic effectiveness. In addition to modifying activity, control of 'essential' water can also have remarkable effects on enzyme stability. For example, chymotrypsin has a half-life of several hours at 100°C in octane compared with only several minutes at 60°C in an aqueous solvent. This effect is attributed to a decrease in conformational mobility resulting from the restricted availability of water, which in turn leads to greater structural stability (Zaks & Klibanov, 1985). Similar effects have been reported for lipases (Zaks & Klibanov, 1984).

An interesting finding from work with enzymes in organic solvents is that the enzyme retains a 'memory' of its chemical environment prior to the intermediate drying phase before transfer into the new solvent. Hence, the activity of the enzyme will vary with the pH of the solution from which it was lyophilized (Zaks & Klibanov, 1985) and therefore the appropriate activity can be 'preprogrammed'. Another opportunity offered by the use of enzymes in organic solvents is the possibility of changing specificity, particularly where hydrophobic interactions between enzyme and substrate are important.

In the area of waste processing the most interesting opportunities for the use of enzymes in non-aqueous solvents concern the treatment of phenolic-containing streams. Traditionally, these would be microbially degraded to prevent pollution; however, tyrosinase can be used directly to oxidize phenols in organic solvents (Kazandjian & Klibanov, 1986). In aqueous solution the oxidized products rapidly polymerize, whereas in chloroform the reaction produces quinones which can be readily recovered.

In his review of the use of enzymes in anhydrous solvents Klibanov (1989) considered the potential of 'solvent engineering', where solvent properties are adjusted to optimize the appropriate reaction. In conjunction with modified enzymes developed using protein engineering techniques, non-aqueous solvent systems offer the potential for a wide range of novel waste recovery processes.

5 CONCLUSIONS

The use of enzymes for the treatment of wastes is still in the early stages of development. However, several extremely successful applications, e.g. starch hydrolysis, have been developed and these point the way to the future. It is clear that economic factors are of paramount importance and any new developments will have to complete with other biotechnological approaches, e.g. microbial culture, and the more traditional chemical methods. Nevertheless, the increasing awareness of the environmental problems that can be caused by the discharge of inadequately treated wastes and the stance taken by regulatory authorities will ensure that waste treatment is a growth industry. It is anticipated that the enzymic treatment of waste will play a significant part in the growth of the industry, particularly as the development of novel enzymes/enzyme systems greatly increases the range of potential applications.

REFERENCES

Adler, E. (1977). *Wood Sci. Technol.*, **11**, 169.

Aiba, S., Kitai, K. & Imanaka, T. (1983). *Appl. Environ. Microbiol.*, **46**, 1059.

Alder-Nissen, J. (1986). *Enzymic Hydrolysis of Food Proteins.* Elsevier Applied Science, London.

Alfani, F. & Cantarella, M. (1987). In *Biotechnology of Waste Treatment and Exploitation*, eds J. M. Sidwick & R. S. Holdom. Ellis Horwood, Chichester, UK, p. 256.

Ammar, M. S. & McDaniel, L. E. (1984). *Zentralbl. Mikrobiol.*, **139**, 61.

Atkinson, T., Scawen, M. D. & Hammond, P. M. (1987). In *Biotechnology*, Vol. 7a, ed. J. F. Kennedy. VCH, Weinheim, Germany, p. 279.

Bailey, J. E. & Ollis, D. F. (1986). *Biochemical Engineering Fundamentals.* McGraw-Hill, New York.

Beguin, P., Grepinet, O., Millet, J. & Aubert, J. P. (1988). In *Proc. 8th Int. Biotechnol. Symp.*, Vol II, eds. G. Durand, L. Bobichon & J. Florent. Société Française de Microbiologie, Paris, p. 1015.

Beppu, T. (1983). *Trends in Biotechnology*, **1**, 85.

Berry, D. R. & Senior, E. (1987). In *Biotechnology of Waste Treatment and Exploitation*, eds J. M. Sidwick & R. S. Holdom. Ellis Horwood, Chichester, UK, p. 209.

Bronowski, K., Sagata, J. & Sturdniarski, K. (1979). *Leder*, **30**, 8.

Brummer, W. & Gunzer, G. (1987). In *Biotechnology*, Vol. 7a, ed. J. F. Kennedy. VCH, Weinheim, Germany, p. 213.

Bucke, C. (1988). In *Principles of Biotechnology*, 2nd edition, ed. A. Wiseman. Surrey University Press, London, p. 143.

Burgess, K. & Shaw, M. (1983). In *Industrial Enzymology,* eds T. Godfrey & J. Reichelt. Macmillan, Byfleet, UK, p. 260.
Buswell, J. A. & Odier, E. (1987). *CRC Crit. Rev. Biotechnol.* **6,** 1.
Cheetham, P. S. J. (1983). In *Principles of Biotechnology,* ed. A. Wiseman. Surrey University Press, London, p. 172.
Chibata, I. & Tosa, T. (1977). *Adv. Appl. Microbiol.,* **22,** 1.
Clanet, M., Durand, H. & Tiraby, G. (1988). *Biotechnol. Bioeng.,* **32,** 930.
Coker, L. E. & Venkatasubramanian, H. J. (1985). *Comprehensive Biotechnology,* **3,** 777.
Dale, J. W. (1988). In *Principles of Biotechnology,* 2nd edition, ed. A. Wiseman. Surrey University Press, London, p. 44.
Daniels, M. J. (1986). In *Process Engineering Aspects of Immobilised Cell Systems,* eds C. Webb, G. M. Black & B. Atkinson. I. Chem.E., Rugby, UK, p. 218.
Daniels, M. J. (1988). *Methods in Enzymology,* **136,** 371.
Demain, A. L. (1983). In *Overproduction of Microbial Products,* eds V. Krumphanzl, B. Sikyta & Z. Vanek. FEMS Symposium No. 13, Academic Press, London, p. 3.
d'Enfert, C., Chapon, C. & Pugsley, A. P. (1987). *Molec. Microbiol.,* **1,** 107.
Denner, W. H. B. (1983). In *Industrial Enzymology,* eds T. Godfrey & J. Reichelt. Macmillan, Byfleet, UK, p. 111.
Durand, H., Soucaille, P. & Tiraby, G. (1984). *Enzyme Microbiol. Technol.,* **6,** 175.
Eriksson, K.-E. & Pettersson, B. (1975). *Eur. J. Biochem.,* **51,** 193.
Estell, D. A., Graycar, T. P. & Wells, J. A. (1985). *J. Biol. Chem.,* **260,** 6518.
Fogarty, W. M. & Kelly, C. T. (1983). In *Microbial Enzymes and Biotechnology,* ed. W. M. Fogarty. Applied Science Publishers, London, p. 131.
Frost, G. M. & Moss, D. A. (1987). In *Biotechnology,* Vol. 7a, ed J. F. Kennedy. VCH, Weinheim, Germany, p. 65.
Fujii, M., Takagi, M., Imanaka, T. & Aiba, S. (1983). *J. Bacteriol.,* **154,** 831.
Fujiwara, N., Tsumiya, T., Katada, T., Hosobuchi, T. & Yamamoto, K. (1989). *Process Biochem.,* **24,** 155.
Gacesa, P. & Hubble, J. (1987). *Enzyme Technology.* Open University Press, Milton Keynes, UK.
Gestrelius, S. & Mosbach, K. (1987). *Methods in Enzymology,* **136,** 353.
Gilkes, N. R., Kilburn, D. G., Langsford, M. L., Warachuk, W. W., Warren, R. A. J., Whittle, D. J. & Wong, W. K. R. (1984*a*). *J. Gen. Microbiol.,* **130,** 1377.
Gilkes, N. R., Kilburn, D. G., Miller, R. C. & Warren, R. A. J. (1984*b*). *Biotechnology,* **2,** 259.
Godfrey, T. & Reichelt, J. R. (1983) in *Industrial Enzymology,* eds T. Godfrey & J. Reichelt. Macmillan, Byfleet, UK, p. 1.
Gonzalez, G., Caminal, G., deMas, C. & Lopez-Santin, J. (1988). *Process Biochem.,* **24,** 62.
Hacking, A. J. (1986). *Economic Aspects of Biotechnology.* Cambridge University Press, Cambridge, UK.
Haemmerli, S. D., Fiechter, A. & Leisola, M. S. A. (1988). In *Proc. 8th Int. Biotechnology Symp.,* eds G. Gurand, L. Bobichon & J. Florent. Société Française de Microbiologie, Paris, p. 1030.
Halling, P. J. (1987*a*). In *Bioreactors and Biotransformations,* eds G. W. Moody & P. Baker. Elsevier Applied Science, London, p. 109.
Halling, P. J. (1987*b*). In *Dechema Biotechnology Conferences,* Vol. 1. Dechema, Frankfurt, p. 35.
Harrison, L. A. (1987). In *Biotechnology of Waste Treatment and Exploitation,* eds. J. M. Sidwick & R. S. Holdom. Ellis Horwood, Chichester, UK, p. 81.
Hirata, H., Negoro, S. & Okada, H. (1985). *Appl. Environ. Microbiol.,* **49,** 1547.
Hoq, M. M., Koike, M., Yamane, T. & Shimizu, S. (1985*a*). *Agric. Biol. Chem.,* **49,** 3171.

Hoq, M. M., Tagami, H., Yamane, T. & Shimizu, S. (1985*b*). *Agric. Biol. Chem.*, **49,** 335.
Jensen, B. F. & Norman, B. E. (1984). *Process Biochem.*, **19,** 129.
Jensen, V. J. & Rugh, S. (1987). *Methods in Enzymology,* **136,** 358.
Joyet, P., Guerineau, M. & Heslot, H. (1984). *FEMS Microbiol. Lett.*, **21,** 353.
Katsunori, M., Youichi, M. & Yoshikawa, S. (1986). Process for preparing fatty acid esters. European patent application 195311.
Katz, B. A. & Kossiakov, A. (1986). *J. Biol. Chem.*, **261,** 15480.
Kazandjian, R. Z. & Klibanov, A. M. (1986). *J. Am. Chem. Soc.*, **107,** 5448.
Keen, N. T., Dahlbeck, D., Staskawicz, B. & Belser, W. (1984). *J. Bacteriol.*, **159,** 825.
Kirk, T. K. & Farrell, R. L. (1987). *Ann. Rev. Microbiol.*, **41,** 465.
Klibanov, A. M. (1989). *Trends in Biochem. Sci.*, **14,** 141.
Klibanov, A. M. & Morris, E. D. (1981). *Enzyme Microbiol. Technol.*, **3,** 119.
Klibanov, A. M., Samokhin, P., Martinek, K. & Berezin, I. V. (1977). *Biotechnol. Bioeng.*, **19,** 1351.
Knowles, J. R. (1987). *Science,* **236,** 1252.
Koch-Schmidt, A. C. & Mosbach, K. (1977). *Biochemistry,* **16,** 2101.
Krusteva, M., Peev, G. & Iotova, L. (1987). *Acta Biotechnol.*, **7,** 93.
Kumakura, M. & Kaetsu, I. (1988). *Process Biochem.*, **23,** 51.
Ladisch, M. R., Lin, K. W., Voloch, M. & Tsao, G. T. (1983). *Enzyme Microbiol. Technol.*, **5,** 82.
Lehtonen, P. O. (1988). In *Proc. 8th Int. Biotechnol. Symp.*, Vol. II, eds G. Durand, L. Bobichon & J. Florent. Société Française de Microbiologie, Paris, p. 1060.
Lilly, M. D., Brazier, A. J., Hocknull, M. D., Williams, A. C. & Woodley, J. M. (1987). In *Biocatalysts in Organic Media,* eds C. Laane, J. Tramper & M. D. Lilly. Elsevier, Amsterdam, p. 3.
Lilly, M. D., Harbron, S. & Narendranathan, T. J. (1988). *Methods in Enzymology,* **136,** 615.
Maiorella, B. L. & Castillo, F. J. (1984). *Process Biochem.*, **19,** 157.
Martinek, K. & Torchilin, V. P. (1988). *Methods in Enzymology,* **137,** 615.
Martinek, K., Levashov, A. V., Klyachko, N., Khmelnitski, Y. L. & Berezin, I. V. (1986). *Eur. J. Biochem.*, **155,** 453.
Marwaha, S. S. & Kennedy, J. F. (1984). *Process Biochem.*, **19,** 79.
Matveeva, I. V., Rogovskikh, I. V. & Puchkova, L. I. (1989). *Khleboprodukty,* 32.
Mellor, J. *et al.* (1983). *Gene,* **24,** 1.
Mercier, C. & Colonna, P. (1988). In *Proc. 8th Int. Biotechnol. Symp.*, Vol. II, eds G. Durand, L. Bobichon & J. Florent, Société Française de Microbiologie, Paris, p. 1042.
Merivuori, H., Siegler, K. M., Sands, J. A. & Montenecourt, B. S. (1984). *Biochem. Soc. Trans.*, **13,** 411.
Messing, R. A. (1985). In *Comprehensive Biotechnology,* Vol. 2, ed. M. Moo-Young, Pergamon Press, Oxford, UK, p. 191.
Miura, Y. & Endo, A. (1962). Process for the preparation of pectolytic enzymes. US Patent 3058890.
Monsan, P. & Combes, D. (1988). *Methods in Enzymology,* **137,** 584.
Montenecourt, B. S. (1983). *Trends in Biotechnology,* **1,** 156.
Oren, A. (1983). *Curr. Microbiol.*, **8,** 225.
Palmer, J. M., Harvey, P. Y. & Schoemaker, H. E. (1987). *Phil. Trans. Roy. Soc.*, A, 494.
Palva, I. (1982). *Gene,* **19,** 81.
Peberdy, J. F. (1987). In *Biotechnology,* Vol. 7a, ed. J. F. Kennedy. VCH, Weinheim, Germany, p. 325.
Perry, L. J. & Wetzel, R. (1984). *Science,* **226,** 555.

Perry, L. J. & Wetzel, R. (1986). *Biochemistry*, **25**, 733.
Phillips, P. A. (1977). Stripping photographic materials. US Patent 4150977.
Piggott, R. P., Rossiter, A., Ortlepp, S. A., Pembroke, J. T. & Ollington, J. F. (1984). *Biochem. Biophys. Res. Commun.*, **122**, 175.
Quaglia, G. B. & Massacci, A. (1982). *J. Sci. Food Agric.*, **33**, 634.
Reichelt, J. R. (1983). In *Industrial Enzymology*, eds T. Godfrey & J. Reichelt. Macmillan, Byfleet, UK, p. 375.
Sarkanen, K. V. & Ludwig, C. H. (1971). *Lignins. Occurrence, Formation, Structure and Reactions*. Wiley-Interscience, New York.
Schoemaker, H. E., Harvey, P. J., Bowen, R. M. & Palmer, J. M. (1985). *FEBS Lett.*, **183**, 7.
Schwartz, R. D. & McCoy, C. J. (1977). *Appl. Environ. Microbiol.*, **34**, 47.
Scopes, R. K. (1982). *Protein Purification: Principles and Practice*. Springer-Verlag, New York.
Snitsar, A. I., Stekol'nikov, L. I., Chernukha, I. M. & Alievska, L. L. (1987). *Myasn. Ind. SSSR*, **10**, 46.
Takizawa, N. & Murooka, Y. (1985). *Appl. Environ. Microbiol.*, **49**, 294.
Taylor, M. M., Diefendorf, E. J., Foglia, T. A., Bailey, D. G. & Feairheller, S. H. (1989). *J. Am. Leather Chem. Soc.*, **84**, 71.
Tien, M. & Kirk, T. K. (1983). *Science*, **221**, 661.
Tien, M. & Tu, C.-P. D. (1987). *Nature*, **326**, 520.
Tomazic, S. J. & Klibanov, A. M. (1988). *J. Biol. Chem.*, **263**, 3092.
Torchilin, V. P. & Martinek, K. (1979). *Enzyme Microbiol. Technol.*, **1**, 74.
Torchilin, V. P., Maksimenko, A. V., Klibanov, A. M., Berezin, I. V. & Martinek, K. (1978). *Biochem. Biophys. Acta*, **522**, 277.
Torchilin, V. P., Trubetskoy, V. S. & Martinek, K. (1983). *J. Mol. Catal.*, **19**, 291.
Ueng, P. P., Volpp, K. J., Tucker, J. V., Gong, C. S. & Chen, L. F. (1985). *Biotechnol. Lett.*, **7**, 153.
Van Brunt, J. (1986). *Biotechnology*, **4**, 611.
van Griethuysen-Dilber, E., Flaschel, E. & Renken, A. (1988). *Process Biochem.*, **23**, 55.
Vasantha, N., Thompson, L. D., Rhodes, C., Banner, C., Nagle, J. & Filpula, D. (1984). *J. Bacteriol.*, **159**, 811.
Webb, E. C. (1984). *Enzyme Nomenclature, 1984*. Academic Press, London.
Wells, J. A. & Powers, D. B. (1986). *J. Biol. Chem.*, **261**, 6564.
Wells, J. A., Ferrari, E., Henner, D. J., Estell, D. A. & Chen, E. Y. (1983). *Nucleic Acid Res.*, **11**, 7911.
Wheatley, A. D. (1987). In *Biotechnology of Waste Treatment and Exploitation*, eds J. M. Sidwick & R. S. Holdom. Ellis Horwood, Chichester, UK, p. 173.
Zaborsky, O. R. (1973). *Immobilized Enzymes*. CRC Press, Cleveland, Ohio.
Zaks, A. & Klibanov, A. M. (1984). *Science*, **224**, 1249.
Zaks, A. & Klibanov, A. M. (1985). *Proc. Natl Acad. Sci. USA*, **82**, 3192.
Zaks, A. & Klibanov, A. M. (1988). *J. Biol. Chem.*, **263**, 3194.

Chapter 3

PROCESSES WITH IMMOBILIZED ENZYMES AND CELLS

SEVERIAN DUMITRIU

Polytechnic Institute of Jassy, Department of Organic and Macromolecular Chemistry and Technology, 23 August St. 11, R-6600 Jassy, Romania

CONTENTS

1 INTRODUCTION

The knowledge base accumulated on enzyme immobilization studies has grown to very large proportions. This wealth of information is one of the primary reasons for the present advances in enzyme engineering. The slower-than-predicted introduction of immobilized enzyme systems into commercial use has been the result of numerous factors, such as the long time required for approval of new processes for use in food applications, the need to control microbial contamination in biological reactor systems, and some enzyme characteristics that limit the economic success of the immobilization process. The engineering of enzymes with better characteristics will overcome some of the problems encountered that have prevented commercial processes from developing.

1.1 Enzyme and Cells Immobilization: Advantages and Disadvantages

The use of freely suspended enzymes has the disadvantages that:

(1) most enzymes are labile under normal operating conditions and therefore have only a very limited life;
(2) since enzymes are water-soluble they are difficult to separate from their substrates and products, and therefore their re-use is difficult, which increases the costs of the process considerably.

The development of an immobilized enzyme process will have the following advantages:

(1) the immobilization of an enzyme to water-insoluble particles has been shown to increase its stability considerably;
(2) the ability to separate the enzyme easily from the products and substrate allows its re-use in a continuous process giving greatly increased economy.

The disadvantages of immobilized enzymes are that the heterogeneous nature of such a catalyst can impose diffusion limitations which can reduce and alter its activity.

The use of immobilized cells in place of the normal batch-grown cells has similar advantages to those of immobilized enzymes:

(1) batch fermentation can be replaced by continuous reactions;
(2) immobilized cells allow the use of considerably higher cell density, hence giving process intensification;
(3) many metabolites or enzymes are active only in resting or stationary phase cells; in an immobilized system cells can be retained in this state.

The disadvantages of immobilized cells are similar to those for enzymes, with the system imposing diffusion limitations.

The advantages and disadvantages of immobilized enzymes compared with immobilized cells depend upon the system involved, but can be summarized as:

(1) the use of immobilized cells removes the need to extract and purify the enzyme;
(2) often, the whole cell system is less sensitive to changes in operating conditions such as pH;
(3) immobilized cells allow a high loading of the support. With isolated enzymes, high loading may reduce activity due to protein–protein interactions;
(4) if the reaction system involved requires many enzymes and cofactor renewal, the immobilized cell system is the best solution.

The disadvantages of the immobilized cell systems are that they impose further diffusional barriers (cell wall), permeabilization may be required, and the maintenance of cell integrity may be difficult.

1.2 Immobilization of Microorganisms or Enzymes?

Many extracellular microbial enzymes are produced in large enough quantities to be used in industrial processes; however, the cost of isolating and purifying intracellular enzymes for commercial processes can dampen the profitability sufficiently for their use to become unattractive. Nonetheless, there are advantages to the isolated enzymes that must be balanced against cost, based upon the nature of the conversion process. Isolated enzymes offer greater purity and the possibility of apparent modified kinetics which, in turn, may yield higher conversion of products less contaminated by miscellaneous materials. Immobilized microorganisms containing the catalyst of interest combine the advantages inherent in the use of immobilized enzymes with those of microbial fermentation processes.

Immobilizing the whole cell should be considered when the extracted enzyme is unstable, since retaining the enzyme within its natural surroundings preserves its stability. Additional considerations for cell immobilization are that no enzyme-catalyzing interfering side reactions be present within the cells (although often these can be deliberately inactivated); that the cells retain greater enzyme activity than is possible with immobilized enzymes; and that the reactor volume per unit of product is much smaller for the continuous, immobilized cells method than for the conventional fermentation process. A conceivable shortcoming could be encountered by the resistance of cell wall and cell membrane to substrate and product transport.

2 IMMOBILIZATION PROCEDURES

2.1 Carriers

In choosing the support material, one should realize that the characteristics of the final immobilized enzyme preparations strongly depend on this material.

Support material ideal for the immobilization of enzymes should have the following characteristics:

(1) large specific surface and sufficient permeability;
(2) relatively high diffusion coefficient for substrate, after immobilization;
(3) presence of functional groups to which the enzyme can be attached under mild conditions;
(4) hydrophilicity;
(5) insolubility;
(6) chemical, mechanical, and thermal stability;
(7) rigid particles of a proper shape;
(8) resistance to microbial breakdown;
(9) regenerability in case of expensive support material;
(10) general recognition as a safe material.

The most important supports usually employed are given in Table 1.

From experience (Porath & Axén, 1976) it appears that hydrophilicity is a very important factor for the preservation of enzymes in a highly active state after their immobilization to the solid support. This fact places hydrophilic gels in a very favorable position among immobilized enzyme matrices. Among synthetic hydrophilic supports, cross-linked polyacrylic acid amide as yet occupies the most important position, while the polysaccharides and some of their derivatives dominate as hydroxylic supports for enzymes. The most important polysaccharide supports are (a) cellulose, (b) dextran, (c) starch and (d) agar. Dextran in cross-linked form, Sephadex, under certain conditions is superior to cellulose and starch. Enzymes fixed to Sephadex exhibit higher relative activity than the corresponding cellulose-bound enzymes (Axén *et al.*, 1967).

Table 1
Carriers for Biomaterials

Nylon
Collagen
Polysaccharide supports: Cellulose, Xanthan, Dextrose, Starch, Agar
Polystyrene
Acrylic copolymers
Agarose and Sephadex
Silicone rubber
Ion-exchange resins
Controlled pore glass and ceramics: Alumina, Silica, Titania
Silica gel
Bentonite
Clay
Magnetic supports

2.2 Methods of Immobilization

In general, six different types of immobilization methods are known (Yamada, 1977) (Fig. 1).

2.2.1 Adsorption and Ion Exchange

In this process, the enzyme is attached to surface-active material, such as alumina, carbon, clay, ion-exchange resins, cellulose, or glass.

Adsorption depends on the balance of interaction between the enzyme, the surface, the solvent and other solutes. Some of these interactions are electrostatic; however, the role of weak forces such as van der Waals forces and H bonding should not be neglected.

For instance, it is expected that;

—the electrostatic enzyme–surface repulsion will have a lower influence if both the enzyme and the support are hydrophobic or if the ionic strength is higher;
—a highly soluble enzyme will nevertheless be adsorbed if its electrical charge is opposite to that of the support.

The surface charge of a solid may be influenced by the composition of the surrounding solution and may be affected by minor constituents of the solid.

The surface area of the support and the pore size distribution are important parameters to consider. They determine:

—the adsorption capacity and thus the amount of enzyme which may be adsorbed per unit reactor volume;
—the accessibility of the enzyme, which may limit the overall reaction rate;
—the stability of the immobilized enzyme.

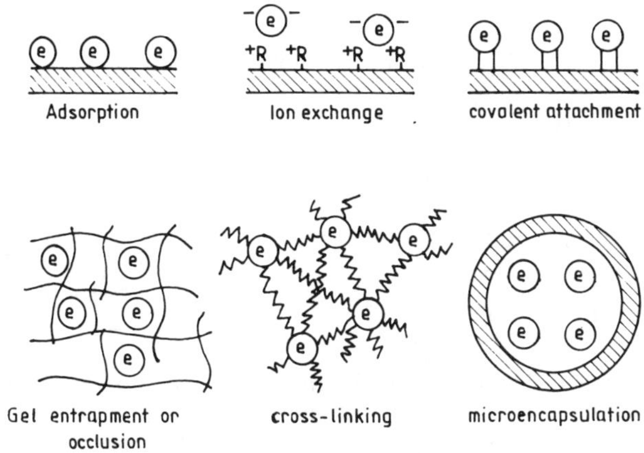

Fig. 1. Immobilization methods.

CELL ADHESION

For whole cells, the term adhesion should be preferred to adsorption, which is typically used for the retention of a molecular entity by a surface.

Adhesion depends on the balance of interactions at the cell-liquid–support interface; the trend expected for the variation of interaction energy is a function of the distance between cell and surface, both bearing a charge of the same sign (Rouxhet, 1985):

—the attractive term is due to dispersion van der Waals forces;
—the repulsive term is due to electrostatic repulsion.

Adhesion of cells to anion exchange resins has been used to immobilize whole cells, though little attention has been paid to selection of the optimum carrier and to control of the amount of cells immobilized per unit area (Rouxhet & Mozes, 1989).

Figure 2 shows the influence of pH on adhesion of *Moniliella pollinis* to a plate (Rouxhet *et al.*, 1984; Rouxhet, 1985). Graph A gives a plot as a function of pH of the amount of cells retained by glass and polystyrene after settling a cell suspension on a plate of each support. The increase of the amount of cells immobilized on glass as pH decreases from 5 to 3·5 is attributed to a decrease of the surface negative charge of both the cells and the support, leading to a reduction of electrostatic repulsion. The decrease of the amount immobilized observed at lower pH is due to a flocculation of the cells, which will be discussed below. The surface charge of polystyrene is not known; however, the shape of the curve leads one to think that cell–support or cell–cell electrostatic repulsions are also important.

We have investigated adhesion of yeast and bacteria, using glass as a model

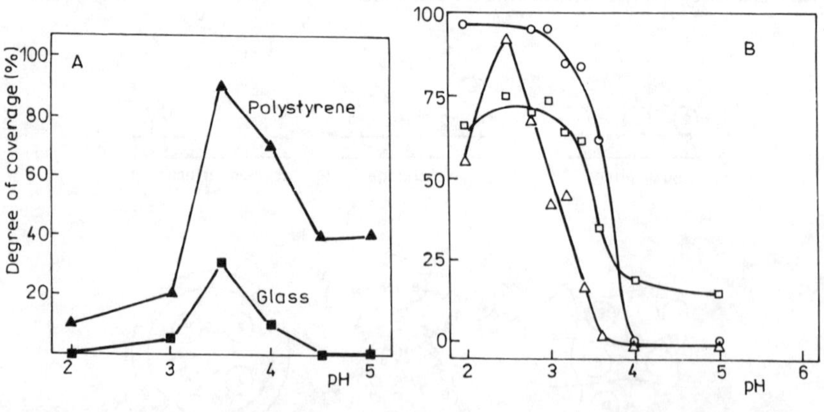

Fig. 2. Illustration of the influence of pH on the adhesion of *Moniliella pollinis* to a support and on cell flocculation. (A) Variation of the amount of cells (% of monolayer) adhering to glass and polystyrene. (B) Variation of parameters related to cell flocculation: △, rate of decrease of turbidity in standard test (arbitrary units); □, volume of sedimented cells ($\mu l/10^8$ cells); proportion of cells involved in flocculation (%). From Rouxhet (1985), by permission of the author.

support. No retention is obtained when none of the component is treated. However, a single, dense and homogeneous layer of cells adhering to the support is obtained by one of the following procedures (Rouxhet *et al.*, 1984, 1987)

—treating the cells by Al^{3+} or Fe^{3+} ions;
—treating the support by Al^{3+} or Fe^{3+} ions;
—coating the support by positively charged colloidal particles of $Al(OH)_3$ or Fe_2O_3.

Adhesion of *Saccharomyces cerevisiae* cells on glass was employed (Van Haecht *et al.*, 1985) for the transformation of glucose waste in ethanol.

MASSIVE RETENTION OF WHOLE CELLS IN A FIXED BED OF SUPPORT PARTICLES

When a cell suspension circulates through a fixed bed of solid particles (wood shavings, crushed bricks), cells may sediment in zones protected from the liquid flow. This provides an immobilization without requiring real adhesion to the support.

FORMATION OF BIOLOGICAL FILMS

The development of a population of microorganisms in contact with a support may lead to a progressive retention of cells on the support and to build-up of a thick film. This is well known for heterogeneous populations of microorganisms encountered in natural systems and in waste water treatment (Mozes *et al.*, 1987; Amory & Rouxhet, 1988*a*, *b*).

Another example is provided by accumulation of *Aspergillus niger* on steel discs; the discs are partially immersed in the culture medium and their rotation ensures a good aeration (Smith *et al.*, 1980).

2.2.2 Covalent Attachment

Nonessential amino acid residues of the enzymes are attached to chemically activated supports, such as cellulose, agarose, glass, or synthetic polymers. This is by far the most intensively studied of the immobilization techniques (Fig. 3).

2.2.3 Cross-linking between Enzyme Molecules or Copolymerization

Enzyme immobilization through reticulation may be performed by inter- and intramolecular reactions with bifunctional agents. Among them the most important are glutaraldehyde, diimidoesters, diisocyanates, bisdiazonium salts, aliphatic diamines, and dithiols.

Schematically, the reactions that may occur are presented in Fig. 4.

If the former case can be avoided by properly selecting the conditions of the reaction (for instance, by decreasing the enzyme concentration in the reaction medium), then it is much more difficult to distinguish between the second and the third variants. Evidently, the success or failure of intramolecular cross-link formation in an enzyme molecule will depend on the extent to which the length

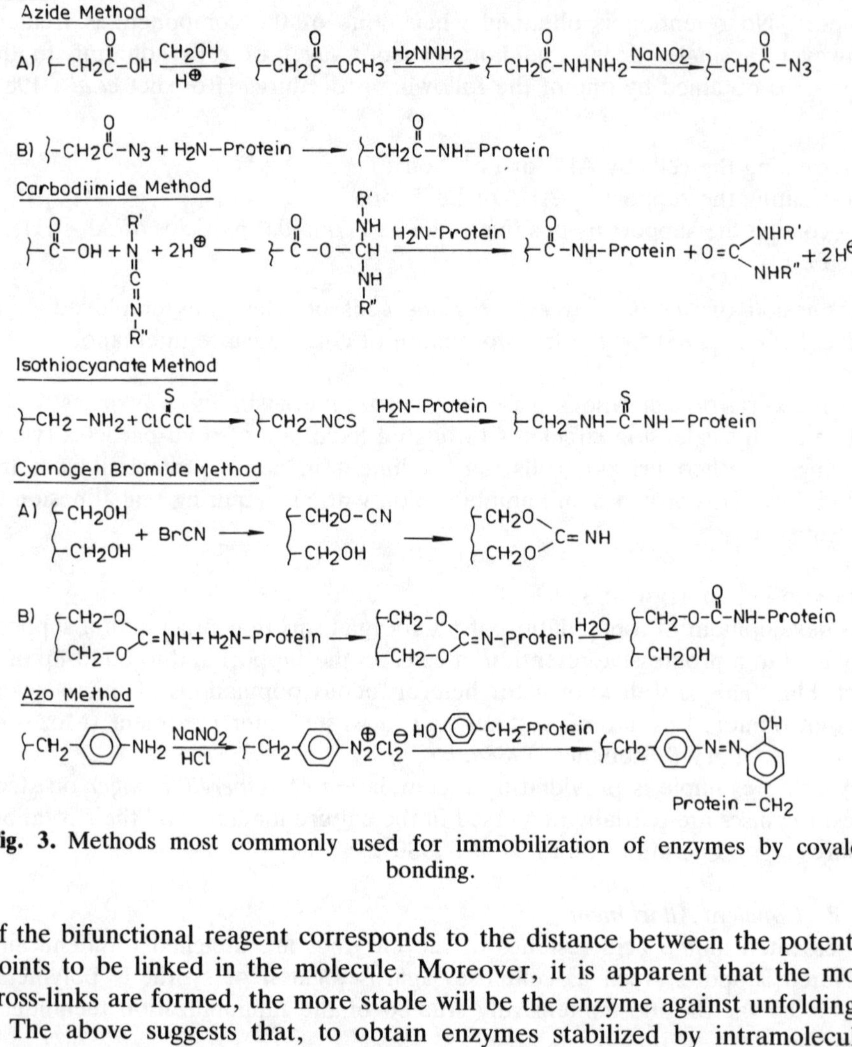

Fig. 3. Methods most commonly used for immobilization of enzymes by covalent bonding.

of the bifunctional reagent corresponds to the distance between the potential points to be linked in the molecule. Moreover, it is apparent that the more cross-links are formed, the more stable will be the enzyme against unfolding.

The above suggests that, to obtain enzymes stabilized by intramolecular crosslinks, one has to try a large number of bifunctional reagents in order to find the optimal one for each enzyme.

Several studies have been undertaken on employing glutaraldehyde. A series of enzymes—such as lactate dehydrogenase (Foster & Thomson, 1973), trypsin (Kazanskaya *et al.*, 1975), chymotrypsin (Gaffield *et al.*, 1973), asparaginase and lysozyme (Payne, 1973), aldolase (Hajdu *et al.*, 1975) and glycogen phosphorylase (Wang & Tu, 1969)—have been modified in reactions of intramolecular reticulations.

These immobilized preparations are generally rather difficult to handle because they are gummy materials or gels. In particulate form they present multiple problems of high pressure drops in a packed-bed reactor and poor fluidizing properties in a fluidized-bed reactor.

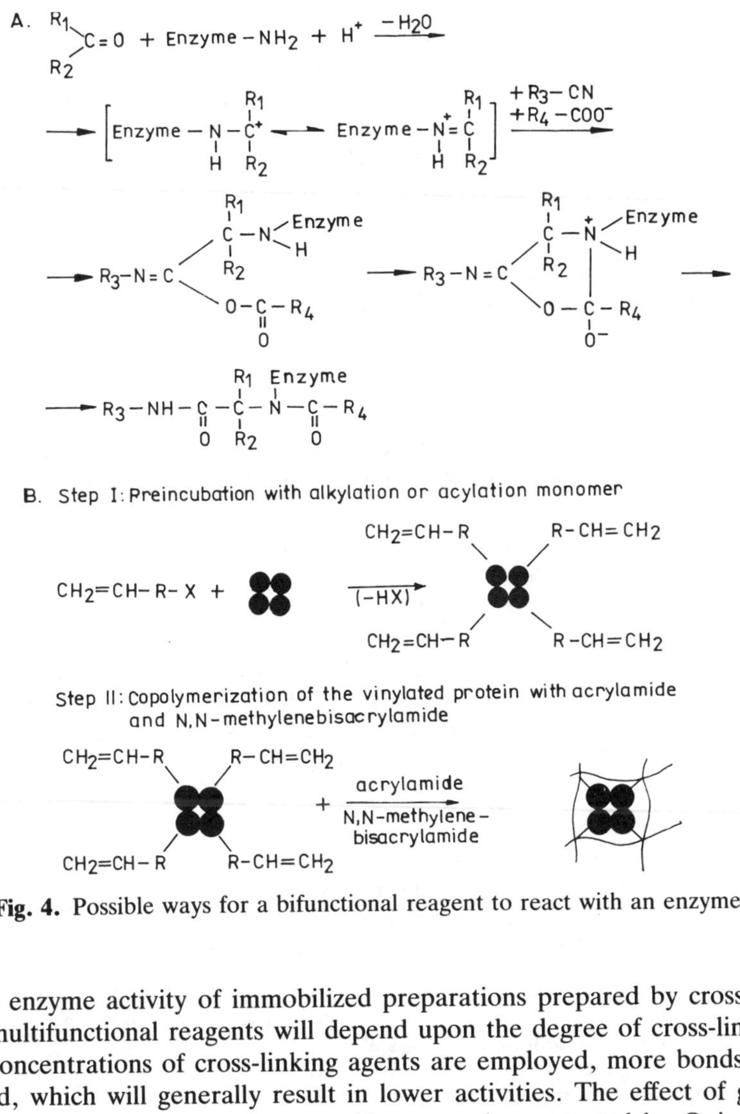

Fig. 4. Possible ways for a bifunctional reagent to react with an enzyme.

The enzyme activity of immobilized preparations prepared by cross-linking with multifunctional reagents will depend upon the degree of cross-linking. If high concentrations of cross-linking agents are employed, more bonds will be formed, which will generally result in lower activities. The effect of glutaraldehyde cross-linking on carboxypeptidase was demonstrated by Quiocho and Richards (1964, 1966).

The preparation of an immobilized trypsin prepared by copolymerization with ethylene and maleic anhydride was described by Bar-Eli and Katchalski (1960). Levin and coworkers (1964) described trypsin preparations formed by copolymerizing the enzymes with peptides.

2.2.4 *Entrapment in a Matrix*

Gels of, for example, polyacrylamides, silica, and starch are formed in the presence of the enzyme. The enzyme is captured within the polymer matrix;

substrates and reaction products move relatively freely. Entrapment is probably the most popular method of immobilization of microbial cells, although other methods have been successfully utilized. In recent years (Sharma *et al.*, 1982), materials like alginic acid and carrageenan have been employed, which form beads of gel when the cell-containing solution of these materials is extruded into salt solutions such as $CaCl_2$.

Cell entrapment in a matrix may be achieved by applying the following methods (Klein & Vorlop, 1985):

—gelation of polymers;
—precipitation of polymers;
—ionotropic gelation of polymers;
—polycondensation;
—polymerization.

Gelation is a temperature-controlled phase transition in polymer–solvent systems, where a homogeneous polymer solution is transformed to a homogeneous gel without change in composition.

Immobilization through polymer precipitation employs concentrated solutions of polymers containing cells that are coagulated by addition of nonsolvent. Solutions of cellulose triacetate, cellulose, polystyrene (Dinelli, 1972; Hackel *et al.*, 1975; Linko *et al.*, 1980) are to be employed.

Ionotropic gelation of polymers is applied to water-soluble polyelectrolyte; when mixed with the appropriate, usually multivalent, counterions, solidification by polysalt formation occurs. Table 2 summarizes those polymer–counterion systems which have been successfully applied to whole immobilization.

Table 2
Ionic Polymers and Related Counterions in the Preparation of Ionotropic Gels for the Entrapment of Whole Cells[a]

Poly-ions		*Counterions*
CO_2^-	Alginate	Ca^{2+}, Al^{3+}, Zn^{2+}, Co^{2+}, Ba^{2+}, Fe^{2+}
	Pectin	Ca^{2+}, Al^{3+}, Zn^{2+}, $Co^{2+\,b}$ and Mg^{2+}
	Carboxymethylcellulose	Ca^{2+}, $Al^{3+\,b}$
	Carboxy-guar-gum	Ca^{2+}, $Al^{3+\,b}$
	Copolystyrene-maleic acid	$Al^{3+\,b}$
PO_3^{2-}	Phospho-guar-gum	Ca^{2+}, $Al^{3+\,b}$
SO_3^{2-}	Carrageenan	K^+, $Ca^{2+\,b}$
	Furcellaran	K^+, $Ca^{2+\,b}$
	Cellulose sulfate	K^+
NH_3^+	Chitosan	Polyphosphates $[Fe(CN)_6]^{4-}$, $[Fe(CN)_6]^{3-}$ Polyaldehydocarbonic acid Poly-1-hydroxy-1-sulfonate-propene-2

[a] Reprinted with permission from Klein & Vorlop (1985).
[b] Also counterions from the corresponding series of the table of elements.

Table 3
Application of Ionotropic Gels for the Entrapment of Whole Cells

Matrix	*Cells*	*Reaction*	*Ref.*
Ca alginate	*Saccharomyces cerevisiae*	Glucose/ethanol	Kierstan & Bucke (1977)
Chitosan alginate	*S. cerevisiae*	Glucose/ethanol	Vorlop & Klein (1983)
Al copolystyrene maleic acid	*Candida tropicalis*	Phenol degradation	Klein *et al.* (1979*b*)
Al alginate	*C. tropicalis*	Phenol degradation	Hackel *et al.* (1975)
Ca, Al carboxymethylcellulose	*C. tropicalis*	Phenol degradation	Klein *et al.* (1979*c*)

Mention must be made of the large range of possible utilizations of these systems in the treatment of residual waters (phenol degradation) and in the production of ethanol, too (Table 3).

Polycondensation reactions were applied in cell entrapment only after obtaining oligomers with reactive functional groups such as hydroxy, amino, epoxy and isocyanate. Table 4 summarizes those systems which have been reported in the literature.

Cells—used in various biochemical processes—have been entrapped by polymerization (Table 5).

2.2.5 Encapsulation or Confinement in a Membrane Structure

The enzymes are entrapped by membranes of various polymers, usually in capsules ranging from 5 to 300 μm in diameter. The enzyme is too large to move through the pores of the semipermeable membrane, but smaller substrate and reaction product molecules can move back and forth.

Chang (1977) has described the encapsulation of enzymes in a variety of membranes. Two of the membranes that he used were cellulose nitrate and nylon.

The most promising of the delivery systems is liposome encapsulation (Bickerstaff, 1987).

The use of *ultrafiltration membranes* allows the fluid to be forced through the

Table 4
Polymer Networks from Polycondensation for the Entrapment of Whole Cells

Matrix	*Cells*	*Reaction*	*Ref.*
Polyurethane	*Arthrobacter simplex*	Steroid conversion	Tanaka *et al.* (1979)
Xerogel®	*Saccharomyces cerevisiae*	Glucose/ethanol	Yamada *et al.* (1980)
Epoxy	*E. coli*	Penicillin G/6-APA	Klein *et al.* (1979*a*)

Table 5
Polymer Networks from Polymerization for the Entrapment of Whole Cells[a]

Matrix	*Cells*	*Reaction*	*Ref.*
Monomers			
Acrylamide	*Umbilica pustulata*	Production of orcinol	Mosbach & Mosbach (1966)
Methacrylamide	*Candida tropicalis*	Phenol degradation	Klein & Schara (1980)
2-Hydroxyethyl methacrylate	*Streptomyces phaerochromogenes*	Glucose/fructose	Kumakura *et al.* (1978)
Oligomers/polymers			
Maleic polybutadiene	*Nocardia rhodocrous*	Steroid transformation	Omata *et al.* (1979)
Activated poly-(propylene glycol)	*N. rhodocrous*	Steroid transformation	Omata *et al.* (1979)
PVA esterified with acrylic acid	*Arthrobacter simplex*	Cortisol/prednisolone	Manecke & Beier (1983)

[a] Reprinted with permission from Klein & Vorlop (1985).

interface, avoiding transport limitations by diffusion. When the reaction involves fragmentation of substances of high molecular weight, the permeability properties of the membrane provide a certain control of the product molecular weight (Rouxhet, 1985).

Devices with an ultrafiltration membrane have been used to allow continuous cofactor regeneration (Wandrey *et al.*, 1984).

Hollow fibers with semipermeable walls allow the filtration area to be increased to a considerable extent. For instance a bundle of fibres may provide an area of 1000 cm^2 for an interval volume of 10 ml. Figure 5 shows various possible modes of using a hollow fibre device (Rouxhet, 1985).

2.2.6 *Formation of Biological Granules*

CELL FLOCCULATION

Cell flocculation is ruled by physical processes similar to those mentioned for cell adhesion. It is very well known in brewing, and provides a direct connection between continuous fermentation and immobilization of a cell used as biocatalyst. Flocculation is the only mode of immobilization well adapted to production requiring a high rate of multiplication.

Flocculation is strongly dependent on the physico-chemical properties of the cell surface and on the composition of the medium. The influence of pH on flocculation of *Moniliella pollinis* is shown in Fig. 5B; aggregates having the size and shape of a rice grain are formed when the pH is close to the isoelectric point.

Flocculation may be promoted by adsorption of various substances (ions, polyelectrolytes, polymers) or by the presence of particles of opposite surface

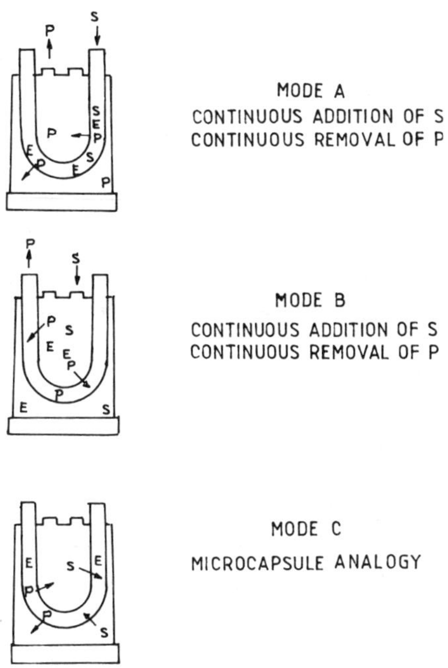

Fig. 5. Schematic representation of possible operational modes using a hollow-fiber device for continuous enzymatic reactions: S, substrate; P, product; E, enzyme. From Rouxhet (1985), by permission of the author.

charge (heteroflocculation), which are brought as such (Kayem & Rouxhet, 1983) or are produced in contact with the cells (precipitation of metallic hydroxides).

FORMATION OF COMPOSITE GRANULES

The process of cell association may be stimulated by the use of a skeleton creating zones protected from attrition by the fluid and favoring accumulation of microorganisms. For instance mesh particles made of stainless steel wire allow accumulation of mycelium into beads with a diameter of about 6 mm and a porosity of the order of 80%, which can be used in a stirred reactor. This method is applicable to bacteria.

Another example is provided by immobilization of cells occurring in a continuous process of anaerobic digestion and methane production (Binot, 1983). After introduction of glass beads (10–100 μm) in the fermentor, formation of a coating on their surface takes place progressively following the main steps already observed in natural environments, for instance in marine fouling (Bitton & Marshall, 1980):

—adsorption of polymeric substances forming a continuous film;
—retention of certain microbial species;
—increase of the number of species and of the biomass held on the surface.

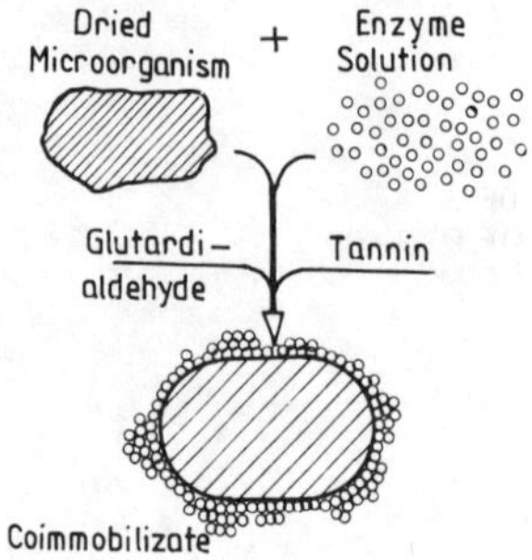

Fig. 6. Coimmobilization of cells and enzymes. Reprinted with permission from Hartmeier (1982).

The particles tend to agglomerate and are maintained together by the growing biomass. The granules formed may reach a size of 1–3 mm and contain a few thousands of glass particles. Within the granule there exists a certain spatial organization with a network of channels which gives a spongy appearance to the granule. The granules are maintained in a fluidized bed and follow a cyclic process of destruction and reconstruction. As compared with a homogeneous stirred digestor, this method of immobilization has allowed:

—obtaining a ratio of 10–100 (instead of 1) between the biomass residence time and the fluid residence time;
—increasing by a factor of about 10 the biomass concentration in the reactor;
—increasing by a factor of about 10 the rate of methane production per reactor volume unit.

2.2.7 Combined Immobilized Enzyme/Cell Systems

According to the schematic presentation in Fig. 6, the dried cells are resuspended in an aqueous solution of the enzyme. The amount of water in the enzyme solution must be limited to such an extent that nearly all the water added is sucked up by the rehydrating cells.

Hahn-Hägerdal & Mosbach (1980) proposed a new method of coimmobilization, in which enzyme molecules are covalently bound on alginates by

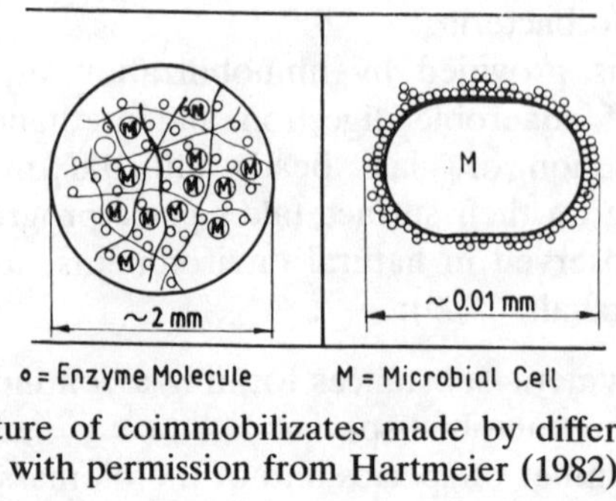

Fig. 7. Size and structure of coimmobilizates made by different methods. Reprinted with permission from Hartmeier (1982).

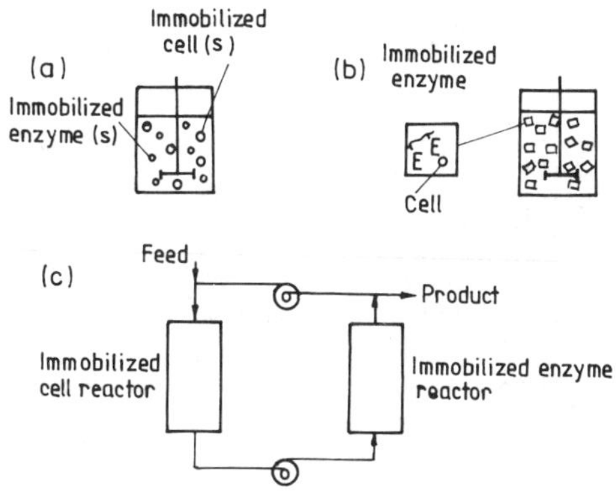

Fig. 8. Immobilized enzyme/immobilized cell reactors: (a) reactor with immobilized enzymes and cells; (b) coimmobilized enzyme and cell system; (c) two-reactor system. Reprinted with permission from Charles & Phillips (1985).

means of carbodiimides. Microbial cells are coentrapped in the enzyme–alginate gel formed with Ca ions. Thus, particles of 2 mm in diameter are to be obtained, while enzyme coupling on the cell surface (Fig. 6) leads to particles of 0·01 mm in diameter (Fig. 7).

Reactors with combined enzyme/cell systems are operating as presented in Fig. 8.

3 REACTORS FOR IMMOBILIZED BIOMATERIAL SYSTEMS

Seven types of immobilized enzyme reactors, schematically presented in Fig. 9, are being employed.

In the *batch reactor,* the initial substrate concentration is very high. The final product concentration is very high, too. Because the total amount of enzyme is initially exposed to high substrate concentrations and to high product concentrations, at the end of the process, this type of reactor is not suitable for reactions with substrate or product inhibition (Pitcher, 1975).

The *continuously stirred tank reactor* (CSTR) is started up as a batch reactor. As soon as a certain degree of conversion has been reached, the substrate is pumped continuously into the reactor and the reaction mixture is simultaneously pumped out of the reactor. This type of reactor is not affected by pH shifts and is operated continuously. Because of the high product concentration, it should not be used in case of product inhibition (Pitcher, 1978).

Analysis of the kinetic behavior of the enzyme immobilized in stirred reactors led to a calculation model for the volume of the curve necessary for attaining the desired conversion degree, in the given conditions (Lilly & Sharp, 1968; Durand & Monsan, 1974).

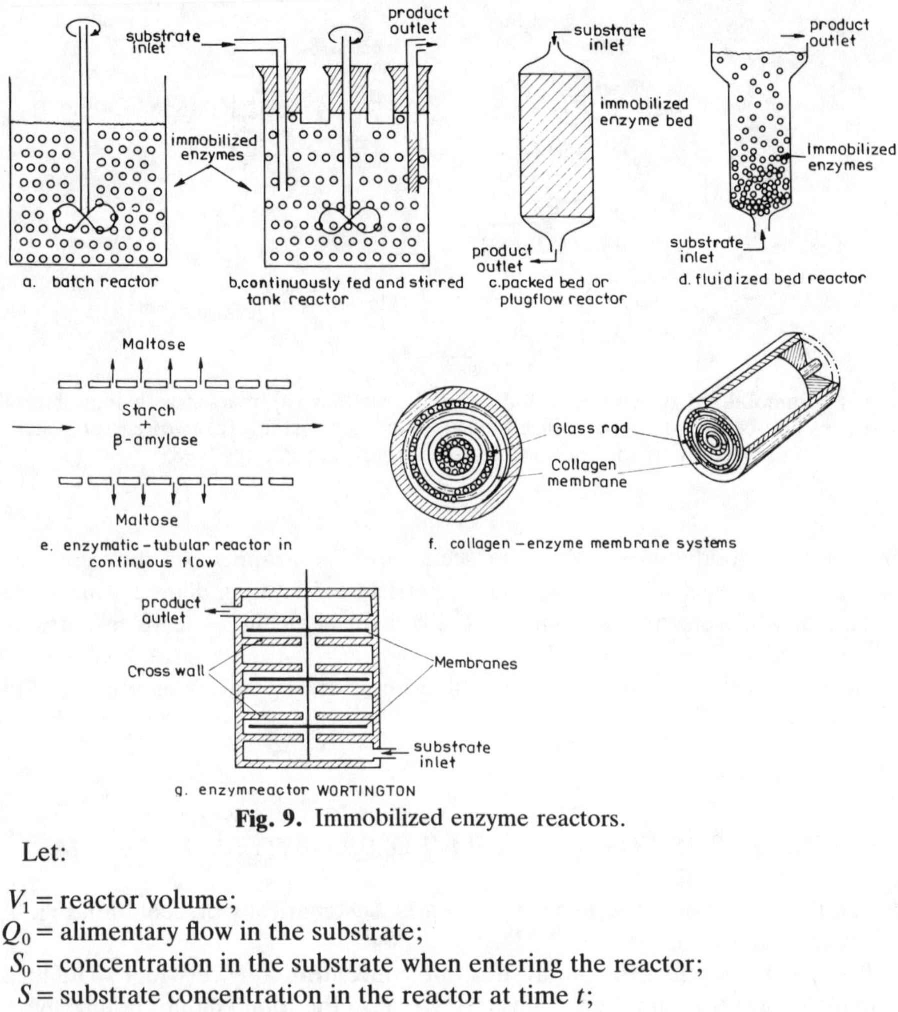

Fig. 9. Immobilized enzyme reactors.

Let:

V_1 = reactor volume;
Q_0 = alimentary flow in the substrate;
S_0 = concentration in the substrate when entering the reactor;
S = substrate concentration in the reactor at time t;

X = conversion degree of the substrate $= \dfrac{S_0 - S}{S_0}$;

K'_m = Michaelis apparent constant (in the presence of inhibitor)

V = rate of the enzymatic reaction $= \dfrac{V_{max} S}{K'_m + S}$

In a *continuously homogeneous reactor,* the enzymatic flow, which depends upon the rate of the catalyzed reaction and upon the reactor's dimensions, equals the difference between the input and output flows:

$$VV_1 = Q_0 S_0 X \tag{1}$$

$$V_1 = \frac{Q_0 S_0 X (K'_m + S)}{V_{max} S} \tag{2}$$

Substituting for S in eqn (2):

$$V_1 = Q_0 X\left[\frac{K'_m}{V_{max}(1-X)} + \frac{S_0}{V_{max}}\right] \tag{3}$$

In the *plug flow* or *packed bed reactor,* the substrate is converted into the product during downflow passage over a column packed with immobilized enzymes. When there is no substrate inhibition and no pH shift, it is very difficult to control the pH in the column without using buffers. In an industrial situation this not generally feasible, but this type of reactor has advantages over the first two types mentioned.

The *fiuidized bed reactor* is theoretically the most attractive, in terms of mass transfer. The substrate is passed through the column in an upflow which has such a velocity that the immobilized enzyme particles are free to move up and down.

In comparing the types of reactors, one more subject should be borne in mind. When the immobilized enzyme reactor is one of the stirred types, the immobilized enzyme particles should be elastic, to avoid crushing. When one needs a packed bed reactor, the immobilized enzyme particles should be relatively rigid; otherwise, pressure drop problems will occur.

If one considers that there is (1) no substrate inhibition, (2) no, or hardly any, pH shift, and (3) an equilibrium reaction in the enzymatic conversion of glucose to fructose, it is not surprising that today quite a large number of plug flow reactors are used for the production of high-fructose corn syrup in industry.

Membrane reactors contain membrane systems representing simple physical barriers retaining the macromolecules in the reaction compartment. The enzyme may be either in a native soluble form or fixed on solid particles. For substrate with high molecular weight, such as starch, Glosset *et al.* (1974) proposed a *tubular reactor* (Fig. 9(e)). The main shortcoming of this type of reactor is the phenomenon of 'polarization', i.e. an accumulation of products with high molecular weight, when entering the reactor, and the blinding of pores.

Several reactors are being equipped with *enzymatically active membranes*; thus, the enzyme is being adsorbed on the surface or entrapped in the support's skeleton. The biocatalytic modulus of Wang and Vieth (1973), Vieth and Venkatasubramanian (1976), is made (Fig. 9(f)) of a collagen and enzyme film, formed by electrode siting, and finally fixed on a cellulose acetate film. The membrane is wrapped on an axis, the various layers being separated by small glass rods.

In *Wortington's enzymreactor* (Fig. 9(g)) (Durand &Monsan, 1974), a series of membrane-bearing disks are fixed on an axis, to be introduced in the reaction medium. This is a multi-floored system, each compartment offering the possibility of being equipped with membranes containing various enzymes.

The reactor performance equations applicable to the real immobilized biomaterial system of interest are one of the pivotal points of information during the system's development. Among other things, these equations permit an estimate of the biomaterial activity in the reactor at a given point in time. Often, it is desirable to arrange to have the reactor behave similarly to either the plug flow or the mixed flow, depending upon the kinetics. Performance equations for the batch, plug flow, and mixed flow reactors, for modified or simple Michaelis–Menten kinetics, have been listed in several reviews (Sümer, 1978). Equation (4) is valid, e.g. for a plug flow system exhibiting product inhibition, as in the case in the hydrolysis of lactose by immobilized lactase:

$$\frac{kE}{F} = S_0\left(1 - \frac{K_m}{K_i}\right)(X_t - X_i) - \left(K_m + \frac{K_m S_t}{K_i}\right)\ln\left(\frac{1 - X_e}{1 - X_i}\right) \quad (4)$$

where:

k = turnover number, a constant whose dimension is a reciprocal of time;
E = total concentration of biomaterial (enzyme), free and bound;
F = Flow rate;
S_0 = initial substrate concentration;
K_m = Michaelis constant with the dimension of concentration; its numerical value corresponds to the substrate concentration at half-maximal reaction rate;
K_i = product inhibition constant;
$X_t = (S_i - S)/S_t$;
$X_i = (S_i - S_0)/S_t$;
S = concentration of the non-bonded substrate at time t;
S_0 = initial substrate concentration;
S_t = total substrate concentrations if all product were converted to substrate;
$X_e = X_i$ at equilibrium.

Some factors that influence the choice of reactor type are as follows.

(1) nature of the biomaterial support;
(2) nature of the substrate to be processed;
(3) reaction kinetics;
(4) cost of reactor construction and operation;
(5) ease of reactor scale-up;
(6) method of regeneration or replacement of the immobilized biomaterial;
(7) operational stability and requirements;
(8) activity per unit volume or productivity required.

Many different geometries for *immobilized cell reactors* have been explored. Geometries include continuously stirred tank reactors (Furui, 1985), fixed beads (Sitton *et al.*, 1980; Young & Dahab, 1982; Lortie & Thomas, 1986), fluidized beds (Dunn *et al.*, 1983; Berk *et al.*, 1984*a*, *b*, *c*; Andrews & Przezdziecki, 1986; Beck *et al.*, 1986; Hogrefe *et al.*, 1986; Sheih & Keenan,

1986; Tanaka *et al.*, 1986; Wallat *et al.*, 1986), spouted beds (Webb *et al.*, 1986), rotating disks (Characklis *et al.*, 1985), hollow fibers (Klei *et al.*, 1981; Miyauchi *et al.*, 1982; Kleinstreuer & Agarwal, 1986; Tharakan & Chan, 1986), and membranes (Furusaki & Miyauchi, 1981; Gekas, 1986). Air-lift fermenters (Weiland *et al.*, 1984) do not seem to have been used for immobilized cells, but have obvious potential. The special case of choice of reactor for immobilized plant cells has been examined (Shuler *et al.*, 1986). Which reactor is preferred in any given situation depends to a considerable extent on the physical properties of the immobilizing medium.

Treatments of residual waters usually employ fixed film reactors, known in three variants: (1) trickling filters; (2) rotating biological contactors; and (3) fluidized beds (Characklis *et al.*, 1985).

Superficially, these reactor configurations appear quite different. In fact, the fluid flow patterns are significantly different. However, the fundamental processes occurring in the biofilm within these different reactor configurations are essentially the same. The heterogeneous reactors (i.e. those requiring two phases) contain active biomass immobilized on the wetted surfaces. The immobilized biomass, herein referred to as the *biofilm,* extracts nutrients from the flowing liquid. As noted before, the flow patterns in the various reactor configurations are different and, as a result, fluid shear stress and mass transfer processes are different at the biofilm–fluid interface.

Consequently, a flow diagram of any model for a fixed film reactor would include two major components: (a) a reactor flow model and (b) a biofilm model (Fig. 10). The reactor flow model will be different for every different configuration. The biofilm model will be the same for all fixed film reactors (Characklis *et al.*, 1985).

Development of a systematic understanding of biofilm processes has been limited by the seeming complexity posed by the interaction of several contributing rate processes. The development and persistence of a biofilm may be described as a net result of the following processes:

(1) transport of material from the bulk fluid to the wetted surface; materials may be soluble (microbial nutrients both inorganic and organic) and particulate (viable microorganisms, other organic, and inorganic particles);
(2) adhesion, attachment and/or adsorption of soluble and particulate material to the wetted surface or the biofilm;
(3) microbial processes within the biofilm which include growth, production formation, decay or maintenance and death or lysis; and
(4) detachment or desorption of soluble and particulate material from the biofilm and its reentrainment in the bulk fluid.

The absence of convenient methods for measuring the biotic and abiotic components has hindered more fundamental process analyses of fixed film

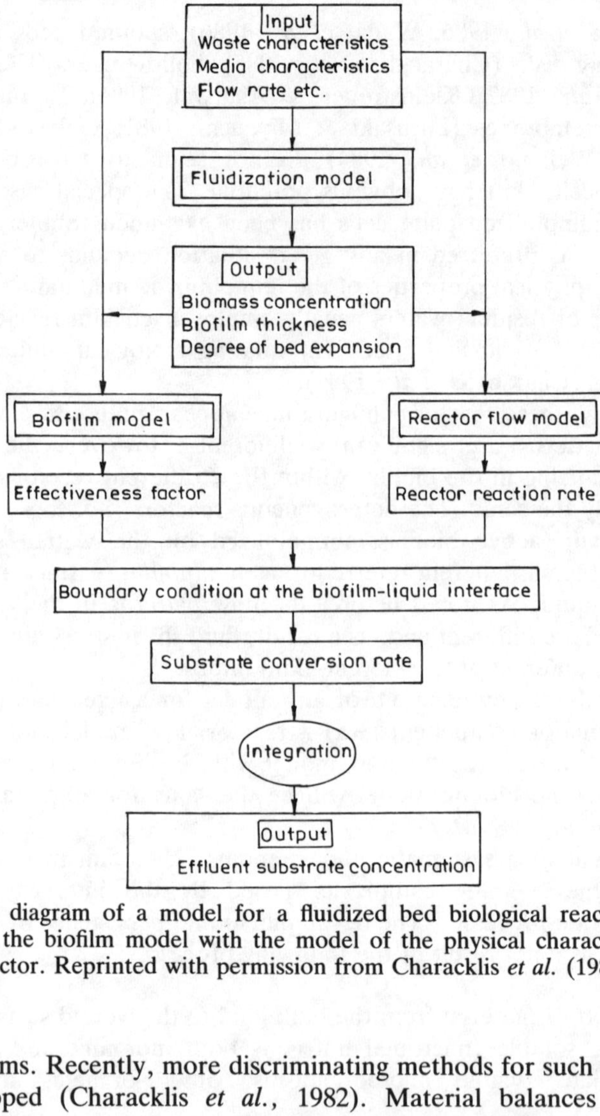

Fig. 10. Flow diagram of a model for a fluidized bed biological reactor illustrating interfacing of the biofilm model with the model of the physical characteristics of the reactor. Reprinted with permission from Characklis *et al.* (1985).

reactor systems. Recently, more discriminating methods for such analyses have been developed (Characklis *et al.*, 1982). Material balances and, hence, methods are necessary for quantifying substrate, product and suspended biomass. In addition, a constitutive equation for the biofilm mass is necessary.

4 DAIRY INDUSTRY WASTES CONVERSION

4.1 Whey Hydrolysis by Immobilized Lactase

It is estimated that nearly 35 million tons of whey are produced annually worldwide. Much of the sweet whey, from the manufacture of hard cheeses, is presently dried and blended in some animal and human processed food products. Utilization of acid whey, a by-product from the manufacture of soft cheeses like cottage cheese, on the other hand, poses a problem due to its acidity. The percent utilization of whey varies from about 60% in the USA and Canada, and about 75% in France and the USSR, to about 95% in Denmark and the Netherlands (Reesen & Strube, 1978). This may leave as much as $10–12 \times 10^9$ pounds in the USA alone that is not properly utilized. Because of its high biochemical oxygen demand and pollution regulations, most of the processed whey is either dried or concentrated. However, the increasing energy prices continue to have impact on the drying costs.

Whey contains 4·5–5% lactose, 0·5–0·8% protein and about 0·7% salts. This whey lactose utilization can be performed by an immobilized biomaterial, or on a catalytic resin (Demaimay, 1977). Other alternatives studied have been those of fermentation to make alcohol and methane (Reesen & Strube, 1978) and crystallization of lactose before protein recovery (Davis, 1972).

4.1.1 Methods of Lactase Immobilization and Properties of the Immobilized Lactase

Lactase (β-galactosidase) hydrolyzes lactose to glucose and galactose.

Lactase immobilization has been performed by employing various methods (Dahlqvist *et al.*, 1973; Olson & Stanley, 1973; Woychik & Wondolowski, 1973; Hinberg *et al.*, 1974; Weetall, 1974), among which microencapsulation (Chang, 1976), hollow-fiber immobilization (Chambers *et al.*, 1976; Prenosil & Hediger, 1985), adsorption on inorganic supports (Pitcher *et al.*, 1976), as well as covalent bondng concomitant with adsorption on methacrylic polymers possessing reactive epoxy groups (Plexazym®) (Plainer *et al.*, 1982), leading to immobilized lactase, have been used industrially. Microencapsulation has been performed in membranes of nitrate cellulose, reticulated with glutaraldehyde (Chang, 1976).

An enzyme-containing solution in contact with the spongy layer of anisotropic membranes will fill the large pores of this layer, and the semipermeable membrane on the thin, inner skin will prevent this enzyme solution from penetrating to the hollow-fiber lumen. Utilizing this behavior, enzymes have been immobilized by Waterland *et al.* (1975). The enzyme is physically absorbed within the macroporous region of the hollow fiber. By applying this technique, a large amount of enzyme is to be immobilized, which is advantageous when considering the production of enzymatic reactors operating at high yields.

Breslau and Kilallen (1976) reported the achievement of 5·96 mg lactase/cm^2 Romicon XM50 membrane surface (117 mg/ml membrane matrix volume) using their new ultrafiltration enzyme-loading technique. At their high loading, the enzyme that is absorbed within the membrane matrix has been found to have gel-like properties. The enzyme is immobilized by pumping enzyme solution into the shell side of a tubular hollow-fiber device. High enzyme loading is achieved by employing high concentrations in the enzyme solution and successive repetitions of the loading process.

Immobilization by entrapment in a polymer-matrix of lactase has been extended considerably. Thus, cellulose and its derivatives (triacetates) have been used in processes of lactase entrapment (Linko *et al.*, 1976; Pastore & Morisi, 1976). Cellulose triacetate fibers with lactase (Pastore & Morisi, 1976) are employed in equipping reactors for processes of lactose hydrolysis from milk. The stability of fiber-entrapped yeast lactase depends on operating temperature, the presence of magnesium ions in the external solution, and bacterial contamination. The enzyme is stable in the range of 0–25°C; it follows that the process temperature, 4–7°C, does not cause a loss of activity. Magnesium ions are necessary for activity.

In the immobilization of lactase through covalent bonds, Brown and Touet (1981) employ polymers based on modified polyacrylamide:

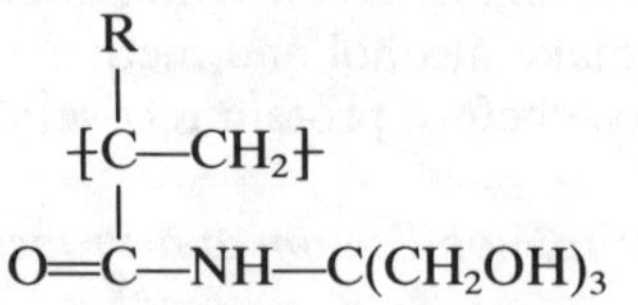

The support polymer is previously activated with BrCN. The polymer pearls thus obtained are used in filling a glass column through which a 5% lactose solution in phosphate buffer has been percolated (0·1M, pH 7, for 6 days, at a temperature of 20°C). The activity–time curve, as plotted in Fig. 11, indicates a sudden decrease of activity of the immobilized lactase in the first 24 h, followed by an increased activity, practically constant for 120 h.

Kay *et al.* (1968) attached lactase to dichloro-s-triazinyl cellulose paper and to chloro-amino-s-triazinyl derivatives of DEAE-cellulose.

Active lactase was immobilized on a polyisocyanate polymer previously applied to Teflon stirring bars (Hustad *et al.*, 1973). Immobilized lactase was active at pH 8·75 and possessed a good stability at this pH, as indicated by subsequent assay at pH 6·5 (Fig. 12). The immobilized lactase (β-galactosidase) was used continuously up to 137·6 h with less sensitivity to glucose inhibition compared to the soluble enzyme.

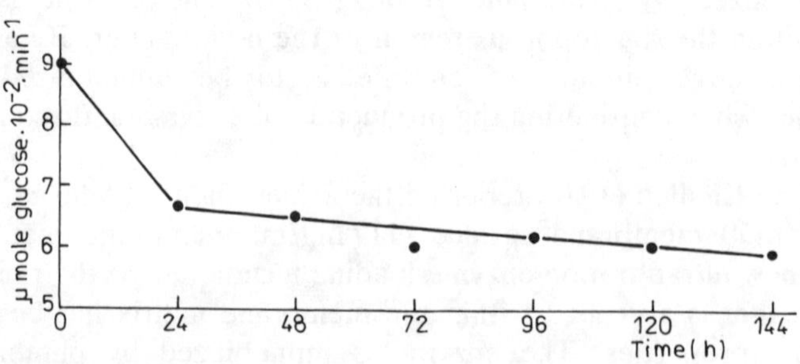

Fig. 11. Continuous hydrolysis of a solution of lactase (5%) through percolation through a column of immobilized β-galactosidase. Reprinted with permission from Brown & Touet (1981).

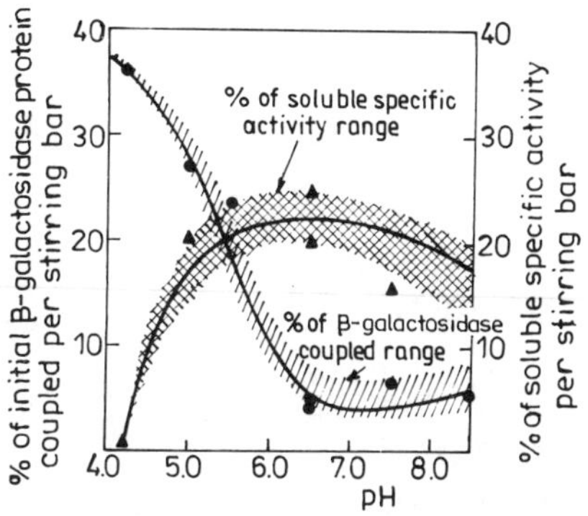

Fig. 12. Effect of pH on efficiency of coupling of β-galactosidase and on percent of soluble specific activity per stirring bar. From Hustad *et al.* (1973), by permission of the authors.

The Michaelis–Menten constants for soluble and immobilized β-galactosidase at pH 6·5 were 13·1mM and 22·1mM (Fig. 13). The immobilized lactase exhibited good stability, retaining 65·6% of the original activity after intermittent use for 118 days.

Plainer *et al.* (1982) employed methacrylic polymers possessing reactive epoxy groups for a one-stage immobilization of lactase, through absorption and covalent binding. The product obtained—known as Plexazym—is hard, having a high reticulation degree and a high pore volume, thus meeting the requirements of an immobilization support. Depending on the lactase source used in the immobilization, several types of Plexazym are known, such as Plexazym LA (lactase from *Aspergillus oryzae*; optimum pH = 4·5, heat stable,

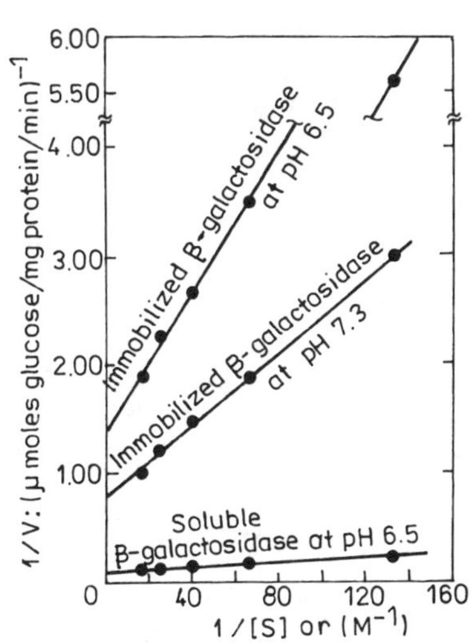

Fig. 13. Lineweaver–Burk plots for immobilized β-galactosidase at ph 6·5 and 7·3 and for soluble enzyme at pH 6·5. From Hustad *et al.* (1973), by permission of the authors.

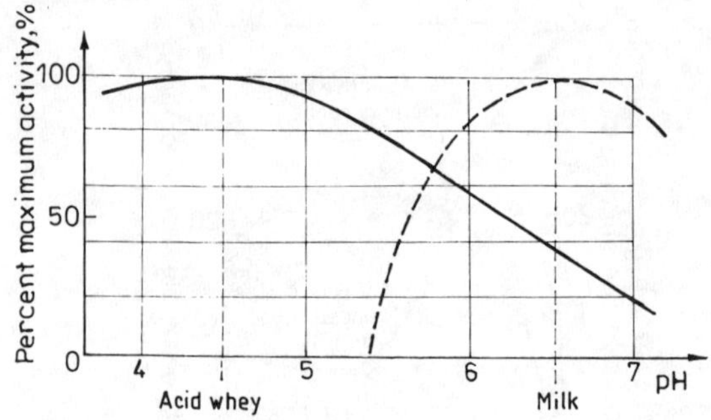

Fig. 14. Influence of pH upon the activity of Plexazym LA and Plexazym LY. From Plainer *et al.* (1982), by permission of the authors.

not sensitive to free calcium ions; large pH range, large utilization); and Plexazym LY (lactase from *Saccharomyces lactis*; optimum pH = 5·6; sensitive towards heat and free calcium; narrow range of utilization). Immobilized lactase from *S. lactis* may be utilized only in neutral media (milk), which is a consequence of the narrow range of optimum pH activity (Fig. 14). The hydrolysis temperature of Plexazym LA is 55°C, which prevents microbial developments, too. With acidic whey (pH ~ 4·5), Plexazym LA may be utilized if employing a temperature range of 30–50°C (Fig. 15). Heterogeneous elements from milk and whey (drops of fatty emulsion and casein micelles) do not hinder the hydrolytic process, due to the favorable morphological properties exhibited by Plexazym. Pore diameter is chosen so that only lactose molecules should enter the support (Fig. 16).

In the case of acidic whey, a productivity of 70 t/kg Plexazym LA is to be attained, for a hydrolysis time of 60 days (Fig. 17).

Lactase enzymes were immobilized on porous inorganic supports by the aqueous silane–glutaraldehyde method (Pitcher *et al.*, 1976). Porous silica bodies, of 30/45 mesh size and 370 Å average pore diameter, were made by Corning Glass Works, using a procedure developed by R. A. Messing. Titania

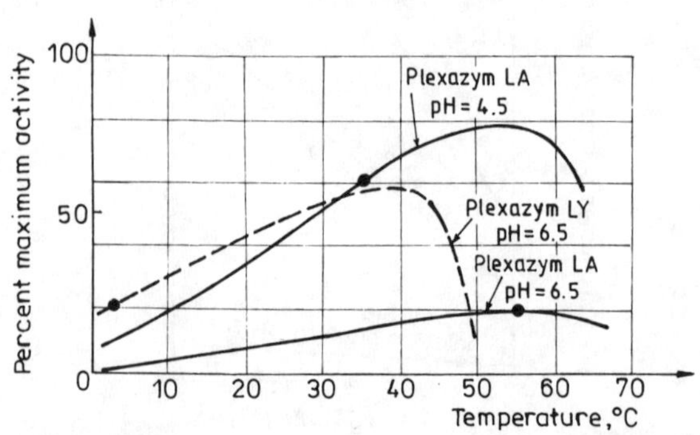

Fig. 15. Influence of temperature upon the activity of immobilized lactases. From Plainer *et al.* (1982), by permission of the authors.

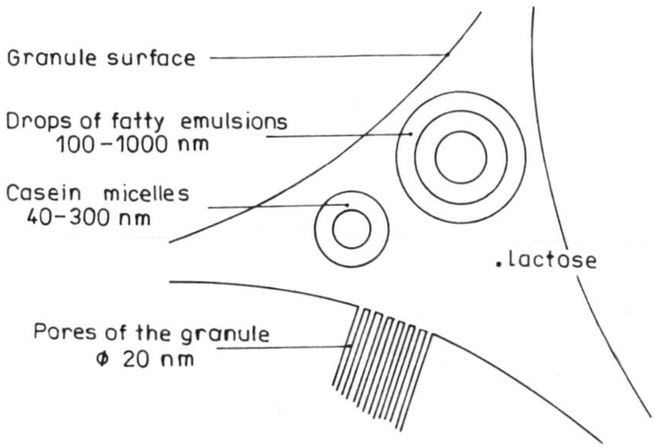

Fig. 16. The pearls–milk system in the hydrolysis of lactose with Plexazym. From Plainer *et al.* (1982), by permission of the authors.

porous bodies were prepared in a similar fashion (400 Å average pore diameter, 1 mm particle diameter). Zirconia-coated controlled-pore glass (CPG) particles, 550 Å average pore diameter and 40/80 mesh particle size, were also used as supports.

The following types of lactase have been employed: Lactase-M, fungal lactase; Lactase-Y, yeast lactase; Lactase-LP, from *Aspergillus niger*; Lactase-W, yeast lactase. The coupling efficiency of these enzymes is given in Table 6. Coupling efficiency is defined as $E_{\mathrm{IME}}/(E_{\mathrm{initial}} - E_{\mathrm{recovered}})$ where E is enzyme activity, i.e. the percentage of enzyme not recovered from the enzyme solution after attachment, that is observed as active IME. For lactase, $E_{\mathrm{recovered}}$ has normally been negligible (Ford & Pitcher, 1974; Weetall *et al.*, 1974).

The pH profiles of Lactase-M (Fig. 18) (Weetall *et al.*, 1974), immobilized as well as soluble, are similar, with a downward shift in pH for the immobilized forms.

A number of IME samples were evaluated for operational durability by continuous use in columns. Activities were calculated from eqn (5) (Ford &

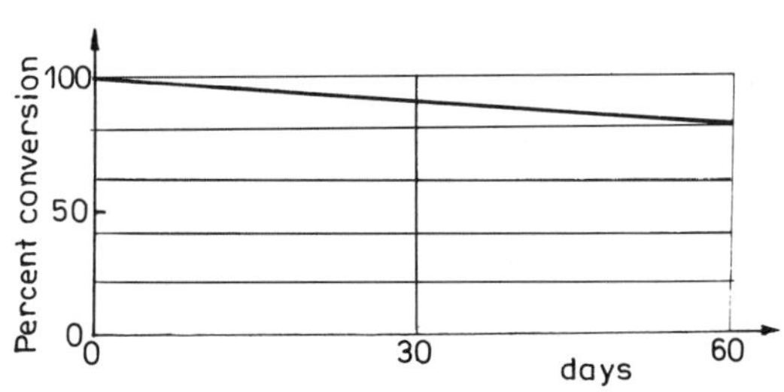

Fig. 17. Utilization of Plexazym LA in the hydrolysis of acidic whey. Fixed layer reactor, 35°C, flow = 50 liter/h, acidic whey (pH = 4·5; 5% lactose), activation every 3 days with a solution of a quaternary ammonium salt. From Plainer *et al.* (1982), by permission of the authors.

Table 6
Coupling Efficiencies[a]

Enzyme	*Carrier*	*Activity offered*	*Units (60°C)/g IME*[b]	*Coupling efficiency (%)*
Lactase-M	ZrO_2–CPG	118	75	64
Lactase-Y	ZrO_2–CPG	963	850	88
Lactase-W	ZrO_2–CPG	1 055	800	76
Lactase-W	TiO_2	1 055	500	47
Lactase-W	SiO_2	1 055	800	76
Lactase-W	SiO_2	1 055	650	62

[a] Reprinted with permission from Pitcher *et al.* (1976).
[b] IME = Immobilized enzyme.

Pitcher, 1974):

$$\frac{E_t}{v} \text{ or } \frac{E}{F} = \frac{1}{k\left(\frac{K_i - K_m}{K_i}\right)(S_0 - S)} + \left(\frac{K_m S_0}{K_i} + K_m\right) \ln\left(\frac{S_0}{S}\right) \quad (5)$$

where:

k = turnover number;
E = amount of enzyme;
F = volumetric flow rate of substrate;
v = volume of substrate concentration;
S = substrate concentation at time t or in the reactor effluent.

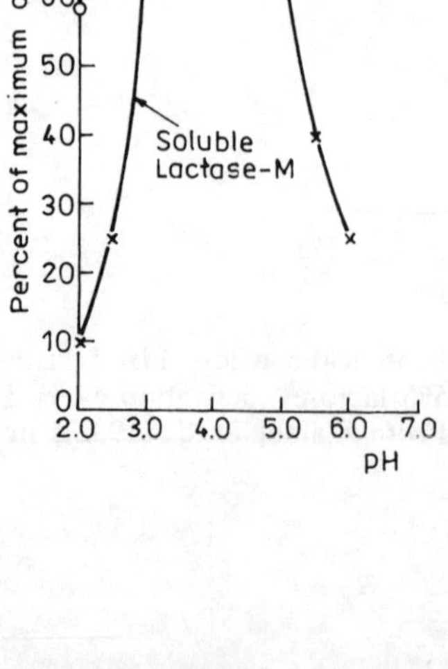

Fig. 18. Lactase-M pH profile at 60°C. Reprinted with permission from Pitcher *et al.* (1976).

4.1.2 Processes of Whey Hydrolysis with Immobilized Lactase

A thorough study of the employment of immobilized lactase in whey hydrolysis has been undertaken by Pitcher *et al.* (1976). On utilizing immobilized lactase on inorganic supports (see Section 4.1.1), the parameters for designing an installation of acidic whey hydrolysis can be settled.

Limitations on internal mass transfer of pore diffusion appear to be minimal, at least up to 60°C with 400 units (40°C) per gram immobilized lactase (IME) (30/45 mesh silica carrier) for the following reason. Assume, as the extreme case, 7–8 kcal/g mol activation energy for TiO_2–Lactase-W. Table 7 gives the result of making the reaction rate entirely diffusion controlled. In that case the apparent activation energy should have been equal to half the intrinsic activation energy plus 1 or 2 kcal/g mol for diffusivity effects. An intrinsic activation energy of 13 or 14 kcal/g mol would be estimated (Pitcher *et al.*, 1976). The mass transfer has been established through the IME film, for columns with heights varying between 6 and 71 cm.

Potential effects of back-mixing can be estimated from a dispersion number calculation described by Levenspiel (1962). The dispersion number is defined as D/uL, where D is the dispersion coefficient, u is the fluid velocity, and L is the bed height. This dispersion number can be related to the Peclet number,

$$N_{Pe} = ud_p/D\omega \quad (6)$$

where d_p = particle diameter and ω = bed void fraction, by

$$D/uL = (N_{Pe})^{-1}(d_p/\omega L) \quad (7)$$

At Reynolds number less than 10 in a packed bed, the Peclet number is approximately equal to 0·5. Even for a small packed bed, 6 cm in height, containing 30/45 mesh particles (0·46 mm in diameter), with $\omega = 0{\cdot}35$ (Pitcher *et al.*, 1976), $D/uL = 0{\cdot}044$.

Prenosil and Hediger (1985) have published an ample study referring to

Table 7
Activation Energies[a]

Enzyme	*Activation energy (kJ/g mole)*
Lactase-Y (soluble)	43·99
Lactase-Y (IME)	47·34
Lactase-M (IME)	28·23
Lactase-W (IME)[b]	32·68
Lactase-W (IME)[c]	50·38

[a] Reprinted with permission from Pitcher *et al.* (1976).
[b] Immobilized on 1 mm TiO_2 particles.
[c] Immobilized on 30/45 mesh SiO_2 particles.

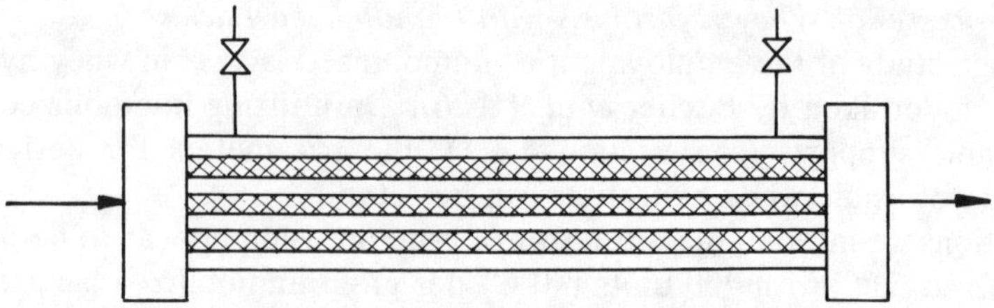

Fig. 19. Continuous hollow-fiber ultrafiltration membrane reactor. Reprinted with permission from Prenosil & Hediger (1985).

modeling of capillary membrane reactors, immobilization of lactase in such reactors, and their utilization in installations of whey hydrolysis.

The obvious advantages of membrane reactors for biological reaction have attracted a host of investigators. Two types of such reactor have been mainly investigated: filtration and diffusion membrane reactors. Our interest has been focused on the diffusion membrane reactor; a hollow-fiber membrane system was chosen in particular for its easy handling and favorable surface–volume ratio.

Let us consider an asymmetric hollow-fiber membrane system such as that shown in Fig. 19. The biocatalyst is kept in the porous structure of the capillary shell side and the substrate solution is fed continuously into the capillary lumen, while the shell side is drained and closed. The substrate diffuses towards the biocatalyst in the porous structure, it reacts, and the product is removed by the opposite way. Figure 20 shows schematically the transport processes in one hollow fiber with the biocatalyst immobilized in the porous structure (region 3) and the substrate flowing in its interior capillary lumen (region 1). Substrate comes into contact with the enzyme by diffusion across

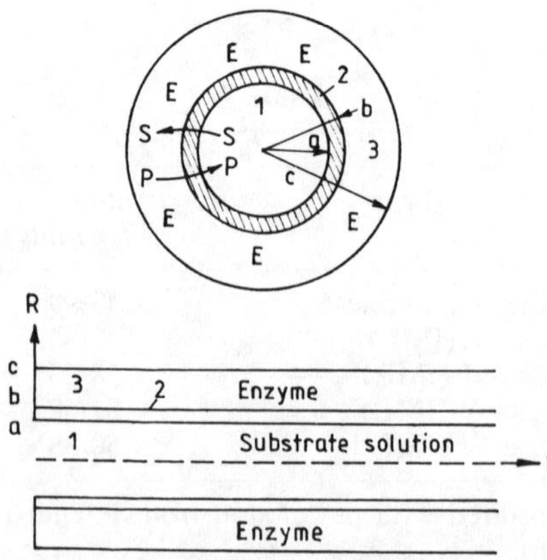

Fig. 20. Geometry and transport processes in one membrane capillary. Reprinted with permission from Prenosil & Hediger (1985).

the membrane skin (region 2). The products of an enzymatic reaction diffuse back again into bulk solution in the lumen.

A steady-state, differential mass balance on elements of fluid in all three regions results in a system of differential equations written in cylindrical coordinates (Prenosil & Hediger, 1985):

$$D_1\left(\frac{\partial^2 C_1}{\partial r^2}+\frac{1}{r}\frac{\partial C_1}{\partial r}\right)=u_z\left(\frac{\partial C_1}{\partial z}\right) \quad \text{Region 1} \qquad (8)$$

$$D_2\left(\frac{\partial^2 C_2}{\partial r^2}+\frac{1}{r}\frac{\partial C_2}{\partial r}\right)=0 \quad \text{Region 2} \qquad (9)$$

$$D_3\left(\frac{\partial^2 C_3}{\partial r^2}+\frac{1}{r}\frac{\partial C_3}{\partial r}\right)=v \quad \text{Region 3} \qquad (10)$$

where D = diffusion coefficient (L^2T^{-1}), C = substrate concentration (ML^{-3}), r and z are radial and axial distances (L), and v = reaction rate ($ML^{-3}T^{-1}$). The terms on the left-hand sides represent the radial diffusion of substrate. The right-hand side of eqn (8) is the convection term, where for laminar flow:

$$u_z = u_0(1 - r^2/a^2) \qquad (11)$$

where u_z and u_0 are the axial and maximum axial fluid velocities (LT^{-1}), and a = internal capillary radius (L). On the right-hand side of eqn (10) is a kinetic expression for the reaction considered. In this case, simple Michaelis–Menten kinetics with product inhibition was used:

$$v = \frac{V_{max} C_3}{C_3 + K_m[1 + (C_0 - C_3)/K_i]} \qquad (12)$$

where V_{max} is the maximum reaction rate ($ML^{-3}T^{-1}$), and K_m and K_i = Michaelis–Menten and inhibition constants (ML^{-3}). Boundary conditions are:

$$C_1(r, z) = C_0 \quad \text{for } z \leqslant 0 \qquad (13)$$

$$(\partial C_1/\partial r) = 0 \quad \text{for } r = 0 \qquad (14)$$

$$D_1(\partial C_1/\partial r) = D_2(\partial C_2/\partial r) \quad \text{for } r = a \qquad (15)$$

$$D_2(\partial C_2/\partial r) = D_3(\partial C_3/\partial r) \quad \text{for } r = b \qquad (16)$$

$$(\partial C_3/\partial r) = 0 \quad \text{for } r = 0 \qquad (17)$$

The model was converted to dimensionless form using the following dimensionless groups:

Dimensionless radial variables:

$$R = r/a \qquad (18)$$

$$\varepsilon = (r - b)/(c - b) \qquad (19)$$

so that $(R, \varepsilon) = (0, 1)$. Further:

$$\beta = b/(c-d) \tag{20}$$

$$Z = z/(N_{\mathrm{Pe}}\, a) \quad \text{Graetz number (dimensionless length)} \tag{21}$$

$$N_{\mathrm{Pe}} = u_0 a/D \quad \text{Peclet number} \tag{22}$$

The Biot number resulted from the analytical solution of eqn (9) for region 2, combined with the boundary conditions, eqns (15) and (16):

$$\mathrm{Bi}_1 = \gamma(D_2/D_1)[1/(2\ln(b/a))] \quad \text{Region 1} \tag{23}$$

$$\mathrm{Bi}_3 = \gamma(D_2/D_3)[1/2\beta\ln(b/a))] \quad \text{Region 3} \tag{24}$$

where γ = partition coefficient, and the Thiele modulus ϕ^2 for the membrane porosity is given by:

$$\phi^2 = \frac{V_{\max}(c-b)^2}{K_{\mathrm{m}} D_3} \tag{25}$$

Dimensionless concentrations and kinetic parameters:

$$C_{\mathrm{i}} = c_{\mathrm{i}}/c_0 \tag{26}$$

$$\xi = x_0/K_{\mathrm{i}} \tag{27}$$

$$\theta = K_{\mathrm{m}}/c_0 \tag{28}$$

The model was solved numerically using an extremely efficient orthogonal collocation method.

The conversion is $X = f(\mathrm{Bi}_1, \mathrm{Bi}_2, \beta, \phi^2, \theta, \xi, Z)$ (Prenosil & Hediger, 1985).

The enzyme lactase was immobilized in the hollow-fiber (Prenosil & Hediger, 1985) module in a back-flush mode under sterile conditions. The procedure was essentially the same for both laboratory and industrial size cartridge, except that for the latter a stronger pump was used to achieve the necessary flow rate of the enzyme solution. The apparatus is schematically shown in Fig. 21. The cross-linking with 1% glutaraldehyde solution and albumin addition procedure was exactly the same as for the laboratory module. The size of a module did not have any influence on the final specific activity per unit membrane area. When not used, the laboratory cartridges loaded with enzyme were stored at 3°C and the industrial cartridge was filled with 0·67% acetic acid solution. For the industrial unit, the flow rate on the lumen side was typically 20–30 liter/min, the entrance pressure being adjusted to $p = 1{\cdot}65$ bar. The process was stopped when the permeate flow at a given pressure reached its initial value as before immobilization.

The apparatus schematically shown in Fig. 21 was used for the continuous hydrolysis experiments. The flow diagram of the pilot-plant is shown in Fig. 22. All the experiments were run at temperature $T = 40$°C and with one cartridge at a time. In most of the laboratory experiments, 5% lactose solution was used as a substrate; in the pilot-plant reconstituted whey was used with slightly

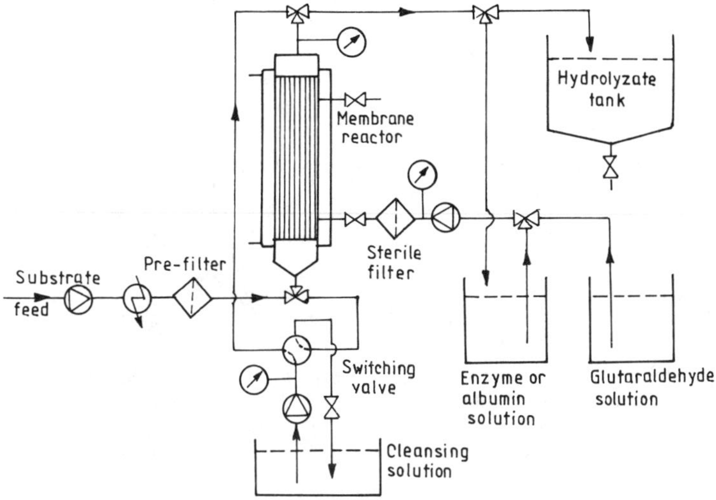

Fig. 21. Hollow-fiber membrane reactor. Reprinted with permission from Prenosil & Hediger (1985).

fluctuating lactose concentration; the pH was controlled and kept at 4·5. Finally, some experiments were done with whole whey and skimmed milk, feeding this substrate directly into the membrane reactor. In the experiments, ϕ^2 and Z were varied by changing the flow rate and the enzyme loading.

The results of the continuous hydrolysis experiments with various enzyme loadings are presented graphically in terms of X as a function of Z or, when

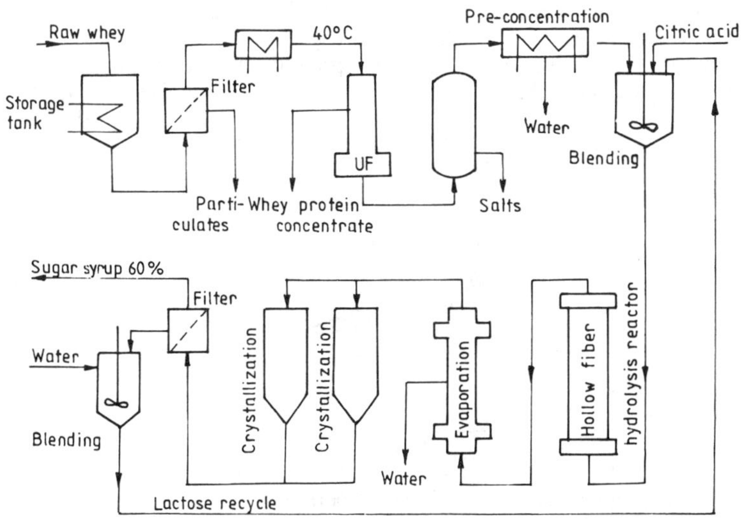

Fig. 22. Flow diagram of the pilot-plant. Reprinted with permission from Prenosil & Hediger (1985).

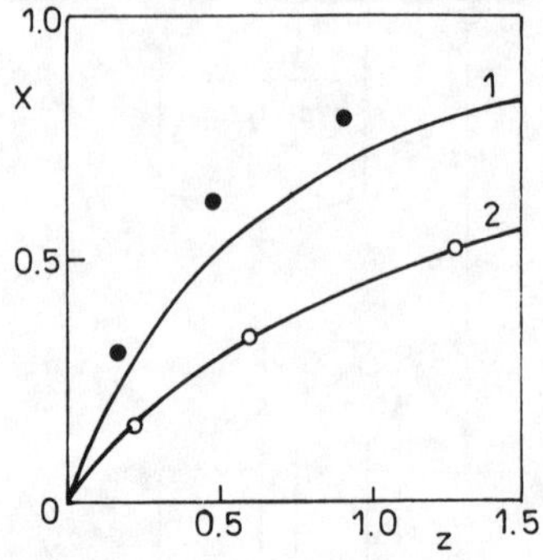

Fig. 23. Model predictions compared with experiments 1 and 2 (see Table 8). Reprinted with permission from Prenosil & Hediger (1985).

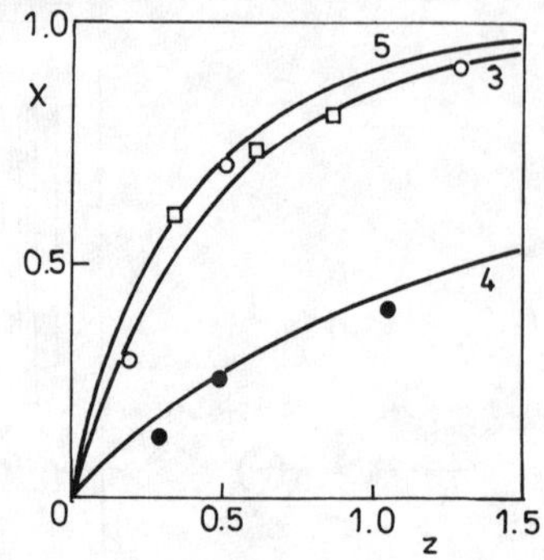

Fig. 24. Model predictions compared with experiments 3, 4 and 5 (see Tables 8 and 10). Reprinted with permission from Prenosil & Hediger (1985).

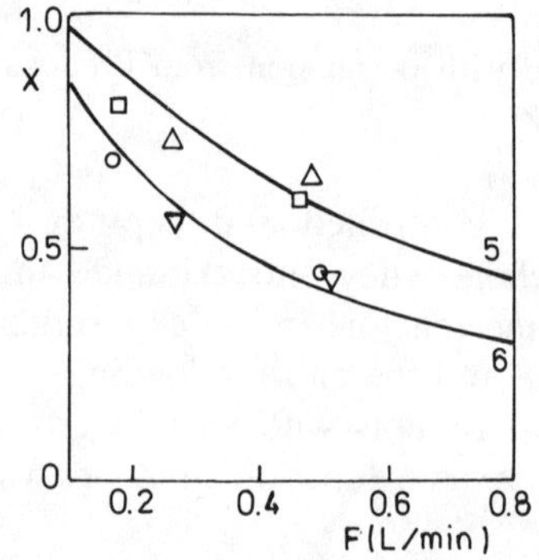

Fig. 25. Model predictions compared with experiments 5 and 6 (see Table 10). Reprinted with permission from Prenosil & Hediger (1985).

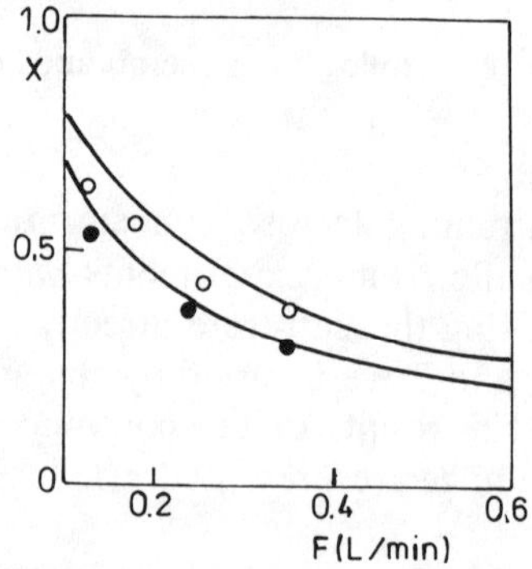

Fig. 26. Model predictions compared with experiments 7 and 8 (see Table 9). Reprinted with permission from Prenosil & Hediger (1985).

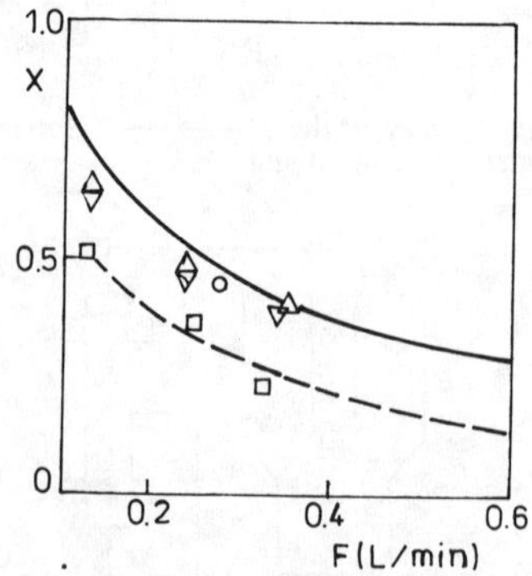

Fig. 27. Model predictions compared with experiment 9 (△); with experiment 10 (▽, 1 week later; ○, feed by reciprocating pump); and with experiment 11 (□, untreated whey, pH = 5·5) (see Table 9). Reprinted with permission from Prenosil & Hediger (1985).

possible, as a function of the flow rate F (Figs 23–27). In the model calculations it was assumed that the ROMICON industrial cartridge had 100% and the AMICON cartridge 80% of the fibers active (Tables 8, 9 and 10).

Figure 28 graphically plots the general diagram of whey bioconversion, applied by Corning Glass Works. The system employs a packed bed reactor in

Table 8
Module AMICON HIP3-20[a]

Experiment no.	*E (mg)*	ϕ^2	*ΔF (ml/min)*	*p (mm H_2O)*	*N*	*Z*	*X*	*Albumin (mg)*
1	102	7·58	0·92	1·9	79	0·908	0·79	—
			1·86	3·6	84	0·479	0·62	—
			5·68	10·2	90	0·168	0·31	—
2	28	2·08	0·80	1·3	97	1·277	0·52	52
			1·74	2·8	100	0·605	0·34	52
			4·88	7·3	108	0·235	0·15	52
3	235	17·64	0·80	1·3	97	1·277	0·90	—
			1·74	3·4	83	0·507	0·70	—
			5·80	8·4	113	0·205	0·29	—
4	24	1·78	0·88	1·6	87	1·045	0·40	59
			1·82	3·5	84	0·486	0·25	59
			3·92	5·6	113	0·305	0·13	59

[a] $C_0 = 146$mM. Reprinted with permission from Prenosil & Hediger (1985).

Table 9
Module ROMICON HF-26·5–43-PMS[a]

Experiment no.	*E (g)*	ϕ^2	*F (liter/min)*	*Z*	*X*	C_0*(mM)*
7	10·06	37·16	0·130	0·356	0·63	126
			0·180	0·257	0·55	126
			0·255	0·182	0·42	126
			0·350	0·132	0·38	126
8	3·99	14·74	0·130	0·356	0·53	121
			0·240	0·193	0·38	121
			0·350	0·132	0·30	121
9	15·03	55·52	0·135	0·343	0·65	134
			0·240	0·193	0·48	134
			0·350	0·132	0·40	134
10	15·03	55·52	0·135	0·343	0·63	130
			0·240	0·193	0·47	130
			0·345	0·134	0·38	130
			0·145	0·320	0·61	130
			0·275	0·168	0·45	130
11	15·10	—	0·130	—	0·51	108
			0·250	—	0·37	108
			0·330	—	0·22	108

[a] $N = N_0$; Albumin = 4 g. Reprinted with permission from Prenosil & Hediger (1985).

Table 10
Module AMICON H5P3-20[a]

Experiment no.	*E (g)*	ϕ^2	*F (liter/min)*	*Z*	*X*	C_o *(mM)*
5	24·12	23·81	0·185	0·861	0·80	124
			0·260	0·612	0·74	124
			0·460	0·346	0·60	124
			0·260	0·612	0·73	124
			0·480	0·332	0·66	124
6	7·63	7·53	0·170	0·937	0·68	127
			0·270	0·590	0·56	127
			0·490	0·325	0·45	127
			0·270	0·590	0·55	127
			0·500	0·318	0·43	127

[a] $N = 0{\cdot}8N_0$; Albumin = 7 g. Reprinted with permission from Prenosil & Hediger (1985).

the downflow mode. The fungal lactase is immobilized on to 30/35 mesh controlled pore silica beads by chemical coupling using the the aqueous silane–glutaraldehyde method. A pilot-scale, fluidized bed reactor system was developed by Coughlin and Charles (1975) and Charles *et al.* (1975) at Lehigh University in the United States. Development studies from Corning Glass Works and others on whey lactose hydrolysis have been documented in the literature (Olson & Stanley, 1973; Stanley & Palter, 1973; Pitcher *et al.*, 1976; Pitcher & Ford, 1978).

5 BIOCONVERSION OF CELLULOSIC WASTES

The potential of cellulose as a renewable raw material cannot be overestimated. It is produced and wasted annually in huge amounts and has several likely uses, if suitable technology can be developed and operated sufficiently cheaply. The main possibilities for its utilization are (1) production of single-cell protein; (2) conversion to glucose which can be used as such or as a feedstock or in other microbial processes, e.g. production of fine chemicals like citric acid or glutamate; and (3) production of liquid or gaseous fuels and solvents.

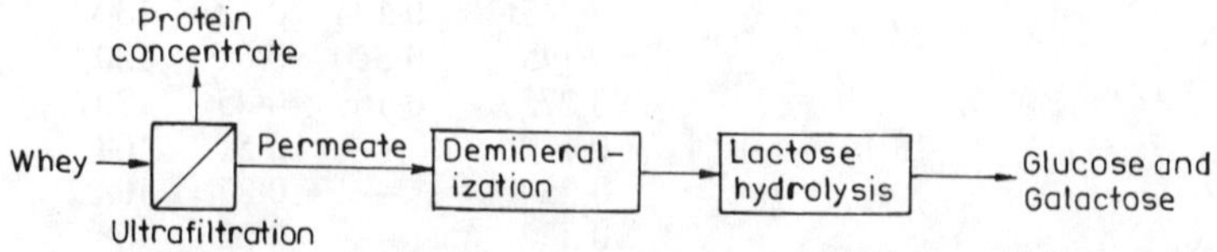

Fig. 28. A scheme for whey utilization. The protein is recovered by ultrafiltration and a sweetener is produced by hydrolyzing the demineralized permeate with immobilized β-galactosidase.

At the moment, starch or simple sugars are, or can be, used for these processes, some of the products of which could replace fossil fuels. The advantage of using cellulose, which currently has little economic value, is that its use would not compete with human nutrition (starch or sugar). Lignin, on the other hand, has little potential for biological exploitation and its presence in wastes is a nuisance. Lignin derivatives in waste waters are possible pollutants. It is fair to say that the use of cellulose in biotechnology is more potential than actual, as critical factors limit cellulose utilization, namely (1) the failure of enzymes to hydrolyze cellulose rapidly and the consequent slow growth of microbes on cellulose; and (2) the high cost of pretreatment to improve the degradability of naturally occurring cellulosic substrates.

5.1 Conversion of Cellulose to Ethanol

Transformation of cellulosic wastes, especially those resulting from paper mills—as cellulosic suspension—through hydrolysis and fermentation, to useful products (glucose, ethanol) represents the most profitable way for their recovery.

Bioconversion of cellulose to ethanol may be achieved through the utilization of coimmobilization systems of cells and enzymes. In this direction, predominant is the utilization of coimmobilization product of cellulose and cells of *Saccharomyces cerevisiae*.

In recent years a new system of soluble immobilized biocatalysts has been utilized to this purpose (Hahn-Hägerdal *et al.*, 1982). Figure 29 illustrates schematically such a system. A two-phase system was used where the biocatalyst and substrate were confined in one phase and the product either evenly distributed between the phases or distributed with some preference for the other phase. The phase systems used were aqueous two-phase systems consisting of dextran T-40 and Carbowax PEG 8000. Based on this principle, reactors have been developed which produce cellulose bioconversion to ethanol (Fig. 30). The lower part is the mixing chamber where the conversion

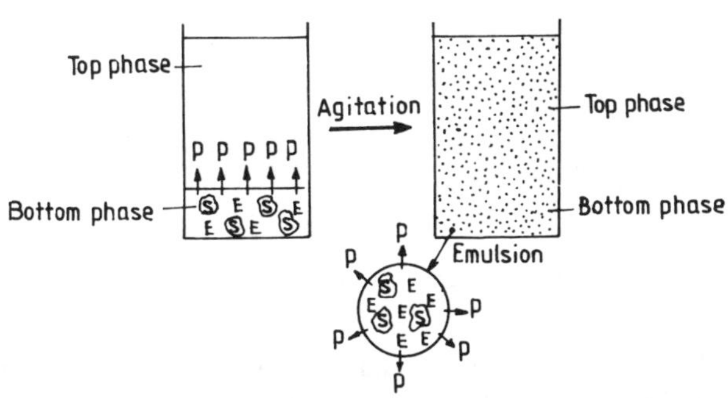

Fig. 29. Principle of soluble immobilized biocatalyst (E) acting on a particular substrate (S) forming products (P) in an aqueous two-phase system. From Hahn-Hägerdal *et al.* (1982), by permission of the authors.

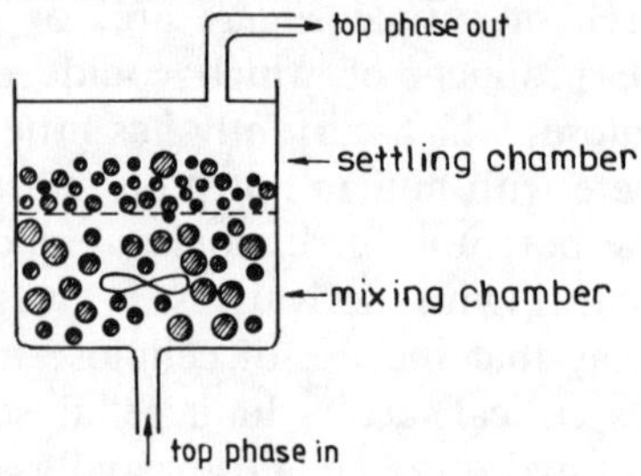

Fig. 30. Reactor for continuous conversion in aqueous two-phase systems. ● Bottom phase. From Hahn-Hägerdal *et al.* (1982), by permission of the authors.

takes place in the bottom phase and the products are extracted into the top phase. The agitation and mixing of the two phases are decelerated by a net above which the bottom phase settles down to the mixing chamber. From the settling chamber the top phase containing the product is withdrawn, the product is removed and the top phase is brought back to the bottom of the reactor. This system was studied in view of converting cellulose to soluble sugars, of fermenting glucose and producing ethanol from cellulose. The results obtained in the cellulose–soluble sugars stage are given in Fig. 31.

The next step, the conversion of glucose into ethanol by *S. cerevisiae,* was also performed both in the phase system and in the buffer system (Fig. 32).

The total reaction sequence cellulose→ethanol was then studied. Here the step of cellulose degradation is coupled to the fermentation step. The results of one such experiment performed in both buffer and phase systems are shown in Fig. 33.

The simultaneous saccharification and fermentation process (SSF), developed by Blotkamp *et al.* (1978), Becker *et al.* (1980) and Emert and Katzen (1980), employed soluble *Trichoderma reesei* cellulase together with *S.*

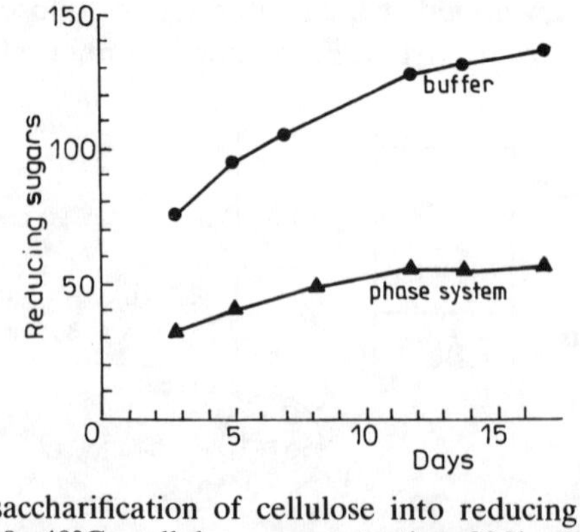

Fig. 31. Enzymatic saccharification of cellulose into reducing sugars (mg/cm³), 0·1M acetate buffer, pH 4·8, 40°C, cellulose concentration 30%. Phase system: 6% (w/w) dextran T40 and 7·5% (w/w) Carbowax PEG 8000 in 0·1M acetate buffer, pH 4·8. Volume ratio of top over bottom phase 3:1. From Hahn-Hägerdal *et al.* (1982), by permission of the authors.

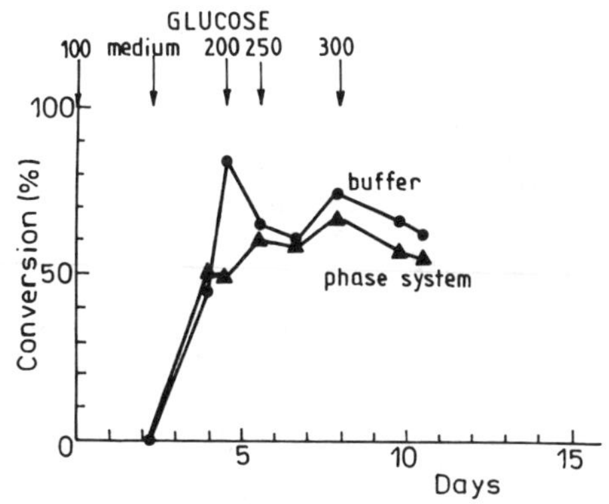

Fig. 32. Glucose fermentation with *Saccharomyces cerevisiae.* 0·1M acetate buffer, pH 4·8, 30°C. Accumulated glucose additions (mg/cm^3) are given on the diagram. Theoretical yield 0·51 mg/cm^3. Phase system is as described in the legend to Fig. 31. From Hahn-Hägerdal *et al.* (1982), by permission of the authors.

cerevisiae in the same vessel to effect a single-step conversion of cellulose to ethanol. On a pilot installation, a productivity of 0·75 g/liter h and a yield of about 53%—as compared with the theoretical values—were obtained. The process was carried out at 40°C, which actually represents a compromise between the optimum established for enzyme and yeasts. For the same mode of operation, 70 g/liter of pulp mill waste was converted to 22 g/liter of ethanol (Becker *et al.*, 1980).

Emert and Katzen (1981) obtained, by applying SSF, a yield ranging between 25 and 40%, a value which increases when employing two-step technologies. They contended that the primary reason for the improvement is the fact that the yeast removed hydrolysis products (e.g. glucose) which

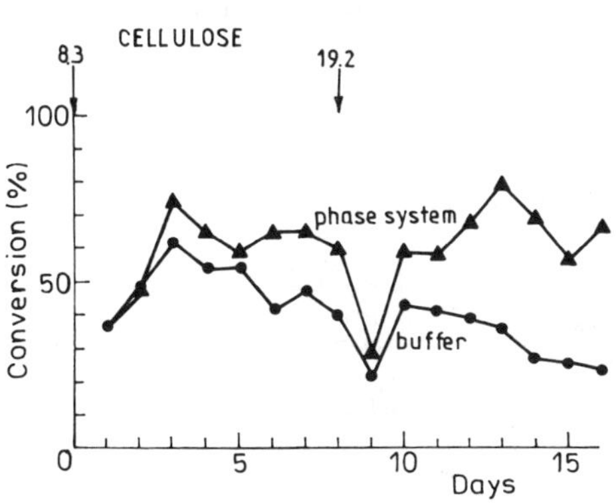

Fig. 33. Bioconversion of cellulose into ethanol with cellulose from *Trichoderma reesei*, cellobiase from *Aspergillus niger* and *Saccharomyces cerevisiae.* Accumulated cellulose additions are given on top of the diagram. Theoretical yield 0·54; same conditions as in Fig. 32. From Hahn-Hägerdal *et al.* (1982), by permission of the authors.

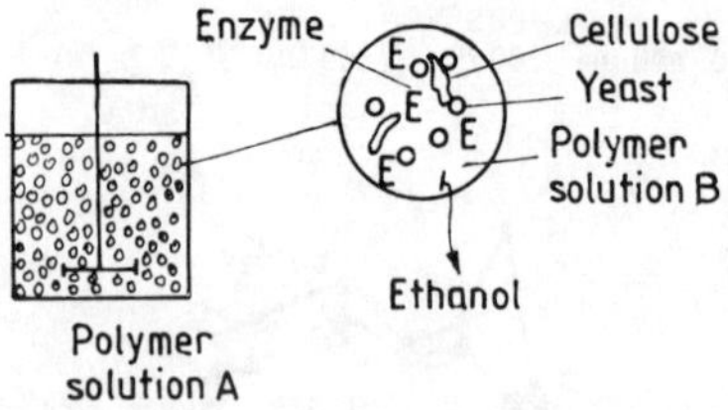

Fig. 34. Direct fermentation of cellulose to ethanol: the use of a two-phase water-soluble polymers system. Reprinted with permission from Charles & Phillips (1985).

inhibited the cellulase enzyme complex. The main problem encountered with the operation of such installations is the inhibition, through ethanol, of cellulolytic activity. A series of authors (Blotkamp *et al.*, 1978) have shown that, up to a concentration of 5% ethanol, no saccharification reduction occurs.

By employing a new technique, that of extraction, Hahn-Hägerdal *et al.* (1981) obtained ethanol from cellulose in only one stage. They exploited the fact that the polymer pair Dextran T-40/Carbowax PEG (and other pairs of water-soluble polymers) formed two phases in aqueous systems. When this two-phase system was agitated, cellulose and yeast tended to remain in the dispersed phase along with the cellulase enzyme (Fig. 34). The retention of cellulase activity in the dispersed phase was due in part to adsorption of the enzyme onto cellulose and in part to the greater solubility of β-glucosidase in the dispersed phase. The combined action of the β-glucosidase and yeast eliminated product inhibition of the cellulase. Ethanol was extracted continuously into the continuous phase and was removed easily by simple decantation.

Yeast immobilization with isomerase made possible cellulose conversion to ethanol (Hahn-Hägerdal *et al.*, 1981). The installation presented in Fig. 35 employs an ultrafiltration reactor, where cellulose is being hydrolyzed to a mixture of glucose and cellobiose, mainly by soluble cellobiase, which is retained by membrane and recycled continuously (Charles & Phillips, 1985).

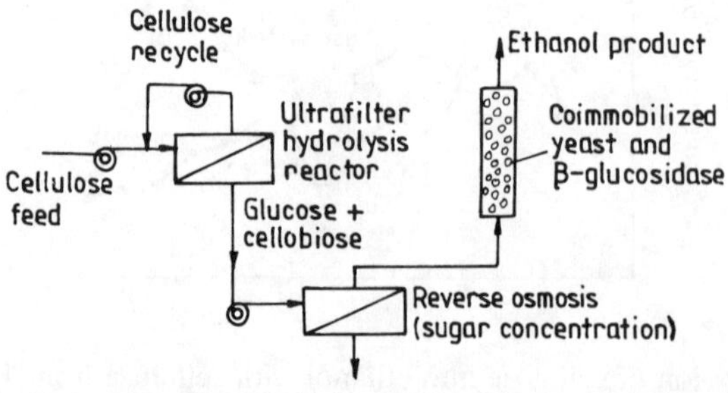

Fig. 35. Enzymatic hydrolysis of cellulose to glucose; the use of membrane reactors and of coimmobilized yeast and β-glucosidase. Reprinted with permission from Charles & Phillips (1985).

The coimmobilized enzyme/cell catalyst comprised a calcium alginate gel-bound mixture of *Saccharomyces cerevisiae* and *Aspergillus niger* β-glucosidase bound covalently to sodium alginate. A continuous reactor operating at a dilution rate of 0·2/h yielded almost 6% ethanol when fed a stream containing 5% glucose and 5% cellobiose; the productivity was 12 g/liter h and the yield was essentially quantitative.

By employing a cellulose hydrolyzate (chemically or enzymatically obtained), Hahn-Hägerdal and Mosbach (1979) transformed it into ethanol by means of a coimmobilization system made up of yeast and β-glucosidase (see Fig. 8(b)). β-Glucosidase covalently bound to sodium alginate hydrolyzes cellobiose to glucose, which is consumed by the yeast coimmobilized through entrapment in the sodium alginate gel, where β-glucosidase is bound through transformation into ethanol. The catalyst remains in stable operation for at least 4 weeks, exhibiting a yield of 80%.

6 HEMICELLULOSE CONVERSION

Hemicelluloses are regarded as wastes resulting from installations of wood prehydrolysis, in view of its subsequent chemical saccharification. These solutions of waste hemicelluloses may be recovered in processes of ethanol production.

6.1 Conversion of Xylose to Ethanol

Recent investigations (Charles & Phillips, 1985) have evidenced that xylose may be converted into ethanol, on employing a cell/enzyme coimmobilization system.

Chiang *et al.* (1981) used *Schizzosaccharomyces pombe* together with isomerase immobilized in whole cells to produce ethanol from xylose: the objective was to develop a process for the conversion of hemicellulose hydrolyzate to ethanol. Using 45 g/liter of yeast and 50 g/liter of enzyme preparation in a single reactor at pH 6 and 30°C, one may obtain about 35 g/liter ethanol in 45 h, from a dilution containing 120 g/liter xylose. The obtained yield is 63%, as compared with the theoretical value, and the productivity is 0·8 g/liter. As yeast operates best at 30°C (pH 4–5), while isomerization occurs at 70°C (pH 7·5), a compromise has been made for a temperature of 30°C and a pH of 6 (Charles & Phillips, 1985).

Jeffries (1981) utilized an installation such as that shown in Fig. 8(c), including two reactors for the production of ethanol from xylose. The immobilized enzyme reactor used xylose isomerase immobilized in whole cells, while the fermenter employed *Candida tropicalis* immobilized in *k*-carrageenin particles. After 144 h, about 41 g/liter ethanol was obtained, with a productivity of 1·2 g/liter h.

For the liquefaction of sugar-beet residues, the installation given in Fig. 36

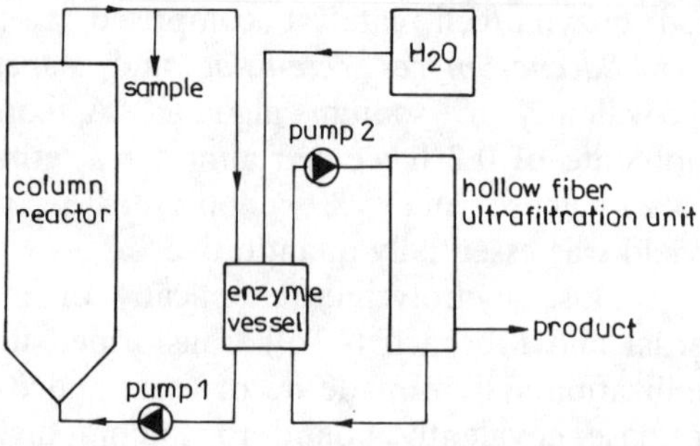

Fig. 36. Reactor system for enzymatic liquefaction of sugar-beet residues. Reprinted with permission from Pilnik (1982).

(Pilnik, 1982) has been proposed. In the reactor, which has run for many days for liquefaction of sugar-beet residues, the enzyme solution circulates through a bed of material by pump (1) and passes through an ultrafiltration cell by pump (2) in which the low-molecular-weight degradation products are separated.

7 BIOCONVERSION OF STARCH WASTES

Residual solutions of starch can be recovered in processes of glucose production, mainly for getting ethanol directly. At the same time, residual solutions of glucose may be utilized as a source of ethanol or to obtain syrup. With this aim in view, a series of immobilized enzymes and cells are being employed.

7.1 Hydrolysis of Waste Starch with Immobilized Glucoamylase

The first investigations, undertaken by Lee *et al.* (1976) and Weetall *et al.* (1976), dealing with glucoamylase (GA) immobilization, employed alkylamine porous silica or porous ceramic as supports, and glutaraldehyde as coupling agent. The activity of the coupling products varied between 2500 and 3000 IU/g, which was sufficient to achieve maximum conversion after 15–20 min residence time in a continuous packed bed reactor, far less than that required for a batch reaction using soluble GA at its most economical level.

Figure 37 presents an industrial pilot for starch hydrolysis to glucose, on using GA immobilized on SiO_2 silanized with glutaraldehyde, showing an activity of 3000 IU/g carrier. A starch solution—preliminarily hydrolyzed with 24 DE (dextrose equivalent)—has been employed. The working temperature was kept at 37°C. The analysis of the product showed a glucose production of 89·6–90·0% by weight of the solid produced and a DE of 92·4–93·0. The enzyme activity during the 70 days of operation showed no detectable change when compared to the original activity.

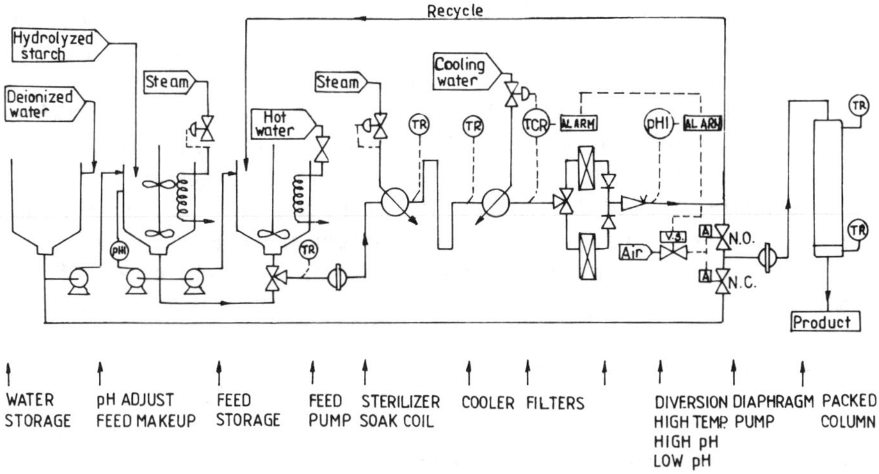

Fig. 37. Process flow diagram of Iowa State glucoamylase pilot plant. Reprinted with permission from Weetall *et al.* (1976).

Recent researches (Simionescu *et al.*, 1984) have shown that GA immobilization on xanthan, through coupling reactions activated by dicyclohexylcarbodiimide, leads to reaction products with an amylolytic activity of 12×10^5 μmol/min g.

7.2 Simultaneous Saccharification and Fermentation of Starch

The direct conversion to ethanol of residual solutions of starch has been discussed by Lee and Long (1974); they employed a continuous fermenter with hollow-fiber ultrafiltration (UF) possessing a recirculating system containing soluble amyloglucosidase (AMG) and *Zymomonas mobilis* (Fig. 38). The process temperature was limited by the use of *Z. mobilis* to 35°C; however, AMG functions best at approximately 60°C, therefore an enzyme dose much higher than normal was required to prevent overall rate limitation by the hydrolysis reaction.

Charles (1983) studied and developed an installation of starch

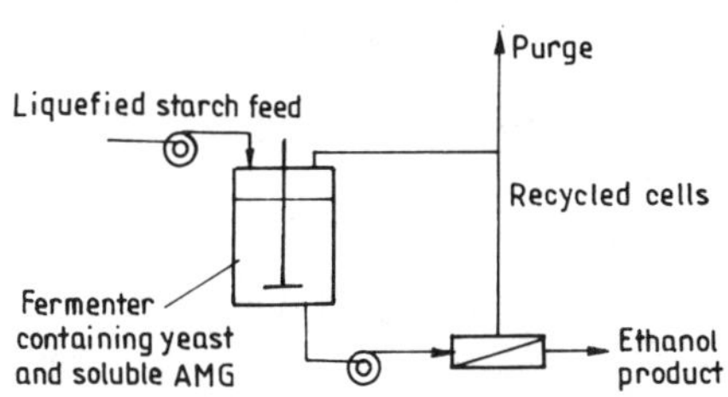

Fig. 38. Continuous ethanol fermentation employing ultrafiltration to retain yeast and AMG. Reprinted with permission from Charles & Phillips (1985).

saccharification/fermentation (S/F) that can be easily adapted for starch wastes. Fermentation occurs at 37°C, while saccharification occurs at 60°C. Two variations of the process were discussed: one used soluble GA, the other used immobilized (IME) GA. In each case, the fermentation broth was circulated constantly, through a hollow-fiber, microporous, cross-flow filter (0·4 μm pores) until hydrolysis was complete. In the soluble enzyme version, dextrins and GA passed through the filter wall and then were reacted at 60°C; in the IME version, only dextrins passed through the filter. Cells, which were retained by the filter, were recycled to the fermenter (Charles & Phillips, 1985).

7.3 Recovery of Waste Glucose Solutions

Waste solutions of glucose result from numerous processes in the food industry, especially from sugar and starch processing factories. In this respect, immobilized glucose isomerase—producing glucose isomerization to fructose—is employed, the sweetness index thus increasing. Several methods of immobilization have been employed, the most important ones, with large industrial applications, being presented in Table 11 (Bucke, 1983; Van Tilburg, 1985).

Table 11
Enzyme Sources and Immobilization Methods Employed for the Commercial Use of Glucose Isomerase[a]

Company	*Enzyme source*	*Immobilization method*
Takasaki	*Streptomyces* spp.	GI fixed in whole cells by heat treatment
Clinton Corn Processing Co.	*Streptomyces* spp.	Cell-free enzyme preparation adsorbed to ion-exchange materials; fibrous and granular forms used
Reynolds Tobacco Corp.	*Arthrobacter* spp.	Whole cells, flocculated, filtered, dried or frozen
Gist-Brocades	*Actinoplanes missouriensis*	Whole cells, entrapped in gelatin, cross-linked with glutaraldehyde; spherical particles
Novo Industri A/S	*Bacillus coagulans*	Cells lyzed, cross-linked with glutardialdehyde, granulated
Corning Glass Works	*Streptomyces* spp.	GI adsorbed to inorganic carrier
Snam Progetti	*Streptomyces* spp.	Cells entrapped in fibers of cellulose acetate
Miles Kali Chemie	*Streptomyces* spp.	Cross-linked whole cells, granulated

[a] Reprinted from Van Tilburg (1985), by courtesy of Marcel Dekker, Inc., New York.

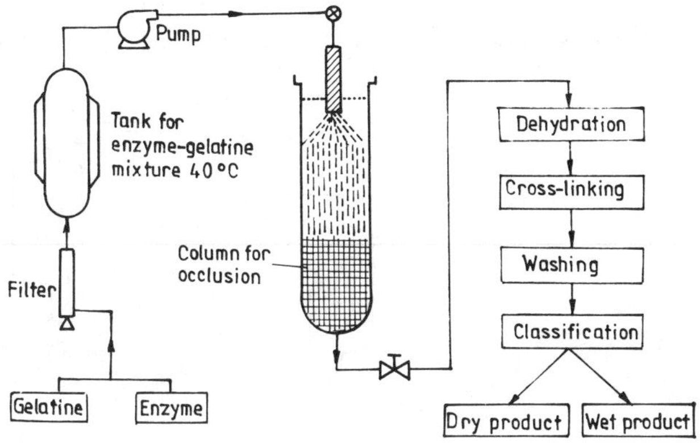

Fig. 39. Immobilization method for glucose isomerase. Reprinted from Van Tilburg (1985), by courtesy of Marcel Dekker, Inc., New York.

A widely utilized method is that applied in the Netherlands (Gist–Brocades) (Bakter Laboratories, 1972; Van Velzer, 1974; Hupkes & Van Tilburg, 1976); it immobilizes glucose isomerase (GI) through gel inclusion, in the presence of glutaraldehyde (Fig. 39). The activity of immobilized GI depends on the following agents:

(1) Metal ions: In the isomerization, two specific metal ions play an important role: Mg^{2+} and Co^{2+}. From studies carried out by Danno (1971), it was concluded that Mg^{2+} ions activate the enzyme and Co^{2+} ions stabilize the enzyme.
(2) Substrate dry substance: For industrial isomerization, the optimum substrate dry substance appears to be about 45% (w/w).
(3) pH: The optimum pH of immobilized glucose isomerase is dependent on temperature, substrate dry substance, and Mg^{2+} concentration. At 60°C, 45% (w/w) dextrose and 3×10^{-3}M Mg^{2+}, the Maxazyme GI—immob. optimum pH is 7·5.
(4) Purity of the substrate: It has been reported (Aschengreen *et al.*, 1979) that purification of the substrate is very important. A single or double treatment with activated carbon followed by ion exchange with a strong cation and a weak anion resin is advised.
(5) Optimum temperature: The optimum temperature employed ranges between 60 and 65°C, in spite of the fact that the immobilized enzyme shows its maximum activity at 90°C (Hupkes & Van Tilburg, 1976).

Utilization of diluted waste solutions of glucose in isomerization processes is not possible, which is why the normal technological flow for glucose isomerization has to be preceded by two preliminary stages consisting of evaporation to a glucose content in solution of 30–35% (Fig. 40) (Poulsen & Zittan, 1976).

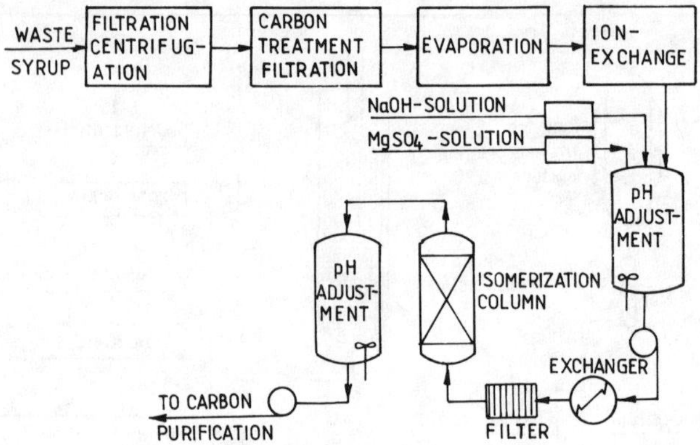

Fig. 40. Production of syrups with a high content of fructose. From Poulsen & Zittan (1976), by courtesy of Academic Press, New York.

8 METHANE BIOCONVERSION OF WASTES

It is generally known that hydrocarbons subjected to conditions of anaerobic digestion are being converted into methane.

Anaerobic digestion, from which methane also results, is performed in four stages, namely hydrolysis, acidogenesis, acetogenesis, and methanogenesis. Methane, as such, is obtained mainly through the following reactions (Fig. 41) (Maës, 1986).

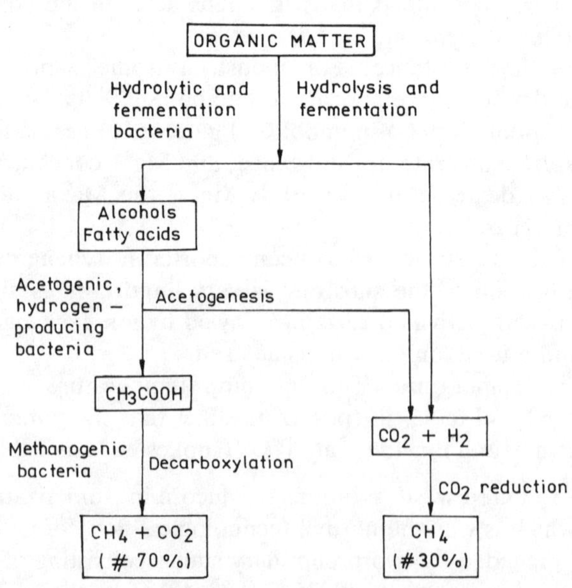

Fig. 41. Simplified scheme of methane biosynthesis.

8.1 Fermentation Procedures

According to the way in which fermenters are supplied, there exist *discontinuous procedures,* characteristic of wastes having solid layers, and *continuous procedures,* employing liquid, pumpable discharges.

Depending on the type of cell cultures, the processes of anaerobic fermentation may be classified into:

(1) cultures of microorganism in free suspension;
(2) cultures of immobilized microorganisms.

Cultures of immobilized microorganisms permit the maintenance of a high concentration of active biomass into the reactor, consequently accepting a strong pollution charge, no dilution or recycling being necessary.

The technique employing immobilized cells involves:

(1) digestion on specially arranged supports in reactors. An ascendant or descendant flow is employed, by passing the discharges through plastic systems or through mineral beds containing immobilized cells;
(2) digestion on granular supports. Microorganisms are fixed on macroporous granular supports, kept in the fluidizing layer.

The main reactors employing immobilized cells for methane production are presented in Fig. 42.

Reactors with fixed mud bed (Fig. 42(a)) contain a layer of granulated mud kept permeating through the upflow of ascending liquid. The biomass may attain 120 g/liter at the base of the fermenter. Such reactors may be applied to residual waters resulting from the agricultural and food industries. The DCO is 2–10 g/liter; the stationing time is 5–6 h; the volumic charge = 10–45 kg DCO/(m^3 day).

In order to overcome the difficulty in recycling the bacteria in the anaerobic contact process, Young and Perry (1969) developed the anaerobic filter concept, which is similar to the aerobic trickling filter process.

In the case of reactors of the anaerobic filter type, in the fixed layer (Fig. 42(b)), the biomass is fixed on a mineral (sand, gravel) or synthetic (PVC and other polymers) support, among which the liquids circulate in upflow or downflow.

These reactors may be used for residual waters having small quantities of solid materials (there is a risk of clogging). The DCO is 2–7 g/liter; the stationing time is a few hours; the volumic charge is 4–8 kg DCO/(m^3 day).

This anaerobic filter saw its first full-scale application in the treatment of wheat starch waste water (Characklis *et al.*, 1985). For carbohydrate-based waste waters this process readily achieves 80–90% removal of COD at space-loadings of 2–4 kg COD/(m^3 day). The most impressive performance is for purification of waste waters containing acetic acid, methanol and formic acid which are direct substrates for the methane bacteria and which have very low biomass yield coefficients. Here the anaerobic filter's outstanding ability to

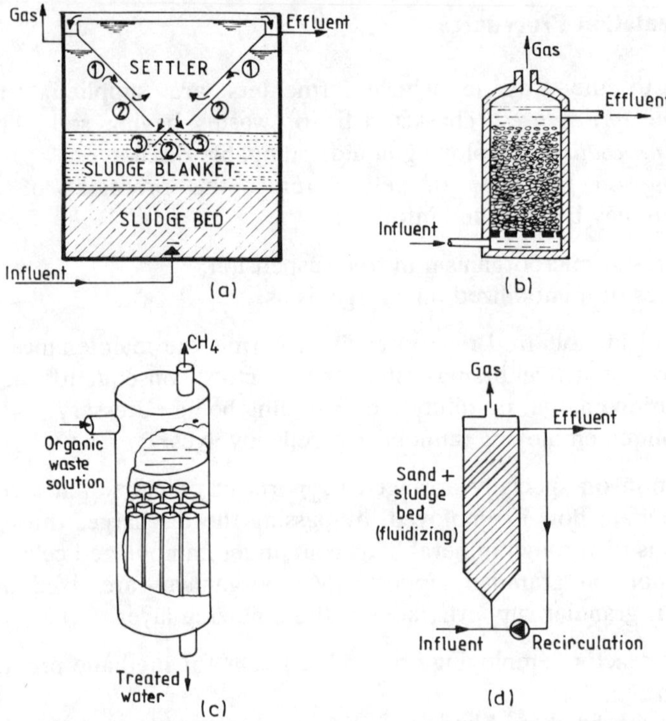

Fig. 42. Common systems of methane fermentation: (a) Scheme of an upflow anaerobic sludge blanket reactor: 1—sludge–liquid mixture inlet; 2—gas screens; 3—settled sludge return opening; (b) Anaerobic filter containing inert support material onto which the microorganisms are attached; (c) Thin film reactor in which the methanogenic bacteria adhere to the inside surface of the tubes; (d) Fluidized bed reactor in which the waste water is passed upwards through a bed of particles carrying the bacteria.

retain biomass permits COD removal of 90% at a space-loading of 10–20 kg COD/(m^3 day) with virtually no surplus sludge production.

Another recent development by Van der Berg and coworkers (1982) is a process similar to that developed by McCarty, the so-called thin film reactor. In this system the acetogenic and methanogenic bacteria adhere to the inside surface of tubes made from red drain-tile clay. Each reactor contains several tubes which have an inside diameter of 5–10 cm and a length of about 140 cm (Fig. 42(c)).

By admitting waste water from above, rather than below as is done with anaerobic filters, problems of channeling and plugging are avoided. The film formed on clay was about 1–3 mm thick and had an activity of about 0·8–1·2 COD/(g day). Maximum methane production rates of 5 cm^3 day were achieved in a reactor with 100 m^2 film support area per m^3 from bean blanching waste water.

This fixed biofilm concept was extended by developing the fluidized bed reactor (Fig. 42(d)). This contains the biomass fixed on fine particles, forming aggregates kept suspended through the upward flowing liquid.

These reactors may be applied to residual waters with a low content of solid materials: the DCO is 5–9 g/liter; the stationing time is 15–30 min.

The process has been successfully employed in the treatment of municipal waste water, and industrial effluents (Hancher, 1978) for nitrogen control as well as the aerobic BOD treatment. Pilot-scale testing has shown that anaerobic fluidized bed reactors are very effective for the treatment of various wastes including dairy, chemical, food processing, soft drink bottling and heat treatment liquors (Hickey & Owens, 1981).

8.2 Wastes Sources and Methods of Bioconversion through Immobilized Methanogenesis Cells

8.2.1 Discharges from Animal Farms

The methods employing immobilized cells have permitted continuous fermentation of pig farm discharges. In France, within the Institut Technique de Porc, the GIDA investigation group has built up a method of fermentation by employing two systems of digestion, permitting the reduction of the stationary time and therefore also of work capacity, and consequently inducing a high yield (Fig. 43).

The functional characteristics of the reactors, along with the effects of methanogenesis upon the discharge composition, are given in Tables 12 and 13. The biological fermentation mass has a significant agricultural value.

8.2.2 Agricultural and Food Wastes

The agricultural and food industries provide important residual effluents. Through their methane fermentation, up to 12% of the energy consumed in the sugar-beet and wine industries is being recovered.

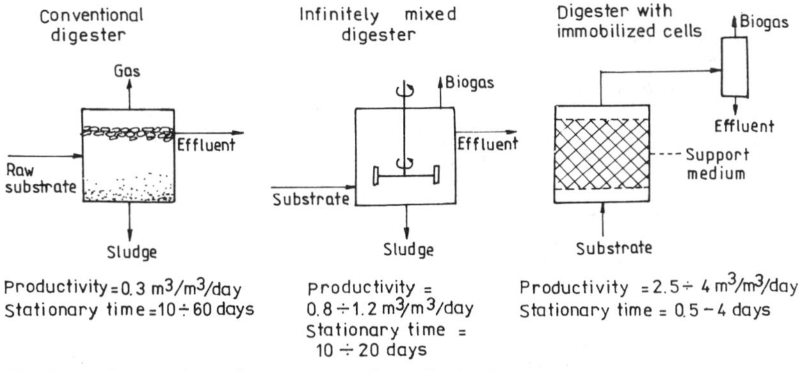

Fig. 43. Procedures of methane processing of pig farm discharges, both conventional and with immobilized cells. Reprinted from Maës (1986), by courtesy of Technique et Documentation Lavoisier, Paris.

Table 12
Functional Characteristics of Reactors[a]

	Digester 'infinite mixture'	Digester 'immobilized cells'
Productivity	1·2 m^3 biogas/m^3 reactor/day	4–5 m^3 biogas/m^3 reactor/day
Composition of biogas	75% CH_4	80–85% CH_4
Stationary time	10 days	3–3·5 days
Self consumed energy	30%	10–20%
Feeding	once per day	continuous
Temperature	37°C	37°C

[a] From Maës (1986), by courtesy of Technique et Documentation Lavoisier, Paris.

Table 13
Effects of Methanogensis upon Pig Farm Discharge Composition[c]

A. *Infinite mixture system*[a]

Parameters	*Concentration (g/liter)*	
	Raw discharge	*Digested discharge*
Total dry materials	50·9	32·1
Organic matter	36·3	19·2
N_2 (Kjeldahl)	4·7	4·4
NH_4^+	2·2	2·5
COD	47·3	26·1
BOD_5	19·6	7·5

[a] Temperature = 35°C; 10 days of hydraulic retention.

B. *Immobilized cells*[b]

Parameters	*(Concentration (g/liter)*		
	Sifted discharge	*Digested discharge*	
		TRH = 3·5	*TRH = 2·8*
Total dry materials	36·2	20·0	18·5
Organic matter	22·5	10·6	10·5
COD	40·3	15·1	13·0
NH_4^+	2·3	2·3	2·5

[b] Temperature = 35°C; TRH = hydraulic retention time (days.
[c] From Maës (1986), by courtesy of Technique et Documentation Lavoisier, Paris.

Second-generation reactors employed in the fermenation of agro-food wastes are equipped with systems of bacterial cells, immobilized on fermentation muds (Albagnac & Verrier, 1983).

Residual wastes resulting from starch enterprises are being processed through methanic fermentation in continuous installations employing mud-entrapped cells (Maës, 1986).

The method developed by the Biothane-USAB employs cells immobilized on mud granules, the installation requiring, a preliminary filtering of the discharges, thus preventing the plugging of the mud granule layer.

Residual waters from fermentation industries may be processed through the methanization in an anaerobic fluidized layer, by applying the Gist–Brocades procedure (Fig. 44) (Ledoux, 1984). According to this Dutch method, the stationary times are low—about 1–3 h. The main stages of the technological process are acidification, methanization, denitrification, and nitrification. The effluent methanized, in two stages, passes through a denitrification column, then through a recirculating nitrification one.

Traditionally, factories producing sugar from beet release large amounts of residual waters that can be recovered through methanization. In the case under study, new technologies, employing cells immobilized as anaerobic filters, have been introduced, the residual waters being circulated downflow (Fig. 45). The installation includes mainly a temperature exchanger device, an 1100 m^3 fermenter, a security flare, and a steam-producing oven.

9 IMMOBILIZED CELLS AND WASTE WATER TREATMENT

Just as cell immobilization has interesting relationships with industrial fermentation, so it does with biological waste treatment. Cells trapped in the slime layers of trickle-bed filters have already been mentioned. In the complementary activated sludge method of waste treatment, it is the flocculation of cells

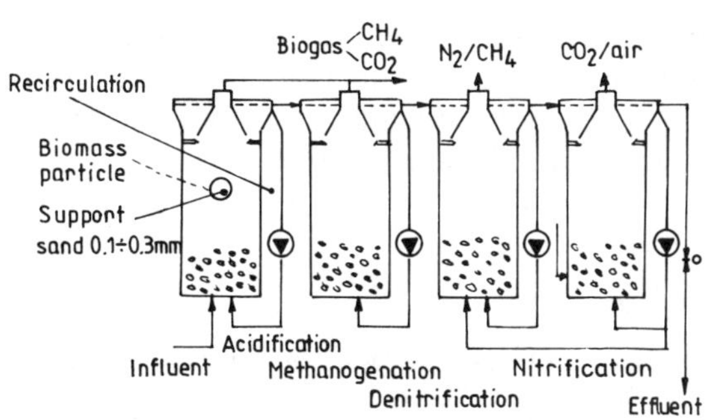

Fig. 44. The Gist–Brocades fluidized layer procedure. Reprinted from Maës (1986), by courtesy of Technique et Documentation Lavoisier, Paris.

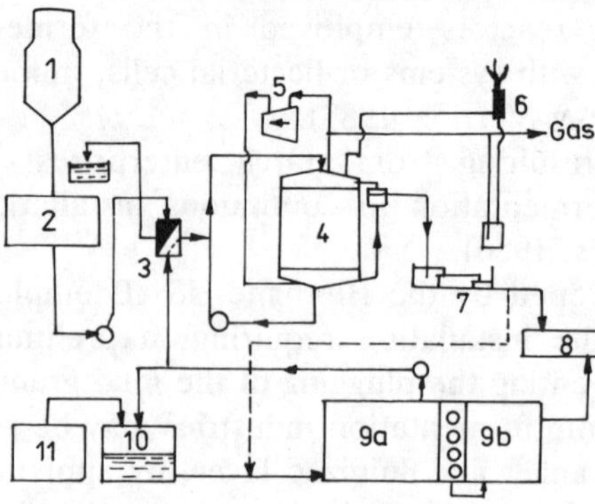

Fig. 45. The SGN procedure, employing immobilized cells. Reprinted from Maës (1986), by courtesy of Technique et Documentation Lavoisier, Paris. 1, Cooler; 2, re-heating circuit; 3, heat transfer; 4, digester; 5, compressor; 6, purge; 7, hydraulic layer; 8, removal; 9a, decantation tank; 9b, stillness tank; 10, beet washing device; 11, plant.

that permits their economic recycling. The enhancement of natural flocculation by the use of synthetic polyelectrolytes can further improve this partial cell immobilization. Several organizations have immobilized cells with a view to removing particular substances from waste water (Dunnil, 1980). At East Hyde Sewage Works, waste water is forced through a sand bed on which species such as *Hyphomicrobium* are deliberately grown with added methanol to cause nitrate reduction. Another plant at Rye Meads, processing 81 000 m^3 of water per day, uses a similar technique, but without methanol addition. Nitrate reductions between 50% and 90% are obtained (Anon; 1977). Mohan and Li (1974) examined the use of *Micrococcus denitrificans* cells encapsulated in liquid membranes for the reduction of nitrate and nitrite. The procedure, involving emulsification in the presence of cells, oil, surfactant and membrane-strengthening additives, gave capsules containing 500–600 cells, which retained 78% of their activity for 120 h, compared with zero remaining activity in 16 h for the free cells. The immobilized cells also tolerated 10^{-4}M mercuric chloride, in strong contrast to the sensitive free cells. It remains to be seen whether the requirement of very low cost can be met by this approach. Immobilized cells have been used in a sensor continuously, to evaluate the biological oxygen demand of waste water (Karube *et al.*, 1977).

REFERENCES

Albagnac, G. & Verrier, D. (1983). *Rapport Institut National de la Recherche Alimentaire,* Lille, CNRS France, p. 1.
Amory, D. E. & Rouxhet, P. G. (1988*a*). *J. Inst. Brew.*, **94,** 79.
Amory, D. E. & Rouxhet, P. G. (1988*b*). *Biochim. Biophys. Acta,* **938,** 61.
Andrews, G. F. & Przezdziecki, J. (1986). *Biotechnol. Bioeng.*, **28,** 802.

Anon. (1977). *New Scientist,* **74,** 401.
Aschengreen, P. H., Helwüg-Nielsen, B., Rosendal, P. & Østergaard, J. (1979). *Stärke,* **31,** 64.
Axén, R., Porath, J. & Ernback, S. (1967). *Nature* (London), **214,** 1302.
Bakter Laboratories, Inc. (1972). British Patent 1274158.
Bar-Eli, A. & Katchalski, E. (1960). *Nature* (London), **188,** 856.
Beck, M., Kiesser, T., Perrier, M. & Bauer, W. (1986). *Can. J. Chem. Eng.,* **64,** 553.
Becker, D. K., Blotkamp, P. Y. & Emert, G. H. (1980). *ACS Div. Fuel Chem.,* **25,** 297.
Berk, D., Behie, L. A., Jones, A., Lesser, B. H. & Gaucher, G. M. (1984*a*). *Proc. 34th Can. Soc. Chem. Eng. Conf.*, Québec City, Canada, Sept.–Oct., p. 279.
Berk, D., Behie, L. A., Jones, A., Lesser, B. H. & Gaucher, G. M. (1984*b*). *Can. J. Chem. Eng.,* **62,** 120.
Berk, D., Behie, L. A., Jones, A., Lesser, B. H. & Gaucher, G. M. (1984*c*). *Can. J. Chem. Eng.,* **62,** 112.
Bickerstaff, G. F. (1987). *Enzymes in Industry and Medicine.* Edward Arnold, London, p. 8.
Binot, R. A. (1983). Biométhanisation en lit fluidisé de microorganismes immobilisés. Thèse de Doctorat en Sciences Agronomiques, Université Catholique de Louvain, Belgium.
Bitton, G. & Marshall, K. C. (1980). *Adsorption of Microorganisms to Surfaces.* John Wiley, New York, p. 439.
Blotkamp, P. J., Takagi, M., Pemberton, M. S. & Emert, G. H. (1978). *AIChE Symp. Ser.,* **74,** 85.
Breslau, B. R. & Kilallen, B. M. (1976). *Enzyme Engineering,* Vol. 3, eds E. K. Pye & H. H. Weetall. Plenum, New York, p. 28.
Brown, E. & Touet, J. (1981). *Markromol. Chem.,* **182,** 305.
Bucke, C. (1983). *Phil. Trans. Roy. Soc., Ser. B,* **300**(1100), 369.
Chambers, R. P., Cohen, W. & Baricos, W. H. (1976). In *Methods in Enzymology,* Vol. XLIV, ed. K. Mosbach. Academic Press, New York, p. 291.
Chang, T. M. S. (1976). In *Methods in Enzymology,* Vol. XLIV, ed. K. Mosbach. Academic Press, New York, p. 201.
Chang, T. M. S. (1977). *Biomedical Applications of Immobilized Enzymes and Proteins,* Vol. 1, ed. T. M. S. Chang. Plenum, New York, p. 69.
Characklis, W. G., Bryers, J. D., Trulear, M. G. & Zelver, N. (1982). *Water Res.,* **16,** 1207.
Characklis, W. G., Bakke, R. & Trulear, M. G. (1985). In *Comprehensive Biotechnology,* Vol. 4, eds M. Moo-Young, C. W. Robinson & J. A. Howell. Pergamon Press, New York, p. 945.
Charles, M., Coughlin, R. W., Paruchuri, E. K., Allen, B. R. & Hasselberger, F. X. (1975). *Biotechnol. Bioeng.,* **17,** 203.
Charles, M. (1983). Final Report, Dept. of Energy Contract. Department of Energy, Washington, DC, p. 1.
Charles, M. & Phillips, J. (1985). In *Comprehensive Biotechnology,* Vol. 2, eds M. Moo-Young, C. W. Robinson & J. A. Howell. Pergamon Press, New York, p. 225.
Chiang, L. C., Hsiao, N. Y., Ueng, P. P., Chen, L. F. & Tsao, G. T. (1981). In *Third Symposium on Biotechnology in Energy Production and Conservation,* ed. C. D. Scott. John Wiley, New York, p. 263.
Coughlin, R. W. & Charles, M. (1975). In *Enzyme Technology Grantees–Users Conference,* ed. E. K. Pye. NTIS PB, p. 107.
Dahlqvist, A., Mattisson, B. & Mosbach, K. (1973). *Biotechnol. Bioeng.,* **15,** 395.
Danno, G., (1971). *Agr. Biol. Chem.,* **35,** 997.
Davis, J. C. (1972). *Chem. Eng.,* **79,** 114.
Demaimay, M. (1977). *Lait,* **57,** 351.

Dinelli, D. (1972). *Process Biochem.*, Aug., 9.
Dunn, I. J., Tanaka, H., Uzman, S. & Denac, M. (1983). *Ann. New York Acad. Sci.*, **413**, 168.
Dunnil, B. P. (1980). *Phil. Trans. Roy. Soc., Ser. B.*, **290**, 409.
Durand, G. & Monsan, P. (1974). *Série Synthèses Bibliographiques*, No. 5, CDIUPA-APRIA, Massy, France.
Emert, G. H. & Katzen, R. (1980). *Chem. Tech.*, **10**, 610.
Emert, G. H. & Katzen, R. (1981). *Biomass as Non-fossil Fuel Source*, American Chemical Society, Chap. 11, Washington, DC.
Ford, J. R. & Pitcher Jr, W. H. (1974). In *Whey Products Conference 'Immobilized Enzyme System'*, 14–18 September, John Wiley, New York, p. 38.
Foster, R. L. & Thomson, A. R. (1973). *Biochim. Biophys. Acta*, **321**, 409.
Furui, A. (1985). *J. Ferment. Technol.*, **63**, 467.
Furusaki, S. & Miyauchi, T. (1981). *J. Chem. Eng. Japan*, **14**, 479.
Gaffield, W., Tamimatsu, Y., Olson, A. C. & Jansen, E. F. (1973). *Arch. Biochem. Biophys.*, **157**, 405.
Gekas, V. C., (1986). *Enzyme Microbiol. Technol.*, **8**, 450.
Glosset, G. P., Cobb, J. T. & Shah, Y. T. (1974). *Biotechnol. Bioeng.*, **16**, 345.
Hackel, U., Klein, J., Meguet, R. & Wagner, F. (1975). *Eur. J. Appl. Microbiol.*, **1**, 291.
Hahn-Hägerdal, B. & Mosbach, K. (1979). Paper presented at the International Congress on Engineering and Food, Food Processing Engineering, Helsinki, Finland, 27–31 August.
Hahn-Hägerdal, B. & Mosbach, K. (1980). In *Food Process Engineering*, Vol. 2, ed. P. Linko & Y. Larinkari. Applied Science, London, p. 129.
Hahn-Hägerdal, B., Andersson, E., Lopez-Leiva, M. & Mattiasson, B. (1981). In *Third Symposium on Biotechnology in Energy Production and Conservation*, ed. C. D. Scott, John Wiley, New York, p. 651.
Hahn-Hägerdal, B., Mattiasson, B., Andersson, E. & Albertsson, P. (1982). *J. Chem. Tech. Biotechnol.*, **32**, 157.
Hajdu, J., Solti, M. & Friedrich, P. (1975). *Acta Biochim. Biophys. Acad. Sci. Hung.*, **10**, 7.
Hancher, C. W. (1978). Report 1978, pp. 1–36, *Energy Res. Abstr.*, **3**(22), Abstr. No. 52142.
Hartmeier, W. (1982). In *Use of Enzymes in Food Technology*, ed. P. Dupuy. Technique et Documentation Lavoisier, Paris, p. 205.
Hickey, R. F. & Owens, R. W. (1981). *Biotechnol. Bioeng. Symp.*, **11**, 399.
Hinberg, I., Korus, R. & O'Driscoll, N. F. (1974). *Biotechnol. Bioeng.*, **16**, 943.
Hogrefe, W., Grossenbacher, N., Cook, A. M. & Hütter, R. (1986). *Biotechnol. Bioeng.*, **28**, 1577.
Hupkes, J. V. & Van Tilburg, R. (1976). *Stärke*, **28**, 356.
Hustad, G. O., Richardson, T. & Olson, N. F. (1973). *J. Dairy Sci.*, **56**, 1111.
Jeffries, T. W. (1981). In *Third Symposium on Biotechnology in Energy Production and Conservation*, ed. C. D. Scott. John Wiley, New York, p. 315.
Karube, I., Matsunaga, T., Mitsuda, S. & Suzuki, S. (1977). *Biotechnol. Bioeng.*, **19**, 1535.
Kazanskaya, N. F., Kost, D. A. & Berezin, I. V. (1975). *Bioorg. Khim.*, **1**, 1387.
Kay, G., Lilly, M. D., Sharp, A. K. & Wilson, R. J. H. (1968). *Nature* (London), **217**, 641.
Kayem, G. J. & Rouxhet, P. G. (1983). *J. Chem. Soc. Faraday Trans. I*, **79**, 561.
Kierstan, M. & Bucke, C. (1977). *Biotechnol. Bioeng.*, **19**, 387.
Klei, H. E., Sundstrom, D. W., Coughlin, R. W. & Ziolkowski, K. (1981). *Biotechnol. Bioeng. Symp.*, **11**, 593.
Klein, J. & Schara, P. (1980). *J. Solid. Phase Biochem.*, **5**, 61.

Klein, J. & Vorlop, K. D. (1985). In *Comprehensive Biotechnology*, Vol. 2, eds M. Moo-Young, C. L. Cooney & A. E. Humphrey, Pergamon Press, New York, p. 203.
Klein, J., Eng, H., Kluge, M. & Vorlop, D. (1979*a*). *IUPAC Proc. Macro Mainz*, **3**, 1547.
Klein, J., Hackel, U. & Wagner, F. (1979*b*). *Am. Chem. Soc. Symp. Ser.*, **106**, 101.
Klein, J., Hackel, U., Schara, P. & Eng, H. (1979*c*). *Angew. Makromol. Chem.*, **76/77,** 329.
Kleinstreuer, C. & Agarwal, S. S. (1986). *Biotechnol. Bioeng.*, **28,** 1233.
Kumakura, M., Yoshida, M. & Kaetsu, I. (1978). *J. Solid Phase Biochem.*, **3,** 175.
Ledoux, H. (1984). CAPPA, Rapport l'Usine Composants Alimentaires pour Panification et Production Associées, Prouvy, France.
Lee, G. K. & Long, M. E. (1974). US Patent 3939041.
Lee, D. D., Lee, Y. Y., Reilly, P. J., Collins, E. V. Jr & Tsao, G. T. (1976). *Biotechnol. Bioeng.*, **18,** 253.
Levenspiel, O. (1962). *Chemical Reaction Engineering*. John Wiley, New York, p. 32.
Levin, Y., Pecht, M., Goldstein, L. & Katchalski, E. (1964). *Biochemistry*, **3,** 1905.
Lilly, M. D. & Sharp, A. K. (1968). *Chem. Eng.*, **215,** 12.
Linko, Y. Y., Pohjola, L., Viskari, R. & Linko, M. (1976). *FEBS Lett.*, **62,** 77.
Linko, P., Poutanen, K., Weckströn, L. & Linko, Y. Y. (1980). *Biochemie*, **62,** 387.
Lortie, R. & Thomas, D. (1986). *Biotechnol. Bioeng.*, **28,** 1256.
Maës, M. (1986). *Dechets Industriels. Mode d'Emploi.* Technique et Documentation Lavoisier, Paris, p. 216.
Manecke, G. & Beier, W. (1983). *Angew. Makromol. Chem.*, **113,** 179.
Miyaauchi, O., Nakamura, K. & Yano, T. (1982). *Agric. Biol. Chem.*, **46,** 2725.
Mohan, R. R. & Li, N. N. (1974). *Biotechnol. Bioeng.*, **16,** 513.
Mosbach, K. & Mosbach, R. (1966). *Acta Chem. Scand.*, **20,** 2807.
Mozes, N. *et al.* (1987). *Biotechnol. Bioeng.*, **30,** 439.
Olson, A. C. & Stanley, W. L. (1973). *J. Agr. Food Chem.*, **21,** 440.
Omata, T., Tanaka, A., Yamane, T. & Fukui, S. (1979). *Eur. J. Appl. Microbiol. Biotechnol.*, **6,** 207.
Pastore, M. & Morisi, F. (1976). In *Methods in Enzymology*, Vol. XLIV, ed. K. Mosbach. Academic Press, New York, p. 822.
Payne, J. W. (1973). *Biochem. J.*, **135,** 867.
Pilnik, W. (1982). In *Use of Enzymes in Food Technology*, ed. P. Dupuy. Technique et Documentation Lavoisier, Paris, p. 425.
Pitcher Jr, W. H. (1975). *Catal. Rev. Sci. Eng.*, 37.
Pitcher Jr, W. H. (1978). *Adv. Biochem. Eng.*, **10,** 1.
Pitcher Jr, W. H. & Ford, J. R. (1978). In *Enzyme Engineering*, Vol. 3, eds E. K. Pye & H. H. Weetall. Plenum Press, New York, p. 483.
Pitcher Jr, W. H., Ford, J. R. & Weetall, H. H. (1976). In *Methods in Enzymology*, Vol. XLIV, ed. K. Mosbach. Academic Press, New York, p. 792.
Plainer, H., Spröbler, B. G. & Uhlig, H. (1982). In *Utilisation des Enzymes en Technologie Alimentaire*, ed. P. Dupuy. Technique et Documentation Lavoisier, Paris, p. 189.
Porath, J. & Axén, R. (1976). In *Methods in Enzymology*, Vol. XLIV, ed. K. Mosbach. Academic Press, New York, p. 19.
Poulsen, P. B. & Zittan, D. (1976). In *Methods in Enzymology*, Vol. XLIV, ed. K. Mosbach. Academic Press, New York, p. 809.
Prenosil, J. E. & Hediger, T. (1985). *Desalination*, **53,** 265.
Quiocho, F. A. & Richards, F. M. (1964). *Proc. Natl Acad. Sci. USA*, **52,** 833.
Quiocho, F. A. & Richards, F. M. (1966). *Biochemistry*, **5,** 4062.
Reesen, L. & Strube, R. (1978). *Process Biochem.*, **13,** 21.
Rouxhet, P. G. (1985). *Cerevisia*, **4,** 167.

Rouxhet, P. G., & Mozes, N. (1989). In *Récents Développements Technologiques dans les Réacteurs a Cultures Fixées,* Nice, 4–6 April. CFRP-AGHTM, Paris, p. 13.
Rouxhet, P. G., Mozes, N., Van Haecht, J. L., Reuliaux, L. & Palm-Gennen, M. H. (1984). In *Third Eur. Congr. Biotechnology,* Vol. I, München, 10–14 September. Verlag Chemie, p. 319.
Rouxhet, P. G., Mozes, N., Hermesse, M. P. & Matta-Ammouri, G. (1987). In *Proc. 4th Eur. Congr. on Biotechnology,* Vol. 2, eds O. M. Neijsael, R. R. Van der Meer & A. M. Luyben. Elsevier, Amsterdam, p. 193.
Sharma, B. P., Bayley, L. F. & Messing, R. A. (1982). *Angew. Chem. Int. Ed. Engl.,* **21,** 837.
Sheih, W. K. & Keenan, J. D. (1986). *Adv. Biochem. Eng. Biotechnol.,* **33,** 131.
Shuler, M. L., Hallsby, G. A., Pyne Jr, J. W. & Cho, T. (1986). *Ann. New York Acad. Sci.,* **469,** 270.
Simionescu, Cr., Dumitriu, S., Popa, M., Dumitriu, M. & Hriţcu, D. (1984). *Colloid & Polym. Sci.,* **262,** 705.
Sitton, O. C., Magruder, G. C., Book, N. L. & Gaddy, J. L. (1980). *Biotechnol. Bioeng. Symp.,* **10,** 213.
Smith, J. E., Berry, D. R. & Kristiansen, B. (1980). *Fungal Biotechnology.* Academic Press, New York, p. 18.
Stanley, W. L. & Palter, R. (1973). *Biotechnol. Bioeng.,* **15,** 597.
Sümer, E. (1978). *Biotechnol. Bioeng.,* **20,** 1853.
Tanaka, A., Jin, N. I., Kawamoto, S. & Fukui, S. (1979). *Eur. J. Appl. Microbiol. Biotechnol.,* **7,** 351.
Tanaka, M., Kawaide, A. & Matsuno, R. (1986). *Biotechnol. Bioeng.,* **28,** 1294.
Tharakan, J. P. & Chan, P. C. (1986). *Biotechnol. Bioeng.,* **28,** 1064.
Van der Berg, L. & Kennedy, K. J. (1982). *Energy Biomass Wastes,* **6,** 401.
Van Haecht, J. L., Bolipombo, M. B. & Rouxhet, P. G. (1985). *Enzyme Microbiol. Technol.,* **27,** 217.
Van Tilburg, R. (1985). In *Starch Conversion Technology,* eds. G. M. A. van Beynum & J. A. Roels. Marcel Dekker, New York, p. 175.
Van Velzer, A. G. (1974). US Patent 3838007.
Vieth, W. R. & Venkatasubramanian, K. (1976). In *Methods in Enzymology,* Vol. XLIV, ed. K. Mosbach. Academic Press, New York, p. 768.
Vorlop, K. D. & Klein, J. (1983). In *Enzyme Technology,* ed. R. M. Lefferty. Springer International, New York, p. 219.
Wallat, J., Monsan, P. & Riba, J. P. (1986). *Biotechnol. Bioeng.,* **28,** 151.
Wandrey, C., Wichmann, R., Berke, W., Morr, M. & Kula, M. R. (1984). In *Third Eur. Congr. Biotechnology,* Vol. I, München, 10–14, September. Verlag Chemie, p. 239.
Wang, J. H. C. & Tu, J. L. (1969). *Biochemistry,* **8,** 4403.
Wang, S. S. & Vieth, W. R. (1973). *Biotechnol. Bioeng.,* **15,** 93.
Waterland, L. R., Robertson, C. R. & Michaels, A. S. (1975). *Chem. Eng. Commun.,* **2,** 37.
Webb, C., Fukuda, H. & Atkinson, B. (1986). *Biotechnol. Bioeng.,* **28,** 41.
Weetall, H. H. (1974). *Anal. Chem.,* **46,** 602.
Weetall, H. H. *et al.* (1974). *Biotechnol. Bioeng.,* **16,** 295.
Weetall, H. H. *et al.* (1976). In *Methods in Enzymology,* Vol. XLIV, ed. K. Mosbach. Academic Press, New York, p. 776.
Weiland, P., Jostmann, T. & Onken, U. (1984). *Chem. Ing. Tech.,* **56,** 244.
Woychik, J. H. & Wondolowski, M. V. (1973). *J. Milk Technol.,* **36,** 31.
Yamada, K. (1977). *Biotechnol. Bioeng.,* **19,** 1563.
Yamada, H., Shimizy, S., Tani, Y. & Hino, T. (1980). *Enzyme Eng.,* **5,** 405.
Young, J. C. & Dahab, M. F. (1982). *Biotechnol. Bioeng. Symp.,* **12,** 303.
Young, J. C. & Perry, L. M. (1969). *J. Water Pollut. Control Fed.,* **41,** R160.

Chapter 4

SOLID SUBSTRATE FERMENTATION—A BIOTECHNOLOGICAL APPROACH TO BIOCONVERSION OF WASTES

O. PAREDES-LÓPEZ & A. ALPUCHE-SOLÍS

Lab. de Biotecnología de Alimentos, Unidad Irapuato, Centro de Investigación y de Estudios Avanzados del IPN, Apdo Postal 629, 36500 Irapuato, Gto, Mexico

CONTENTS

1 INTRODUCTION

Solid substrate fermentation processes have been used by man for many centuries. The term 'solid substrate fermentation' (SSF) describes the microbiological transformations on and/or within the particles of solid matrix (solid substrate) where the liquid content, bound with them, is at the level corresponding to the water activity (a_w) assuring growth and metabolism of cells, but not exceeding the maximal water-holding capacity of the solid matrix (Cannel & Moo-Young, 1980*a*; Durand *et al.*, 1988; Paredes-López & Harry, 1988). In other words, in SSF the moist water-insoluble solid substrate is fermented by microorganisms in the absence or near-absence of free water, resulting in semisolid or solid fermentation systems (Hesseltine, 1977*a*). 'Solid state' and 'solid substrate' fermentations are terms used by different workers but they are essentially the same. On the other hand, in liquid state fermentations (LSF), the substrate is solubilized or suspended as fine particles in a large volume of water. SSF have been used long before the microbiological or biochemical processes involved were understood. Various SSF processes with histories reaching far back in time are still practiced today (Moo-Young *et al.*, 1983; Steinkraus, 1983*a*). These traditional SSF systems deal with fermented foods (e.g. tempeh in Indonesia, miso in Japan, pozol in Mexico); mold-ripened cheeses; starter cultures for fermented brews; and silage and compost. Some of the first information on fermented oriental food preparations dates back to 1000 BC (Cannel & Moo-Young, 1980*a*).

The koji process may be considered the prototype of SSF. At present, SSF processes are used on a commercial scale for the production of different types of fermented foods, particularly in Asia and in some developing countries (Aidoo *et al.*, 1982). In addition to these types of foods, such processes have been used successfully in recent years for large scale production of protein-enriched feed from starchy wastes, single cell protein (SCP) from a variety of wastes, fungal metabolites, and bioconversion of plant/animal/domestic wastes into useful products (Aidoo *et al.*, 1982; Tengerdy, 1985).

There are abundant organic wastes in most countries of the world (Fig. 1). The net productivity of dry biomass due to photosynthesis by plants on the earth is estimated to be close to 150×10^9 t per year (Bassham, 1975; Rajarathnam & Bano, 1989), of which about 11% is represented by the primary production of materials with food/feed properties. Wastes of various kinds (e.g. unedible parts of plants, wastes generated during the harvesting season, wastes during handling and processing), suitable for consideration as

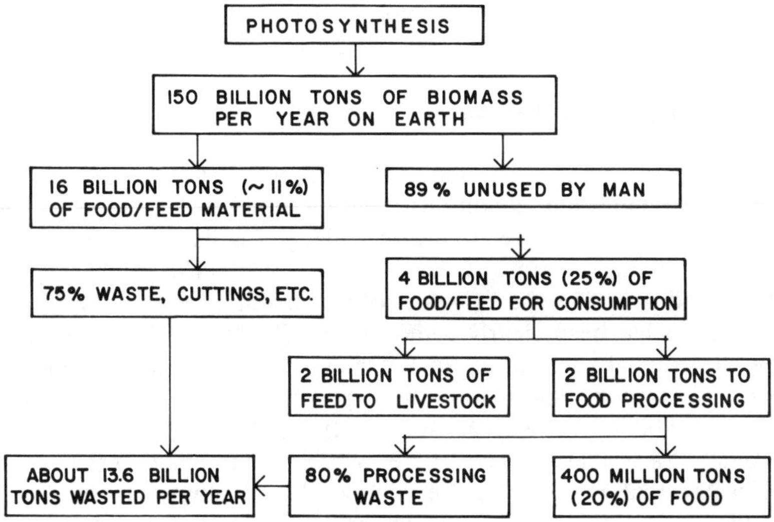

Fig. 1. Biomass produced on earth by photosynthesis and annual availability of wastes.

sources of raw materials, may amount to more than 13×10^9 t per year. The potential for the transformation of these wastes via SSF into valuable products is remarkable.

The attraction of SSF comes from its closeness to the natural way of life for many microorganisms and from its simplicity (Tengerdy, 1985; Hesseltine, 1987). A good example of these assessments is a compost pile where multiple microbial cultures grow (e.g bacteria, actinomycetes, yeasts, fungi). Another attraction of SSF is the low volume of water present in the media per mass of substrate which can substantially reduce the space occupied by the fermentor without severely sacrificing the yield of products, and the cost of water removal at the end of fermentation is minimized. Aeration and mixing requirements may also be easily met (Hesseltine, 1977*a, b*; Moo-Young *et al.*, 1983). This type of fermentation requires relatively low capital investments, and reduced energy inputs; also it is especially suitable for growing mixed cultures of microorganisms where symbiosis stimulates better growth and productivity (Bushell & Slater, 1981). These advantages can outweigh the disadvantages of SSF in relation to LSF, which are the lower product yields and the slowness of the fermentation. They also present heat build-up problems, bacterial contamination, problems in scale-up, and restrictions for growth measurement, and are more difficult to control due to the lack of adequate sensors and efficient solids-handling techniques, especially for continuous operations. This chapter focuses on the fundamental principles of SSF, both scientific and technological, and it also pays attention to the various opportunities to convert agroindustrial wastes by SSF into useful products.

2 CRITICAL FACTORS FOR MICROBIAL GROWTH ON SOLID SUBSTRATES

2.1 Water Activity and Moisture

Water potential is directly related to the energy state of water in organic substrates. At equilibrium conditions between liquid, solid and vapor phases, the water potential may be related to the relative humidity of the vapor phase or to the a_w of the substrate (Gervais *et al.*, 1988). The influence of water potential, conditioned by the a_w values of the medium, on the development of microorganisms has been clearly demonstrated (Mossel, 1975). It is important to know the optimal value of a_w for individual physiological phenomena, such as microbial growth, sporulation and production of main primary and secondary metabolites (Grajek & Gervais, 1987). Thus, it is very important to know the a_w of a system at a given time, which is an estimate of the proportion of free water that is available for biological and physicochemical activity. During SSF, substrate polymers are degraded into smaller molecules and CO_2 while metabolites and water are produced. This means that composition and structure of the medium changes during fermentation together with its sorption properties (Paredes-López & Harry, 1988).

Perhaps the most widely sought parameter with regard to the effects of a_w on microorganisms is that relating to the minimal level for growth. Theoretically, this point defines the limit below which a microorganism or group of microorganisms can no longer reproduce (Beuchat, 1981; Troller, 1987). The specificity and uniqueness of minimal levels can be illustrated in Table 1. Bacteria usually are capable of subsisting at higher a_w ranges than are yeasts and molds. The minimum a_w at which molds and yeasts are capable of growing is about 0·60. This corresponds to a water content of 10–20%, and it can be concluded that SSF may take place at moisture levels between 10–20% and 80%. The exploitation of combinations of environmental factors together with lowered a_w levels is common in industrial microbiology. Generally, as the minimal a_w for growth of a microorganism is approached, alteration of other environmental factors (e.g. pH) will have a greater impact on growth rates (Troller, 1987). Several studies have been conducted on the interacting effects of a_w for growth and for production of metabolites (Table 2) (Norholt *et al.*, 1977, 1979; Narahara *et al.*, 1982; Grajek & Gervais, 1987; Oriol *et al.*, 1988). The optimum a_w values for microbial growth are usually lower than those required for production of the respective metabolites.

2.2 Temperature

Temperature control is intimately connected with water activity control, partly because of the dependence of thermal diffusion on substrate moisture content, but partly because water evaporation is the greatest source of cooling that SSF systems may have (Tengerdy, 1985). Heat build-up causes evaporative water

Table 1
Water Activity and Growth of Microorganisms in Various Substrates

Range of a_w	*Microorganisms inhibited by lowest a_w of this range*	*Substrate/food*
1·000–0·950	*Escherichia, Pseudomonas, Shigella, Bacillus, Clostridium perfringens*, some yeasts	Fresh fruits and vegetables, meat, fish, milk, breads, substrates containing up to 40% sucrose or 7% sodium chloride
0·950–0·910	*Lactobacillus, Salmonella, Clostridium botulinum*, some molds, yeasts	Some cheeses, cured meat, coffee pulp, substrates containing 55% sucrose or 12% sodium chloride
0·910–0·870	Many yeasts (*Candida, Hansenula, Torulopsis*)	Dry cheeses, fermented sausage, substrates containing 65% sucrose or 15% sodium chloride
0·870–0·800	Most molds, most *Saccharomyces*	Most fruit juice concentrates, most cereal flours, pulses containing 15–17% moisture, syrups
0·800–0·750	Most halophilic bacteria	Glacé fruits, marmalade
0·750–0·650	Xerophilic molds (*Aspergillus chevalieri, A. candidus*)	Molasses, raw cane sugar
0·650–0·600	Osmophilic yeasts (*Saccharomyces rouxii*), few molds (*Aspergillus candidus, Wallermia sebi*), *Saccharomyces bisporus*	Dried fruit and vegetable wastes containing 15–20% moisture, honey
0·500	No microbial proliferation	Spices containing 10% moisture, starchy substrates containing 12% moisture
0·400	No microbial proliferation	Grain hulls, egg powder containing 5% moisture
0·300	No microbial proliferation	Bread crusts containing 3–5% moisture
0·200	No microbial proliferation	Dried fruit and vegetable wastes containing 5% moisture, whole milk powder containing 2–3% moisture

Table 2
Water Activity for Growth and for Metabolite Production in Solid Substrate Fermentation

Microorganism	*Product/ metabolite*	*Optimum growth* a_w	*Production*		*Reference*
			Minimum a_w	*Optimum* a_w	
Aspergillus oryzae	Koji	>0·900	0·942	0·975	Narahara *et al.* (1982)
Trichoderma viride TS	Polygalacturonase	<0·995	—	0·995	Grajek & Gervais (1987)
	D-xylanase	<0·995	—	0·995	
	β-glucosidase	>0·980	—	0·960–0·980	
Aspergillus niger	Microbial protein	0·980			Oriol *et al.* (1988)
Aspergillus flavus	Aflatoxin	>0·800	0·830–0·870	—	Norholt *et al.* (1977)
Penicillium cyclopium	Ochratoxin	>0·810	0·870–0·900	—	Norholt *et al.* (1979)

loss, stops vegetative growth and induces protective but nonproductive sporulation of the molds. It has been reported (Chahal, 1983) that about 59% of heat generated during SSF is used to evaporate water produced during the metabolic activity while the remaining 41% is left in the substrate to be dissipated. This amount of energy, if not dissipated immediately, as it is released, will reduce productivity or may kill the organism (Hayes, 1977). Different methods (e.g. forcing large quantities of air through the fermentor, circulation of water in a jacket around the fermentor) are used to remove the heat from the fermentors and to maintain the temperature in the range 25–32°C, which is the usual temperature range employed in SSF (Lonsane *et al.*, 1985).

2.3 pH

Although pH is one of the critical factors, the monitoring and control of pH during fermentation is not usually attempted in SSF (Lonsane *et al.*, 1985; Paredes-López & Harry, 1989). Most fungi are able to grow in a wide pH range of 5–8 (Chahal, 1983). Some of the substrates (e.g. straws, grains) have good buffering capacity and there may not be any need to adjust the pH during SSF. This advantage is, therefore, exploited in the initial adjustment of the pH of the solids in the range 4·5–5·0 during moistening by using water at the desired pH level. However, local changes in the pH of the agglomerates produced cannot be checked and result in low productivity in unagitated fermentors. In relation to the effect of pH on productivity, Chahal (1985) reported that maintenance of pH around 4·5 is very crucial for cellulase production with *Trichoderma reesei.*

2.4 Aeration and Oxygen Transfer

In a solid state culture, aeration is needed for three important functions:

(1) To maintain aerobic conditions, because the partial pressures of O_2 and CO_2 in the gas environment of a SSF are critical factors for growth and product formation. Increased aeration stimulated α-galactosidase and invertase production in SSF of wheat bran with *Aspergillus awamori.* Similarly, aflatoxin production by *A. flavus* was enhanced considerably when the aeration was increased (Silman, 1980). In contrast, ochratoxin biosynthesis by *A. ochraceus* in a rotating drum fermentor stopped when the air flow rate was greater than 0·1 liter/kg min (Lindenfelser & Ciegler, 1975).

(2) To control substrate temperature because, in comparison with submerged culture, the low moisture environment in SSF creates difficult conditions for heat transfer and temperature regulation

(3) To regulate water content/a_w of the substrate, CO_2 and volatile metabolites produced during metabolism. As was described previously,

the a_w markedly influences the cell growth and the metabolism in SSF. Narahara *et al.* (1982) and Grajek & Gervais (1987) reported that production of enzymes was strongly affected by the a_w of the substrate.

Ulmer *et al.* (1981) found that biomass production by *Chaetomium cellulolyticum* in SSF of manure fibers was not oxygen limited but depended on the ventilation of CO_2. Inhibitory effects of CO_2 under conditions of adequate O_2 supply were also observed in submerged fermentations (Paredes-López *et al.*, 1974). In a SSF of rice with *Aspergillus oryzae,* Bajracharya & Mudgett (1980) have found that growth and biomass accumulation was insensitive to pO_2 but was sensitive to pCO_2 in a closed gas system.

As it has been pointed out by Lonsane *et al.* (1985), although SSF systems are highly aerobic in nature, except those in composting/ensiling and some food fermentations (Paredes-López & Harry, 1988), little information is available on the mechanism of O_2 transfer. This mechanism is probably similar to that which takes place in hydrocarbon fermentation (Prokop *et al.*, 1971) and hence may occur from the gas phase directly and also from the O_2 dissolved in the water which keeps the substrate moist, though the contribution of the latter will be negligible.

2.5 Mixing

Mixing is an additional control parameter used in connection with aeration. Mixing of the fermenting mass has beneficial effects like providing new surfaces to aeration, distribution of inoculum, promotion of homogeneity and of growth on individual particles of the substrate, and prevention of aggregate formation and of localized changes (Hesseltine, 1977*a*; Moo-Young *et al.*, 1983). The speed of mixing or rotation of the fermentor is dictated by the same factors as those governing the rate of aeration.

3 MICROBIAL GROWTH PATTERNS AND CONTROL OF FERMENTATION

3.1 Microbial Types and Inoculum

Many microorganisms can grow on solid substrates, but only filamentous fungi can grow to a significant extent in the absence of free water (Moo-Young *et al.*, 1983). Bacteria and yeasts will grow on solid substrates at the 40–70% moisture levels, such as in composting, anaerobic and aerobic ensiling, but the growth and propagation of single cell organisms always require free water. Thermophilic bacteria grow predominantly in the first stage of composting. Lactic acid bacteria grow anaerobically and aerobically in ensiling processes causing souring and flavour changes (Tengerdy, 1985). Yeasts usually grow on solid substrates in symbiosis with other microorganisms in composting, ensiling and some industrial SSF processes involving fruit wastes and other sugar-containing natural substances.

Among the filamentous fungi, three classes have gained practical importance in SSF: (1) the Phycomycetes such as the genera *Mucor* and *Rhizopus*; (2) the Ascomycetes with the genera *Aspergillus* and *Penicillium*; and 3) the Basidiomycetes, especially white rot fungi, among them certain edible mushrooms. The former two classes grow naturally on fruits and grains and the latter on decaying lignocellulose, wood, straw, and other forestry and agricultural wastes.

For many SSFs inoculum can be prepared and used in the same fashion as in liquid media. Either preformed inoculum or dry spores in a carrier may be used to inoculate the substrate (Hesseltine, 1987). However, in the case of fungi used commercially that do not sporulate in culture, such as *Agaricus, Lentinus, Pleurotus* and *Volvariella,* the inoculum has to be prepared as spawn (Rajarathnam & Bano, 1989). In this procedure, the fungus is grown on solid particles such as rye kernels and the infected kernels are dispersed evenly throughout the substrate.

3.2 Fungal Growth Patterns and Growth Rate

3.2.1 Growth Patterns

In nature, bacteria grow best only when in a liquid phase or at least when the nutrients are in free water. Likewise, single-celled fungi, the yeasts, grow well when the nutrients are in a soluble form. Growth of such microorganisms in SSF becomes difficult where substrate carbon is not available in soluble form (Chahal, 1983). A limited success in converting lignocellulosic wastes into animal feed by using bacteria (e.g. *Cellulomonas* sp., *Alcaligenes faecalis*) or yeasts (e.g. *Aureobasidium pullulans, Candida utilis*) has been reported in semi-solid state fermentations (Han & Anderson, 1975). Low protein yields in the final product might be attributed to the fact that in such systems the unicellular organisms (bacteria and yeasts) were unable to penetrate deep into the tissue for complete utilization of the substrate.

On the other hand, filamentous fungi grow on solid surfaces such as wood, seeds, roots and leaves of plants, much as roots grow in the soil. High protein contents in the finished product have been recorded in SSF of cellulosic materials with *Chaetomium cellulolyticum* (Abdullah *et al.*, 1985). High protein content in the final product has been attributed to the fact that the hyphae of this fungi have the power to penetrate deep into the intercellular and intracellular spaces for better utilization of the substrate. In SSF experiments on growth of *Rhizopus oligosporus* on hard-to-cook common beans used as substrate, it was found that the long tubular body of the fungal filament grows alongside the solid particle, using the available surface liquid film as a source of moisture and nutrients, and penetrating cracks and pores in the substrate with special appendages to scavenge for further nutrients (Fig. 2) (Paredes-López, O., unpublished results). Filamentous fungi grow differently in LSF than in SSF. In a stirred fermentor, the fungi multiply by mycelial fragmenta-

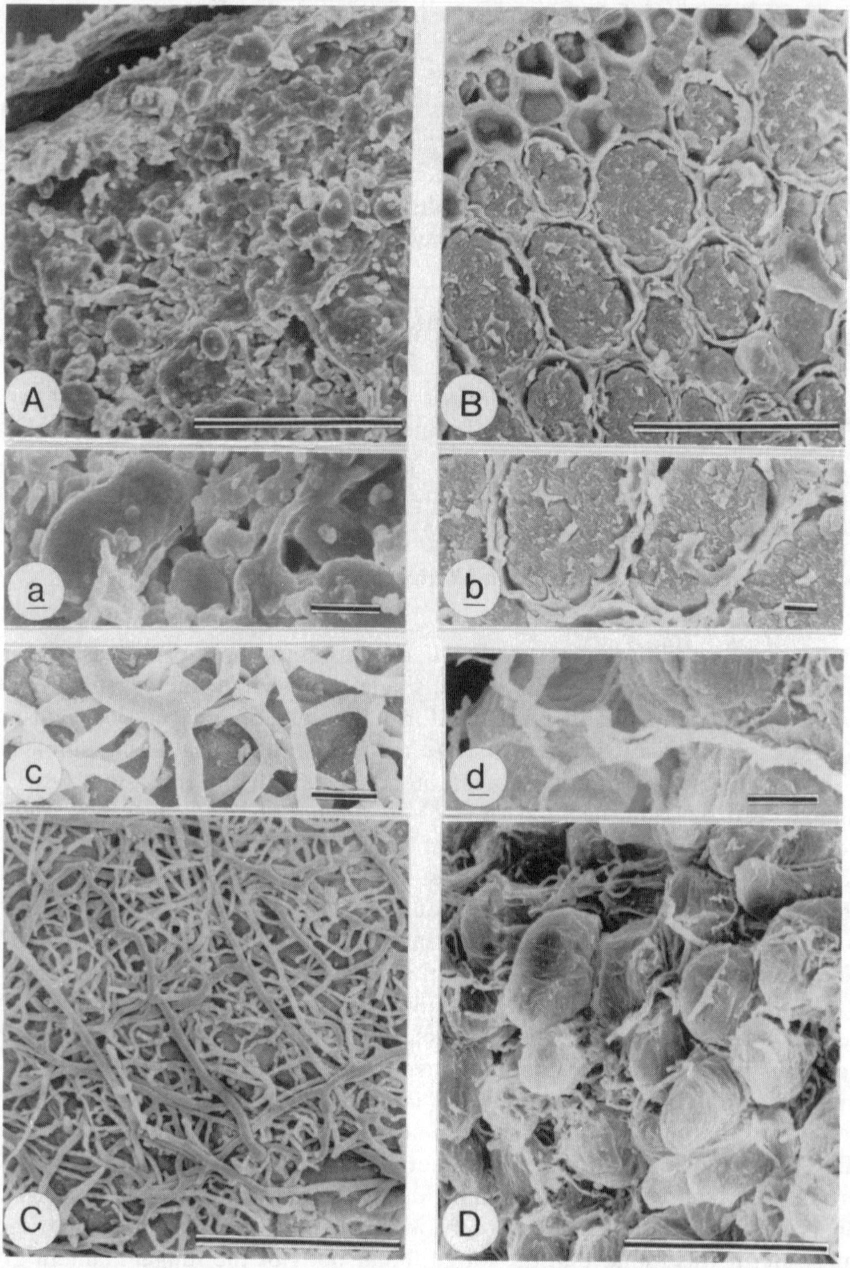

Fig. 2. Scanning electron micrographs of samples from the solid state fermentation using *Rhizopus oligosporus* and hard-to-cook common bean as substrate for preparation of a tempeh-like product. A, B = Different sections of the uninoculated cotyledons of common bean; C = growth of the fungus on the surface of substrate after 48 h of fermentation; D = growth of the fungus in the internal part of substrate after 48 h of fermentation. Bar size for A, B, C and D samples = 100 μm. a, b, c and d = Insets of corresponding samples (bar size of all insets = 10 μm).

tion. On an agar plate, fungi grow typically by apical extension in a radial fashion toward an increasing concentration gradient of nutrients. Growth rate is linear and depends on the width of the peripheral growth zone where hyphal growth is exponential and approximately equal to the specific growth rate (0·11–0·15/h) observed in liquid cultures (Moo-Young *et al.*, 1983). The low packing density of mycelia in the internal part of the substrate (Fig. 2) has been suggested to be partially due to geometric limitations (Laukevics *et al.*, 1985). Laukevics *et al.* (1985) reported that the available space per unit mass is 6–10 times smaller in SSF than in LSF. Such conditions set an upper theoretical limit on the achievable biomass concentration in SSF.

Georgiou and Shuler (1986) presented a relatively simple computer model for the growth of a mold colony on a solid surface with a defined medium utilizing glucose. Unlike submerged cultures the model needs to account for both cellular differentiation and the spatial heterogeneity in the system. Results of the simulation studies suggest that mass transfer limitations are at least partially responsible for the proliferation of differentiated structures on solid substrates as compared to liquid cultures. The major limitation of the model may well be the assumption that simple saturation kinetics with external nutrient control can be used for a dynamic system.

3.2.2 Growth Rate

One of the disadvantages of SSF is the technical difficulty for growth rate assessment in a direct way. Most of the standard procedures used in LSF for biomass measurement are of limited value or are indeed useless. The content of biomass has been estimated by measuring some of the mycelial components such as protein, DNA and RNA. Corrections need also to be made for the measured component present in the substrate. However, not much attention has been given to the fact that qualitative and quantitative changes of these components take place in actively growing cultures. Outstanding differences in protein content may be given by the standard assays at the various developmental stages of the culture (Paredes-López *et al.*, 1988). Some of the indirect methods reported for cell growth estimations are based on O_2 uptake (Sato *et al.*, 1983; Oriol *et al.*, 1988), CO_2 evolution (Narahara *et al.*, 1982; Sato & Yoshizawa, 1988), glucosamine or chitin measurement (Aidoo *et al.*, 1982; Narahara *et al.*, 1982) and infrared techniques (Durand & Chereau, 1988).

For fungi, the average mass doubling time in LSF is 3 h. In SSF of beet pulp with *Aspergillus oryzae* a maximum specific growth rate (μ) of 0·10/h was observed (Czajkowska & Ilnicka-Olejniczak, 1988). However, Oriol *et al.* (1988) reported a μ of 0·45/h for SSF of cassava by *A. niger*. For some yeasts, a μ of 0·47/h in LSF has been reported (Paredes-López *et al.*, 1976) and of 0·48/h in SSF (Sato *et al.*, 1988).

3.3 Control by Physical and Nutritional Factors

As mentioned above, the partial pressures of O_2 and of CO_2 in the gas environment of SSF may influence microbial growth and product biosynthesis.

High pO_2 increased amylase production in SSF of rice with *Aspergillus oryzae* and increased pCO_2 deteriorated amylase production in the same fermentation (Bajracharya & Mudgett, 1980). Mitchell *et al.* (1986) developed a model solid substrate to simplify studies of the basic aspects to control SSF. The substrate consisted of starch and other nutrients embedded in a κ-carrageenan gel matrix. The model system was designed to mimic the growth of *Rhizopus oligosporus* on cassava tubers, a process which may be used for protein enrichment of this substrate for use as animal feed. A large improvement in protein content was obtained by simultaneously decreasing the particle size of the substrate and increasing the nitrogen content of the fermentation medium (Mitchell *et al.*, 1988). Other physical factors (e.g. pH, temperature, mixing) may be used to control the SSF.

In SSF the restriction effects of nutritional factors may be very severe due to the limited diffusion rate of the substrate and the limited access of microorganisms to the substrate (Moo-Young *et al.*, 1983). The C/N ratio is an important indicator in SSF for nutritional regulation of growth. For composting an ideal C/N ratio is 20 (Cannel & Moo-Young, 1980*b*). The C/N ratio of garbage from North America ranges from 20 to 70 because of its high paper content. In this situation another source of nitrogen (e.g. sewage sludge) may be added to speed up the fermentation process. Nitrogen starvation favors lignin degradation whereas higher nitrogen levels favor celluose depolymerization. Fermentation media with a high C/N ratio are optimal for secondary metabolite formation (Hutter, 1982).

4 REACTORS FOR SOLID SUBSTRATE FERMENTATION

The design of SSF systems has been for the most part highly empirical in nature. The absence of accurate and reliable experimental data for such systems is the major reason for the lack of sound knowledge in this area (Aidoo *et al.*, 1982; Moo-Young *et al.*, 1983). The information is very scarce on the instrumentation used to monitor fermentation processes (e.g. dissolved oxygen electrodes, pH measurement). At the pilot-plant and large-scale fermentation of moist solid substrate, some factors need to receive special consideration for the construction or selection of the fermentation equipment (Durand & Chereau, 1988; Durand *et al.*, 1988):

(1) inoculation, sampling and transfer techniques must be simple;
(2) homegeneity of culture medium is important to prevent particle agglomeration and the settling of substrate during fermentation;
(3) agitation suitable for the substrate, not destructive for the microorganism;
(4) appropriate aeration rate and prevention of heat build-up problems;
(5) equipment sterilization;
(6) monitoring and control devices of fermentation parameters;

(7) handling (empty, refill, clear out, maintenance) must be simple;
(8) materials used for construction of the fermentation vessel;
(9) low capital costs and operating expenses.

There are different types of reactors presently used for SSF. These can be summarized as follows.

4.1 Tray Fermentor

Traditionally, most SSFs are conducted in shallow trays where the substrate is spread in about 10 cm layers or in specially formed cakes. Heat build-up is avoided in such shallow trays (Tengerdy, 1985). The fermentor is humidified by humidifiers or by forcing in humid air. This equipment is used with success in small-scale operations at the pilot or commercial level (Lonsane *et al.*, 1985); it gives a final product uniform in nature and with high enzymatic activity. It is widely used in the traditional oriental food preparation schemes, such as in the koji process (Steinkraus, 1983*b*; Paredes-López *et al.*, 1987). However, the operation of this equipment is labor intensive. The need for a large area is another disadvantage of tray fermentors.

4.2 Rotating Drum Fermentor

The rotating drum fermentors are equipped with an inlet and an outlet for air. The air inlet tube may almost reach the bottom of the drum. The drum may be provided with baffles or may be sectioned into four parts. The drum is placed on a shaft rotating on two half-bearings and is powered by a variable speed drive motor. The unit is provided with a port for charging the fermentor, water addition, inoculation and sampling (Lindenfelser & Ciegler, 1975). For the most effective agitation, baffles permit the substrate to be elevated and subsequently dropped at some point during each revolution (Hesseltine, 1977*a*). Microbial growth in these fermentors appears prolific and uniform. Operation is simple and cleaning fast. They present some problems such as microbial contamination, medium aggregation and heat build-up. They have been used successfully for ochratoxin production by *Aspergillus ochraceus* (Lindenfelser & Ciegler, 1975). Pou *et al.* (1987) used a rotating reactor to increase the protein content of an acid-pretreated bagasse pith with *Trichosporum penicillatum*.

4.3 Packed Column Fermentor

The packed column reactor uses as substrate sugar-rich wastes such as pineapple residues (González *et al.*, 1985), and granulated starchy materials such as cereal grain residues, cassava (Raimbault & Alazard, 1980) and sweet potato residues (Yang, 1988). These substrates, loosely packed into the columns, are inoculated with molds and fermented for various times. The final product is used for animal feed purposes. Also, Cochet *et al.* (1988) reported the use of a tubular reactor in a semi-solid state fermentation to produce ethanol.

4.4 Auger Tube Fermentor

The auger tube fermentor has been designed for a batch and continuous operation (Gibbons & Westby, 1988; Gibbons, 1989). It consists of: (1) a nonported steam pasteurization chamber to destroy bacterial contaminants; (2) an inoculum port; and (3) an auger tube that simultaneously conveys and mixes the substrate under fermentation. The horizontal auger measures 15·2 cm diameter by 470 cm long. The fermentation chamber is surrounded by a plastic heating/cooling jacket. This fermentor has been used to produce ethanol from sweet sorghum stalks and Jerusalem artichoke tubers. Starchy wastes may be processed as well in this reactor.

4.5 Helical Screw Fermentor

According to Tengerdy (1985), a promising new design is the helical screw fermentor. Its form of operation avoids compacting the substrate, does not damage the mycelia by shearing forces, and provides thorough mixing. Tengerdy and his group have tested a 15 liter reactor, coupled to a microcomputer-regulated temperature and moisture control device for batch and continuous cultivation. Tests have been run with the fungus *Chaetomium cellulolyticum* and wheat straw.

4.6 Fluidized Biomass Fermentor

The productivity of a microbial process is a strong function of the number of microorganisms or of microbial biomass per unit of volume of bioreactor and per unit of time. In general, the volumetric concentration of microbial biomass in any type of SSF bioreactor is small (Laukevics *et al.*, 1985). In a fluidized biomass reactor the fluidizing medium is air or an inert gas (Mishra *et al.*, 1982; Brauer, 1988). The advantages observed for this reactor at laboratory level are (1) high biomass concentration; (2) intensive mixing of all ingredients in the system; (3) optimal mass transfer; (4) continuous operation (substrate input and product removal); (5) reduction of inhibitory effects by final products; and (6) ease of control. Despite the fact that this type of reactor has a high productivity because of its high cell density compared to submerged cultures, Mishra *et al.* (1982) found a relatively long generation time of 14–17 h for growth of *Saccharomyces cerevisiae*. However, Sato *et al.* (1988) studied the production of ethanol with a thermophilic yeast in a SSF system with circulation of CO_2, and found that this procedure had, compared with the conventional SSF, a higher fermentation efficiency and a shorter fermentation time. Further studies on growth rate kinetics and scaling-up of the fluidized system are necessary.

4.7 Miscellaneous Types

A new laboratory scale reactor (2·5 kg dry-matter capacity) was designed and constructed to allow control of temperature and moisture level of the substrate without agitation (Fig. 3). A fermentor unit is composed of two reactors. Each cylindrical vessel contains a removable basket with a perforated bottom (cylinder of 40 cm diameter and 60 cm height). Forced aeration is accomplished by means of thermostated air injected at the bottom of the reactor. By bubbling in a water bath and after heating, the thermostated air allows the regulation of temperature and moisture content of the medium during the cultivation. The two reactors have strictly the same conditions for temperature and relative humidity of the inlet air. This system has a great advantage in that experiments can be made in duplicate for studying certain parameters without varying the environmental conditions. Previous experience (Durand *et al.*, 1988) with this type of equipment has demonstrated that mechanical agitation is not necessary and that there is a uniform temperature distribution.

The development and investigation of high-efficiency SSF bioreactors is one of the fundamental problems biochemical engineers have to solve. By increasing efficiency of the reactor the downstream processes will be much simplified, and consequently energy and costs will be reduced. Based on the

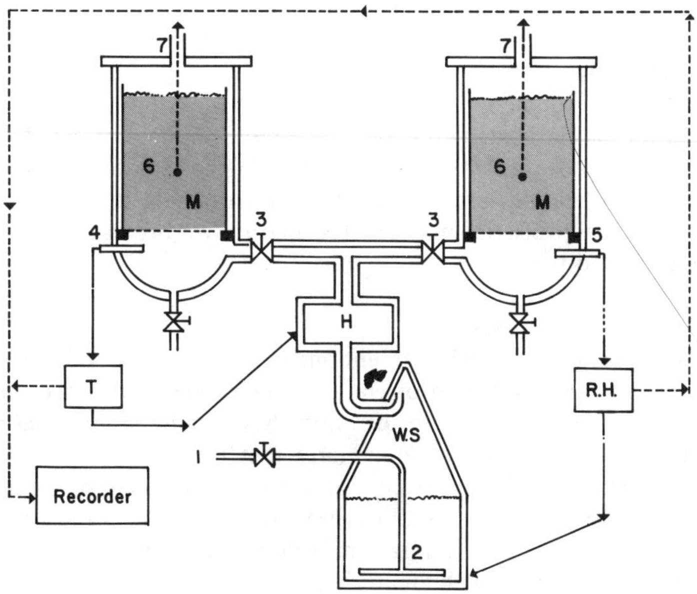

Fig. 3. Schematic representation of a fermentor unit composed of two reactors for solid state fermentation. H = heating device; M = individual reactors with perforated basket; RH = relative humidity control; T = temperature control; WS = water system for air humidification. 1 = air inlet; 2 = air sparger; 3 = valves for air flow regulation; 4, 5 = probes for temperature and relative humidity, respectively; 6 = probe for temperature of sample under fermentation; 7 = air exhaust.

findings of the fundamental research, the following requirements for high efficiency reactors may be established:

(1) high rate of microbial conversion per unit of biomass;
(2) identical conditions for enzymatic/microbial reactions in all volumetric segments of the reactor (e.g. uniform distribution of microorganisms, temperature, pH, substrate concentration);
(3) low energy requirements;
(4) uniform distribution of energy/oxygen in the reactor;
(5) small volume and site requirement of reactor;
(6) simplicity of design;
(7) applicability for batch and continuous operation;
(8) possibility of integration into production systems.

Finally, in a few words, a modern solid substrate fermentor should be able to mix large amounts of substrate for a uniform fermentation, allow maximum automation and scale-up of the process, allow continuous operation, and facilitate fermentation controls (Laukevics *et al.*, 1984; Tengerdy, 1985; Durand *et al.*, 1988).

5 FERMENTATION PROCESSES AND COMPOSITIONAL CHANGES

5.1 SSF Processes

Figure 4 shows the basic steps involved in a biotechnological process, in solid state conditions, to use wastes from different origins. Some of these steps are optional (e.g. sterilization, product isolation) according to the objectives of the particular SSF. SSF processes may be classified in two broad categories, which are briefly reviewed here.

5.1.1 Typical SSF Processes

In nature solid substrates, such as plant and animal wastes, undergo microbial colonization and transformation by various microbiological processes. Therefore, studies on solid-state cultures of molds have traditionally been restricted to substrates comprising mostly polymeric materials (e.g. starch, cellulose) with a given water holding capacity in their porous matrix. Ralph (1976) described these fermentations as those in which a solid material serves as the main source of nutrients for an organism which intimately associates itself with the surface and interstices of the material.

5.1.2 SSF Processes on Inert Solid Support

Several attempts have been made to grow fungi on solid inert materials impregnated with a liquid nutrient. Ralph (1976) described these processes as those in which a nutritionally inert solid support provides some advantage to

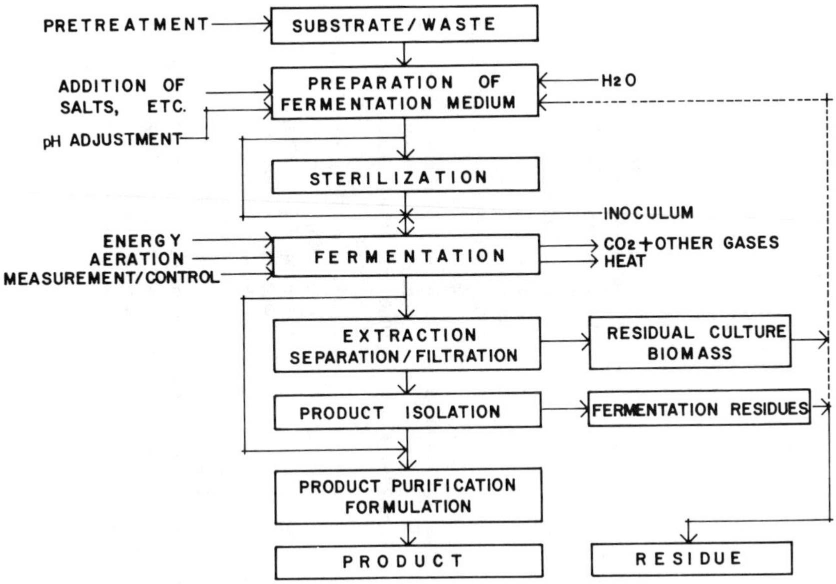

Fig. 4. General description of the fundamental steps of a biotechnological process for the transformation of wastes from different origins.

the organism in respect of access to nutrients. It has been found that production rates of amylase with *Aspergillus oryzae* cultured on vermiculite impregnated with a starch solution were higher than those in submerged cultures (Aidoo *et al.*, 1982). This type of fermentation has also been investigated by Barrios-González *et al.* (1988) and more recently by Larroche and Gros (1989). Aidoo *et al.* (1982) and Larroche and Gros (1989) reported that fermentation with inert porous particles allows (1) the use of low-molecular-weight carbohydrates directly available to the microorganism; (2) the handling of high concentrations of substrate; (3) direct determinations of biomass at any given time as in LSF; and (4) simplified product recovery.

5.2 Some Currently Practiced SSF Processes

Table 3 describes the basic characteristics of some of the most important SSF processes, most of which may use wastes from plant and animal origin. Oriental-type foods have always been very popular in several Eastern countries. Pozol is one of the SSF products with the longest tradition in Latin America. Lactic acid fermentations have a strong potential to use various agroindustrial wastes. Mushroom cultivation is steadily gaining popularity and economic viability. Animal feed production involves the use of an array of wastes and some of these processes are practiced on a commercial scale. Processes for ethanol (Gibbons, 1989), enzymes (Blain, 1975; Grajek, 1987; Ramesh & Lonsane, 1987) and specialty biochemicals (Hang, 1988) production appear to have a very sound economic future.

Table 3
Typical Solid Substrate Fermentation Processes

Microorganisms	*Oxygen need*	*Process*	*Products*	*References*
Mixed fungal cultures and fungal monocultures	Aerobic	Koji process	Oriental food and beverages	Steinkraus (1983*b*); Paredes-López *et al.* (1987)
Mixed microbial flora	Aerobic	Dough formation	Mexican food (pozol)	Ulloa-Sosa (1974)
Saccharomyces cerevisiae/Lactobacillus sanfrancisco	Aerobic/anaerobic	Bread dough formation	Bread	Aidoo *et al.* (1982)
Composting flora, spawn culture, mushroom cropping	Aerobic	Mushroom cultivation	Edible mushrooms	Präve *et al.* (1987); Rajarathnam & Bano (1989)
Mixed lactic acid bacteria	Aerobic/anaerobic	Ensiling of crops and animal excreta	Preserved and enriched animal feed	Moo-Young *et al.* (1983); Iñiguez-Covarrubias *et al.* (1989)
Fungal monocultures	Aerobic	Animal feed production	Converted lignocellulose	Ulmer *et al.* (1981); Laukevics *et al.* (1985); Hatakka *et al.* (1989)
Mixed microbial cultures and fungal monocultures	Aerobic	Animal feed production from fruit and vegetable wastes	Protein-rich feedstuff	González *et al.* (1985); Rolz *et al.* (1988); Yang (1988)
Mixed microbial flora	Aerobic/anaerobic	Composting	Soil conditioner	Rajarthnam & Bano (1989)
Mixed microbial flora	Anaerobic	Methane production	Methane	Molnar & Bartha (1989)
Yeast	Anaerobic	Alcohol production	Alcohol	Sato & Yoshizawa (1988)
Fungal monoculture	Aerobic	Enzyme production	Various specific enzymes (e.g. amylases, cellulases, proteases)	Tengerdy (1985)
Fungal monocultures	Aerobic	Specialty biochemicals	Various biochemicals (e.g. mycotoxins, organic acids, antibiotics)	Lindenfelser & Ciegler (1975); Hang (1988); Yang & Ling (1989)
Acetobacter species	Aerobic	Vinegar production	Vinegar	Aidoo *et al.* (1982)

5.2.1 Food Production

Lactic acid fermentation is an ancient process whereby a varied group of bacteria ferment carbohydrates producing lactic acid as the major end product. This type of fermentation is used for the production of dairy products, sauerkraut, bread, meat and silage. In particular, SSF fermentations of legumes, cereals and starchy substrates have been associated in many regions of the world with the activity of lactic acid bacteria (Fukushima, 1985; Paredes-López & Harry, 1988). Several processes have been developed to produce fermented foods involving a fermentation by lactic acid bacteria (e.g. *Pediococcus halophilus, Klebsiella pneumoniae*) followed by a fungal fermentation, using as substrate wastes and by-products of plant origin (Steinkraus, 1983*a*; Paredes-López & Harry, 1988).

Tempeh is a generic term that is applied to foods that have been fermented by a filamentous fungus to obtain a compact cake of substrate particles knitted together by the mycelium (Steinkraus, 1983*a*). Tempeh is a food of Indonesian origin and is now produced at the industrial level in various countries (Harrigan, 1985). It is a food most commonly prepared from soya; however, this fermentation technique is extremely versatile and has been applied to a wide variety of plant wastes and by-products (Paredes-López & Harry, 1988; Paredes-López *et al.*, 1990). One of the most important steps is the soaking of the substrate during which, at ambient temperatures in the tropics, lactic acid bacteria ferment the solutes leached from the substrate to lactic acid (Fig. 5). The consequent acidification prevents the growth of spoilage microorganisms that may impede the growth of the fungus used in the second fermentation step (Harry, 1988).

There are significant amounts of fresh bananas in the banana producing regions (e.g. Central America) which are unacceptable to both internal and external markets. The use of these wastes has become an important matter for economic and ecological reasons. The natural pH of banana purée (4·8–5·0) requires a drastic heat sterilization process followed by aseptic canning. This preservation process has a relatively small market. An alternative may be the fermentation process developed by De Porres *et al.* (1985) for the preparation of a purée by lactic acid bacteria, which may be associated more with yoghurt than with natural banana purée (Fig. 6). Enzymatic browning of the ripe, peeled fruit is prevented by blanching in boiling water before pulping. Lactic bacteria inoculum is added to the purée, which reduces pH below 4·5 in a short time and produces an acceptable flavor. The fermented purée is suitable for use in banana flour specialty products. This technology may be implemented by cooperatives in rural areas.

Traditional SSFs of Latin America have received scarce attention as compared to food fermentations of Asia (Paredes-López & Harry, 1988). Lactic acid fermentations of maize in the solid state have been used in Mexico at village level for centuries (Ulloa-Sosa, 1974). The transformation of the substrate gives an end product, termed pozol, with improved nutritional properties and with promising characteristics for industrialization (Cravioto *et al.*, 1955).

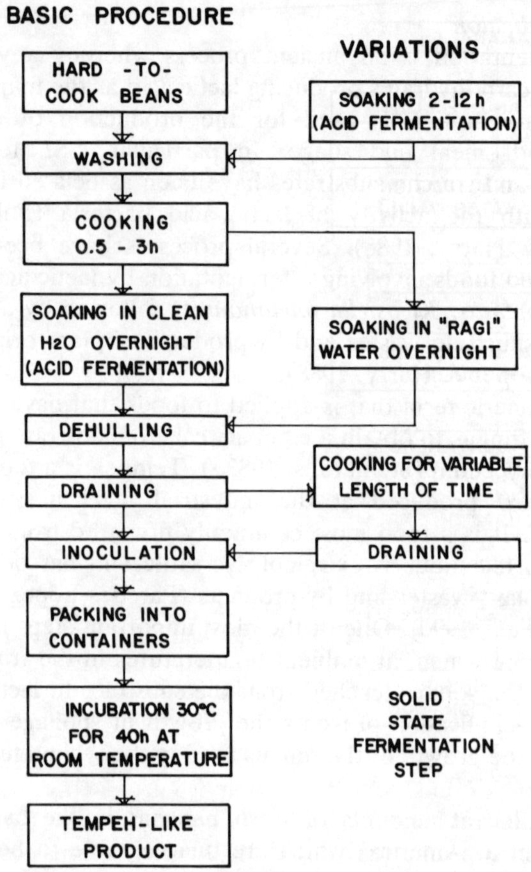

Fig. 5. Technological procedure to produce a tempeh-like food using *Rhizopus oligosporus* and hard-to-cook common beans as substrate.

The demand for edible fungi has risen greatly in the last decade and has been satisfied by an expansion of the production of cultivated mushrooms using a variety of agricultural and lignocellulosic wastes as substrate (Dare *et al.*, 1988; Rajarathnam & Bano, 1989). The mushroom species cultivated in North America, Latin America and Europe are mostly *Agaricus bisporus*, *A. bitorquis*, *Agaricus spp* and *Pleurotus ostreatus*, whereas in Asia are cultivated *Lentinus edoides* and *Volvariella volvacea* (Präve *et al.*, 1987). The biotechnology for modern cultivation of mushrooms includes two basic steps: (1) preparation of suitable composts; and (2) growth of the mycelium and fructification (Fig. 7). After 8–9 days the compost is packed into boxes and subjected to pasteurization. This kills animal pests and harmful microorganisms. The free ammonia present in the compost, which is toxic for fungi, is eliminated by thermophilic microorganisms. After inoculation and mycelial growth, the compost is covered with a mixture of soil, peat and chalk. Yields of

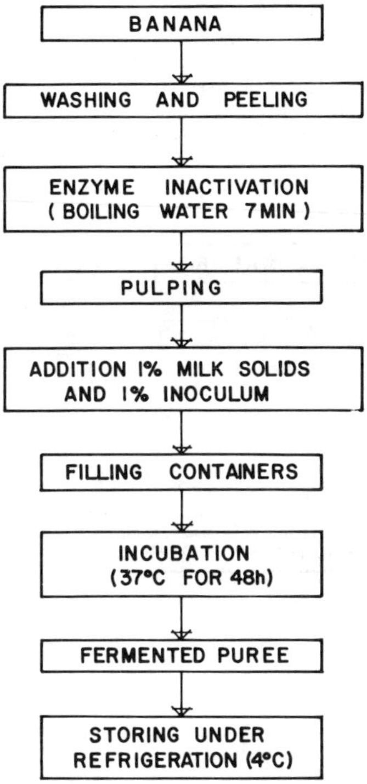

Fig. 6. Utilization of fresh banana wastes and lactic bacteria as inoculum, under solid substrate fermentation, to produce a fermented purée to be used in yoghurt-like products.

0·5 to 1 kg of mushrooms per kg of compost dry matter may be obtained. Besides being a delicacy, the mushrooms are also an important source of food protein for human consumption (Gray, 1970). The residual compost could be used as an upgraded form of ruminant feed (Gujral *et al.*, 1987), as a source of degradatory enzymes or as a fertilizer (Rajarathnam & Bano, 1989).

5.2.2 Feed Production

The process of ensilage is traditionally used for animal feeding in various regions of the world and is an alternative to the drying of green crops. This process involves the controlled fermentation of green crops such as grass, maize plants and other available by-products and wastes from agricultural activites (Carrizales & Ferrer, 1984). Molasses are sometimes added to promote fermentation and increase palatability. As the fermentation proceeds the temperature rises and the product becomes more acid. The acidity prevents the growth of many putrefactive organisms, and as the oxygen is consumed strict aerobe microorganisms are inhibited, but facultative and strict anaerobes continue to grow. At first, species such as *Escherichia coli* and *Aerobacter* spp. predominate, but these are replaced by lactobacilli and streptococci as pH declines. Lactic, butyric, propionic and acetic acids are synthesized giving flavors very acceptable to animals.

One of the agroindustrial wastes available in some tropical countries is the

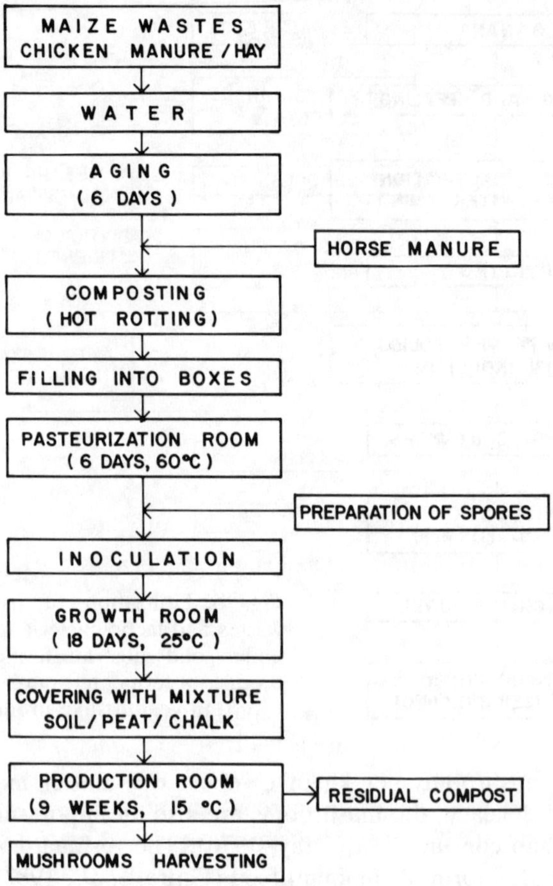

Fig. 7. Mushroom production using wastes from plant and animal origin.

coffee pulp from wet processing plants. Approximately half of the world coffee harvest is processed by the wet method in which the coffee berry is subjected to mechanical and biological operation in order to separate the bean or seed from the exocarp (skin), mesocarp (mucilagenous pulp) and the endocarp (parchment) (Rolz *et al.*, 1988). The skin and most of the pulp is separated in the pulpers. This fraction represents about 40% of the weight of the fresh fruit and presently is underutilized, causing serious pollution problems. The use of coffee pulp as an animal feed has been mentioned as an attractive possibility; however, such utilization is limited by antiphysiological factors naturally occurring in the material (Peñaloza *et al.*, 1985). Figure 8 describes the basic steps for ensiling and for SSF, under controlled conditions, of coffee wastes. The most important changes in composition are given in Table 4. Both processes appear to promise improvement in the nutritive features of coffee wastes for monogastric animal feeding (Carrizalez & Ferrer ,1984; Peñaloza *et al.*, 1985).

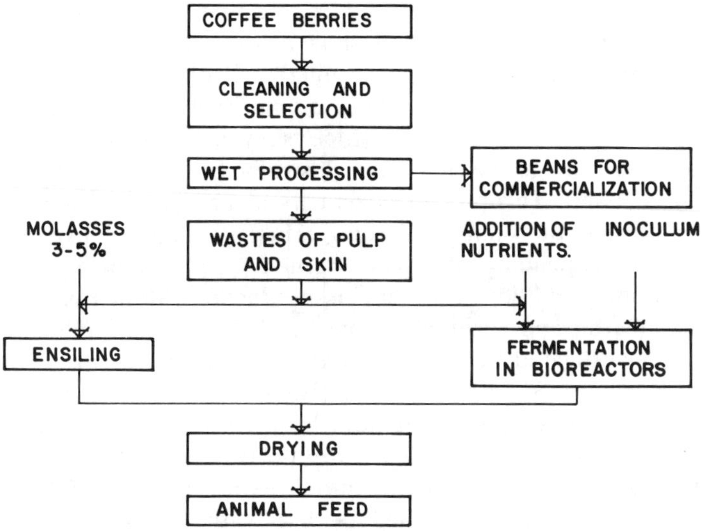

Fig. 8. Ensiling and solid substrate fermentation, under controlled conditions, of coffee pulp and skin.

Ensiling is also an economical method of preserving and rendering animal excreta silages safe from potentially pathogenic microorganisms (Iñiguez-Covarrubias *et al.*, 1989). Animal wastes represent one of the most underutilized resources. Utilization of animal excreta as feed can alleviate pollution problems, decrease feed costs and increase the supplies of available nitrogen and essential mineral resources. The practice of feeding unprocessed wastes

Table 4
Effect of Ensiling and Fermentation under Controlled Conditions on Chemical Composition of Coffee Pulp (% Dry Basis)

Component	*Ensiling*		*SSF under controlled conditions*[a]	
	Fresh pulp	*Ensiled product*	*Fresh pulp*	*Fermented product*
Protein ($N \times 6{\cdot}25$)	11·6	13·6	10·9	36·0
Crude fiber	15·2	17·2	23·7	13·1
Ether extract (fat)	5·0	2·2	2·2	2·4
Ash	6·7	8·1	5·0	12·6
Free-nitrogen extract	61·5	58·9	58·2	35·9
Caffeine	0·95	0·82	0·68	0·46
Cell wall constituents	—	—	31·4	23·5
Hemicellulose	—	—	0·98	0·05
Acid detergent fiber	—	—	30·5	23·4
Cellulose	—	—	18·6	12·0
Tannins	—	—	2·3	2·8

[a] SSF = Solid substrate fermentation.

may be a potential health hazard as excreta may contain agents harmful to human and animal health. The ensiling process suppresses the activities of undesirable microorganisms. The ensiling technology has been reported to eliminate pathogenic bacteria such as *Salmonella* spp., *Mycobacterium* spp. and *E. coli* from beef cattle wastes (McCaskey & Wang, 1985), and to reduce the viability of bovine coccidia and clostridia from animal wastes (Iñiguez-Covarrubias, 1989). This process is actually being used at commercial level in various countries for ruminant feeding. However, further exploitation of its great potential is pending. Another alternative for manure is the use of SSF under anaerobic conditions for methane production (Molnar & Bartha, 1989).

The use of fruit and vegetable wastes in SSF for feed production has been thoroughly investigated by several workers. Wastes from citrus (Rodríguez *et al.*, 1986; Nicolini *et al.*, 1987), banana (Baldensperger *et al.*, 1985) and potato (Yang, 1988) industrial processing have been converted into feed supplements of high nutritional value. González *et al.* (1985) reported that during pineapple processing each tonne of fresh fruit produces not less than 500 kg of wastes. These wastes may be transformed by SSF with *Trichoderma viride* into a valuable feed product. SSF is also a useful technology for enriching the feed value of lignocellulosic materials (Abdullah *et al.*, 1985; Hatakka *et al.*, 1989). In relation to costs associated with SSF, Durand and Chereau (1988) estimated that at industrial scale for the whole process the most significant cost is that for drying.

5.2.3 Ethanol Production

Various SSF processes for ethanol production using sugar-rich substrates have been studied. For this purpose, crops such as fodder beets (Gibbons & Westby, 1988), sugar beets (Cochet *et al.*, 1988), sweet sorghum and Jerusalem artichoke (Gibbons, 1989) have been tested. These SSF processes appear to be economically competitive compared to traditional LSF procedures based on corn, sugar beets and sugar cane substrates. The use of SSF for ethanol production from sugar-rich wastes may be a promising alternative as well. The increased volumetric productivity and savings in distillation costs make the SSF process attractive (Tengerdy, 1985).

5.2.4 Enzyme Production

It is estimated that SSF will have truly significant economic importance in high volumetric productivity systems where recovery expenses are considerably reduced or even obviated (Aidoo *et al.*, 1982). Table 5 shows some of the most important enzymes produced by SSF. For enzyme production, SSF may be attractive due to its low-level technology, the high product concentration and reduced cost of dewatering (Laukevics *et al.*, 1984; Macris *et al.*, 1987). Hatakka *et al.* (1989) postulated that protein enrichment of lignocellulosic wastes may be economically feasible if at the same time wood-rotting fungi are used for the production of extracellular enzymes (e.g. cellulases, lignases). In these SSF processes, lignin could be selectively removed, the protein content

Table 5
Production of Enzymes by Solid Substrate Fermentation

Enzymes	*Microorganisms*	*Substrates*	*References*
Amylases	*Aspergillus oryzae*	Broken rice kernels	Narahara *et al.* (1982)
	Bacillus megaterium	Wheat bran	Ramesh & Lonsane (1987)
Amyloglucosidase	*Rhizopus* sp.	Wheat bran	Blain (1975)
Cellulases	*Trichoderma reesei*	Wheat straw	Chahal (1983, 1985)
		Wood residues	Hatakka *et al.* (1989)
	Thermoascus aurantiacus	Sugar-beet pulp	Grajek (1987)
	Neurospora crassa	Wheat straw	Macris *et al.* (1987)
Pectinases	*Penicillium capsulatum*	Sugar-beet pulp	Considine *et al.* (1987)
	Trichoderma viride	Sugar-beet pulp	Grajek & Gervais (1987)
Proteases	*Aspergillus oryzae*	Wheat bran	Nakadai & Nasuno (1988)
	Aspergillus oryzae	Broken rice kernels	Narahara *et al.* (1982)
D-xylanase	*Trichoderma viride*	Sugar-beet pulp	Grajek & Gervais (1987)
β-glucosidase	*Penicillium capsulatum*	Sugar-beet pulp	Considine *et al.* (1987)
	Aspergillus phoenicis	Sugar-beet pulp	Deschamps & Huet (1984)

of the residue increased, and extracellular enzymes extracted from the residue prior to use as animal feed. SSF is also a very valuable technology for the production of crude enzyme preparations for the food industry (Deschamps & Huet, 1984; Considine *et al.*, 1987; Nakadai & Nasuno, 1988).

5.2.5 Specialty Biochemicals

Various procedures for production of specialty biochemicals by SSF have been reported. The increased volumetric productivity and resulting cost reductions in product recovery make the SSF a viable alternative to the LSF. Some of these biochemical products are mycotoxins (Norholt *et al.*, 1977), organic acids (Lakshminarayana *et al.*, 1975; Kumar & Lonsane, 1987) and antibiotics (Yang & Ling, 1989) (Table 6). Agricultural wastes may be used as substrate for the production of these metabolites.

Table 6
Biochemicals Produced by Solid State Fermentation

Microorganisms	*Substrates/ wastes*	*Biochemical products*	*References*
Aspergillus flavus	Rice, corn	Aflatoxin	Norholt *et al.* (1977)
Aspergillus ochraceus	Wheat	Ochratoxin	Lindenfelser & Ciegler (1975); Hesseltine (1977*a*); Norholt *et al.* (1979)
Aspergillus niger	Sugar cane bagasse, apple pomace	Citric acid	Lakshminarayana *et al.* (1975); Hang (1988)
Aspergillus sp.	Sucrose	Koji acid	Aidoo *et al.* (1982)
Gibberella fujikuroi	Wheat bran	Gibberellic acid	Kumar & Lonsane (1987)
Streptomyces viridifaciens	Sweet potato residue	Tetracycline	Yang & Ling (1989)
Penicillium chrysogenum	Sugar cane bagasse (inert support)	Penicillin	Barriose-González *et al.* (1988)

6 ADVANTAGES AND DISADVANTAGES AND FUTURE PROSPECTS OF SOLID SUBSTRATE FERMENTATION

6.1 Advantages and Disadvantages

There are many advantages of SSF processes for bioconversion of wastes over the conventional LSF systems at both laboratory and industrial scales (Aidoo *et al.*, 1982; Laukevics *et al.*, 1984; Lonsane *et al.*, 1985; Hesseltine, 1987; Paredes-López & Harry, 1988). The most attractive advantages of SSF may be the high fermentor volumetric productivity, and the reduction in product recovery expenses, pollution problems and total cost of production. There also are some important disadvantages of SSF (Aidoo *et al.*, 1982; Paredes-López & Harry, 1988), such as (1) the limited type of organisms which can grow at reduced water activities, namely fungi, some yeasts, and some bacteria and streptomycetes; (2) technical problems in controlling the heat generated during fermentation; (3) low product yields; (4) the slowness of fermentation; and (5) the difficulty of using reliable devices to measure/control some of the fermentation parameters (e.g. moisture, temperature, pH, oxygen, product concentration).

6.2 Future Prospects

As the availability of agricultural land declines and pollution problems increase, the bioconversion of wastes from plant and animal origin through SSF into valuable products is becoming more pressing than ever in both developed and developing countries. Food is becoming more expensive in terms of basic inputs such as water and energy. The increasing use of wastes and agroindustrial residues to produce SSF foods and feeds may optimize the indigenous resources, increase the availability of nutritious products and reduce pollution problems. A combination of anaerobic and aerobic fermentation procedures appears also to be an attractive way of increasing the storability, sensory properties and nutritional value of starchy residues. Technological procedures for composting will be used increasingly for agricultural, industrial and municipal solid waste disposal. SSF is expected to have in the near future a significant economic impact in the production of metabolites of relatively high price per unit of weight (e.g. enzymes, organic acids, antibiotics).

More research is required for the development of new SSF bioreactors with improved performance in the utilization of wastes and residues. Also, more attention should be given to SSF processes based on the use of solid inert material, impregnated with suitable media, to improve the control of some fermentation parameters and reduce recovery expenses of metabolites. Research into the genetic improvement of fermentative microorganisms for more efficient utilization of a variety of substrates, and conversion into more valuable products, is pending.

ACKNOWLEDGMENTS

Research on fermentation was supported by Organización de los Estados Americanos and Consejo Nacional de Ciencia y Tecnología-México. The authors are grateful to Dr A. Cárabez, Inst. de Fisiología Celular, UNAM, and Jaquelina González, Unidad Irapuato, for their competent technical assistance.

REFERENCES

Abdullah, A. L., Tengerdy, R. P. & Murphy, V. G. (1985). *Biotechnol. Bioeng.*, **27,** 20.
Aidoo, K. E., Hendry, R. & Wood, B. J. B. (1982). *Adv. Appl.Microbiol.*, **28,** 201.
Bajracharya, R. & Mudgett, R. E. (1980). *Biotechnol. Bioeng.*, **22,** 2219.
Baldensperger, J., Le Mer, J., Hannibal, L. & Quinto, P. J. (1985). *Biotechnol. Lett.*, **7,** 743.
Barrios-González, J., Tomasini, A., Viniegra-González, G. & López, J. (1988). *Biotechnol. Lett.*, **10,** 793.
Bassham, J. A. (1975). In *Cellulose as a Chemical and Energy Resource,* Biotech. Bioeng. Symp. No. 5, ed. C. R. Wilke. Interscience, New York, p. 9.
Beuchat, L. R. (1981). *Cereal Foods World,* **26,** 345.
Blain, J. A. (1975). In *The Filamentous Fungi,* eds J. E. Smith, D. R. Berry & B. Kristiansen. Edward Arnold, London, p. 210.
Brauer, H. (1988). *Bioproc. Eng.*, **3,** 11.
Bushell, M. E. & Slater, J. H. (1981). *Mixed Culture Fermentations,* Special Publ. No. 5. Society of General Microbiology, Academic Press.
Cannel, E. & Moo-Young, M. (1980*a*). *Process. Biochem.*, **15**(3), 2.
Cannel, E. & Moo-Young, M. (1980*b*). *Process Biochem.*, **15**(4), 24.
Carrizales, V. & Ferrer, J. (1984). In *Symposium on the Importance of Lactic Acid Fermentation in the Food Industry,* Mexico City, Mexico.
Chahal, D. S. (1983). In *Foundations of Biochemical Engineering,* eds H. W. Blanch, E. T. Papoutsakis & G. Stephanopoulos. Am. Chem. Symp. Series 207, Washington, DC, p. 421.
Chahal, D. S. (1985). *Appl. Environ. Microbiol.*, **49,** 205.
Cochet, N., Nonus, M. & Lebeault, J. M. (1988). *Biotechnol. Lett.*, **10,** 491.
Considine, P. J., Hackett, T. J. & Coughlan, M. P. (1987). *Biotechnol. Lett.*, **9,** 131.
Cravioto, O. R., Cravioto, Y. O., Massieu, G. & Guzmán, J. (1955). *Ciencia* (Mexico), **15,** 27.
Czajkowska, D. & Ilnicka-Olejniczak, O. (1988). *Acta Biotechnol.*, **8,** 407.
Dare, P. H., Clarke, T. A. & Chu-Chou, M. (1988). *Process Biochem.*, **23,** 156.
De Porres, E., de Arriola, M. C., García, R. & Rolz, C. (1985). *Lebensm.-Wiss. u. Technol.*, **18,** 379.
Deschamps, F. & Huet, M. C. (1984). *Biotechnol. Lett.*, **6,** 55.
Durand, A. & Chereau, D. (1988). *Biotechnol. Bioeng.*, **31,** 476.
Durand, A., de la Broise, D. & Blachère, H. (1988). *J. Biotechnol.*, **8,** 59.
Fukushima, D. (1985). *Food Rev. Int.*, **1,** 149.
Georgiou, G. & Shuler, M. (1986). *Biotechnol. Bioeng.*, **28,** 405.
Gervais, P., Bensoussan, M. & Grajek, W. (1988). *Appl. Microbiol. Biotechnol.* **27,** 389.
Gibbons, W. R. (1989). *J. Ferment. Bioeng.*, **67,** 258.
Gibbons, W. R. & Westby, C. A. (1988). *Biotechnol. Lett.*, **10,** 665.

González, E. E., Vernon, E. J. & Moctezuma, A. R. (1985). *Ind. Environ.*, **8**(4), 19.
Grajek, W. (1987). *Appl. Microbiol. Biotechnol.*, **26**, 126.
Grajek, W. & Gervais, P. (1987). *Enzyme Microbiol. Technol.*, **9**, 658.
Gray, W. D. (1970). *The Use of Fungi as Food and in Food Processing*. CRC Press, Cleveland, Ohio.
Gujral, G. S., Bisaria, R., Madan, M. & Vasudevan, P. (1987). *J. Ferment. Technol.*, **65**, 101.
Han, Y. W. & Anderson, A. W. (1975). *Appl. Microbiol.*, **30**, 930.
Hang, Y. D. (1988). *Biotechnol. Lett.*, **10**, 421.
Harrigan, W. F. (1985). *Food Flav. Ingr. Proc.*, **10**, 32.
Harry, G. I. (1988). Desarrollo de procedimientos fermentativos para la elaboración de tempe a partir de frijol común y su incorporación en algunos alimentos. MSc thesis, Unidad Irapuato, CIEA-Inst. Politécnico Nal., Irapuato, Mexico.
Hatakka, A. I., Mohammadi, O. K. & Lundell, T. K. (1989). *Food Biotechnol.*, **3**, 45.
Hayes, W. A. (1977). *Composting*. The Mushroom Growers' Association, London.
Hesseltine, C. W. (1977*a*) *Process Biochem.*, **12**(6), 24.
Hesseltine, C. W. (1977*b*). *Process Biochem.*, **12**(7), 29.
Hesseltine, C. W. (1987). *Internat. Biodeter.* **23**, 79.
Hutter, R. (1982). In *Bioactive Microbial Products: Search and Discovery*, eds J. D. Bullock, L. J. Nisbet & D. T. Winstanley. Academic Press, London, p. 112.
Iñiguez-Covarrubias, G. (1989). Fermentación de estiércol de cerdo para la obtención de un alimento para rumiantes. PhD thesis, Inst. Investigaciones Biomédicas, UNAM, Mexico, D.F.
Iñiguez-Covarrubias, G., Franco-Gómez, M. J. & Ruiz-Carrera, V. (1989). *Biol. Wastes*, **28**, 293.
Kumar, P. K. R. & Lonsane, B. K. (1987). *Biotechnol. Lett.*, **9**, 179.
Lakshminarayana, K., Chaudhary, K., Ethiraj, S. & Tanyo, P. (1975). *Biotechnol. Bioeng.*, **17**, 291.
Larroche, C. & Gros, J. B. (1989). *Process Biochem.*, **24**, 97.
Laukevics, J. J., Apsite, A. F., Viesturs, U. E. & Tengerdy, R. P. (1984). *Biotechnol. Bioeng.*, **26**, 1465.
Laukevics, J. J., Apsite, A. F., Viesturs, U. E. & Tengerdy, R. P. (1985). *Biotechnol. Bioeng.*, **27**, 1687.
Lindenfelser, L. A. & Ciegler, A. (1975). *Appl. Microbiol.*, **29**, 323.
Lonsane, B. K., Ghildyal, N. P. Budiatman, S. & Ramakrishna, S. V. (1985). *Enzyme Microbiol. Technol.*, **7**, 258.
Macris, B. J., Kekos, D., Evangelidou, X., Galiotou-Panayotou, M. & Rodis, P. (1987). *Biotechnol. Lett.*, **9**, 661.
McCaskey, T. A. & Wang, Y. D. (1985). *J. Dairy Sci.*, **68**, 1401.
Mishra, I. M., El-Temtamy, S. A. & Schügerl, K. (1982). *Eur. J. Appl. Microbiol. Biotechnol.*, **16**, 197.
Mitchell, D. A., Greenfield, P. F. & Doelle, H. W. (1986). *Biotechnol. Lett.*, **8**, 827.
Mitchell, D. A., Doelle, H. W. & Greenfield, P. F. (1988). *Biotechnol. Lett.*, **10**, 497.
Molnar, L. & Bartha, I. (1989). *Biol. Wastes*, **28**, 15.
Moo-Young, M., Moreira, A. R. & Tengerdy, R. P. (1983). In *The Filamentous Fungi*, Vol. IV, eds J. E. Smith, D. R. Berry & B. Kristiansen. Edward Arnold, London, p. 117.
Mossel, D. A. A. (1975). In *Water Relations of Foods*, ed. R. B. Duckworth. Academic Press, London, p. 347.
Nakadai, T. & Nasuno, S. (1988). *J. Ferment. Technol.*, **66**, 525.
Narahara, H., Koyama, Y., Yoshida, T. & Taguchi, H. (1982). *J. Ferment. Technol.*, **60**, 311.
Nicolini, L., von Hunolstein, C. & Carilli, A. (1987). *Appl. Microbiol. Biotechnol.*, **26**, 95.

Norholt, M. D., Van Egmond, H. P. & Paulsch, W. E. (1977). *J. Food Protect.*, **40,** 778.
Norholt, M. D., Van Egmond, H. P. & Paulsch, W. E. (1979). *J. Food Protect.*, **42,** 485.
Oriol, E., Raimbault, M., Roussos, S. & Viniegra-González, G. (1988). *Appl. Microbiol. Biotechnol.*, **27,** 498.
Paredes-López, O. & Harry, G. I. (1988). *CRC Crit. Rev. Food Sci. Nutr.*, **27,** 159.
Paredes-López, O. & Harry, G. I. (1989). *J. Food Sci.*, **54,** 968.
Paredes-López, O., Vital-Bori, G. & Camargo-Rubio, E. (1974). *J. Ferment. Technol.*, **52,** 592.
Paredes-López, O., Camargo-Rubio, E. & Ornelas-Vale, A. (1976). *Appl. Environ. Microbiol.*, **31,** 487.
Paredes-López, O., Harry, G. I. & Montes-Rivera, R. (1987). *Biotechnol. Lett.*, **9,** 333.
Paredes-López, O., Guevara-Lara, F., Schevenin-Pinedo, M. L. & Montes-Rivera, R. (1988). *Plant Foods Hum. Nutr.*, **39,** 137.
Paredes-López, O., Harry, G. I. & González-Castañeda, J. (1990). *J. Food Sci.*, **55,** 123.
Peñaloza, W., Molina, M. R., Gómez, R. & Bressani, R. (1985). *Appl. Environ. Microbiol.*, **49,** 388.
Pou, J., Vittori, N. Fernández, M. J. & Garrido, J. (1987). *Interferón Biotecnol.*, **4,** 121.
Präve, P., Faust, U. & Sittig, W. (1987). *Fundamentals of Biotechnology*. VCH, Weinheim, Germany.
Prokop, A., Erickson, L. E. & Paredes-López, O. (1971). *Biotechnol. Bioeng.*, **13,** 241.
Raimbault, M. & Alazard, D. (1980). *Eur. J. Appl. Microbiol. Biotechnol.*, **9,** 199.
Rajarathnam, S. & Bano, Z. (1989). *CRC Crit. Rev. Food Sci. Nutr.*, **28,** 31.
Ralph, B. J. (1976). *Food Technol. Austr.*, **28**(7), 247.
Ramesh, M. V. & Lonsane, B. K. (1987). *Biotechnol. Lett.*, **9,** 323.
Rodríguez, J. A. *et al.* (1986). *Acta Biotechnol.*, **6,** 253.
Rolz, C., de León, R. & Arriola, M. C. (1988). *Biol. Wastes,* **26,** 97.
Sato, K. & Yoshizawa, K. (1988). *J. Ferment. Technol.*, **66,** 667.
Sato,K., Nagatani, M., Nakamura, K. & Sato, S. (1983). *J. Ferment. Technol.*, **61,** 623.
Sato, K., Miyazaki, S., Matsumoto, N. & Nakamura, K. (1988). *J. Ferment. Technol.*, **66,** 173.
Silman, R. W. (1980). *Biotechnol. Bioeng.*, **22,** 411.
Steinkraus, K. H. (1983*a*). In *Handbook of Indigenous Fermented Foods,* ed. K. H. Steinkraus, Marcel Dekker, New York, p. 131.
Steinkraus, K. H. (1983*b*). In *The Filamentous Fungi,* Vol. IV, eds J. E. Smith, D. R. Berry & B. Christiansen. Edward Arnold, London, p. 171.
Tengerdy, R. P. (1985). *Trends Biotechnol.*, **3,** 96.
Troller, J. A. (1987). In *Water Activity: Theory and Applications to Food,* eds L. B. Rockland & L. R. Beuchat. Marcel Dekker, New York, p. 101.
Ulloa-Sosa, M. (1974). *Bol. Soc. Mex. Mic.*, **8,** 17.
Ulmer, D. C., Tengerdy, R. P. & Murphy, V. G. (1981). *Biotechnol. Bioeng. Symp. Ser.*, **11,** 449.
Yang, S. S. (1988). *Biotechnol. Bioeng.*, **32,** 886.
Yang, S. S. & Ling, M. Y. (1989). *Biotechnol. Bioeng.*, **33,** 1021.

Chapter 5

COMPOSTING PROCESSES

S. P. MATHUR

*Land Resource Research Centre, Central Experimental Farm, Agriculture Canada, Ottawa, Ontario, Canada K1A 0C6**

CONTENTS

* Presently on Secondment to Environmental Issues Working Group, Office of the Associate Deputy Minister, Agriculture Canada, Ottawa.

1 INTRODUCTION

Recently heightened concerns about global warming, depletion of stratospheric ozone, environmental pollution, wildlife, shrinking biodiversity, health, food safety, land degradation, pressures on nonrenewable resources and intergenerational justice, have produced the Sustainable Development ethos expressed by the Brundtland Commission in its report. 'Our Common Future'. In practical terms, the new paradigm promotes the 4R strategy of Reduce, Re-use, Recover and Recycle wastes to conserve resources and reduce pollution by emulating, adopting and maximizing beneficial processes of an ecosystem. An ecosystem is maintained in essence by the cycling of carbon and mineral nutrients so that the losses of nutrients from the system are not greater than the relevant mineral ions gained through soil weathering and the nitrogen fixed by biological agents or the electrochemical force in lightning.

The carbon is cycled through the food chain of herbivores, omnivores, carnivores and soil-inhabiting decomposer organisms, because without it photosynthesis will cease within a decade or so for lack of atmospheric CO_2. The beneficial biological activity involved in carbon cycling also helps to control pathogens, pests and parasites through antagonism and competition in a balanced ecosystem. A portion of the carbon is maintained in soil humus, an amphoteric heteropolycondensate of carbon, nitrogen, sulphur and phosphorus compounds that is relatively biostable (Paul & Clark, 1989). Humus shelters and nourishes soil organisms while it buffers and moderates physical, chemical and biological processes and attributes of soil.

When man first cultivated land and removed portions of the organic produce as food or materials, the loss of nutrients and carbon was not a serious problem as long as agriculture was confined to the river valleys and deltas which recouped the losses through the alluvial silt deposited by flood waters, or when the slash and burn strategy was used on forest lands at low frequency (Altieri, 1987). Agriculture was sustainable, particularly on lands out of the two aforesaid areas, only where the organic wastes were recycled to maintain soil humus and fertility. The organic wastes, however, ideally were not returned to the cultivated land in the raw state as that would have been unhygenic for both plants and man, and would have overloaded the system's capacity to cycle the carbon in the wastes in the near absence of on-site herbivores present in a balanced ecosystem.

Processing of the raw wastes in composite mixtures in some ancient cultures reduced bulk, dissipated part of the carbon, sanitized the wastes and conserved nutrients (Howard, 1943). Agriculture was thus sustained in China, where the connection between disease and fecal matters was first realized, for 5000 years, and in the Minoan civilization of Greece to a lesser extent. In contrast, according to Hughes (1980), the Mayas of middle America, who had no domestic animals, exhausted the lands they cultivated, and had to move and resettle repeatedly, for lack of the cycling of organic matter and nutrients through manuring.

Awareness of the oriental practice of recycling wastes through the composting process was gained strikingly through Dr Kings' account of agriculture there in the Far East, (King, 1911).

Krishna Murthy (1978) and Howard (1943) have traced the history of development of the scientific principles of composting, mostly in India, culminating in the aerobic Indore, and the partially aerobic Bangalore, methods. Indore and Bangalore are cities in the erstwhile Holkar and Mysore states, the Maharajahs of which employed researchers such as Howard, Wad, Jackson and Acharya, during the British colonial rule of India.

With the advent of cheap chemical fertilizers, and due to other economic and social factors, recycling of organic matter through composting lost ground to the 'artificial manures'. Liebig is credited with initiating the chemical approach by discovering that plants absorb nutrients only as mineral ions. Liebig did, however, state that 'there is but one manure which permanently keeps up the fertility of land, and that is farmyard manure' (Krishna Murthy, 1978). The rediscovery of this type of thinking has reawakened interest in composting which had in the interim held the interest mostly of the avid home gardener and the sanitary engineer, almost irrespective of the labour in the former and cost in the latter, while food processing wastes and manures were being disposed contaminatively or buried in landfills to pose insidious, slow but definite threat to the environment.

This chapter therefore deals with composting processes, with special reference to a novel passively aerated windrow system used for seafood wastes and manure slurries.

2 DEFINITION AND PRINCIPLES OF COMPOSTING

2.1 Definition

Composting is the biological conversion of waste materials, under controlled conditions, into a hygienic, humus-rich, relatively biostable product that conditions soils and nourishes plants.

Composting and the product compost is thus defined by the multiple objectives it achieves together, not singly. For example, the sanitization of wastes by radiation or biostabilization by desiccation will not produce humus or conserve nutrients. The addition of soluble nutrients to organic wastes and their joint application to land will neither sanitize the wastes nor produce all the possible humus, as the leachable nutrients may not remain with the organic matter at the surface. Decomposition of certain wastes alone, e.g. bark or sawdust, may generate sanitizing heat due to oxidation of organic matter without producing the nutrient-rich compost beneficial to plants. Organic matter merely left to rot also does not produce compost any more than a rotting grape on the table produces champagne. Composting indeed is akin to the making of other biological products like beer, wine and cheese, and therefore needs to be achieved under controlled conditions of environment and materials.

Many definitions of composting can be found in the literature, varying in their focus, priorities and specificities. For example, Zucconi and De Bertoldi (1987) proposed the definition that composting is a controlled bioxidative process that (1) involves a heterogeneous organic substrate in the solid state; (2) evolves by passing through a thermophilic phase and a temporary release of phytotoxin; and (3) leads to production of carbon dioxide, water, minerals and stabilized organic matter.

2.2 Principles

In essence, composting involves oxidative dissipation of part of the carbon in the waste to CO_2 and H_2O while the plant nutrients are assimilated and mineralized by organisms. Some by products of the carbon oxidation, bioresistant compounds or their derivatives, secondary metabolites and microbial biomass fractions, are condensed and polymerized into humus through the presence of reactive free radicals.

The heat generated during the oxidation of carbon, particularly by heat-loving (thermophilic) organisms, antagonistic compounds produced by decomposer organisms, and the competitive ability of a versatile consortium of heterotrophic organisms, combine to kill plant and animal pathogens or at least to suppress them to acceptable levels. The survival capacity of pathogens is limited in composts due to their narrow requirements of specific substrates and temperatures for optimal activity. It is therefore obvious that conditions in the composting mass must foster optimal bioxidation by a suitable group of organisms for an adequate period.

2.3 The Nature of Compostables

Any biodegradable or biological material can be composted, alone if it has the right physical and chemical properties but more commonly with other suitably matched material(s). All materials of biological origin contain, in various proportions, water-soluble organic compounds; proteins; lipids (fats and waxes); structural or storage polysaccharides (starches, cellulose, hemicelluloses; chitin, glycogen, etc.); and, in the case of plants, phenylpropanoid compounds and their polymers such as lignin. As animals and their intestinal flora do not digest all of these fully, the same are also present in excreta along with their products and microbial biomass. The protein-rich bacterial biomass may account for as much as 50% of the carbon in the fecal matter (see Mathur *et al.*, 1990*b*). Animal excreta, particularly urine, is therefore a valuable source of N in compost-mixes, although animal manure slurries pose special problems for lack of litter and high moisture content (Gray *et al.*, 1987; Mathur *et al.*, 1989, 1990*a,b*).

Composting of slaughterhouse wastes, meat leftovers, small bones, blood, seafood processing wastes and other animal matters can be accomplished by the use of special care and practices (Mathur *et al.*, 1986; Dalzell *et al.*, 1987).

Vegetable materials left over from extraction of juices, oils, fibres and pulps are valuable as compost materials (Krishna Murthy, 1978; Chen & Hadar, 1987; Pudelski, 1987). Residues of tea and coffee processing and residues from microbe-based beverage and pharmaceutical industries belong to the same group.

Putrescible town and home refuse are, of course, major sources of compostable materials (Gotaas, 1956; Dalzell *et al.*, 1987; De Bertoldi *et al.*, 1987), as are cullings, weeds, fallen leaves and grass clippings.

Wastes from wood-based industries tend to be low in N and P and therefore are composted best in combination with other materials or with N and P fertilizers. Wood wastes have phenolic compounds and polymers which release bioinhibitory phenols, terpenes and tannins during decomposition. When ammoniacal-N is added or released into the wood wastes, e.g. from urea or urine, the slightly alkaline pH in presence of air causes neutralization and autooxidation of phenols to produce semiquinone free radicals and hydroxyquinones which polymerise, by themselves, or through enzymatic catalysis, into insoluble humus-like polymers, thus removing the bioinhibitory compounds from an active role (Mathur & Farnham, 1985).

A later section (5) focuses on seafood wastes and manure slurries.

2.4 Requirements of Optimal Composting

The enzymology and microbiology of composting parallel those of organic matter decomposition in soils and therefore the optimum conditions required also have some similarities. Optimization is achieved not by having only one or two conditions at their best but by manipulating the whole set of conditions to be near the most favourable set achievable under the circumstances.

2.4.1. C/N Ratio

The activity of microorganisms involved in composting is directed at synthesizing microbial protoplasm which contains about 50% C, 5% N and 0·25–1% P on a dry weight basis (Alexander, 1977). Since the atomic weights of C and N are so close, the C/N ratios on atomic and percent bases are 8 and 10 respectively which, not coincidentally, match the ratio in mature humus (Kononova, 1966). The assimilation and maintenance of every atom of C in microbial biomass require the oxidation of about two atoms of C for energy. Consequently, Waksman (1938) established that the microbes overall utilize about 30 parts of carbon for each part of nitrogen. Since these are approximate values, and not all the C and N in a compost mixture are bioavailable, nor is all the mass converted to humus or microbial biomass, the optimal ratios for different materials have been found to vary from 26 to 35 (Gotaas, 1956). Satisfactory results are obtainable mostly with C/N ratios between 30 and 40 (University of California, 1953).

Table 1 presents the C/N ratios of various materials.

When the C/N ratio is too narrow N will be lost from the system as ammonia. If the C/N ratio is too wide the synthesis of biomass, and its breakdown for new formation of protoplasm, will be repeated many times before a stable state is reached, slowing down the composting process.

As the word compost itself implies, different materials have to be mixed (composited) or layered to achieve an optimal C/N ratio, and thus optimal composting.

Obviously, for reasons similar to some of the above, a C/P ratio of 100–200

Table 1
Approximate Nitrogen Content and C/N Ratios of Some Compostable Materials (dry weight basis)

Materials	*%N*	*C/N*
Urine	15–8	0·8
Blood	10–14	3
Mixed fish scrap	6·5–10	—
Filleted fish scrap	8·2	4
Crab scrap	8·2	3·5
Lobster scrap	4·6	—
Night soil	5·5–6·5	6–10
Sewage sludge	5·0–6·0	6–8
Animal manures (without litter)	1·7–3·7	11–25
Farmyard manure (average, with litter)	2·15	14
Grass clippings (average)	2·4	19
Green vegetable wastes, weeds	2·0–3·7	11–20
Cereal straw	0·3–1·05	45–130
Combined refuse	1·05	34
Rotted sawdust	0·25	208
Raw sawdust	0·11	511
Peat (horticultural)	1·0	50
Newspaper	Negligible	Infinity
Paper sludge	Negligible	Infinity

Sources: Gotaas, 1956; Mathur *et al.*, 1988*a*, 1990*a*, *b*.

is also desirable, but this is less critical than the C/N ratio as most organic materials contain sufficient P for the purpose.

2.4.2. pH

As most biological fluids are balanced in their cationic and anionic ions near neutrality, pH 7·0, that is also the ideal for soils and a composting mass. At a pH much higher than 7·0, the aqueous medium of activity will contain dissolved base ions at a concentration higher than those in the microorganisms, causing an osmotic loss of water from their vital fluids. The excess of the OH^- ions will also cause loss of the ammonium ions as ammonia, and hydroxylation of essential biological elements like Cu and Zn, rendering them amenable to precipitation as insoluble mixed carbonates. At the other extreme, mass action of the excessive H^+ ions at low pH can cause decomplexation and desorption of essential base ions, e.g. Ca and Mg from organisms, and toxic metal ions like Al, Mn and Cu from minerals and organic matter.

The pH of manures is usually slightly above 7·0; that of plant materials and most household garbage is between 5·0 and 7·0, more towards the lower figure if the material is partly putrified as decomposition initially releases organic acids. On the other hand, if the compostable material is high in substances like proteins, urea, uric acid and chitin, enzymatic hydrolysis initially will release ammonia, thus raising the reaction to alkalinity.

Mixtures or alternating layers of acidogenic and ammonifying materials do not pose a problem provided the interlayer and overall pH is between 6·5 and 7·5, which is optimal for most decomposer organisms. Since composting is a dynamic biochemical process, marginal fluctuations in reaction will occur temporarily and spatially without affecting the overall process.

For most materials, therefore, pH adjustments are not necessary, but exceptions do occur. Wood wastes and sludges from pulp and paper mills may have low pH of 5–6 and a high buffering capacity. As these materials also have a wide C/N ratio, necessary addition of ammonia-releasing compounds like blood, urine or urea can be made to neutralize the acidity until base cations are released by decomposition of the wood waste. In the interim, loss of ammonia can be prevented by placing a 100–150 mm-thick layer of acidic peat or mature compost as an envelope over the composting mass. The other example is of seafood wastes or other food processing wastes where alkaline materials may have been used in the factory. The alkalinity can be neutralized by use of peat or acidic wood materials in layers or mixtures.

Amelioration of pH can also be achieved by using sulphur to create acidity, and lime to neutralize acidity. In so doing, one must take into account the buffering capacities of the materials involved, and pretest different rates of application and their effect over time.

2.4.3. Physical State (Particle Size)

Various shredders, raspers, rotary drums, or hammer mills are sometimes used to break down the materials into a suitable physical state.

Composting involves actions of enzymes, microorganisms and oxygen. Therefore the smaller the size of the material acted on, the greater is the surface area and easier the contact between various reactants, except perhaps oxygen. Oxygen and the water produced and present do not move freely in narrow interstitial spaces between small particles that pack together tightly, nor does air diffuse easily into large particles to a distance greater than 2·5 cm. Consequently, the composting is optimal in a mixture of particles of various sizes in which the largest is less than 5 cm in its greatest dimension. Ideally, as in a soil, both micro- and macropores are needed between particles so that water can be retained in the former to provide a medium for the microorganisms, while excess water, the CO_2 produced, and the necessary air can move easily in the latter.

2.4.4 Moisture Content

The main site of microbial activity in soil or composts is the thin film of water on the surface of particles. Water is essential also for dissolving and transporting the nutrients and substrates that the organisms can absorb only as solutions. Motile organisms and free enzymes also move through aqueous medium. However, for optimal aerobic activity, as in soils, the interstitial spaces, voids or pore spaces between particles need to be occupied by and accessible to both water and air.

The water absorption capacity of soil particles and materials varies widely so that a sandy medium with 50% moisture content will be nearly waterlogged while peat is air-dry at a moisture content equal to 50% of moist or 100% of dry weight. In soil science, therefore, it has long been known that maximum aerobic microbial activity occurs when the soil moisture content is two-thirds of its water holding capacity (WHC) at $\frac{1}{3}$ bar suction (Waksman, 1938). WHC in effect is influenced both by hygroscopicity of the particles and by the proportion of macro- and micropores in the soil. Surprisingly, no such criterion seems to have been applied to composts. Research findings for the appropriate water content on a moist weight basis therefore vary from 45 to 90% for different materials (Gotaas, 1956; Poincelot, 1975; Biddlestone *et al.*, 1987), with 50–60% being found to be suitable for most materials. Thostrup (1985) observed that 70–80% moisture in solid manures of a bulk density equal to 300 kg/m^3 gave the best results. The author of this chapter has also effectively composed peat-based composts with 70–90% moisture content and paper mill sludges with 70% water. Pieces of wood in the sludge, and peat have the necessary structural strength due to their fibrosity. Peat and the sludge have low bulk densities of 60–300 kg/m^3. In contrast, any material that tends to form a tight mat may not compost well at higher than 60% moisture content.

The area of moisture content nonetheless needs more research to seek a relationship between the water retention capacity of the compost-mix and the optimal moisture content for it. In any case, one should also be aware of the changes that might occur in water relationships in the composting process, the system of aeration used, and the climate. In warm climates the evaporative loss of moisture may necessitate periodic irrigation of the compost-mix. Some materials like paper and seaweeds may become soggy on wetting or decomposition. Frequent or continuous mixing of the hot compost-mix may change the structure of the mass and cause excessive moisture loss.

2.4.5 Air Spaces and Aeration

A substantial portion of the organic carbon is oxidized to CO_2 and H_2O. Jeris and Regan (1973) found that each gram of volatile organic matter in municipal refuse required 144 mg of O_2. Their results suggested that a minimum of 30% free air space should be maintained for a wide variety of composting mixtures. Thostrup (1985) on the other hand obtained best results with 70% free air space in solid manures. As the air within a composting mass may be highly enriched in CO_2, it is important to consider the oxygen content itself. Wiley and Spillane (1962) reported that the proportion of oxygen 38 cm inside a pile was 18·6% but fell to as low as 1–2% at 61 cm. It has been suggested that a minimum concentration of 5% oxygen in the gas phase is essential for high-temperature (>45°C) composting. Biddlestone *et al.* (1987) have recommended 10–18% oxygen in the gas phase of the matrix, although maintaining those levels may involve energy-consuming mechanical aeration or turning to an extent that may be deleterious to structure and retention of H_2O and NH_3. Thostrup's (1985) experiment revealed that the O_2 content of exhaust air should be not less than 10%, although even 7% is not limiting. It would seem that the proper criterion for monitoring and maintaining oxidative conditions should be the redox potential in the compost-mix; and we are attempting to test the feasibility of using this criterion, and the oxygen repletion rate.

It is important to maintain a proper redox condition to avoid the malodour-generating anaerobiosis where reduced sulphur and nitrogen compounds like sulphides and amines are produced. That and the lack of heat-sanitization were indeed among the reasons why partial or total anaerobic composting generally fell out of favour. Composting methods (see section 4) therefore now vary mostly on the basis of the manner in which the aerobic state is achieved and maintained. The optimal is considered to be a supply of 0·6–1·8 m^3 air/day/kg volatile solids during the thermophilic stage of composting.

2.4.6 Temperature

The temperature of a compost-mix is important in several perspectives. As heat is generated by oxidation of organic matter it is symptomatic of progress of the process. Microbial processes of thermophiles and chemical reactions are faster at higher temperatures but beyond 70°C the organisms and many exoenzymes are inactivated and 'thermo killed', causing the process to stall. High temperatures are also conducive to excessive evaporative loss of water and odour emanation (Finstein & Miller, 1985; Van der Hoek & Oosthoek, 1985). Retention of heat is necessary not only for fostering thermophilic microbial activity but also for sanitizing the waste by 'pasteurization-type' killing of pathogens. Retention and continual generation of heat are influenced by ambient temperatures, configuration and size of the compost mass, its insulating property, and the chemical nature of its constituents.

The heats of combustion of the three most common organic constituents of biomass—proteins, carbohydrates and lipids—range from 9 to 40 kJ/g with lipids yielding about twice as much heat per unit weight as the other two. Wiley (1957) found that 75% of the 22 000–26 500 BTU of heat released on oxidation of 1 kg volatile solids into CO_2 and H_2O was derived from the lipid fraction. Pulverized waste was experimentally determined to release in an 8–10 day period 7 kJ/g of volatile solids. It is therefore not surprising that within 2–5 days of initiation a composting mass attains a temperature of >40°C which inhibits mesophilic (moderate-temperature-loving) organisms, causing organic acids, the intermediate products of their metabolism, to accumulate. Above 40°C the mesophiles are killed, except some spores, and the thermophiles start to dominate the population. These increase the temperature to beyond the 60°C that is lethal to most of the fungi in the compost. Above 60°C the decomposer activity is carried out mainly by actinomycetes such as *Streptomyces* whose spores generate the good-earth smell of a mature compost. The process may also be affected in the above succession as proteins in the original mix or the killed biomass are hydrolyzed by enzymes to produce ammonia. At the high temperature, a combination of the ammonia or amino compounds, sugars or phenols may theoretically cause Maillard-type reactions to produce brown humus-like polymers, although this has apparently not been demonstrated.

The rate at which the heat is lost at this stage is determined by the size and shape of the compost mass. Readily available substrates such as starches, sugars, lipids and protein are the main 'fuel' at this stage until the temperature falls below 60°C to allow fungi and actinomycetes to attack cellulose, hemicellulose and lignin. The need to maintain the temperatures at high levels has been perceived to be due to their lethality to pathogenic microorganisms and ova of annelid parasites and nuisance insects. The earlier perception that the higher the temperature the greater is the decomposition has been altered by the realization that few organisms are active at above 70°C. Based on various criteria, optimal temperatures for decomposition of different materials have been found to range from 48°C to 71°C (Poincelot, 1975). Recently, heightened awareness of (a) the dependence of pathogen inactivation on both the period and temperature of exposure rather than the latter alone, (b) the role of antagonistic organisms and bioinhibitory properties of certain intermediate products of decomposition, and (c) the need for odour control, as well as detailed observations of maximal decomposition at near 55°C, have seemingly led to a consensus that 55–60°C is the optimal temperature range (Bollen, 1985; Finstein & Miller, 1985; Lopez-Real & Foster, 1985). Significantly, Lopez-Real and Foster (1985) found that the plant pathogens they tested survived higher temperatures *in vitro* than in the compost.

Of special importance to agriculture is the presence of seeds of weeds in farm wastes. Certain seeds survive or even need passage through animal tracts while others are gained by manures on exposure to air during storage in dumps or lagoons. Field application of raw farm wastes consequently often increases weed infestation. Aerobic composting of the wastes kills or inactivates the weed seeds (Gotaas, 1956; Dalzell *et al.*, 1987).

2.4.7 Heap Size

It is well documented that a minimum height of 1·5 m and width of 2·5 m is necessary to retain enough heat in a composting mass to promote the desirable thermophilic activity (Biddlestone *et al.*, 1987), although Mathur *et al.* (1985, 1990*a*) showed that a height of 1 m was sufficient when the medium of composting is peat and mixtures of peat with fibre-containing manure, both of high thermal-insulation capacities.

The length of the heap and its total size will also depend on the aeration and agitation system used. In any case, an important fact to remember is that large heaps tend to allow pockets of anaerobic conditions due to compression by the overlying burden, and should be avoided.

2.4.8 Odour and Fly Control

Decomposition of organic material naturally may release ammonia and aliphatic acids under aerobic conditions and foul-smelling H_2S, mercaptans, diamines, methyl amines, and skatoles under anaerobic conditions. Some of these products are cadaverine and putrescine which need no elaboration of their implications as they are related to the words cadaver and putrefaction.

Besides the aesthetic undesirability of the aforesaid compounds, their presence on the compost mass surface attracts nuisance insects such as fruit, vinegar and house flies. The eggs laid by these insects, and their subsequent hatching in compost bags or beds, render such composts unattractive to vendors, purchasers and users of the product, and pose some health risks due to vectoring of diseases by some insects.

The classic control of these two problems, used since ancient times in China, was a mud plaster of 5–10 cm in thickness, often containing binding agents such as chopped rice straw (Dalzell *et al.*, 1987). Similar control is possible by the use of 10–15 cm of peat, mature compost, chopped straw, or hay (e.g. Mathur *et al.*, 1985, 1990*a*,*b*). The latter two materials may also help shed rainwater when the heaps are sloped.

The control of odour by avoiding high heat has already been dealt with in the previous section.

2.4.9 Additives (Optional)

The first step in compost-mix formulation is the matching of different materials for their diverse C/N ratios, water absorption capacity, moisture content, and physical structure to obtain optimal conditions for the composting process (Gotaas, 1956). In many systems air is added continuously or periodically by mixing, natural ventilation or the use of fans (see section 4).

Liming of compost-mixes, except to mitigate biologically limiting acidity, may hasten decomposition of acidogenic materials but is generally inadvisable because the high pH generated by the lime and the base elements released on decomposition are conducive to loss of ammonia (Poincelot, 1975).

Addition of sources of bioessential mineral elements, where they are lacking, may promote microbial activity. Theoretically, addition of degradative exoenzymes may also hasten decomposition. However, the natural microflora of most wastes are capable of producing the biocatalysts quite rapidly. Some autolytic enzymes may also be already present in the biomass being degraded.

Addition of sources of N and P may be desirable when the materials to be composted lack these elements markedly, such as when bark, sludge from pulp and paper mills, and certain wastes of food processing are to be composted. For example, urea or ammonium nitrate, preferably the former, can be used for composting certain wood wastes lacking in N (Pudelski, 1987), but steps need to be taken to avoid loss of ammonia as the urea is hydrolyzed quite rapidly in the compost by the almost ubiquitous enzyme urease.

The addition of soluble phosphate compounds in composts was researched by Acharya (1954), originally to increase plant availability of the fertilizer-P in soils of high phosphate-fixation capacity. Subsequently, sparingly substantiated claims were made that phosphate addition suppresses denitrification (NO_3^- reduced to N_2) and promotes biological nitrogen fixation by fulfilling the microbial need for P.

Addition of insoluble phosphorus compounds is also made to solubilize unavailable forms of it from plants such as soft and hard (sedimentary and igneous) rock phosphate, and basic slag from old-type steel mills. The heat, high concentrations of CO_2, Ca-chelating properties of acidic intermediate products of decomposition and humus, as well as the anions CO_3^{2-}, SO_4^{2-} and NO_3^-, may help mobilize the insoluble phosphate materials. The literature on this subject was recently reviewed by Mathur *et al.* (1987) who reported on the composting of an igneous rock phosphate.

Another type of additive is represented by the starters and inocula touted by many commercial enterprises, to activate the composting process due to their high levels of enzymes, 'special vitamins and hormones needed by compost organisms', and decomposer organisms. Any ecosystem at equilibrium has all the niches covered by the type of organisms it can sustain in their most suitable form. Addition of any extraneous organisms is both unnecessary and an added stress on the normal population. Therefore any beneficial effects of such additives to wastes that already contain normal saprophytes, and are open to natural seeding by soil and dust, are unlikely to be due to the microbial inocula except where the wastes are low in population due to some pretreatment (Gotaas, 1956; Poincelot, 1975). Experiments with inoculations of composts therefore give neutral, positive or negative results (Nakasaki *et al.*, 1985). This does not contravene the evidence of rapid and beneficial decomposition of straw alone by cellulolytic fungi and N-fixing organisms, as the latter may hasten cellulose hydrolysis by consuming the product, glucose (Biddlestone *et al.*, 1987). When the readily available sources of energy for the microorganisms are extremely low, inclusion of soluble carbohydrates in sugar-related industry by-products or materials which yield soluble carbohydrates readily, e.g. seaweeds and starchy wastes, may hasten initiation of the thermophilic phase (Krishna Murthy, 1978; Mathur *et al.*, 1986; Dalzell *et al.*, 1987).

The best additive for a compost-mix is old but moist mature compost to provide a suitable starting population throughout the composting mass and the bioavailable minor elements essential to life.

3 CHEMISTRY AND BIOLOGY OF THE COMPOSTING PROCESS

The multitude of reactions and alterations that occur during composting are analogous to those that biotransform organic detritus in soil to humus (Paul & Clark, 1989) except for the thermophilic stage, and therefore the speed of the bioconversion. Also, the dismutation or physical breakdown of the plant residues in or on soil by the soil fauna (centipedes, millipedes, springtails, mites, nematodes, earthworms, etc.) is limited in composting to the initial and maturation stages, due to the characteristic heat-intolerance of cold blooded animals. This, of course, does not apply to the mild temperature composting of wastes specifically by earthworms.

At the initial stage, mesophilic microorganisms that occur naturally on soil, dust particles and waste begin the decomposition process in which sugar-utilizing fungi and acid-producing bacteria are the most active in depleting soluble sources of carbon in the compost mix. The oxidation of the sugars, amino acids and alcohols generates heat. Temperatures beyond 45°C inactivate or kill most mesophiles, except their spores. At this stage the more readily accessible polysaccharides such as starch, chitins and pectins, proteins and lipids may be hydrolyzed by exoenzymes amylase, chitinase and pectinase, protease, and lipase to release sugars, amino sugars, uronic acids, amino acids, alcohols and fatty acids. These fuel thermophilic fungal activity until the temperature goes beyond about 60°C when the heat-tolerant actinomycetes begin their domination.

Many bacteria and fungi produce exoenzymes (extracellular enzymes) to hydrolyze insoluble amylose and amylopectins outside the parent cells to produce soluble compounds they can absorb and use intracellularly. Alpha-amylases cleave amylose and amylopectins into glucose directly while β-amylase produces the dimer maltose. Both maltose and dextrins are generated from amylopectins. Pectins like other hemicelluloses are copolymers of hexoses, pentoses and, in some cases, uronic acid. Their complete decomposition involves several enzymes called pectinases. A polysaccharide of special interest is cellulose, a major constituent of plant materials and municipal wastes (due to their paper content). The degradation of cellulose by bacteria and fungi requires not one but three major enzymes as a complex. Endo-β-1,4-glucanase hydrolyzes insoluble crystalline polymers of cellulose randomly to release glucose and cellobiose. The smaller chains of glucose are then the substrate for the exo-β-1,4-glucanase which cleaves off cellobiose dimers from the nonreducing end of the polymer chain. The disaccharide cellobiose is hydrolyzed by cellobiase into glucose. Thermophilic bacteria and fungi utilize cellulose mostly during the maturation process and while the temperature is all below about 60°C. At higher temperatures, actinomycetes dominate. They are specifically active in decomposing hemicelluloses.

A polymer that does not seem to have been researched extensively in this context is chitin (poly-*N*-acetyl glucosamine), and its homologue poly-*N*-acetyl muramic acid found in bacterial cell walls. Chitin is the main component of exoskeletons of crustaceans and of fungal cell walls. Lack of research in this area is surprising because fecal excreta contain as much as 50% of their carbon in microbial biomass and detritus while insects are commonly present in many urban and industrial wastes. Chitin decomposition is of special importance to composting of shellfish wastes. Extracellular chitinase enzymes do occur in soils (Mathur, 1982).

Proteins are the major N-containing polymer in living bodies. Proteolytic enzymes hydrolyze the peptide links between amino acid monomers of the protein polymer. Microbial proteases, pronase and subtilisin, depolymerize proteins by cleavage of terminal amino acids. Decomposition of most proteins in soils and composts should be fairly rapid except when the protein is highly insoluble. Keratin, the S-rich highly cross-linked insoluble protein in hair, feathers, hooves and horns is degraded, albeit slowly, by actinomycetes and some fungi. The microbial proteins produced in soils and composts may resist decomposition due to their protective tanning by free phenolic compounds and humic substances, and that is the reason for the desirable slow-release nature of part of the N in composts.

Monomers of both lipids and proteins are easily utilized by soil organisms. Lipids are depolymerized by the lipolytic enzymes lipases, except when the lipids are water-insoluble waxy materials. Aliphatic compounds of long-chain fatty acids have lately been recognized to persist in soils where they may have a role in waterproofing soil aggregates (e.g. Preston *et al.*, 1986).

Another material that resists decomposition is lignin as it is not a specific polymer of identical repeating units, and thus is not degraded by specific enzymes. Also its phenylpropanoid units or their phenolic products tend to be biostatic. In the latter stages of the thermophilic range, when fungi begin to return to the core from the cooler periphery, they transform and break by attrition the layers of lignin that encrust cellulose and hemicellulose fibres. The oxidized by-products of lignin attrition are semiquinones, hydroxyquinones and their derivatives, the amino quinones. All these free radicals condense with each other and with amino acids, peptides and other N compounds to form the recalcitrant humic substances. The phenolic compounds that contribute to humus formation may also be microbial metabolites or products of other aromatic substances in the detritus.

The amino acids and other N-compounds are converted to ammonia which, in turn, is oxidized to nitrates by a specific group of mesophilic acid-intolerant autotrophic organisms. The reader is referred to Paul and Clark (1989) for the microbiology and biochemistry of soils, and to Poincelot (1975) for those of composts.

4 THE TECHNOLOGY OF COMPOSTING

This section outlines a number of aerobic 'hot' composting methods but excludes wormicomposting and anaerobic systems. In wormicomposting (Edwards *et al.*, 1985) the main objective is to harvest the nutrients as earthworms which then have to be used as hog, fish or fowl feed to recycle the nutrients in the waste. The composts cannot be allowed to be hot enough to kill the worms. The whole multistep system requires intensive management and extremely careful environmental controls.

The anaerobic or partly anaerobic systems were designed mainly to prevent loss of ammonia from night soil, or to produce methane ('gobar gas', biogas) which is a subject in itself.

The aerobic liquid composting system (LiCom) is discussed in the subsection on manure slurries.

Aerobic systems of solid composting can be classified into open systems and confined, in-vessel, or container systems. The former are based on the Indore method of Howard and Wad (1931), severally modified and refined.

4.1 Open Systems

4.1.1 The Indore Method

The Indore method requires a heap of trapezoidal cross-section. The heap is about 2 m wide at the bottom and 1·4 m wide at the top, with sloping sides 1·3 m high, and a minimum of 2 m in length. The heap contains, starting with carbonaceous wastes at the bottom, 20 cm of carbon-rich and 10 cm of nitrogen-rich materials in layers. A depression at the top, in hot climate areas, traps rainwater. The heap is covered with soil or hay as thermal insulation. Where drainage channels are provided, the liquid is added back to the pile. The pile is turned over at 6- and 12-week intervals until the compost is mature. The Indore method also works when the heaps are built in partially filled trenches (allowing space for mixing and turning), as long as aeration ducts and drainage channels are provided.

4.1.2 The Simple Turned Windrow

A compost windrow should be a minimum of 1·5 m high and 2·5 m wide, of any convenient length. The materials are mixed and placed manually or mechanically. At the farm scale, such heaps can be built by a modified solid manure spreader which can also effect mixing and shredding of the materials. If the outside of the windrow does not have to be subjected to sanitizing heat, such as when composting fallen leaves, no steps are necessary for remixing and relaying the materials (turning) or for forcing air into the mass. This type of composting can, however, take a long time to be completed, up to two summers in temperate climate areas. To promote faster decomposition and better hygiene, and to further break the materials physically, any of many different types of turning equipment can be used. For certain types of mixtures of good structural strength, vertical aeration channels (chimneys) can be created by using rods which are pushed in, jiggled, and then withdrawn (Biddlestone *et al.*, 1987). This may be repeated at the first few turnings.

Advantages of the windrow systems are the low capital costs, faster product maturation compared to the nonaerated static piles, and generally good quality of the product. Disadvantages are the cost of turning, where practiced; odour emanation from some wastes not covered with mature compost or other suitable deodorizing envelopes; vermin; the high requirements for land; loss of ammonia and dispersal of allergenic spores and dust during turning; and vulnerability to adverse conditions of temperature and precipitation. Protection from rain, however, can be achieved with simple tent or free-standing hat-type structures where the site is protected by windbreaks (see, e.g., Mathur *et al.*, 1986, 1988*b*). It has been argued that in temperate climate areas mature compost prepared by any method has to be stored anyway, and protected from rain to prevent leaching of nutrients in piles of limited height (to avoid anaerobic pockets created by compression), because compost is not applied on farmland which is unsuitably wet, frozen or covered by crop plants of more than a certain height.

4.1.3 The Passively Aerated Windrow (PAW) System

This system will be dealt with in a later section (5.1.3).

4.1.4 Static Piles

A mixture of solid wastes is stockpiled in the open air and turned occasionally for aeration. This low technology method is considered to waste both ammonia and energy. Both of these disadvantages have been substantially mitigated by mechanical forcing of air through perforated pipes at the bottom of the static pile by (a) continuous air suction, or (b) continuous air blowing, or (c) alternate blowing and sucking, or (d) intermittent air blowing to keep the temperature below 60°C.

The static pile system with an air suction component was created by the US Department of Agriculture staff at Beltsville, for the US Environmental Protection Agency, to compost a mixture of sewage sludge and wood chips. Undecomposed wood chips are screened out and reused in a new pile (Epstein *et al.*, 1976). Air is sucked through continuously, at a rate, into the blower, of about $0{\cdot}2\ m^3/min/1000\ kg$ of the composting mass to maintain an oxygen concentration of nearly 15% in the mixture. The inevitable odour in the exhaust air has to be filtered through piles or towers of mature compost. Several operations in the USA are based on this system.

The sucking of air provides the least amount of oxygen in the centre of the

Table 2
Confined Composting Systems

System	*Configuration*	*Aeration*	*Agitation*
Dana	Horizontal flow	Forced	Continuous
Earp–Thomas	Vertical flow	Forced	Continuous
Ewesen	Horizontal flow	Forced	Continuous
Fairfield	Mixed flow	Forced	Continuous
Paygro	Walled container	Forced	Intermittent
Peabody	Vertical flow	Forced	Intermittent
Tollemache	Walled container	Natural and forced	Intermittent

pile where it is needed most, does not aid desiccation, and allows temperatures to exceed 60°C—the maximum tolerated by most lignolytic and cellulolytic fungi. These problems were avoided in a system formulated by a team at Rutgers University, New Jersey, (Finstein *et al.*, 1983). The Rutgers system involves blowing air through the compost mass in response to a thermistor and fan both controlled by a microcomputer to maintain temperatures below 60°C. To overcome the problem that temperatures in the compost pile may not be high enough everywhere to sanitize the wastes, several systems now provide for sucking of air through the pile in the primary phase and then reversing the air flow in conjunction with temperature control.

4.2 In-vessel (or Reactor Confined) Systems

Several such systems (Table 2) have been developed since the first in-vessel method was designed by Becarri and modified by Verdier and Bordas (see Gotaas, 1956).

The main advantages of these systems are rapidity of the process (10–14 days), low land requirements, complete process control, and consistency of end-product. In many cases, part of the composting process and all of the maturation occurs in stockpiles outside the enclosed systems.

In-vessel systems were reviewed by Gotaas (1956), Gray *et al.* (1973) and De Bertoldi *et al.* (1985). The main feature of these systems is that the compost materials are mixed and the mass advanced and aerated automatically and mechanically. They are costly to install and operate, and need intensive and skilful management.

5 COMPOSTING OF CERTAIN PROBLEMATIC WASTES WITH THE NOVEL PASSIVELY AERATED WINDROW (PAW) SYSTEM

5.1 Seafood Wastes

As indicated by Table 3, reproduced from Swanson *et al.* (1980), commercial fishing generates large proportions of waste. The waste may be 80% or greater

Table 3
United States Seafood Processing Solid Waste

Category	*Average 1972–76 landings (10^6 kg)*	*Product yield (%)*	*Solid waste (%)*	*Peak season*
Fish:	1 850 (1977)			
Bottom fish	238 (1972)	20–24[a]	55–75	Summer
Catfish	72 (1975)	63[b]	37	Fall to spring
Halibut	10	90[c]	10	May to September
Menhaden	877	28–36[d]	15[a]	June to September
Anchovies	115	28–36[d]	15[e]	June to September
Salmon	105	62–68[f]	16–27	July to August
		65–80[b]	21–23	
Tuna	181	40–50[g]	49–59	Summer
Shellfish:	516 (1977)			
Clams	46[h]	10–15	82–90	Year round
Oysters	24[h]	10–15	82–90	Fall and winter
Crabs:				
Blue	61	9–16	51	Spring
Dungeness	10	17–27	50–60	December to March
Alaska King	43	25–36	57	August to September
Tanner	27	10–20	50–60	January to May
Shrimp	171	12–20	65–85	Varies with location
		80[i]	15	

[a] Filleting process; fish flesh operations may achieve higher yields. [b] Farm raised catfish processing. [c] Fresh/frozen processing. [d] Rendering process; both fish meal and fish oil are considered final products. [e] Represents concentrated fish solubles which are commercially marketed. [f] Canning process. [g] Does not include red meat (8–10% of tuna) which is processed for pet food. [h] Excluding shell weight; total harvested weight is 5–10 times this value. [i] If beheaded at sea.
Source: Swanson *et al.*, 1980.

when landed catches are judged unfit as food, or when only eggs (roe) are harvested from herring (*Clupea harengus*) or other fishes. Filleting of fish of the cod family (Gadidae) discards 40–60% of the total landings. About one-third of fish cut into steaks, or canned, is culled. The scrapped exoskeletons and other parts of processed shellfish like the American lobster (*Homarus americanus*), snow crab (*Chinocetes opilio*) and shrimp constitute 30–80% of their total weights. Marketing of whole fishes generates little waste but nearly all catches contain some fishes of unsuitable state, size or species (by catch). Some fishes are crushed in dragnets.

Consequently about 140 000–200 000 t of unutilized waste is generated yearly solely from cod family fishes in Canada alone. There may be more elsewhere, as Canada lands only 5% of the global catches of bottom-dwelling ground fishes of the cod family and 2% of all related marine products. For example, seafood processing produces about 2 million metric tonnes of wastes in the USA (Swanson *et al.*, 1980).

5.1.1 The Problem

Environmental concerns and regulations in North America have made it costly (US $15–50 per ton) to dispose of fish wastes by dumping on open seas, or by burial in land (landfilling). Due to the nuisance of malodours and scavengers, and the delayed pollution it causes in ground and surface waters, landfilling of fish wastes is likely to be prohibited in more and more jurisdictions.

Fish wastes include whole waste fish, offal that contains viscera, and fish scrap that is the residue of filleting. The scraps (racks) contain skin, heads, fins, tails and backbones. Finfish wastes are thus rich in proteins and bones, and may contain some oil. Dogfish (*Squalus acanthius*) gurry, being from a shark, has no bones but contains cartilagenous heads, the tough sharkskin, tails and viscera, all in a slurry. Shellfish wastes are high in chitin (poly-*N*-acetyl glucosamine) and Ca- and P-rich crustations, and contain some meat. The finfish and shellfish wastes therefore contain up to 8% N, 7% P and 15% Ca.

Direct use of the fish wastes for land manuring, or landspreading, is deterred by the uniquely obnoxious odours of putrefying fish. Such use even in areas where it is permitted is therefore limited mainly to internment in soil at or before planting (as once practiced by American Indians). The fish proteins, lipids, and chitin are easily hydrolyzable by abiotic (nonliving) soil exoenzymes (Mathur, 1982), and by autolytic enzymes in dead cells, even under conditions suboptimal for microbial growth. The fish lipids being rich in polyunsaturated fatty acids oxidize rapidly, releasing volatile products with obnoxious rancid odours. Aerobic decomposition of fish wastes in soil and ordinary hot composts is therefore so rapid (and malodorous) that much of the ammonia produced by abiotic and exothermic enzymatic hydrolysis and deamination is vaporized—more so, because the Ca present in the wastes raises the pH of the medium to the alkalinity that promotes ammonia loss.

Even the inclusion in fish composts of mildly acidic materials with low stable buffering capacity (e.g. river muds, any wood wastes, citrus and banana wastes) may not conserve most of the escaping NH_3, nor provide a naturally well-aerated matrix. Slowing the degradation of fish wastes by restricting vapour escape and air supply limits the loss of NH_3 from anaerobic cold composts and waterlogged soils, but anaerobic decomposition of flesh produces foul-smelling and poisonous compounds.

5.1.2 A Solution

A team of researchers in Canada in 1983 felt that fish wastes should be composted aerobically by being mixed with and enveloped in a material that (a) has a wide C/N ratio; (b) is acidic and hydrophilic enough to trap NH_3 in solution; (c) has high capacities for adsorbing and complexing cations like NH_4^+ and Ca^{2+}; (d) is fluffy enough to be well aerated so that malodours of anaerobic decomposition are not created but acidic SO_4^{2-} and NO_3^- ions that help dissolve bone phosphates are generated (oxidatively); (e) deodorizes any malodours produced even transiently; (f) provides heat insulation; and (g) though biodegradable, will not sustain thermophilic microbial activity by itself so that the composts can mature early in a short Canadian summer without having to be turned or energetically aerated.

It was surmised that horticultural sphagnum (blonde) peat and light brown fuel peat could meet these requirements well. In addition, peats have high market acceptability as soil conditioners although by themselves they are generally poor sources of fertilizer elements. One potential constraint could be the occurrence of biostatic phenolics in some peats. These inhibitors, however, are largely deactivated by Ca and proteins.

In the context of soil erosion and degradation caused and accompanied by loss of soil organic matter, it is noteworthy that peat is known to be 5–8 times more effective in building soil humus (Janssen, 1984) than uncomposted manure, crop residues and green manuring. Also, a market exists for high-priced, limed and fertilized peat for use in greenhouses, house plants, and home gardens. Fish peat composts could be homologous substitutes for or a component of such high-priced enriched peat formulations, or could be prepared for use on farms with access to peat and fish wastes.

5.1.3 The PAW System

The rationale of the PAW system was fairly simple. The air warmed by the heat generated in aerobic decomposition in the middle of a pile or windrow naturally rises upwards and outwards due to the density and temperature gradients. The warm air will be replaced by cooler air mostly from or through other parts of the compost, setting up a mostly internal circulation until the air is uniformly warm and oxygen deficient (but rich in CO_2), thus creating anaerobic pockets.

If the pipes with perforations all around them, used in static Beltsville or Rutgers systems and others, were present at the base, most of the unaided air circulation would still be internal. This is interrupted in windrow systems by turning which also accomplishes subjection of the outer material to the sanitizing heat of the compost mass interior. Mathur *et al.* (1985) felt that both needs for turning had to be avoided in the case of seafood wastes to forestall emanations of malodours and ammonia and amines that formed very rapidly due to the autolytic enzymes present in the waste. The need for turning was obviated by (a) basal placement of pipes that had holes only on the top side and with both their ends open to the ambient air, and (b) use of peat (that is already sanitary) as the base and envelope of the compost heap. The strategically situated pipes and the perforations in them naturally direct fresh oxygen-rich cooler air to where it is needed the most. The rate of air-intake is controlled in effect by the heat generation and thus the oxygen demand of the decomposer microflora.

For wastes other than from seafood, the peat can be replaced by strippable sanitary wastes such as straw, hay or wood chips, or preferably by mature composts. Both peat and mature compost can trap ammonia. Although the cation exchange capacity of peat (125 meq/100 g) is about five times that of a compost, its bulk density is about one-fifth that of a mature compost. Therefore a unit volume of the two will have similar buffering capacities. Because the envelope is continuously in the mesophilic temperature range, it should support the heat-intolerant nitrifying organisms which would persistently oxidize the ammonia to NO_3, thus replenishing the ammonia-neutralizing capacity of the envelope throughout the process.

The PAW system is employable for both short and long windrows. For the short, the compost heaps are 1 m high, trapezoidal in cross-section, with base and top planes of 3 m × 2 m, and 2 m × 0·3 m, respectively. A basal 10 cm thick layer of peat is laid in a fluffy state on the ground. Two 3 m long standard ABS soil pipes, 10 cm in diameter with perforations 1·2 cm in diameter, are placed lengthwise on the peat layer about 0·6 m from the margins. The two parallel rows of perforations at 5 cm intervals are about 10 cm apart on the two sides of the apex. Such pipes are routinely used for discharging and spreading effluents from septic tanks in North America.

5.1.4 Composting with the PAW System

A layer of fish scrap or crab scrap is spread on the basal layer and covered with peat or seaweeds and then peat or mixtures of peat with any seafood waste according to the formulation for the different composts. A suspension of old mature seafood waste compost can be sprinkled on the layer or mixed into a compost layer as inoculum. There are three layers of each of the nonpeat materials in a heap or a suitable mixture heaped to a total height of about 90 cm. The heap is then covered on the top and sides with a 10–15 cm thick layer of peat that is lightly tapped to ensure complete coverage.

More details of the system and its monitoring were given by Mathur *et al.* (1986) for the short windrows and by Mathur *et al.* (1988*a*) for the long windrows.

For building long windrows for commercial-scale production, a 20 cm thick basal layer of fluffy peat is laid on which are placed the reusable ABS plastic ventilation pipes (with holes on the top side) as above, but in this case they are across the 3 m basal width of the windrows, at intervals of 60 cm.

A mixture of peat with wastes of shellfish, or finfish, is placed on the pipes to a height of about 90 cm at which stage the width is about 1·5 m. At this height, another series of pipes are placed across the width at 60 cm intervals at points above the space between the lower row of ventilation pipes. Another 30 cm thick layer of waste and peat mixture is placed on top of the second row of pipes. The width at the top is 35 cm. Finally a 15–20 cm thick envelope of peat is provided on all sides of the windrows. The length of the windrows can be any distance depending on the site and availability of the fish wastes, but is generally 15 m (50 ft). Peat of about 35% dry matter content is generally 40% of the compost mix by volume, nearly half of which is used in the base and envelope layers. Acidic peats of low or high fibrosity, that is both the brown low-grade fuel peat and the horticultural blonde or white sphagnum moss, can be used.

The PAW system is being used by some farmers in Atlantic Canada where the seafood waste, peat and the humus-depleted sandy soils of poor fertilizer-retention capacity occur together. Six commercial concerns are also involved. Three of them market the product in bags after screening (First Organic Farms, Aquaterre Inc., Lameque Peat Moss). The inclusion of seaweeds to provide potassium and readily degradable polysaccharides was later abandoned as being unnecessary and costly.

Table 4 describes the properties of the peats used and the seafood wastes. Seafood wastes from the following sources have been successfully composted by the PAW system, to the author's personal knowledge: cod (scraps with or without evisceration), flounder, sand shark (dogfish), redfish, herring (whole and scraps), capelin, sole, snow crab (queen crab), shrimps and lobster.

Table 4
Properties of Some Materials Used in the PAW System for Composting of Seafood Wastes and Peat

Property	*Peat I*	*Peat II*	*Peat III*	*Dogfish gurry*	*Mixed fish scrap*	*Crab scrap*	*Lobster scrap*
pH	3·5	3·65	3·73				
Cation exchange capacity (meq/100 g)	155·0	150·0	148·0				
Moisture (%)	65·0	65·0	67·0	73·0	50·0	35·0	45·0
Organic matter	98·5	97·0	94·0		65·5	56·3	
Carbon (%)	50·1	59·8	46·1		38·7	27·2	
Nitrogen (%)	1·05	0·91	1·1	5·49	8·2	8·1	4·6
Phosphorus (%)	0·01	0·035	0·04	0·92	6·1	6·1	3·5
Potassium (%)	0·008	0·013	0·11	0·53	0·37	0·32	0·5
Calcium (%)	0·9	0·17	0·20	1·40	10·19	14·91	
Magnesium (%)	0·10	0·11	0·17	0·14	0·27	0·90	
Iron (%)	0·08	0·23	0·16	0·22	0·02	0·09	
Aluminium (%)	0·08	0·08	0·33		0·01	0·03	
Boron (μg/g)	UD	19·0		16·0	UD	UD	
Manganese (μg/g)	35·8	130·0	109·4	38·0	31·4	43·2	
Zinc (μg/g)	3·9	2·8	25·6	42·0	100·4	43·7	
Copper (μg/g)	1·5	0·8	5·7	15·0	12·4	17·5	
Chromium (μg/g)	UD	UD		2·0	UD	UD	
Nickel (μg/g)	1·1	UD	1·71	<0·1	3·8	3·4	
Cadmium (μg/g)	UD	UD		<0·1	UD	UD	
Mercury (μg/g)	UD	UD			UD	UD	
Lead (μg/g)	1·2	UD	14·70	2·0	8·7	2·8	
Sodium (%)	0·03	UD	0·07	0·66	<1·0	<1·0	

UD = Undetectable. Sources: Mathur *et al.*, 1986, 1988*a*.

Table 5
Periods (in Days) of Temperature Regimes of Interior of Composts of Some Seafood Waste Composts with Peat

Seafood source	*Warming to above control peat temp.*	*Rising to 45°C*	*Maintained between 45°C and 60°C*	*Cooling to 30°C*	*Cooling to control peat temp.*	*Total days*
Cod + sole	4	6	30	49	17	106
Crab	1	1	17	39	10	68
Cod	2	4	26	48	15	95
Cod + crab	1	3	42	28	18	92
Redfish	3	2	50	34	16	106

Source: Mathur *et al.*, 1988*a*.

Temperatures in the interior of the short windrows rise within 2 or 3 days, attaining the thermophilic range of 45–60°C within 10 days even at ambient temperatures of 4–10°C (Mathur *et al.*, 1986). The thermophilic temperatures are maintained for about 2 weeks and ambient temperatures attained within a total of 60 days. Higher temperatures, if attained, can be subdued by temporary plugging of the open ends of the ventilation pipes, but that has never been found to be necessary for either the long or the short windrows. The composts do not reheat on turning or stacking as the peats used themselves decompose little in the process and never support thermophilic activity on their own, even when neutralized to pH 7·0 with lime (Mathur *et al.*, 1988*b*). The composts, however, should be matured, and protected from rain to avoid nutrient leaching, for a period of at least a few weeks to allow further nitrification. The commercial-scale long windrows take longer to reach maturity (Table 5).

The quality of the composts compares favourably with standards and other composts (Tables 6 & 7) (Mathur *et al.*, 1986, 1988*a*; Preston *et al.*, 1986; Mathur & Johnson, 1987).

5.2 Manure Slurries

Modern methods of high-density raising of pigs, poultry and, at times, other animals in shelters with little or no use of litter, generally result in collection and storage of animal excreta as slurries. The slurry management systems with their open or covered outdoor lagoons or storage tanks were evolved in North America and Western Europe primarily to replace manual labour with power when energy was cheap, and animal farms were distinctly away from residential areas. Now, the adverse environmental impact of the slurry storage and disposal invite restrictions from agencies and litigations from nonagricultural neighbours (Viets, 1974; Loehr, 1977; Jewell *et al.*, 1984; Voorburg, 1988).

Table 6
Properties of Some Seafood Waste Composts Prepared in Maine, USA

Property	*I-A Flounder redfish*	*I-B Dogfish*	*I-C Herring, crab*	*I-D Herring, crab, lobster*	*I-E Herring*
pH (in H_2O)	7·18	6·51	7·49	5·97	7·15
Moisture (%)	44·3	69·8	29·0	54·8	60·7
Organic matter (%)	91·3	79·5	89·2	91·5	92·4
Nitrogen (%)	3·84	4·62	5·12	3·38	4·47
Phosphorus (%)	0·46	0·57	0·85	0·22	0·29
Potassium (%)	0·22	0·41	0·46	0·25	0·37
Calcium (%)	0·96	0·72	1·20	0·21	0·12
Cadmium (μg/g)	0·22	0·31	0·21	0·22	0·25
Chromium (μg/g)	1·38	2·59	2·10	2·20	2·00
Copper (μg/g)	1·54	2·75	9·90	7·40	4·00
Nickel (μg/g)	1·70	3·10	0·62	0·77	1·40
Lead (μg/g)	5·10	11·00	7·40	8·20	5·40
Mercury (μg/g)	0·15	<0·01	<0·01	<0·01	<0·01

Source: Mathur *et al.*, 1988*a*.

Table 7
Properties of the Peat Composts Prepared near Lameque, New Brunswick, Canada (±standard error of means)

Property (on dry basis)[a]	*II-1 Cod + sole*	*II-2 Crab*	*II-3 Cod*	*II-4 Cod + crab*	*II-5 Redfish*
pH (in H_2O)	7·99 ± 0·35	7·75 ± 0·13	7·83 ± 0·03	7·72 ± 0·03	7·90 ± 0·29
Salinity (%)	0·9 ± 0·1	1·0 ± 0·0	0·9 ± 0·1	0·9 ± 0·1	0·4 ± 0·2
Conductivity (mmho)	1·44 ± 0·21	1·76 ± 0·20	1·43 ± 0·24	1·41 ± 0·07	0·94 ± 0·17
Moisture (%)	61·6 ± 1·7	60·0 ± 1·4	64·8 ± 2·8	60·9 ± 0·9	59·1 ± 3·6
Organic matter (%)	66·5 ± 1·5	63·4 ± 0·9	71·4 ± 5·5	65·0 ± 4·9	63·4 ± 12·8
Nitrogen (%)	3·52 ± 1·03	2·61 ± 0·09	2·84 ± 0·72	2·70 ± 0·23	2·74 ± 0·93
Phosphorus (%)	1·97 ± 0·65	1·75 ± 0·54	0·88 ± 0·19	1·06 ± 0·26	1·15 ± 0·03
Potassium (%)	0·34 ± 0·06	0·25 ± 0·01	0·31 ± 0·04	0·26 ± 0·04	0·24 ± 0·04
Calcium (%)	2·46 ± 1·88	3·52 ± 1·22	0·76 ± 0·18	1·54 ± 0·22	1·31 ± 0·47
Magnesium (%)	0·60 ± 0·31	0·60 ± 0·12	0·27 ± 0·27	0·19 ± 0·04	0·15 ± 0·06
Iron (%)	0·45 ± 0·04	0·36 ± 0·08	0·45 ± 0·15	0·35 ± 0·07	0·49 ± 0·28
Zinc (μg/g)	37·0 ± 3·0	36·0 ± 6·0	25·0 ± 4·0	32·0 ± 11·0	26·0 ± 5·0
Copper (μg/g)	7·8 ± 0·4	17·0 ± 3·0	8·6 ± 0·7	7·7 ± 0·8	6·9 ± 0·5
Manganese (μg/g)	200·0 ± 20·0	190·0 ± 10·0	180·0 ± 40·0	270·0 ± 110·0	220·0 ± 80·0
Molybdenum (μg/g)	1·8 ± 0·1	<1·8 ± 0·1	<2·0 ± 0·2	<1·7 ± 0·1	<1·7 ± 0·1
Boron (μg/g)	590 ± 60	520 ± 10	<590 ± 60	550 ± 70	480 ± 40

[a] Except % moisture
Source: Mathur *et al.*, 1988*a*.

5.2.1 The Problem

The C/N ratio in the urine-containing slurries is so low that during decomposition much of the N is lost as it is in excess of the requirement for all the microbial biomass synthesis possible from the C present (Witter & Lopez-Real, 1987). Also much of the N is in forms readily degraded by exothermic enzymatic hydrolysis. The microbial proteins, urea and uric acid thus yield foul-smelling aliphatic amines, sulphides and NH_3 even under conditions suboptimal for microbial activity (Miner & Hazen, 1969, 1977; Elliott *et al.*, 1971; Mathur, 1982). The aerobic microbial activity that transforms the malodorous compounds into NO_3^-, SO_4^{2-}, CO_2 and H_2O is impaired by lack of oxygen and by low temperature. The microbial activity that does occur anaerobically produces CH_4, amines, NH_3, CO_2, and the toxic and malodorous sulphides including H_2S. High concentrations of the CO_2 promote retention of the NH_3 as the unstable $(NH_4)_2CO_3$ even at neutral to slightly alkaline pH. But the high viscosity needed to occlude and maintain CO_2 at the suitably high concentrations curtails the slurry's capacity to retain amines, H_2S and other sulphides in aqueous solution, thus heightening the problem of malodours. On the other hand, the NH_3 lost from manures, and the nitrogen oxides from partly aerated systems, promote eutrophication of water bodies and the acidification of soils in more than the immediate vicinity as the NH_3 is oxidized to nitrates on soil and water (Schroder, 1985; Voorburg, 1988). Even the highly viscous and malodorous slurries would lose ammonia rapidly during handling and application to soil, as the CO_2 is degassed to the atmosphere from the solution. The less odorous, dilute, NH_3-poor slurries, on the other hand, are costly to transport and apply.

Consequently, the available treatment systems for animal slurries probably do not eliminate all adverse environmental impacts as they cannot both conserve the N and contain the malodours at all stages.

5.2.2 Attempted Solutions

Many attempts have been made to mitigate the aforesaid problems with manure slurries (Vogtmann & Besson, 1978; Jewell *et al.*, 1984; Gray *et al.*, 1987; Voorburg, 1988). These include the feeding of herbal and deodorizing materials to animals (Ingram *et al.*, 1973) and curtailing of microbial activity by addition of chlorine, lime or permanganate to the slurries, both without much success (Loehr, 1977; Hammond *et al.*, 1968; Burnett & Dondero, 1970).

Biogas production from the slurries produces effluents which are not necessarily poor in P or free of pathogens (Chen *et al.*, 1984; Jewell *et al.*, 1984; Voorburg, 1988). The eutrophication caused by such waters can be curtailed by oxidative secondary treatment, P removal by alum, and production of feed algae in certain climates (Clark, 1965; Thostrup *et al.*, 1987). The sludges from both the biogasification and secondary aerobic treatment may still need to be dried and hygienically buried.

The so-called LiCom system of liquid hot composting of slurries by energy-powered aeration, without the addition of carbon-rich materials, can curtail malodours but not the loss of NH_3 (Nebiker, 1976; Witter & Lopez-Real, 1987). When the difficult process succeeds, the N and P in the slurries should be converted to water-soluble nitrates and phosphates which make the effluents eutrophic. The same would occur, albeit to a lesser extent, when carbon-rich materials such as chopped straw are added because the additions raise the C/N ratio to only about 12, not the >35 necessary to conserve the N (Grant, 1975; Terwelleger & Crauer, 1975; Nebiker, 1976; Jewell *et al.*, 1984; Dalzell *et al.*, 1987; Witter & Lopez-Real, 1987). At the same time, even the limited addition of the straw renders the mixture difficult to aerate and homogenize. The application of slurries to layers of woody materials on the field surface to effect sheet composting cannot prevent the loss of NH_3 during application, nor lead to rapid hygienic maturation of the composts.

Another kind of liquid aerobic composting of slurries alone involves the use of the heat generated for 'biodrying' the manure into a microbial-cell-rich powder that can be fed to cattle (Jewell *et al.*, 1984). Nearly two-thirds of the available protein in the untreated manure is destroyed in this system while 0·06% of the potential pathogens survive.

In systems where the solid part of the slurries is removed by centrifugation and dried, while the liquid part is aerated and chemically treated, the primary objective is hygiene, not conservation and recycling of the plant nutrients, nor the saving of costs (Thostrup *et al.*, 1987; Voorburg, 1988). However, cost seems to have ceased to be the main criterion due to the necessity of meeting environmental agency requirements. More stringent rules may be expected, as a recent study has revealed that the N lost from manures may be contributing more to soil acidification in the Netherlands than all the SO_2 produced there by power generation and automobiles. This may be relevant to many other areas of intensive animal production, e.g. Denmark (Schroder, 1985). Therefore, the most recent system in the Netherlands is of separation of the solid, drying of its cake, and evaporation of the filtrates (Voorburg, 1988). However, the end product is still manure, albeit dry. If the effluents are used for biomethanation and the solids as horticultural media (Chen *et al.*, 1984; Raviv *et al.*, 1987), the threat of pathogen and nutrient pollution is not necessarily eliminated.

The objective of conserving the N is unlikely to be fully met by bulking of the slurries for aerobic overground composting with materials of low buffering capacity such as straw and sawdust (e.g. Environment Canada, 1984; Biddlestone *et al.*, 1985; Gray *et al.*, 1987). The low water absorption capacity of sawdust (Hong *et al.*, 1983) would cause the composts to be poor in nutrients. Also, wood-based composts take long to mature properly due to alternate and differential decompostion of the highly biodegradable cellulose and the recalcitrant lignin (which yields bioinhibitory phenols). For the straw, special maçhinery is needed for shredding it prior to mixing with slurry (Gray *et al.*, 1987). And yet foul-smelling leachates are produced as the matrix decomposes. Return of the leachates, forced aeration, and/or repeated mixing would allow emission of both NH_3 and malodours.

The economics, ease of operation, and environmental aspects of producing earthworms and 'compost' from slurry sediments, and methane from the liquid, are also not totally satisfactory (Fieldson, 1985), although the system is sophisticated to a fault.

The above is in accord with the following statements, both made in the context of animal manure slurries: 'Unfortunately, a great deal of technology has been applied to the management of animal wastes . . . without significant impact on the total problem' (Jewell *et al.*, 1984), 'There exists not a simple treatment system that can overcome this situation' Voorburg, 1988).

In short, the above approaches have been to *treat* the material to meet point-source environmental regulations, or to mask or dilute the offensive aspects, not to recycle the elements in an environmentally benign manner.

5.2.3 Composting Slurries with the PAW System

A need was felt for testing the feasibility of composting liquid manure with a material that:

(1) deodorizes (e.g. peat biofilters—Furusawa *et al.*, 1984);
(2) absorbs high proportions of water;
(3) is rich enough in exchangeable H^+ ions to neutralize the ammonia and then the cations released by decomposition, thus preventing the loss of ammonia by remaining slightly acidic throughout the composting process;
(4) has capacity for adsorbing anions, thus retarding the leaching of NO_3^- and PO_4^- when added to soil (Moore *et al.*, 1981);
(5) is fluffy enough to provide thermal insulation and enough replaceable air to prevent anaerobic production of malodours;
(6) can also be used as a litter and/or a floating cap on the slurries;
(7) is biodegradable but, unlike straw and wood wastes, not capable of sustaining thermophilic microbial activity by itself so that the compost can mature early in a short Canadian summer; and
(8) is devoid of relevant pathogens and weed seeds so that an outer layer of the material does not have to be subjected to sanitizing heat, thus obviating the need for turning the composts inside-out while still hot which would require energy and permit loss of NH_3 and dispersal of allergens and pathogens.

Horticultural (blonde) sphagnum peat meets these requirements besides being attractive, even at a relatively high price, to home gardeners, landscapers, turf agronomists, and nurserymen for its physical properties and exceptionally high capacity for enhancing soil humus, which is 4–8 times that of straw (Janssen, 1984). Peat also has considerable capacity for detoxification of metals by their complexation into nonbioavailable forms (e.g. see Mathur & Farnham, 1985; Dumontet & Mathur, 1989).

Sphagnum peats have bulk densities of about 0·07 g/cm^3, cation exchange capacities of nearly 130 meq/100 g, and pH between 3 and 4, and can adsorb up to 20 times their weight in water, and adsorb NH_3 up to 3% of their weight even in an air-dry state (Bulganina *et al.*, 1983; Mathur & Farnham, 1985).

As Mathur *et al.* (1989) have shown, part of the peat can be first used as a floating cap on the slurries to curtail malodours, prevent nuisance insects, and trap the escaping ammonia.

Mathur *et al.* (1988*b*, 1989, 1990*a*) tested the feasibility of composting slurries from manures from sheep, poultry, hogs and dairy cattle by the short windrow PAW system using peat as base and envelope of a mixture of the peat and a slurry (Table 8). The total ratio of slurry to peat varied from 6 to 8 in the first charging of the bulking agent. It was found that liming the peat with 1% Ca was not necessary.

The limed or unlimed amphoteric peat of acidic pH adsorbed the NH_3 (see Table 9) and any other malodorous compounds, and its fibrosity (bulk density 0·06 g/cm^3) supported aerobic thermophilic decomposition in which neither malodorous amines nor sulphides are produced significantly. No detectable NH_3 or H_2S emanated from the compost piles. Temperatures greater than 45°C were reached in the composts within 1–3 days and maintained above 45°C below the 10-cm envelope for 2–5 weeks without further mixing of the composts or any mechanical aeration (Table 10). When the composts were remixed and reheaped, 7 weeks after initial formulation, no reheating occurred, suggesting that the composts were 'biostable' as is peat. Germination tests confirmed the 'biostability'. The mature composts had water holding capacities nearly equal to that of the original peat, suggesting that the air-dried

Table 8
Properties of the Peat and of the Animal Manure Slurries Used
(on dry matter basis unless indicated otherwise)

Property	*Poultry manure*	*Dairy cow manure*	*Sheep manure*	*Peat*
Moisture (% of dry matter)	2 870	1 186	1 566	100·0
Dry matter (% moist wt)	3·4	7·8	6·0	50·0
Organic matter (%)	68·4	80·7	63·9	96·0
Carbon (%)	34·6	40·0	31·3	43·7
Total nitrogen (%)	3·52	2·09	2·27	0·8
NH_4 nitrogen (mg/kg)	5 050	740	2 650	300
NO_3 nitrogen (mg/kg)	118	51	252	67
pH	6·52	8·35	7·41	3·0
C/N ratio	9·8	19·1	13·8	53·3
Phosphorus (%)	2·90	1·08	0·59	0·04
Potassium (%)	2·89	3·40	2·28	0·05
Calcium (%)	5·9	1·7	4·2	0·1
Magnesium (%)	0·74	0·53	0·63	0·1
Copper (mg/kg)	50·4	52·9	23·4	2·8
Iron (%)	0·43	0·15	0·25	0·1
Manganese (mg/kg)	380	120	115	78
Zinc (mg/kg)	970	210	440	9

Table 9
Average pH of the Outer Surface and Middle of the Compost Piles, at Time Zero and 1 Week Later

Treatment	*Outer surface*		*Middle of the pile*	
	Time-0	*1 week*	*Time-0*	*1 week*
Peat alone	3·2 ± 0·1	3·2 ± 0·1	3·3 ± 0·1	3·3 ± 0·0
Peat + lime	4·2 ± 0·1	4·3 ± 0·1	4·3 ± 0·1	4·3 ± 0·0
Peat + sheep manure	3·2 ± 0·0	4·2 ± 0·1	5·4 ± 0·1	5·4 ± 0·0
Peat + sheep manure + lime	3·2 ± 0·0	4·4 ± 0·1	5·4 ± 0·1	5·6 ± 0·0
Peat + cow manure	3·2 ± 0·0	4·3 ± 0·1	5·9 ± 0·1	6·1 ± 0·0
Peat + cow manure + lime	3·2 ± 0·1	4·5 ± 0·1	6·3 ± 0·1	6·4 ± 0·0
Peat + poultry manure	3·2 ± 0·1	5·8 ± 0·1	6·9 ± 0·1	7·0 ± 0·0
Peat + poultry manure + lime	3·3 ± 0·1	6·5 ± 0·1	7·6 ± 0·1	7·8 ± 0·0

Table 10
Periods (in Days) of Temperature Regimes of Interior of Composts

Temperature regime	*Source of manures in the composts*		
	Sheep	*Dairy cows*	*Poultry*
Warming to 45°C	1	1	1
Warming to above 55°C	4	4	4
Maintained between 45° and 65°C	14	22	36
Cooling to 30°C	19	15	8
Cooling to control peat temperature (20° ± 3°)	13	6	4
Total composting period	46	43	48

peat–manure composts can be recharged with manure slurries. The properties of the composts (Table 11) suggested that they could be marketed and used as a substitute for or a supplement to limed and fertilized peat generally used in gardening, landscaping and greenhouse culture.

6 CRITERIA OF COMPOST MATURITY

Effective composting transforms organic wastes into a useful product which is rich in available forms of plant nutrients, poor in readily biodegradable

Table 11
Properties of the Composts Prepared from Mixtures of Animal Slurries and Peat, with or without Lime Addition (±standard error of means, on dry weight basis)

Property	*Poultry manure + peat*		*Dairy manure + peat*		*Sheep manure + peat*	
	Without lime	*With lime*	*Without lime*	*With lime*	*Without lime*	*With lime*
pH	5·67 ± 0·05	5·83 ± 0·06	5·25 ± 0·19	5·26 ± 0·06	4·73 ± 0·05	5·81 ± 0·18
WHC[a]	420 ± 15	360 ± 30	540 ± 30	480 ± 25	630 ± 30	560 ± 40
Moisture (% dry matter)	245 ± 16	175 ± 19	304 ± 21	281 ± 13	340 ± 20	343 ± 19
Organic matter (%)	63·5 ± 2·4	59·9 ± 1·9	74·7 ± 0·3	65·0 ± 2·1	77·7 ± 1·5	76·9 ± 1·6
Carbon (%)	28·7 ± 1·4	28·1 ± 34·9	34·6 ± 0·1	30·9 ± 0·6	36·1 ± 1·2	36·4 ± 0·8
Total N (%)	1·71 ± 0·05	1·75 ± 0·10	1·83 ± 0·01	1·79 ± 0·20	1·71 ± 0·05	1·57 ± 0·06
NH_4 nitrogen (μg/g)	4 159 ± 85	4 114 ± 243	4 393 ± 85	1 690 ± 253	2 697 ± 119	227 ± 135
NO_3 nitrogen (μg/g)	1 054 ± 142	1 800 ± 145	822 ± 45	1 745 ± 320	2 271 ± 119	3 225 ± 198
Total P (%)	1·18 ± 0·10	1·20 ± 0·21	0·87 ± 0·10	0·89 ± 0·11	0·41 ± 0·05	0·40 ± 0·05
HCl-soluble P (μg/g)	9 372 ± 200	9 508 ± 190	2 901 ± 170	2 516 ± 190	2 927 ± 111	3 133 ± 180
C/N ratio	16·8 ± 1·8	16·1 ± 1·1	18·9 ± 0·1	17·2 ± 0·4	21·1 ± 0·9	23·1 ± 0·6
Potassium (%)	1·38 ± 0·1	1·48 ± 0·0	1·70 ± 0·10	1·69 ± 0·1	1 55 ± 0·9	1·58 ± 0·1
Calcium (%)	3·9 ± 0·3	4·8 ± 0·2	1·7 ± 0·2	2·9 ± 0·1	1·80 ± 0·1	2·3 ± 0·1
Magnesium (%)	0·87 ± 0·06	1·14 ± 0·01	0·62 ± 0·01	0·81 ± 0·05	0·59 ± 0·05	0·47 ± 0·01
Copper (μg/g)	28·1 ± 0·5	27·1 ± 0·1	24·1 ± 0·4	23·1 ± 0·6	12·8 ± 0·1	11·1 ± 0·1
Iron (%)	1·08 ± 0·08	0·95 ± 0·05	0·59 ± 0·10	0·60 ± 0·01	0·71 ± 0·01	0·69 ± 0·01
Manganese (μg/g)	360 ± 21	410 ± 10	190 ± 10	230 ± 20	145 ± 12	130 ± 10
Zinc (μg/g)	150 ± 11	170 ± 9	130 ± 9	120 ± 11	79 ± 10	82 ± 10

[a] Water holding capacity at $\frac{1}{3}$ bar suction, % of dry weight.

carbon, devoid of phytoinhibitory substances, and relatively free of plant pathogens. Subjectively, a mature compost should be dark brown or black; granular, spongy or fibrous in feel, and smell mouldy or rather like good earth reminiscent of the aroma from a sunbaked garden patch that has first received a summer shower (due to the release of spores of thermophilic *Streptomyces*). In objective analysis, attributes of a compost should indicate that the purposes of the process have been achieved. A lack of reheating of the compost mix indicates that the amount of readily biodegradable C sources is insufficient to support thermophilic activity.

6.1 C/N Ratio

The C/N ratio of a mature compost should ideally be about 10 but this is hardly ever achievable due to the presence of recalcitrant organic compounds, or materials which resist decomposition due to their physical or chemical properties. For example, peat does not decompose readily when added to mineral soils because even 75% moisture content in the total mass may be held too strongly physicochemically to be bioavailable. Lignin, on the other hand, resists decomposition because it does not have identical monomers and therefore no specific enzyme depolymerizes it, and because intermediate phenolic products of lignin decomposition tend to be biostatic or bioinhibitory. The critical point about the C/N ratio is therefore whether decomposition of the mature compost in soil will release mineral nitrogen, or cause competition with plants for the N in soil solution. C/N ratios of 14–20 in mature composts are therefore acceptable as long as their further decomposition is slow and does not require additional N from the soil. Gotaas (1956) refers to studies where composting a material with a C/N ratio of 78 produced a biostable product with a C/N ratio of 35.

6.2 Absence of Plant Inhibitors

Perhaps the best indicator of whether a compost is biomature is the absence of bioinhibitory aliphatic acids (Dveleeschauwer *et al.*, 1982) and phenolics which can be assayed chromatographically or by seed germination tests. Small seeds placed on a filter paper soaking in a water extract of the underlying compost in a Petri dish should germinate to the same percentage level as those placed on the paper soaked in pure water (e.g. Mathur *et al.*, 1986).

6.3 Absence of Human Pathogens

For microbiological properties, Zucconi and De Bertoldi (1987) proposed that a compost should be considered hygienic if a 100 g sample contains no salmonellae, no infective parasitic ova and no more than 5×10^4 fecal coliforms and 5×10^5 fecal streptococci. In certain specific types of composts, other microbiological criteria may be advisable. For example, Mathur and Johnson

(1987) tested fish waste composts for the toxins produced by pathogenic microbes by using tissue cultures of Y-1C mouse adrenal tumour cells, vero (monkey kidney cells), CHO (Chinese hamster ovary cells); and the Suckling Mouse Test. These tests ensured that neither the pathogens nor the microbial toxigens that actually cause the diseases were present at significant levels in the mature composts. Some of the toxins were present in the raw materials and immature composts.

6.4 Other Criteria

Other criteria one may use are related to the actual rate of decomposition of the compost and the amount of soluble humus present in the compost. The former may be limited in value due to the presence of bioinhibitory metals or organic compounds in a compost, and the latter by the presence of metals that complex the humic substances into insoluble substances. Preston *et al.* (1986) therefore subjected the whole composts of seafood wastes to NMR spectroscopy, to find that the proteins, lipids and chitin present in raw seafood wastes were not present in mature composts and that the C, N and P in the composts were distributed as in humus from peat or soils. NMR spectroscopy, where feasible, may be a highly reliable means of testing compost maturity.

7 USES OF COMPOSTS

As implied earlier the historical use of composts has been as a soil conditioner and fertilizer, to restore and maintain soil humus and productivity. Humus shelters and nourishes soil organisms, subdues moisture stress, stores and supplies plant nutrients, buffers reaction, and is generally responsible for soil tilth. Compost application is the backbone of organic farming and some definitions of 'organically-produced foods' that fetch premium prices, particularly from consumers concerned about and vulnerable to agrichemical residues, insist on the practice. Other major uses of composting are landscaping, turf production and stabilization of steep slopes in vineyards and orchards.

The composts used in mushroom culture are not the bioconverted mature product dealt with in this chapter but mostly bioavailable-carbon-rich mixtures of various materials composited for the specific purpose. These mixtures may include the biologically produced composts, in mature or immature state.

Recent impetus to the use of composts in or as media for greenhouse production of seedlings, flowers and vegetables comes from evidence that soil-borne pathogens are naturally controlled by their antagonists in mature composts (Hoitink *et al.*, 1987). Pudelski (1987) has recently described such horticultural use of composts, particularly in greenhouses where the heat generated in maturing compost at a lower layer of the on-ground growth medium helps save the energy used for heating the glasshouses in temperate climates. Some studies have indeed been made to examine the feasibility of using the heat produced during composting (Thostrup *et al.*, 1987), and the CO_2 generated by the decomposition, to enhance photosynthesis in greenhouses.

Hoitink and Kuter (1984) predicted that the biological agents of pest control, particularly fungi and bacteria antagonistic to plant pathogens, may be grown and marketed in a compost medium. This may become a significant industrial market for composts as more enterprises that supply organic farmers and pesticide manufacturing companies respond to the new consumer and market conditions favourable to alternative agriculture with minimal or no chemical inputs.

Composts can be used as a base for biofilters to trap and biotransform foul odours from animal houses, abattoirs and industry, into harmless products. The nitrates and phosphates produced in the compost biofilters can be leached and the leachate used for algae culture, or for irrigation of field and greenhouse crops.

Organic wastes have been used in aquaculture both to feed detritus-eating fishes (e.g. tilapia, carp) directly and to increase the production of phytoplankton, the main natural fish food algae. Composts are better than the raw wastes for this purpose as the former create much lower demand for biological oxygen in the water as it is already biostable. Consequently, composts do not significantly deplete the dissolved oxygen needed by the fish. Banerjee and Srinivasan (1983) demonstrated the advantage of compost application to fish ponds in India, and Edwards *et al.* (1983) in Africa, for tilapia culture.

8 SUMMARY

Composting is the managed bioconversion of waste into a hygienic soil conditioner and fertilizer. Heightened awareness of it is due to the new ethos of sustainable development that blends economic and environmental considerations with a sense of intergenerational justice. Restoration and maintenance of soil humus conserves resources and represents one of the most essential and beneficial ecological processes.

To compost, one needs to match different materials to achieve ideal conditions of C/N ratio, pH, moisture content and redox. To optimize the oxidation needed to dissipate carbon and generate the sanitizing heat, one needs to mix and relay the material, or use various strategies of moving air through the composting mass or of moving the mass through air, in or out of enclosed systems. The costly mechanization and intensive skilful management required in high-technology methods have been the main reasons for the virtual limitation of the practice of composting to home gardens and to sanitization of residues containing pathogen-laced human wastes.

A new and inexpensive system of composting, the Passively Aerated Windrow (PAW) System, harnesses the heat produced during decomposition to naturally move air through the mass via specific types of pipes placed suitably at or near the base of the windrow (elongated triangular or trapezoidal piles). The pipes are open-ended and have perforations only on the top side. The need to invert the outside areas of the piles into the centre of the mass, to sanitize them with the heat there, is obviated by the use of hygienic materials such as mature compost as the base and the envelope of the compost. The PAW system has been successfully used for seafood wastes and peat, and for manure slurries bulked with peat.

The criteria for a mature compost reflect achievement of the objectives of composting—the production of a biostable, hygienic, humus-rich product that benefits plants.

A multitude of chemical and biological processes combine to transform wastes into humus, microbial biomass, waxy compounds, and mineral nutrients like NH_4^+, NO_3^-, $P_2O_5^-$ and SO_4^{2-}.

Composts can be used to condition soils, fertilize plants, stabilize slopes; to grow mushrooms, nursery plants, ornamental or vegetable crops in greenhouse; as biofilters of odours from certain industries; as a feed for detrivorous fishes in aquaculture and a promoter of plankton growth; and as a medium for mycorrhizae, and organisms used for their antagonism to plant pathogens.

REFERENCES

Acharya, C. N. (1954). *Indian Farming,* **4,** 9.

Alexander, M. (1977). *Introduction to Soil Microbiology*. John Wiley, New York.

Altieri, M. A. (1987). *Agroecology*: *The Scientific Basis of Alternative Agriculture.* IT Publications, London, p. 227.

Banerjee, R. K. & Srinivasan, K. V. (1983). *Agric. Wastes,* **7,** 209.

Biddlestone, A. J., Gray, K. R. & Cooper, D. J. (1985). *Environ. Health,* **93,** 67.

Biddlestone, A. J., Gray, K. R. & Day, C. A. (1987). In *Environmental Biotechnology,* eds C. F. Forster & D. A. J. Wase. Ellis Horwood, Chichester, UK, p. 136.

Bollen, G. J. (1985). In *Composting of Agricultural Wastes,* ed. J. K. R. Gasser. Elsevier Applied Science, London, p. 282.

Bulganina, V. N., Kuznetsova, L. M., Mandelbaum, L. M. & Menchinova, A. A. (1983). In *Recent Technologies in the Use of Peat,* eds G. W. Luttig & E. Schweizerbart. Verlagsbuchhandlung (Nagele U. Obermiller), Stuttgart, p. 51.

Burnett, W. E. & Dondero, N. C. (1970). *Trans. Am. Soc. Ag. Eng.,* **13,** 221.

Chen, Y. & Hadar, Y. (1987). In *Compost*: *Production, Quality and Use,* eds M. DeBertoldi, M. P. Ferranti, P. L'Hermite & F. Zucconi. Elsevier Applied Science, London, p. 71.

Chen, Y., Inbar, Y., Raviv, M. & Dovrat, A. (1984). *Acta Hort.,* **150,** 553.

Clark, C. E. (1965). *J. Sanit. Engg Div. ASCE,* **91,** 27.

Dalzell, H. W., Biddlestone, A. J., Gray, K. R. & Thurairajan, K. (1987). *FAO Bulletin,* **56,** 177.

De Bertoldi, M., Vallini, G. & Pera, A. (1985). In *Composting of Agricultural and Other Wastes,* ed. J. K. R. Gasser. Elsevier Applied Science, London, p. 27.

De Bertoldi, M., Ferranti, M. P., L'Hermite, P. & Zucconi, F. (eds) (1987). *Compost*: *Production, Quality and Use.* Elsevier Applied Science, London, p. 853.

Dumontet, S. & Mathur, S. P. (1989). *Soil Biol. & Biochem.,* **21,** 431.

Dveleeschauwer, D., Verdonck, O. & Van Asshe, P. (1982). In *Composting, Theory*

and Practice for City, Industry and Farm, eds Staff of *Compost Sci. Land Utilization.* J. G. Press, Emmaus, Pennsylvania, 54.
Edwards, P., Polprasert, C., Pacharaprakiti, C., Rajput, V. S. & Suthirawut, S. (1983). *Agriculture,* **32,** 409.
Edwards, C. A., Burrows, I., Fletcher, K. E. & Jones, B. A. (1985). In *Composting of Agricultural and Other Wastes,* ed. J. K. R. Gasser. Elsevier Applied Science, London, p. 229.
Elliott, L. F., Schuman, G. E. & Viets, F. G. Jr (1971). *Soil Sci. Soc. Amer. Proc.,* **35,** 752.
Environment Canada (1984). *Amélioration et démonstration du procède de compostage du lisier du porcs.* Report by Roche Environment Associés Ltée, Groupe-Conseil, Ottawa, Canada.
Epstein, E., Willson, G. B., Burge, W. D., Mullen, D. C. & Nekri, N. K. (1976). *J. Water Pollut. Control Fed.,* **48,** 688.
Fieldson, R. S. (1985). In *Composting of Agricultural and Other Wastes,* ed. J. K. R. Gasser. Elsevier Applied Science, London, p. 243.
Finstein, M. S. & Miller, M. C. (1985). In *Composting of Agricultural and Other Wastes,* ed. J. K. R. Gasser. Elsevier Applied Science, London, p. 13.
Finstein, M. S., Miller, F. C., Strom, P. F., MacGregor, S. P. & Psarianos, K. M. (1983). *Bio/Technology,* **1,** 347.
Furusawa, N., Togashi, I., Harai, M., Shoda, M. & Kubota, H. (1984). *J. Ferment. Technol.,* **62,** 589.
Gasser, J. K. R. (ed.) (1984). *Composting of Agricultural and Other Wastes.* Elsevier Applied Science, London.
Gotaas, H. B. (1956). *Composting Sanitary Disposal and Reclamation of Organic Wastes.* World Health Organization, Monograph series, No. 31, Geneva.
Grant, F. A. (1975). *Proc. 3rd Int. Symp. on Livestock Wastes,* Amer. Soc. Ag. Eng., p. 497.
Gray, K. R., Biddlestone, A. J. & Clark, R. (1973). *Process Biochem.,* **8**(10)**,** 11, 30.
Gray, K. R., Biddlestone, A. J., Garner, P. J. & Vere, A. E. (1987). In *Proc. Int. Conf. on Straw—a Valuable Raw Material.* Cambridge, UK, **1,** 227.
Hammond, W. C., Day, D. L. & Hansen, E. L. (1968). *Agricultural Engg,* June, 340.
Hoitink, H. A. J. & Kuter, G. E. (1984). *Biocycle,* May/June, 40.
Hoitink, H. A. J., Chen, W., Trillas-Gay, M. I. & Chung, Y. R. (1987). In *Compost: Production, Quality, and Use,* eds M. De Bertoldi, M. P. Ferranti, P. L'Hermite & F. Zucconi. Elsevier Applied Science, London, p. 414.
Hong, J. H., Matsuda, J. & Ikenchi, Y. (1983). *Trans. Amer. Soc. Ag. Eng.,* **26,** 533.
Howard, A. (1943). *An Agricultural Testament.* Oxford University Press, London.
Howard, A. & Wad, Y. D. (1931). *The Waste Products of Agriculture: Their Utilization as Humus.* Oxford University Press, London.
Hughes, E. G. (1980). In *Handbook of Organic Waste Conversion,* ed. M. W. M. Bewick. Van Nostrand Reinhold, New York, p. 109.
Ingram, S. H. *et al.* (1973). *J. Animal Sci.,* **36,** 207.
Janssen, B. H. (1984). *Plant & Soil,* **76,** 297.
Jeris, J. S. & Regan, R. W. (1973). *Compost Sci.,* **14,** 10.
Jewell, W. J. *et al.* (1984). *High temperature stabilization and moisture removal from animal wastes for by-product recovery.* Report for the Co-operative State Research Service, US Department of Agriculture, under project No. SEA/CR 616-15-168.
King, F. H. (1911). *Farmers of Forty Centuries or Permanent Agriculture in China, Korea and Japan.* Harcourt Brace & Co., New York, p. 441.
Kononova, M. M. (1966). *Soil Organic Matter.* Pergamon Press, Oxford, UK.
Krishna Murthy, R. (1978). *A Manual on Compost and Other Organic Manures.* In Today and Tomorrow Printers and Publishers, New Delhi.
Loehr, R. C. (1977). *Pollution Control for Agriculture.* Academic Press, New York.

Lopez-Real, J. & Foster, M. (1985). In *Composting of Agricultural and Other Wastes*, ed. J. K. R. Gasser. Elsevier Applied Science, London, p. 291.
Mathur, S. P. (1982). *Trans. 12th Congr. Int. Soc. Soil Sci.*, **1,** 125.
Mathur, S. P. & Farnham, R. S. (1985). In *Humic Substances in Soil, Sediment and Water,* eds G. R. Aiken, D. M. McNights, R. L. Warshaw & P. MacCarthy. John Wiley, New York, p. 53.
Mathur, S. P. & Johnson, W. M. (1987). *Biol. Agric. Hort.*, **4,** 235.
Mathur, S. P., Levesque, M. P., Dinel, H. & Daigle, J.-Y. (1985). In *Proc. Int. Peat Soc. Symp. 85,* Rivière-du-Loup, Québec International Peat Society, Helsinki, p. 279.
Mathur, S. P., Daigle, J.-Y., Levesque, M. & Dinel, H. (1986). *Biol. Agric. Hort.*, **4,** 27.
Mathur, S. P., Proulx, J. G., Levesque, M. & Sanderson, R. B. (1987). In *Agrogeology in Africa.* Commonwealth Sci. Council Tech. Publ. Ser. No. 226, p. 129.
Mathur, S. P., Daigle, J.-Y., Brooks, J. L., Levesque, M. & Arsenault, J. (1988*a*). *Biocycle,* **29,** 44.
Mathur, S. P., Patni, N. K. & Levesque, M. P. (1988*b*). *Can. Soc. Agric. Eng. Annual Meeting,* Paper No. 88-123.
Mathur, S. P., Proulx, J. G. & Daigle, J.-Y. (1989). In *Proc. 1989 Int. Peat Soc. Symp.* Québec City, eds R. P. Overend & J. K. Jeglum. International Peat Society, Helsinki.
Mathur, S. P., Patni, N. K. & Levesque, M. P. (1990*a*). *Biol. Waste,* **34,** 323.
Mathur, S. P., Schnitzer, M. & Schuppli, P. (1990*b*). *Biol. Agric. & Hort.*, **7,** 153.
Miner, J. R. & Hazen, T. E. (1969). *Trans. Amer. Soc. Ag. Eng.*, **12,** 772.
Miner, J. R. & Hazen, T. E. (1977). In *Soils for Management of Organic Wastes and Waste Waters,* eds L. F. Elliott & F. J. Stevenson. Amer. Soc. Agron., Madison, Wisconsin, p. 379.
Moore, J. J., Fahy, L. & Harrington, A. (1981). *Irish J. Environ. Sci.*, **1,** 71.
Nakasaki, K., Sasaki, M., Shoda, M. & Kubota, H. (1985). *Appl. Environ. Microbiol.*, **49,** 724.
Nebiker, H. (1976). *Chemische Rundschau,* **47,** 607.
Paul, E. A. & Clark, F. E. (1989). *Soil Microbiology and Biochemistry.* Academic Press, San Diego, California.
Poincelot, S. P. (1975). *Biochemistry and Methodology of Composting.* Connecticut Agric. Expt. Sta. Bulletin 754.
Preston, C. M., Ripmeester, J. A., Mathur, S. P. & Levesque, M. (1986). *Can. J. Spectroscopy,* **31,** 63.
Pudelski, T. (1987). In *Compost: Production, Quality and Use,* eds M. De Bertoldi, M. P. Ferranti, P. L'Hermite & F. Zucconi. Elsevier Applied Science, London, p. 20.
Raviv, M., Medina, S., Chen, Y., Inbar, Y. & Geler, Z. (1987). In *Compost: Production, Quality and Use,* eds M. De Bertoldi, M. P. Ferranti, P. L'Hermite & F. Zucconi. Elsevier Applied Science, London, p. 377.
Schroder, H. (1985). *Agriculture, Ecosystems and Environment,* **14,** 279.
Swanson, G. R., Dudley, E. G. & Williamson, K. J. (1980). In *Handbook of Organic Waste Conversion,* ed. M. W. M. Bewick. Van Nostrand Reinhold, New York, p. 281.
Terwelleger, A. R. & Crauer, L. S. (1975). *Proc. 3rd Int. Symp. on Livestock Wastes,* Amer. Soc. Ag. Eng., p. 501.
Thostrup, P. (1985). In *Composting of Agricultural and Other Wastes,* ed. J. K. R. Gasser. Elsevier Applied Science, London, p. 167.
Thostrup, P., Jespersen, L. M. & Jensen, P. E. B. (1987). In *Compost Production, Quality and Use,* eds M. De Bertoldi, M. P. Ferranti, P. L'Hermite & F. Zucconi. Elsevier Applied Science, London, p. 147.

University of California (1953). *Reclamation of Municipal Refuse by Composting.* Technical Bulletin No. 9, Berkeley, California.
Van der Hoek, K. W. & Oosthoek, J. (1985). In *Composting of Agricultural and Other Wastes,* ed. J. K. R. Gasser. Elsevier Applied Science, London, p. 271.
Viets, F. G. (1974). *Fed. Proc.,* **33,** 1178.
Vogtmann, H. & Besson, J. M. (1978). *Compost Sci.,* Jan.-Feb., 15.
Voorburg, J. H. (1988). *Pig production and environment.* Inst. of Agric. Engineering, Wageningen, The Netherlands, Xnr 11890, 14.
Waksman, S. A. (1938). *Humus.* Williams and Wilkins Co., Baltimore, Maryland.
Wiley, J. S. (1957). In *Proc. 12th Industrial Wastes Conf.,* Purdue University, USA. *Engineering Bulletin,* **17,** 596.
Wiley, J. S. & Spillane, J. T. (1962). *Compost Sci.,* **2,** 18.
Witter, E. & Lopez-Real, J. M. (1987). *Biol. Agric. & Hort.,* **5,** 1.
Zucconi, F. & De Bertoldi, M. (1987). In *Compost: Production, Quality and Use,* eds M. De Bertoldi, M. P. Ferranti, P. L'Hermite & F. Zucconi. Elsevier Applied Science, London, p. 30.

BIOCONVERSION APPLICATIONS

Chapter 6

BIOPROCESSING OF AGRO-RESIDUES TO GLUCOSE AND CHEMICALS

V. S. Bisaria

Biochemical Engineering Research Centre, Indian Institute of Technology, Hauz Khas, New Delhi-110 016, India

CONTENTS

1 INTRODUCTION

Vast quantities of agricultural and agro-industrial residues that are generated as a result of diverse agricultural practices represent one of the most important energy-rich resources. Accumulation of this biomass in large quantities every year results not only in deterioration of the environment but in a loss of potentially valuable material which can be processed to yield food, fuel, feed and a variety of chemicals. A few of these potential residues are listed in Table 1. While cellulose, hemicellulose and lignin are the major constituents of these lignocellulosic residues, minor quantities of protein, pectin, soluble sugars, vitamins and minerals are also present. The composition of these residues,

Table 1
A Few Potential Biodegradable Agricultural and Agro-industrial Residues

Source	*Residues*
1. Agricultural crops (rice, wheat, corn, maize, etc.)	Rice straw, wheat straw, maize stalks, castor stems, tapioca stem, banana stem, coconut stem, cotton sticks, corn cobs, bamboo dust, barley straw, guar straw
2. Agro-industrial wastes and by-products:	
—Rice milling industry	Rice husk, rice bran
—Sugar industry	Bagasse, pith
—Cotton processing industry	Cotton linters, cotton seed hulls, cotton gin wastes
—Jute industry	Jute sticks, jute mill wastes
—Sawmill industry	Sawdust, wood chips
—Coconut industry	Coconut husk, shell and pith

which depends on the age of the plant, is also different in different parts of the same crop such as stalks, stems, straws, hulls and cobs. The composition of a few lignocellulosic residues of agricultural origin is summarized in Table 2.

The complete utilization of lignocellulosic residues continues to remain an intractable problem. Effective chemical procedures exist which can fractionate the lignocellulose into its major constituent components and further hydrolyze them into its essential building blocks (i.e. glucose from cellulose, pentoses and hexoses from hemicellulose, and phenylpropanoid units from lignin). However, the chemical processes, besides being energy-intensive, require expensive corrosion-resistant equipment, extensive washing and disposal of chemical wastes. Severe conditions of chemical treatments may also result in

Table 2
Composition of a Few Lignocellulosic Residues of Agricultural Origin

Agro-residue	*Cellulose (%)*	*Hemicellulose (%)*	*Lignin (%)*
Wheat straw	30	50	15
Rice straw	35	25	17
Corn cobs	45	35	15
Corn stalks	35	25	35
Guar straw	28	23	16
Maize straw	32	28	13
Cane bagasse	33	22	14
Oat straw	49·3	25	14–22
Hardwood	45·8	30·7	20·3
Softwood	43·8	24·5	29·5

Sources: Hawley *et al.*, 1986; Bisaria *et al.*, 1987*a*; Parisi, 1989.

partial degradation of desired end-products to toxic byproducts (e.g. furfural in acid hydrolysis). Enzymatic treatment of the residues, on the other hand, results in specific release of end-products and can be achieved at moderate conditions of pH and temperature, obviating the need for expensive equipment. However, it is not realistic at present to consider the enzymatic processes as the competing ones, because of their low catalytic efficiencies, especially towards crystalline cellulosic substrates. A future possibility may lie in the development of a hybrid process which uses a chemical treatment to accomplish the fractionation of the residues and an enzymatic treatment for conversion of a specific component (such as cellulose) to desirable product (e.g. glucose).

In a few cases, it may be more important and economical to recover a useful product, which may be a minor constituent of the whole crop, before processing the lignocellulose for bioconversion. This seems to be applicable to a number of 'botanochemical' crops which can be processed to give useful products such as soluble polyphenols, whole plant oils, isoprene polymers, paper making fibres, animal feed and fermentation substrates. A few botanochemical crops and their components which can be extracted for various applications have been reported (Buchanen *et al.*, 1980; Dale, 1987).

Because of the drawbacks associated with acid hydrolysis of lignocellulosics, rather extensive research efforts have been directed towards their enzymatic hydrolysis. The enzymes catalyzing the breakdown of cellulose (i.e. cellulases), hemicellulose (i.e. hemicellulases such as xylanases, mannanases, etc.) and lignin (i.e. lignin peroxidase, manganese peroxidase and other oxidases) possess high specificity and act at moderate temperatures of 30–50°C. A number of fungi and bacteria produce these enzymes in extracellular fluid which can be employed in various bioconversion processes to get useful products. A few of the current and potential applications of cellulases are mentioned in Table 3. At present, they find their greatest use in the removal of crude fibres of cell walls for various end-uses in different industries. Cellulases, in many cases, are used together with cell separating enzymes (CSE) which possess protopectinolytic activity. This enzyme removes protopectin present as intercellular substance in vegetable tissues and exposes the cell walls which can be degraded by cellulases. One of the potential applications of hemicellulase enzymes lies in the pretreatment of kraft pulp in the paper industry. Traditionally the residual lignin (which imparts brown color) is removed in a multistage bleaching procedure using chlorination and extraction stages, resulting in severe environmental pollution due to formation of numerous chlorinated phenols and dioxines. Use of xylanase preparation in the pretreatment step is reported to substantially reduce the requirement of chlorine and to elevate the brightness without adversely affecting the strength and yield (Viikari *et al.*, 1986). Various potential applications of lignin peroxidase and related enzymes have also been reported in biopulping, biobleaching, treatment of kraft bleaching effluents, removal of environmental pollutants, improving the feed value of lignocellulosic residues, modification of lig-

Table 3
Some Current and Potential Applications of Cellulase Enzymes Produced on Cellulosic Residues

Industry	*Applications*
1. Food	—to increase soaking efficiency and homogeneous water absorption of rice
	—removal of external soybean coat to produce unicellular soybean. This prevents coloration of fermented soybean foods such as soysauce and miso
	—isolation of proteins from soybean and coconut
	—isolation of starch from corn and sweet potato
	—gelatinization of seaweeds (such as *Laminaria, Phaeophyceae*) to increase digestibility
	—extraction of agar-agar from seaweeds (such as *Rhodophyceae, Gelidium*)
	—extraction of various components of green tea such as tannin, caffeine, saponin and aroma
	—to digest ball-milled lignocellulose to obtain lignin with very large surface area, which can be used as a food additive
2. Brewing	—purification of water soluble α-glucan from milled barley
3. Wine and fruit juice	—to break down cellulose and other insoluble plant materials in fruit pulps
4. Paper and pulp	—increasing the tensile strength of high alpha-sulfite pulp
	—to convert waste fiber in primary clarifier sludge into sugars
5. Feed	—as a supplement in feed for cattle and poultry
6. Alcohol/solvent	—to hydrolyze cellulose into glucose which can be fermented to ethanol, butanol, 2,3-butanediol, etc.
7. Other uses:	
(1) Research and development	—elucidating the structure of complex polysaccharides
	—to produce protoplasts of higher plants
(2) Medical	—as digestive aids in anti-flatulence tablets
(3) Pollution abatement	—clarification of sludge to improve the efficiency of digestion in a septic tank

Sources: Toyama, 1969; Ghose & Pathak, 1973; Mandels, 1985; Jurasek *et al.*, 1987, Stark & Yin, 1987.

nochemicals for use as adhesives and resins, and in specific biotransformations (Kirk *et al.*, 1980; Crawford, 1981; Kirk & Farrell, 1987; Kaushik & Bisaria, 1989).

Before using any bioprocessing step for utilization of cellulose, it is important to realize that cellulose constitutes generally less than half of the total lignocellulose and, therefore, any bioconversion process based on cellulose utilization alone is not likely to be attractive and economical. This is because of (a) the lost revenue from hemicellulose or lignin (or other constituents), and (b) the added waste disposal costs. Hemicellulose and lignin and their degradation products find a number of uses which are listed in Table 4 along with those of cellulose and glucose. Cellulose utilization should accordingly be linked to the effective use of the lignocellulose constituent of highest possible product value, so that glucose production does not carry the entire processing and raw material costs. In the utilization of lignocellulosic wastes, therefore, an integrated approach may aim at removal of hemicellulose (which is easily recoverable by mild acid hydrolysis) and conserving the value of lignin and protein which have higher base value than sugars. It was shown (Dale, 1987), for example, that the various protein products available from integrated conversion of alfalfa were approximately twice as valuable as ethanol derived from hydrolysis and fermentation of alfalfa cellulose.

The range of products that can be derived from the agricultural residues are many, depending on the type of bioprocessing steps employed for their conversion. Figure 1 outlines the preferred major processing routes for

Table 4
Application of Products Derived from Cellulose, Hemicellulose and Lignin

Products	*Applications*
I. From cellulose	
1. Glucose	Ethanol
	Fructose (sweetener)
	Energy (alcohol, ATP)
	SCP—Protein concentrate
	—Yeast glycan
	—Nucleic acids
	—Autolyzates
	Chemicals
	2,3–butanediol
	Acetone/butanol
2. Cellobiose/cellodextrins	Research chemicals
3. Ethanol	Chemicals
	Solvents
	Beverages
	Biological energy (ATP)
	Petrochemical synthesis (via ethylene)
	Gasoline diluent and octane enhancer
	Food and feed
II. From hemicellulose	
1. Xylose	Ethanol
	Xylitol (sweetener)
	SCP
2. Furfural and 2,3-butanediol	Chemicals and solvents (such as methyl ethyl ketone)

(continued)

Table 4—*contd.*

Products	*Applications*
III. From lignin	
1. Polymeric lignins	Adhesive
	Dispersant
	Complexing agent
	Stabilizer
	Antioxidant
	Precipitant
	Deflocculant
	Coagulant
	Pesticide carrier
	Emulsion stabilizer
	Soil conditioner
	Rubber reinforcement
	Phenol-formaldehyde resin and extender
	Urea-formaldehyde resin and extender
	Polyurethane foam
2. Chemical intermediates	Phenols
	Cresols
	Catechols
	Vanillin
	Vanillic acid
	Syringaldehyde
	Syringic acid
	Dimethyl sulfide
	Dimethyl sulfoxide
	Methyl mercaptan
	Ferulic acid
	Lignosulfonates (additives in stabilizers, grinding aids, binders)
	Butanol–lignin slurry (furnace and diesel fuel)
IV. From lignocellulose	
(Untreated or pretreated)	Edible mushrooms
	animal feed
	Methane (biogas)
	Fertilizer

Sources: Bisaria & Ghose, 1981; Hawley *et al.*, 1986.

conversion of these residues into food (mushrooms), fuel (methane), feed and chemicals. Consideration of each one of them is clearly out of the scope of the present article. The discussion is, therefore, confined to the production of glucose, ethanol and a few chemicals of industrial importance, which can be derived from cellulosic residues. It is worth noting that amongst the various microbial or enzymatic processes utilizing these residues which have been exploited commercially, the ones involving mushrooms are the only processes which are commercially viable and utilize either unmodified (as in the case of *Pleurotus* or *Volvariella* cultivation) or partially modified (as in the case of

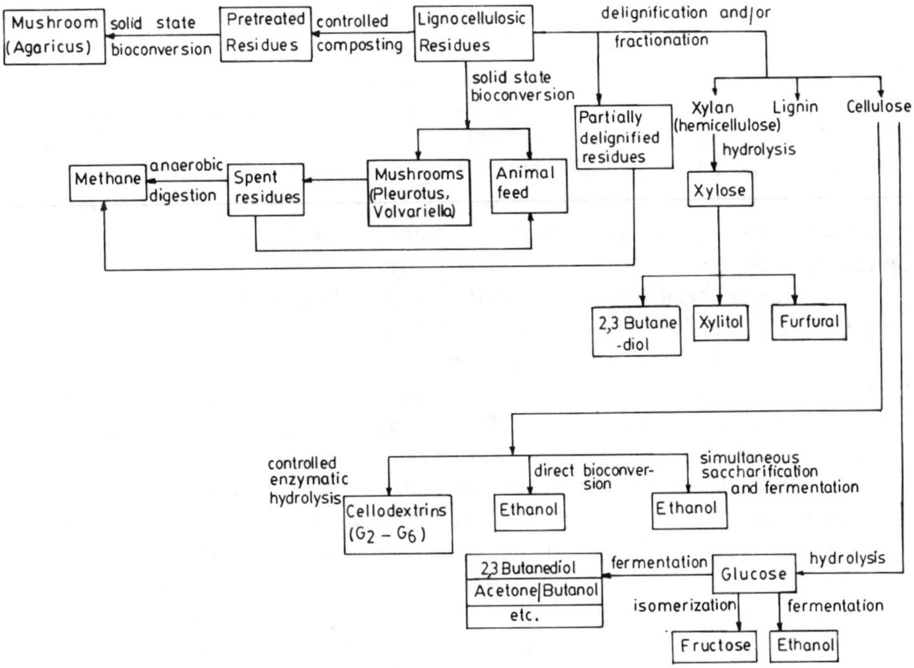

Fig. 1. The preferred major processing routes for bioconversion of agro residues into food, feed, fuel and chemicals.

Agaricus) residues to convert them into food (Zadrazil & Grabbe, 1983; Bisaria *et al.*, 1987*b*).

2 CHARACTERISTICS OF CELLULOSIC MATERIALS AND CELLULOLYTIC ENZYMES

2.1 Cellulosic Materials

2.1.1 Structural Features

As mentioned above, cellulose is one of the three major components of agricultural residues. Cellulose is a linear polymer of D-glucose residues (building blocks), held together by β-1,4-glucosidic linkages. The number of D-glucose units in a cellulose molecule may be as large as 15 000. The average degree of polymerization (DP) may vary from about 50 to a few thousand depending on its source and prior treatment.

The β-1,4-linkage causes the polyglucose molecules to adopt a fully extended linear configuration which is attained by flipping of each glucose unit 180° relative to the preceding one. The glucosidic linkage acts as a functional group, and this along with the hydroxyl groups mainly determines the chemical properties of cellulose. Also, the intramolecular hydrogen bonds stabilize each cellulose chain. The linear cellulose molecules pack together to form rodlike structures known as elementary fibrils or protofibrils or micelles, which are stabilized by intermolecular hydrogen bonds between adjacent cellulose chains. Micelles, the smallest structural unit, are packed into microfibrils which are each about 10 nm in diameter and up to several micrometers in length. About 10 microfibrils in turn are packed together to give rise to fibers or macrofibrils (approximately 50 nm in diameter) which can be seen by an optical microscope.

Lignocellulosic residues also contain hemicellulose and lignin. It has been proposed (Fengel, 1971; Fengel & Wegner, 1984) that the microfibrils, which are about 12 nm in cross-section and about 30 nm in length, are surrounded by hemicelluloses and lignin in wood. The elementary fibrils are also surrounded by monolayers of hemicelluloses. The precise manner in which the native cellulose molecules are aggregated to form the elementary fibril structure is not yet well understood. In the fringe-micellar model, the extended cellulose crystallites have been segmented into crystalline and paracrystalline (or amorphous) regions, whereas in the folding-chain model, each cellulose chain is not extended through a series of several crystalline and paracrystalline regions but is predominantly folded within a single cellulose crystallite (Chang, 1971; Scallan, 1971; Rowland & Roberts, 1972). Pretreatment of cellulosic materials, aimed at degrading the protective layers of hemicelluloses and lignin around elementary fibrils and microfibrils, is essential to rapid enzymatic hydrolysis.

The molecular structural features of cellulose, the elementary fibril and the microfibril are important for enzymatic degradation of cellulose. The proportion of crystalline and amorphous regions ranges from 50 to 90% in different materials; about 70% crystallinity exists in native cellulose. Generally, the amorphous regions in the fringe micelles are attacked first in a partially crystalline cellulose fibril, leading to an enrichment of crystalline regions which are, in turn, gradually solubilized after loosening of their peripheral parts. The ability of the cellulolytic microorganisms or that of enzymes to hydrolyze cellulose varies greatly with the nature of the cellulosic materials.

The second major constituent of lignocellulosic biomass is hemicelluloses, the amorphous polysaccharides made up of xylose, galactose, mannose, arabinose, glucose and their uronic acids (Dekker, 1985). These occur as heteroglycans containing different types of sugar residues, often as short mono- or oligosaccharide appendages linked to the main backbone chain. The proportions of these pentoses and hexoses vary widely in the hemicelluloses of different biomass origin. Coniferous woods, for example, have a higher proportion of glucomannans and deciduous woods a higher proportion of xylans and arabinoxylans. Agricultural residues, though more complex in hemicellulose composition, in general have an important bearing on further processing of their enzymatic hydrolysis product; mannose, for example, is readily fermentable by ordinary yeasts whereas xylose and arabinose can be utilized only by a few yeasts. Further, hemicelluloses appear to be associated with both cellulose and lignin in the microfibrillar structure (Fengel & Wegner, 1984).

Lignin, the third major constituent of lignocellulosic residues, is an amorphous three-dimensional polymer composed of three phenylpropanoid units—coniferyl alcohol, sinapyl alcohol and *p*-coumaryl alcohol (Crawford, 1981). The monomers in the lignin molecule are not linked in the same way. Some monomers may be involved in several linkages, others in only one or two, and the intermonomeric bonds may link the monomers at different positions through C—C, C—O ether (or sometimes ester) bonds. Overall, the lignins are heterogeneous and much-branched polymers of indefinite structure and size; their molecular weights range from a few thousand to over a million. The proportion of the three monomeric residues varies with the species. The lignin in conifers is richer in coniferyl alcohol-derived residues whereas that from deciduous species has more sinapyl alcohol-derived residues. As a consequence, the lignin in conifers is more condensed and less tractable to degradation than the lignin in hardwoods. Agricultural residues have relatively less lignin but are relatively high in *p*-coumaryl derivatives.

2.1.2 Physical and Chemical Constraints on Enzymatic Hydrolysis of Cellulose

The susceptibility of cellulosic substrates to enzymatic hydrolysis is determined largely by their accessibility to cellulolytic enzymes (Tanaka *et al.*, 1988*a*). Direct physical contact (adsorption) of cellulase enzymes on to the cellulosic substrate is a prerequisite to enzymatic degradation (Ghose & Bisaria, 1979; Castanon & Wilke, 1980; Kyriacou *et al.*, 1989), and hence any structural feature that limits the accessibility of cellulosic substrates to enzymes will retard its susceptibility to hydrolysis. These features are listed in Table 5 and have been elaborately discussed by Cowling (1975) and Fan *et al.* (1980).

2.2 Cellulolytic Enzymes

Cellulases are synthesized by a large number of microorganisms which include fungi, actinomycetes, gliding bacteria and true bacteria. While many bacteria utilize cellulose by cell-bound enzymes, fungi secrete many of these enzymes in the growth medium. The extracellular cellulases from the fungi *Trichoderma reesei* (Mandels, 1975; Nevalainen *et al.*, 1980; Ryu & Mandels, 1980), *T. koningii* (Wood, 1985; Wood *et al.*, 1988), *T. viride* (Gritzali & Brown, 1979), *Penicillium pinophilum* (Wood *et al.*, 1988), *Sclerotium rolfsii* (Lachke & Deshpande, 1988), *Fusarium solani* (Wood & McCrae, 1977), *Phanerochaete chrysosporium* (Eriksson & Wood, 1985), *Talaromyces emersonii* (Moloney *et al.*, 1985), and *Neocallimastix frontalis* (anaerobe from bovine rumen) (Pearce & Bauchop, 1985; Wood *et al.*, 1988) are amongst the best known examples of cellulases that can degrade the native crystalline cellulose. Most other extracellular cellulases from fungi have little action on crystalline cellulose but

Table 5
Structural Features of Cellulosic Residues that Determine their Susceptibility to Enzymatic Degradation

1.	Moisture content of the fiber (water-swollen fibers significantly increase surface area by opening up the capillary structure)
2.	Capillary structure of cellulosic fibers (this determines the surface area of cellulosic fibers as defined by size, shape and surface properties of the microcrystalline capillaries within the fiber)
3.	Crystallinity of cellulose
4.	Unit cell dimensions of cellulose
5.	Conformation and steric rigidity of anhydroglucose units within the crystalline regions
6.	Degree of polymerization of cellulose
7.	The nature of the substances with which cellulose is associated (lignin, hemicelluloses, minerals and extraneous materials)
8.	The nature, concentration and distribution of substituent groups (methyl, ethyl, carboxymethyl, hydroxyethyl) on cellulose molecule (the substitution affects the solubility and crystallinity of the cellulose and also, at a high concentration, activity of the enzyme)

Sources: Cowling, 1975; Fan *et al.*, 1980.

can hydrolyze amorphous celluloses and soluble derivatives of cellulose such as carboxymethyl cellulose (CMC). The fungal cellulases do not seem to form physical complexes with each other, but they act in strong synergy where cooperative action is greater than the sum of the action of the individual enzyme components (Wood, 1985).

All cellulases cleave the same chemical bond, viz. the β-1,4-glucosidic bond, but there is variation in the microenvironment of these bonds in the cellulosic substrates, as the same cellulose molecule contains both crystalline and amorphous structure. Moreover, as mentioned above, the presence of lignin and hemicellulose in cellulosic materials further complicates the cellulose matrix. Moreover, as cellobiose is the repeating unit in the cellulose crystallite, the consecutive glucosidic linkages between glucose residues are quite different stereochemically (Wood *et al.*, 1988). This complexity of structure is probably one of the reasons why the cellulolytic microorganisms produce a family of different cellulases with different substrate specificities. It is also worth mentioning that the specificities of many of the cellulases may be much broader than the present nomenclature and traditional classification would suggest.

Fractionation of cellulase proteins present in the culture filtrate of many of the fungi mentioned above have revealed that there are at least three major types of enzymes involved in hydrolysis of native cellulose into glucose. These are:

—Endo-β-1,4-glucanases (endo-β-1,4-D-glucan 4-glucanohydrolase, EC 3.2.1.4)

—Exo-β-1,4-glucanase, present as cellobiohydrolase (exo-β-1,4-D-glucan 4-cellobiohydrolase, EC 3.2.1.91). Some cellulase systems also contain exo-β-1,4-D-glucan 4-glucohydrolase (EC 3.2.1.74) as a minor constituent
—β-glucosidase (EC 3.2.1.21)

On the basis of the mode of action of these components on various cellulosic substrates, the following generalizations about their specificities can be made (Wood, 1988; Bisaria & Mishra, 1989).

Cellobiohydrolases degrade cellulose by removing cellobiose from the non-reducing end of the chain, and have very limited action on substituted celluloses such as CMC. They readily hydrolyze 'swollen' partially degraded amorphous cellulose and cellodextrins. They are, with a few exceptions, considered to be inactive against crystalline cellulose but exhibit highly cooperative synergistic action in the presence of endoglucanases. The endoglucanases hydrolyze CMC or swollen cellulose in a random fashion, as a result of which there is a rapid decrease in chain length together with a slow increase in reducing groups. They also act on cellodextrins and convert them to cellobiose and glucose. β-Glucosidases hydrolyze the degradation products of cellulose (i.e. cellobiose and cellodextrins) to glucose. In spite of these well-known specificities of the cellulase components, there is still no clear picture of events that take place during hydrolysis of crystalline cellulose. This is primarily due to the complex properties of the enzymes and also the inherent difficulties in analyzing the various insoluble products of cellulose degradation. Accordingly, one finds in the literature that the action of 'purified' cellobiohydrolase on cotton and Avicel™ cellulose from a variety of fungal sources has been reported to be both extensive (Chanzy *et al.*, 1983; Nummi *et al.*, 1983) and negligible (Wood & McCrae, 1979). A detailed discussion on the mode of action of cellulase components on solubilization of cotton and other crystalline cellulose can be found in the literature (Bisaria & Ghose, 1981; Mandels, 1981; Wood, 1985; Enari & Niku-Paavola, 1987; Coughlan & Ljungdahl, 1988; Wood, 1988; Wood *et al.*, 1988).

It is also important to realize that additional enzymes besides cellulases may also be required for degradation of cellulose. Thus, whereas the purified and reconstituted mixture of cellobiohydrolases, endoglucanases and β-glucosidases from the culture filtrates of *Penicillium pinophilum, Fusarium solani, Trichoderma koningii* and *Talaromyces emersonii* are capable of synergistically degrading cellulose to glucose, those from *Trichoderma reesei, Phanerochaete chrysosporium, Polyporus adustus* and *Myrothecium verrucaria* require certain oxidative enzymes also for hydrolysis of cellulose (Eriksson, 1981). In *T. reesei,* lactonase (EC 3.1.1.17) has been shown to act synergistically in solubilizing cellulose (Bruchmann *et al.*, 1987). Cellobiose oxidase (EC 1.1.99.18) and cellobiose:quinone oxidoreductase (EC 1.1.5.1) have also been reported to be involved in cellulose degradation in a number of cellulolytic organisms (Dekker, 1980; Eriksson, 1981; Coudray *et al.*, 1982; Morpath & Jones, 1986).

3 BIOCONVERSION TO INDUSTRIAL PRODUCTS

3.1 Glucose

3.1.1 Pretreatment of Lignocellulosic Residues

Since glucose is an easily fermentable sugar, it remains one of the most sought-after products of enzymatic degradation of cellulosic materials. However, a number of factors in terms of the characteristics of both the substrate and the enzyme limit the hydrolysis of cellulose to glucose to an appreciable extent. The physical and chemical constraints imposed by cellulosic substrates have been mentioned in section 2.1. In order to overcome these constraints and to enhance the susceptibility of the residues to enzyme action, pretreatment is an essential prerequisite. Numerous pretreatment methods capable of significantly enhancing the rate and extent of hydrolysis of pure celluloses as well as celluloses found in biomass materials have been described in the literature. These methods along with the underlying phenomenon used for increasing the susceptibility and representative examples are summarized in Table 6. Most pretreatment processes have substantial energy requirements depending on the severity of the process. Severe mechanical and thermochemical processes, in general, have energy inputs which, when combined with the energy requirements of the product separation, can make cellulose bioconversion processes very uneconomical (Kosaric *et al.*, 1983; Klyosov, 1986). The mechanical pretreatments give fine substrates which result in higher slurry concentrations or higher bulk density, thereby reducing the reactor volume. However, lignin remains associated to act as a significant inhibitor to enzyme accessibility. Chemical pretreatments have been used extensively to remove lignin and to modify the structure of lignocellulose. Conventional pulping processes, however, such as kraft, sulfite and soda processes, which remove lignin and preserve the quality of cellulose for paper manufacture, are not used for bioconversion because of their very high cost. The most common chemical pretreatment has been caustic swelling. Pretreatment with caustic soda results in increased surface area due to swelling and disruption of lignin structure. Acid pretreatment uses dilute acids to remove hemicelluloses by hydrolysis. Oxidizing agents such as ozone and hydrogen peroxide attack lignin in preference to cellulose. They also disrupt the association between the carbohydrate polymers and lignin, producing a residue with enhanced susceptibility to cellulases.

Organosolv pretreatments involve the use of an aqueous solvent (ethanol, butanol, etc.) in the presence of a catalyst. The catalyst and water hydrolyze many of the carbohydrate bonds in the hemicellulose as well as the lignin–carbohydrate and lignin–lignin bonds. The solvent dissolves lignin while cellulose remains as a solid. Organic acids that are liberated from the substrate during the pretreatment process accelerate delignification and result in concurrent solubilization of the hemicelluloses. The organosolv pretreatment processes are attractive because the solvents can be recovered and recycled.

Also, this process separates the lignocellulose into a solid cellulose, a solid lignin which has undergone few condensation reactions, and a liquid containing hemicellulose hydrolyzate (Sarkarsen, 1980). Fan *et al.* (1981*b*) performed a cost analysis of various chemical pretreatment methods and found that caustic pretreatment is a potential candidate for large-scale development based on pretreatment cost, hydrolysis rate and its sugar yield. Chemical pretreatments, however, have serious disadvantages in terms of need of specialized corrosion-resistant equipment, extensive washing, and disposal of chemical wastes. They also give rise to very low bulk density of pretreated product, and therefore suspensions of even 4–5% are too thick to agitate.

Cellulose solvents pertaining to the four major categories of (a) strong mineral acids, (b) quaternary ammonium bases, (c) nonaqueous organic solvents such as dimethyl sulfoxide (DMSO) and dimethyl formamide (DMF), and (d) metal complexes have also been employed for pretreatment of lignocellulosics. They act by reducing crystallinity and disrupting lignin–carbohydrate association, resulting in dissolution of cellulose and hemicellulose. Some of these solvents transform native cellulose (Cellulose I) to the Cellulose II form which has been found to be highly susceptible to enzyme action (Hamilton *et al.*, 1984).

Biological pretreatments are primarily concerned with the use of fungi, especially the white-rot ones, and bacteria that can degrade lignin preferentially. The complete selective removal of lignin has not so far been possible as the microorganisms, unable to utilize lignin as a growth substrate, degrade cellulose and/or hemicellulose for their growth. Cellulase-less mutants of *Sporotrichum pulverulentum* have been isolated which partially delignify wood without affecting its cellulose (Eriksson *et al.*, 1980). However, these mutants cause a significant loss of hemicellulose or externally added glucose while degrading lignin. Biological pretreatments suffer from the serious drawback of lengthy (2–5 weeks) pretreatment time (Hatakka, 1983; Srivastava, 1989). Pretreatment processes for enzymatic solubilization of lignin (Leisola & Fiechter, 1985) and hemicellulose (Ghose & Bisaria, 1979) have also been reported. However, they seem to be too expensive due to the low activity and low stability of the lignin peroxidases (Kaushik & Bisaria, 1989).

Out of the combined pretreatment processes, the steam-explosion process seems to be the most promising. Steam pretreatment at high pressure (4 MPa) and temperature (185–240°C) results in the partial decomposition of some of the hemicellulose to uronic acids and other acids (mainly acetic acid) which catalyze the depolymerization of the hemicellulose and lignin. After steaming, the residue can either be explosively discharged by a sudden drop in pressure or emptied out after a gradual 'bleed-down' (Wood & Saddler, 1988). This type of pretreatment is attractive because it allows the selective fractionation of all the three main constituents, viz. cellulose, hemicellulose and lignin, of a wide range of hardwoods and agricultural residues (Saddler *et al.*, 1982*a*). While the recovered lignin and hemicelluloses can be suitably converted into useful products, cellulose can be enzymatically hydrolyzed with very high yields.

Table 6
Pretreatment Methods for Enhancing the Susceptibility of Cellulose to Cellulase Action

Pretreatment method	*Underlying phenomenon*	*Representative examples*	*References*
1. Mechanical	Utilize shearing and impacting forces to yield a fine substrate of low crystallinity index and high specific surface	Ball milling, two-roll milling, hammer milling, colloid milling, vibro energy milling, extrusion	Ghose (1969); Mandels *et al.* (1974); Tassinari & Macy (1977); Millet *et al.* (1979); Fan *et al.* (1981*a*).
2. Physical	Increase in pore size and partial hydrolysis of hemicelluloses (steaming); extensive depolymerization (irradiation); depolymerization, oxidation or dehydration (pyrrolysis)	Steaming, wetting, pulping; freezing/thawing; ν-irradiation; pyrrolysis	Millet *et al.*, (1976); Neese *et al.* (1977); Macdonald & Mathews (1979); Han *et al.* (1981); Han & Ciegler (1983)
3. Chemical	Removal of lignin, and/or hemicellulose, modification of the structure of lignocellulose, reduction in crystallinity, increase in surface area	Swelling agents (NaOH, NH_3); dilute acids (HCl, H_2SO_4, H_3PO_4); oxidizing agents (peracetic acid, O_3, H_2O_2); gases (SO_2, NO_2, ClO_2); organosolv (methanol, ethanol, butanol, glycerol, dioxane, phenol,	Nesse *et al.* (1977); Grethlein (1978); Wilke (1978); Binder *et al.* (1980); Ben-Ghedalia & Miron (1981); Gould (1984); Hamilton *et al.* (1984); Neely (1984); Chou (1986); Grohmann *et al.* (1986); Lynd &

		ethylene glycol in the presence of a catalyst such as Lewis acids); cellulose-dissolving solvents: concentrated mineral acids (H_2SO_4, HCl), ammonia-based solvents (NH_3, hydrazine), aprotic solvents (DMSO, sulfur oxides) and metal complexes (cadoxen, cuoxam)	Grethlein (1987); Chum *et al.* (1988)
4. Biological	Preferential removal of lignin, modification of lignocellulose structure	White-rot fungi; soft-rot fungi; bacteria; cellulaseless mutants of white-rot fungi	Hatakka (1983); Srivastava (1989)
5. Enzymatic	Selective removal of lignin, hemicelluloses	Lignin peroxidase enzymes; xylanases	Ghose & Bisaria (1979); Viikari *et al.* (1986); Kaushik & Bisaria (1989)
6. Combined pretreatments	Depends on the combination of pretreatments used, e.g. steam explosion loosens the cell wall structure, increases surface area, reduces DP of cellulose, and separates the individual components	Steam explosion/RASH; ammonia-freeze-explosion; high-temperature milling; alkali plus ball milling; SO_2 plus steaming; NO_2 plus irradiation	Dale & Moreira (1982); Saddler *et al.* (1982*a*); Dekker & Wallis (1983); Brownell & Saddler (1986, 1987); Schultz *et al.* (1988)

It has been shown that rapid decompression does not greatly improve the accessibility of cellulose (Brownell & Saddler, 1986, 1987). Based on this, a process termed Rapid Steam Hydrolysis (RASH) has been developed which also continuously extracts the gaseous and soluble products (Schultz *et al.*, 1989). Steam treatment under alkaline conditions enhances accessibility of the cellulose (Dekker & Wallis, 1983). Treatment of both hardwood and softwood with SO_2 prior to explosion also resulted in a residue with increased susceptibility to cellulases (Clark & Mackie, 1986). The agricultural residues and a few hardwoods can also be made highly susceptible by subjecting them to the ammonia-freeze-explosion (AFEX) process which treats them with liquid ammonia under pressure followed by sudden release (Dale & Moreira, 1982).

3.1.2 Production of Cellulolytic Enzymes

In the enzymatic conversion of lignocellulosic materials to liquid fuels, the cost of cellulase enzyme has been shown to have a major impact on overall economic viability of the process. It has been computed that nearly 51% of the sugar cost or 40% of the ethanol cost is attributed to cellulase production (Andreotti *et al.*, 1981). Various approaches have been suggested as a possible way of lowering the enzyme-associated costs, including recycling of the enzymes (Deshpande & Eriksson, 1984; Isaacs, 1984; Puls *et al.*, 1985) in the production process of useful metabolites from cellulose. Although complete cellulase is secreted in extracellular fluid by a number of fungi mentioned in section 2.2, *Trichoderma reesei* has been subjected to extensive studies for improving the yield and productivity of cellulases synthesized by it. This has been achieved by using various approaches, such as (a) genetic improvement through mutagenesis by overcoming the control mechanisms of cellulase biosynthesis: this has resulted in several hyperproducing mutants of *T. reesei* (Montenecourt & Eveleigh, 1979; Durand *et al.*, 1988); (b) optimization of fermentation conditions (Sternberg & Dorval, 1979); (c) affecting the physiology by non-inducing sugars (Nanda *et al.*, 1986); (d) use of different reactor configurations (Allen & Andreotti, 1982); and (e) use of mixed fungal cultures (Duff *et al.*, 1987; Panda *et al.*, 1987).

Cellulase synthesis in *Trichoderma* is governed by both induction as well as catabolite repression. While various soluble inducers such as cellobiose, sophorose, lactose and cellobionolactone induce cellulase synthesis, the insoluble substrate cellulose has been the best inducer in the sense that a complete array of cellulase enzymes, capable of degrading crystalline cellulose most effectively, are synthesized only in its presence (Gong *et al.*, 1979; Sternberg & Mandels, 1979; Bisaria & Mishra, 1989). The inductive formation of cellulases and β-glucosidase is also subject to catabolite repression by glucose and other readily metabolizable substrates. Therefore, most studies on its production have been carried out on insoluble carbon sources such as Solka floc(TM), steam-exploded biomass and even untreated agricultural residues (Gonzalez *et al.*, 1986; Doppelbauer *et al.*, 1987).

By overcoming the regulatory mechanisms that control the synthesis of cellulases, largely through mutagenesis, several hyperproducing mutants of *T. reesei* have been obtained. The genealogy of such mutants developed independently in many laboratories is given by Durand *et al.* (1988). The increased cellulase activity in some hyperproducing mutants such as MCG 80 has been attributed to enhancement of secretion activities of the mutants. In almost all cases except for a few Cetus mutants (Shoemaker *et al.*, 1981) and Rutgers' mutant, Rut-P37 (synonym RL-P37) (Eveleigh, 1981; Sheir-Neiss & Montenecourt, 1984), the specific cellulase activity (enzyme units per milligram of extracellular protein) has remained the same. This suggests that the improvements in the efficiency of fungal cellulases are possible by specifically affecting the regulation of the secretory pathway. It should be noted that the specific activity of the best cellulase producers, in terms of filter paper units, is approximately 1·3 IU/mg protein while the commercial amylase preparations have specific activities of about 100 IU/mg protein.

In recent years, there have been numerous reports on cloning and expression of cellulase genes from *T. reesei* and other cellulolytic organisms in *Escherichia coli, Saccharomyces cerevisiae* and other hosts (Knowles *et al.*, 1987; Beguin *et al.*, 1988). The aim of these studies is to understand the gene structure and expression which would be helpful in obtaining the enzyme components in pure form and in quantity for effecting solubilization of native cellulose structure in lignocellulosic residues (Knowles *et al.*, 1987). For a detailed discussion on the factors which govern the biosynthesis and secretion of cellulase enzymes in various mutants and recombinant cells, a recent review by the author (Bisaria & Mishra, 1989) is available.

In spite of the low specific activity of most cellulases, several studies have been carried out in batch, continuous, fed-batch and immobilized cell reactors under varying substrate and environmental conditions to maximize the enzyme titer (i.e. activity in IU/ml) and productivity (IU/liter h). A few of these studies on production characteristics of cellulase fermentations are summarized in Table 7. As seen from the table, there is wide variation in the activity and productivity values, which may be due to a number of factors such as the strain, growth conditions, concentration and type of inducers, the method of fermentation, and also the assay procedure used for activity determination (Ghose, 1987). However, in a recent study on the effects of strain and fermentation conditions on production of cellulase by three strains, viz. QM6a, QM9414 and Rut-C30, of *T. reesei* in continuous and fed-batch cultures, Allen and Roche (1989) reported that when glucose repression was removed by carbon limitation, the parent strain QM6a could produce as much as and even higher activity than that of mutants on cellulose hydrolyzate. The hydrolyzate contained the strong inducer sophorose besides glucose, xylose, cellobiose, gentibiose, and laminaribiose. Keeping in view the similarity of the cellulase systems of these strains, these results are significant for further genetic improvement studies, as they suggest that the mutations may not affect the cellulase genes directly but probably affect the mechanisms controlling the

Table 7
Cellulase Production by Various Strains of *Trichoderma reesei* in Reactors of Different Configurations[a]

Process	*Strain*	*Particulars of inducer(s) and control variables*[b]	*Activity (IU/ml)*[c]	*Productivity (IU/liter h)*	*References*
1. Batch	D 1–6	6% Cellulose, 150 liter fermentor	11 (192 h)	57	Ghosh *et al.* (1982*a*)
	MCG-77	6% Cellulose, bi-level pH control/biotin supplement	11·5 (120 h)	88	Gallo (1982)
	Rut C-30	as above	13·3 (120 h)	95	Gallo (1982)
	MCG-80	3% Cellulose	17·2 (96 h)	142	Gallo (1982)
2. Semicontinuous	E-12	(a) 6·3% Avicel, 7 liter fermentor	8·6 (180 h)	47·8	Ghose, T. K., 1981; pers. comm.
	E-12	(b) 5·2% Solka floc BW 200 (initial conc.), 5 liter fermentor, pH cycling between 3·2 and 5·4. Semicontinuous feeding of cellulose	17·5 (293 h)	97·3	Ghose, T. K., 1981; pers. comm.
3. Continuous	MCG-77	Lactose, 2-stage culture, 15 liter fermentors, $D = 0{\cdot}026$–$0{\cdot}028\ h^{-1}$ in second stage, run for 1350 h	2·6	90	Ryu *et al.* (1979)
	MCG-80	5% Lactose, $D = 0{\cdot}028\ h^{-1}$, pH 3·5, 5 liter fermentor	6·0	168	Allen & Andreotti (1982)
4. Repeated fed batch	MCG-80	2% Lactose (initial conc.), pH 3·5, 5 liter fermentor	10·0 (145 h)	100	Allen & Andreotti (1982)
5. Fed batch	Rut C-30	Solka floc BW 200, 5 liter fermentor, addition rate 2·5 g cellulose per hour, total final addition of 100 g of cellulose per liter	22·4 (160 h)	133	Hendy *et al.* (1984)
	QM 6a	Cellulose hydrolyzate of BW 200, 5 liter fermentor	11·8 (290 h)	55	Allen & Roche (1989)
	MCG-80	Cellulose hydrolyzate of BW 200, 5 liter fermentor	9·7 (290 h)	—	Allen & Roche (1989)
6. Continuous (immobilized cells)	Rut C-30	Lactose, celite II support, loading 80 g/liter	1·93	134	Montenecourt *et al.* (1981)

[a] A few more production results can be found in Kosaric *et al.* (1983).
[b] D = dilution rate (h^{-1}).
[c] Time in parentheses indicates time of maximum activity in batch cultures.

synthesis and secretion of cellulase proteins. Although the cellulases do not find application in the enzymatic hydrolysis of cellulose to produce glucose, they are used in a number of other processes in various industries (see Table 4).

Cellulolytic fungi other than *Trichoderma* which produce complete cellulase enzyme systems (see section 2.2) have also been subjected to mutagenic treatment for isolation of hyperproducing mutants. One such potential mutant seems to be *Penicillium pinophilum* JAB/3B/15, which produced higher cellulase as well as β-glucosidase activities compared to that of *T. reesei* Rut C-30 in a shake flask (Brown *et al.*, 1987). However, since extensive studies on this and other similar strains have not been conducted in large scale bioreactors under controlled environmental conditions, the full biosynthetic potential of all such seemingly better (than *Trichoderma*) strains remains to be seen.

A number of bacteria belonging to the genera *Clostridium, Cellulomonas, Bacillus, Thermomonospora, Ruminococcus, Bacteriodes, Erwinia, Acetovibrio, Microbispora* and *Streptomyces* have also been studied for their cellulase enzyme system (MacKenzie & Bilous, 1982; Hiltner & Dehority, 1983; Giuliano & Khan, 1984; MacKenzie *et al.*, 1984; Ghangas & Wilson, 1987; Lee & Pack, 1987; Beguin *et al.*, 1988; Chippaux, 1988; Miller *et al.*, 1988; Yablonsky *et al.*, 1988; Demain & Wu, 1989) and extensive studies have been conducted at the genetic level in a few organisms to understand their regulatory mechanisms (Lee & Pack, 1987; Beguin *et al.*, 1988; Miller *et al.*, 1988). In several cases, it has been found that though the purified cellulases possessed specific activities equal to or higher than those of fungal origin, they produced smaller amounts of cellulase enzymes than fungi. Furthermore, relatively few of the bacteria produce an extracellular cellulase that can completely hydrolyze crystalline cellulose. An extensive survey of properties and genetics of cellulases of bacterial origin has recently been made by Robson and Chambliss (1989). Thus, in the context of glucose production through enzymatic hydrolysis of cellulose, the cellulases of fungal origin and especially those of *Trichoderma* spp. have been employed.

3.1.3 Production of Glucose through Enzymatic Hydrolysis of Cellulose

A comparison of the salient features of the acidic and enzymatic processes (Table 8) for hydrolysis of cellulose would reveal that the enzymatic route is preferred, especially due to its specificity, mild reaction conditions and absence of side products. Although none is economical at present, significant breakthroughs, e.g. in terms of improved enzyme properties, are still possible in the enzymatic route. A closer examination of the properties of *Trichoderma* cellulases, which have been used in almost all large-scale hydrolysis investigations, will reveal its drawbacks and consequently the need for having or designing better cellulose hydrolyzing agents.

The enzyme properties that affect the hydrolysis of crystalline cellulose are (a) catalytic efficiency, (b) thermal stability, (c) adsorption, (d) end-product

Table 8
Salient Features of Acid and Enzymatic Catalysis of Cellulosic Materials for Glucose Production

Feature	*Acid catalysis*	*Enzymatic catalysis*
1. Specificity	Nonspecific catalysts hydrolyze cellulose; may also degrade lignin	Highly specific
2. Rate of hydrolysis	High	Low
3. Reaction conditions	Harsh conditions (high temperature and pressure; corrosion-resistant equipment)	Mild conditions (atmospheric pressure, temperature 40–60°C and pH 4·8–7·0 depending on the enzyme source)
4. Effect on DP	Considerable depolymerization without appreciable weight loss	Slight decrease in DP with appreciable weight loss
5. Catalyst requirement to achieve same degree of hydrolysis	Very high compared to enzyme (approx. 10^5 times)	Very much less
6. Cost of catalysts	Not generally high (require catalyst recovery for costly chemicals)	High (approx. 50% of the total cost) (requires recovery and/or recycling)
7. Side reactions	Generally results in unwanted side products (e.g. furfural from hemicellulose)	No side reaction, products of hydrolysis can be directly utilized for subsequent fermentations
8. Yield of glucose	Overall yield is low due to degradation	Overall yield is high primarily due to effective pretreatment step

Sources: Gilbert & Tsao, 1983; Kosaric *et al.*, 1983.

inhibition and (e) shear inactivation. Table 9 shows the quantitative values of the properties of a few endoglucanases from *Trichoderma* and a few other organisms. The catalytic efficiency of endoglucanases on crystalline cellulose is very low, being of the order of 0·051 s^{-1} for the most active component (Table 9). Even against filter paper (a pretreated substrate), which is used for measuring its activity, its catalytic constant k_{cat} is only approximately 1·0–1·2 s^{-1} considering its specific activity of 1·2 IU/mg protein and molecular weight of 50 000 Da. Glucoamylase, on the other hand, has a k_{cat} of 58 s^{-1} with specific activity about 100 times higher (≅100 IU/mg protein). However, the same value is shown by endoglucanase against soluble cellulose CMC. This indicates that catalytic activity of cellulase is low for crystalline cellulose, due to the inherent structural features of cellulose outlined in section 2.1. Table 9 also reveals that *T. reesei* cellulases also suffer from the following disadvantages under actual operating conditions of cellulose hydrolysis: (a) low thermostability which requires replenishment of the lost activity at prolonged usage; (b) high degree of product inhibition which results in cessation of

Table 9
Properties of Endoglucanase from *Trichoderma* and a Few Other Sources

Enzyme source	*Adsorption constant on crystalline cellulose* K_A *(liter/g)*	*Thermal stability at 65°C*[a] $t_{1/2}$*(min)*	*Inhibition constant against cellobiose* K_i *(g/liter)*	*Catalytic efficiency against crystalline cellulose at 40°C*[b] $k_{cat}(s^{-1})$
1. *Trichoderma reesei*	0·02–2·8	14–40	51–83	0·027 & 0·051;[c] 1·6(AC); 58(CMC)
2. *Trichoderma viride* (Meiji Seika)	0·6 & 8·5	20	31	—
3. *Myceliophthora thermophila*	2·7	10 200	35	0·013; 1·4(AC)
4. *Aspergillus wentii*	0·02 & 0·75	150	116	—
5. *Aspergillus* sp. (NOVO)	0·045	65	86	—
6. *Clostridium thermocellum*	0·01–5·5	40–36 000	87–114	—

[a] Thermal stability of enzyme was determined in the absence of substrate.
[b] AC represents amorphous cellulose.
[c] Where more than one value is given, they represent the values exhibited by different multiple forms of the enzyme from the same source or by the enzyme from different strains.
Source: Klyosov, 1989.

hydrolysis at increased concentration of glucose and cellobiose; (c) inactivation by shear; and (d) low adsorption of some components on crystalline cellulose which results in its washing-away in, for example, a flow-through reactor for continuous hydrolysis. The implications of these properties of cellulase on hydrolysis of crystalline cellulose under process conditions have been elaborately discussed by Klyosov (1989). It seems that none of the cellulases discovered so far meet the requirement of these properties for 'biotechnological' applications, though a few of them do have some desirable properties (e.g. thermal stability of cellulases from *Clostridium thermocellum* and *Myceliophthora thermophila*). There is, therefore, a need to develop cellulases from either *Trichoderma* or other organisms which possess these properties. The desirable attributes of the cellulases and those of the organisms producing them are summarized in Table 10.

In spite of the shortcomings of the cellulases available so far, several reports have appeared on their application on the hydrolysis of crystalline and/or pretreated cellulose in reactors of different configurations, with a view to maximizing the yield of sugars, especially glucose. The important parameters in the hydrolysis such as the yield and concentration of sugars, the duration of hydrolysis and the enzyme loading are all interrelated and obviously depend on both the characteristics of cellulosic residue and the enzyme. Yields are higher in dilute systems due to decreased inhibition by sugars. High enzyme loadings can increase the yield and sugar concentration to some extent but will require

Table 10
Desirable Attributes of Cellulases and of the Organisms Producing Them

Attributes	*Comments*
1. Ability to produce complete cellulase system in quantity (hypercellulase producers)	Hyperproducing strains of *Trichoderma reesei* such as Rut C-30, RL-P37 and E-12 are available
2. Cellulase synthesis on soluble sugars (constitutive mutants)	Constitutive strains of *T. reesei* (e.g. C-5) are available. Still considerable scope for improvement
3. Synthesis not subject to catabolite (glucose) repression (catabolite repression resistant mutants)	Catabolite repression resistant mutants such as *T. reesei* D1-6 and Rut C-30 are available
4. High specific activity	Enzymes of high specific activity (e.g. those from RL-P37) can significantly improve the economics of cellulose hydrolysis
5. High catalytic efficiency against crystalline cellulose	All cellulases available so far seem to have very low catalytic efficiency. Not much information available on this important hydrolysis parameter
6. Thermostable at hydrolysis temperature	A few cellulases (e.g. from *Clostridium thermocellum* and *Myceliophthora thermophila*) are very stable
7. High specific adsorption on crystalline cellulose	Cellulases with tighter binding on crystalline cellulose are required (Klyosov *et al.*, 1986)
8. Decreased susceptibility of enzyme to inhibition by cellobiose and glucose (end-product inhibition resistant mutants)	Cellulase (especially the cellobiohydrolase component) is severely inhibited by cellobiose (Berghem *et al.*, 1975). Inhibition constants need to be 2–3 orders of magnitude higher (Klyosov, 1989)
9. Stable against shear forces	Shear resistant cellulases will be useful where agitation must be provided to suspend the solid cellulose in hydrolysis reactor (Reese & Ryu, 1980)

Note: At present, none of the four properties mentioned against 5, 6, 7 and 8 seem to be possessed by any cellulase from a single source.

higher quantities of the costly enzyme (Sattler *et al.*, 1989). It has been found that enzyme loadings of more than, say, 20 FPU/g cellulose do not significantly enhance the rate and extent of cellulose hydrolysis (Mandels *et al.*, 1981). A more recent study on process economics of separate hydrolysis and fermentation for ethanol production showed that higher savings are attained with enzyme loading of approximately 12 FPU/g cellulose (Wright *et al.*, 1986). Addition of supplemental β-glucosidase, preferably in immobilized form, to a cellulase preparation (which is deficient in β-glucosidase) can also significantly enhance the hydrolysis of cellulose (Sternberg *et al.*, 1977; Fadda *et al.*, 1989); the optimum ratio of β-glucosidase to filter paper cellulase is approximately

1·25–1·5:1 (Srivastava, 1985). The addition of enzymes such as xylanase in hydrolyzing lignocellulosic residues at increased rates can also be beneficial (Ghose & Bisaria, 1979). Cooperative action of esterases and xylanases on enhanced degradation of acetyl xylan has also been reported (Biely *et al.*, 1986).

The economics of the hydrolysis process can also be enhanced if the enzymes could be stabilized, recovered and reused. Thermal stability of the cellulase and β-glucosidase has been reported to increase by treatment of cellulase at pH 10·0 and by chemical modification of β-glucosidase with the cross-linking reagent glutaraldehyde (Woodward *et al.*, 1981). Cellulase can be recovered by employing an ultrafiltration membrane system and then reused but the system seems to be impractical due to high costs (Ghose & Kostick, 1970; Sundstrom *et al.*, 1981; Tan *et al.*, 1987; Tanaka *et al.*, 1988*b*). The most promising system for recovery of the enzyme seems to be through adsorption of cellulase on cellulose due to its high affinity for the substrate (Wilke *et al.*, 1976; Bisaria & Ghose, 1978; Klyosov, 1986). A novel approach has recently been developed by Taniguchi *et al.* (1989) which makes use of a soluble immobilized preparation of cellulase for cellulose hydrolysis. The immobilized preparation, which was made by using enteric coating polymer Eudragit L, can exist in both soluble and insoluble forms, depending on the pH of the medium, and hence can be recovered easily by simply changing the pH and reused to hydrolyze fresh cellulosic feed.

Several types of reactor configurations have been used for enzymatic hydrolysis of cellulose. These include batch and continuous reactors, the fluidized bed reactor using immobilized cellulase, and the ultrafiltration membrane reactor (Ladisch *et al.*, 1983; Gusakov *et al.*, 1987). The use of the attrition bioreactor in which simultaneous wet milling and enzymatic hydrolysis take place in the same vessel has been found to enhance the hydrolysis rate compared to that achieved in the normal stirred batch reactor (Deeble & Lee, 1986). A novel approach to enzymatic hydrolysis has been reported by Tjerneld *et al.* (1986) who combined an aqueous two-phase system and ultrafiltration for enzyme recycling during a semicontinuous cellulose hydrolysis process. These authors achieved a sugar concentration of 7·4% using 8·0% cellulose slurry with no intermittent additions of cellulase for a period of 50 days. This seems to be the longest time reported for any cellulose hydrolysis process using only the initially added enzymes, indicating the high stability of the cellulases in aqueous two-phase systems. For the product concentration the highest reported value seems to be by Chen and Gong (1982) who obtained a solution containing 19% glucose and 1% cellobiose from hydrolysis of 20% cellulose within 72 h. It must be emphasized, however, that almost all studies which have reported a conversion of, say, more than 80% have employed pretreated cellulose of one type or the other. Although several reactor configurations have been proposed, none of them could be discarded as being inappropriate, and the ultimate reactor design will be governed by the characteristics of both the enzyme and the substrate.

3.2 Ethanol

There are essentially three major routes for the production of ethanol from cellulosic residues using biochemical means. In the conventional two-step process, cellulose is enzymatically hydrolyzed to glucose which can subsequently be fermented to ethanol using either the yeasts (*Saccharomyces*) or bacteria (*Zymomonas*). The technology for fermentation of glucose into ethanol by the yeasts is well developed and can be used for cellulose hydrolyzates as long as inhibitory agents are not present in the hydrolyzate.

The single-step processes for conversion of cellulose into ethanol can be of two types depending on the use of enzyme or the bacteria. In the process employing cellulase enzymes, the saccharification of cellulose to glucose and subsequent fermentation of glucose to ethanol by yeast or bacteria take place in the same vessel and the process is known as the simultaneous saccharification and fermentation (SSF) process. The processes employing bacteria make use of either a single cellulolytic ethanologen (monoculture fermentation) or two organisms, a cellulolytic organism and an ethanologen (co-culture fermentation). It must be mentioned that pretreated cellulose has normally been used in all these processes due to the resistance of native cellulose to enzymatic or microbial attack (see section 2.1).

Several reports have appeared on the SSF process which utilizes the cellulase enzyme (primarily from *Trichoderma* sp.) and yeast or bacterial species for bioconversion of cellulose into ethanol. In view of the properties of cellulases (section 2.2), the following advantages of the SSF process are clear (Takagi *et al.*, 1978; Pemberton *et al.*, 1980; Ghose *et al.*, 1984; Szczodrak, 1989):

(1) Enhanced rate of cellulose hydrolysis due to removal of sugars which inhibit cellulase activity.
(2) Lower enzyme loadings.
(3) Higher product yields.
(4) Limitation of initial cellulose concentration can also limit the concentration of ethanol to a level which is not inhibitory to the yeast or the cellulase.
(5) Decreased requirement of aseptic conditions.
(6) Considerable reduction in process time.
(7) Use of a single reactor.

However, in spite of these advantages, the SSF process must fulfill the following requirements to produce ethanol at competitive prices (Ghose & Tyagi, 1979; Beja da Costa & Van Uden, 1980; Bailey *et al.*, 1982):

(1) Compatible conditions of saccharification and fermentation.
(2) High ethanol yield and rate of production.
(3) High ethanol tolerance of the culture.
(4) Minimal inhibition of the enzyme by ethanol.
(5) Optimal level of β-glucosidase in cellulase to minimize accumulation of cellobiose.

(6) Resistance of the culture to the lytic effects of cellulase.
(7) Resistance of cellulase to proteolytic enzymes of the ethanologen.

A number of studies on SSF have been reported in which the yeasts, *Candida brassicae, Saccharomyces cerevisiae* and *S. carlsbergensis* have been widely used to produce ethanol from cellulose hydrolysis mixture (Takagi *et al.*, 1978; Pemberton *et al.*, 1980; Blanco *et al.*, 1982; Ghosh *et al.*, 1982*b*; Saddler *et al.*, 1982*b*). The bacterium *Zymomonas mobilis* and a few other yeasts have also been studied in the SSF process for ethanol production (Viikari *et al.*, 1981; Saddler *et al.*, 1983; Ghose *et al.*, 1984; Spangler & Emert, 1986). Kinetic studies conducted on the ethanol fermentation by *Z. mobilis* and *S. carlsbergensis* indicate the superiority of *Z. mobilis*. This includes specific growth rate (2·4 times higher), specific ethanol production rate (2·9 times higher) and specific glucose uptake rate (2·6 times higher). Further, whereas the ethanol yield of the bacterium is slightly higher, its ethanol tolerance is lower than that of the yeast (Rogers *et al.*, 1980; Esser & Karsch, 1984; Sahm, 1989). It has been reported that for continuous cell recycle systems operating at the same glucose input concentration of 100 g/liter, the bacterial ethanol productivity (120 g/liter h) was 4·1 times that of the yeast and that the highest productivity achieved by a yeast system was only 68% of that obtained with *Z. mobilis* (Rogers *et al.*, 1980). On the other hand, Khorakiwala *et al.* (1986) found the yeast *S. cerevisiae* to be the preferred organism to *Z. mobilis* for ethanol production in a membrane recycle bioreactor. With 100 g/liter glucose feed the yeast resulted in ethanol productivity of 100 g/liter h with essentially complete glucose utilization, whereas *Z. mobilis* gave ethanol productivity of only 45–55 g/liter h. Various approaches have been used to increase the volumetric productivity of ethanol. These approaches include techniques of keeping cells in the bioreactor by using immobilization and recycle systems (Guidoboni, 1984; Kolot, 1984; Khorakiwala *et al.*, 1986; Kanellaki *et al.*, 1989). Other approaches employed were directed at keeping the ethanol concentration low in the reactor by removing it through solvent extraction or an enriched vapor stream (Wilke *et al.*, 1983; Ghose *et al.*, 1984; Mairoella *et al.*, 1984; Crabbe *et al.*, 1986; Mulder & Smolders, 1986; L'Italien *et al.*, 1989).

As mentioned above, one of the most important requirements for a successful SSF process is the compatibility of the saccharification and fermentation system with respect to temperature, pH and substrate concentration. Since the optimum temperature of *T. reesei* cellulase activity is 50°C while that for anaerobic fermentation using *S. cerevisiae* or *Z. mobilis* is approximately 30°C, a compromise between these two temperatures must be made. Since the cellulase activity is still quite high at 40–45°C, the use of thermotolerant yeasts or bacteria would have obvious advantages in an SSF process. A few of the potential advantages and disadvantages associated with the use of thermotolerant/thermophilic ethanologens are summarized in Table 11. Isolation and use of a few thermotolerant yeasts which have been used along with cellulase in an SSF process at 40–45°C have been reported (Ghose *et al.*, 1984;

Table 11
Potential Advantages and Disadvantages of Thermotolerant/Thermophilic Ethanologens

Advantages
—High catabolic activity at optimum growth temperatures leading to shorter fermentation time and higher productivity
—Ability to ferment wide range of substrates
—Easier maintenance of long-term anaerobic conditions due to low solubility of oxygen at higher temperatures
—Greater solubility of substrates at higher temperatures
—Lower energy requirements to agitate the fermentation broth due to decreased viscosity at higher temperatures
—Easier recovery of ethanol using, for example, continuous ethanol removal processes such as Vacuferm
—Wide pH and temperature optima, in general
—Less chances of contamination (especially by pathogens)
—Lower energy inputs to heat fermentors (if necessary) compared to their cooling

Disadvantages
—Lower ethanol tolerance resulting in low ethanol concentration
—Lower substrate tolerance
—Formation of by-products such as acetic acid
—Extensive evaporation of fermentation medium
—For aerobic thermophiles, oxygen avilability is low
—Greater stress on hardware (e.g. probes)
—Limited fundamental knowledge of physiology and biochemistry

Sources: Wiegel, 1980; Kosaric *et al.*, 1983; Slapack *et al.*, 1987.

Szczodrak & Targonski, 1988; Spindler *et al.*, 1989). Szczodrak and Targonski (1988) screened 58 yeast strains for their ability to produce ethanol, and reported that their best isolate, *Fabospora fragilis* CCY 51-1, was able to accumulate 56 g/liter ethanol from 140 g/liter glucose at 43°C in less than 48 h. Using a cellobiose fermenting yeast, *Candida lusitaniae* Y-5394, and *Saccharomyces uvarum* in a mixed culture SSF process, Spindler *et al.* (1989) reported better performance of the mixed culture than that of either single constituent culture, as it combined the cellobiose fermenting capability with the high ethanol tolerance and rapid glucose fermentation of conventional *Saccharomyces* species. However, the fermentation took a long time: 8 days to accumulate 40 g/liter ethanol in the mixed culture SSF process.

If the cellulase preparation is deficient in β-glucosidase and a non-cellobiose fermenting yeast is used, it will result in accumulation of cellobiose which is a strong inhibitor of cellulase enzyme (Sternberg *et al.*, 1977). This can be overcome by supplementation of cellulase with higher β-glucosidase activity enzyme from another source such as *Aspergillus niger* (Szczodrak, 1988) or by using cellulase from β-glucosidase hyperproducing mutant of *T. reesei* (Szczodrak, 1989). When a cellulase preparation from a β-glucosidase hyperproducing mutant of *T. reesei* was employed with *Kluyveromyces fragilis* in an

SSF process with 10% pretreated wheat straw, the ethanol concentration increased from 2·5% (w/v) to 3·4% (w/v) in 24 h instead of 48 h (when the parent *T. reesei* cellulase was used) (Szczodrak, 1989). The highest productivity of ethanol in an SSF process using a thermotolerant yeast seems to have been reported by Ghose et al. (1984). Using cellulase from *T. reesei*, *β*-glucosidase from *Aspergillus wentii* and *Candida acidothermophilum* at 40°C, and operating the SSF process under vacuum recycling with intermittent substrate feeding, cellulose utilization of 98% and ethanol productivity of 4·5 g/liter h was achieved (Ghose *et al.*, 1984).

Cellulose can also be converted directly to ethanol by using a cellulolytic ethanologen such as *Clostridium thermocellum* (Ng *et al.*, 1977; Brener & Johnson, 1984; Saddler & Chan, 1984). This bacterium has a low tolerance to fermentation products; the inhibitory effects of ethanol and organic acids on cell viability, growth and cellulase activity have been reported (Herrero, 1983; Kundu *et al.*, 1983). As a result *C. thermocellum* does not produce more than 0·5% ethanol from cellulose. However, since this bacterium is the only one amongst many thermophiles which produces a true cellulase system, and has many advantages in fermentation processes (Table 11), a large body of scientific literature has been devoted to various aspects of this thermophile such as characterization of the cellulase system, genetics and molecular cloning of its various cellulase genes (Beguin *et al.*, 1988; Lamed & Bayer, 1988; Demain & Wu, 1989).

Although *C. thermocellum* breaks down cellulose and hemicellulose to glucose, cellobiose and xylose, it is unable to utilize xylose. Furthermore, cellobiose and glucose also inhibit its cellulase activity, resulting in low ethanol yields. Thus, by co-culture of these highly cellulolytic bacteria with ethanol-producing bacteria capable of fermenting xylose, cellobiose and glucose, the advantages of the SSF process can be realized and ethanol can be produced in higher yields. Still, this single-step mixed culture process employing cellulolytic and ethanologenic bacteria for conversion of cellulose to ethanol suffers from two major drawbacks: (a) ethanol toxicity due to low ethanol tolerance, and (b) the rate-limiting step of cellulose hydrolysis. Some of the co-cultures employed for direct conversion of cellulose to ethanol include *C. thermocellum* with *C. thermohydrosulfuricum* (Ng *et al.*, 1981; Zeikus *et al.*, 1981; Ng & Zeikus, 1982), *C. thermocellum* with *C. thermosaccharolyticum* (Avgerinos *et al.*, 1981; Wang *et al.*, 1983), *C. thermocellum* and *Thermoanaerobacter ethanolicus* (Ljungdahl & Wiegel, 1981; Ljungdahl *et al.*, 1981) and *C. thermocellum* with *Z. mobilis* or *Z. anaerobia* (Saddler *et al.*, 1981; Saddler & Chan, 1982). Attempts to increase the ethanol tolerance of these bacteria have resulted in the isolation of a few mutants (Avgerinos *et al.*, 1981; Carreira & Ljungdahl, 1983; Tailliez *et al.*, 1989). Most of them are, however, still unable to produce more than approximately 3% (w/v) in co-culture processes using cellulosic substrates (Avgerinos *et al.*, 1981). The physiological aspects of thermophilic bacteria and ethanol tolerance have been elaborately discussed in a recent book by Stewart and his co-workers (Slapack *et al.*, 1987).

Notwithstanding the disadvantages associated with thermophilic bacteria (such as ethanol tolerance and stability of high-ethanol-yielding strains), a recent study (Lynd, 1989) on economic evaluation of thermophilic bacteria for ethanol production shows that they might compete with the available yeast technology if more efficient technologies to recover ethanol from dilute solutions are applied.

3.3 Other Chemicals

Most of the studies on solvent production through bioconversion of lignocellulosic residues have concentrated on ethanol. However, solvents such as acetone and butanol can also be produced from a variety of hexose and pentose sugars present in the hydrolyzates of lignocellulosic substrates. Like ethanol, these solvents and 2,3-butanediol find numerous applications in chemical syntheses.

3.3.1 Acetone–Butanol

This fermentative pathway is much more complicated than that of ethanol and is a branched one. Depending on energy status of the cell and the fermentation conditions such as pH and substrate level, the cells produce either the solvents or the acids. It has been shown (Monot *et al.*, 1982; Votruba *et al.*, 1986) that at low pH (*c.* 4·5) and high glucose concentration, the main products are acetone and butanol, whereas only acids are produced at high pH and low glucose concentration by the strict anaerobe *Clostridium acetobutylicum*. The concentration of resulting solvents (acetone, butanol and ethanol) normally does not exceed 18 g/liter due to the inhibitory effect of butanol, although improvement of strains through recombinant DNA technology has resulted in accumulation of solvents up to 23 g/liter (Parisi, 1989). Approaches to increase the solvent concentration and productivity have included the use of continuous and immobilized reactors, nutrient cycling, liquid–liquid extraction, adsorption, and hollow-fiber fermentor–extractor (Dadgar & Foutch, 1986; Soni *et al.*, 1986*a,b*; Chakrabarti, 1989; Park *et al.*, 1989; Shukla *et al.*, 1989). The use of an immobilized cell trickle bed reactor resulted in butanol and acetone concentrations of 8·82 and 5·22 g/liter with butanol and total solvents yields of 19·4 and 34·1% respectively from 60 g/liter initial glucose concentration. The solvent productivity, which seems to be the highest reported so far, was 4·2 g/liter h, 10 times higer than that obtained in batch fermentation using free cells and 2·76 times higher than that of an immobilized CSTR (Park *et al.*, 1989). The potential of *C. acetobutylicum* and *Klebsiella pneumoniae* for converting biomass substrates into acetone, butanol, ethanol and 2,3-butanediol has been investigated by Yu and Saddler (1987). Using a combined hydrolysis and fermentation approach, the production of butanol and acetone was 3·4 and 1·0 g/liter respectively from 50 g/liter xylan, and 6·5 and 3·1 g/liter respectively from 50 g/liter Solka floc(TM) cellulose (Yu & Saddler, 1987). The low butanol yields from xylan were attributed to lower efficiency of xylose utilization. Chakrabarti (1989) made use of immobilized cells of *Clostridium* *saccharobutylacetonicum* in a continuous reactor and obtained butanol and acetone concentrations of 5·45 and 4·15 g/liter respectively and solvent productivity of 2·33 g/liter h by using 40 g/liter reducing sugar concentration in cellulose hydrolyzate. The physiological consequences of end-product inhibition by butanol and various possible approaches to overcoming this inhibition have recently been discussed by Lindley and Durand (1989).

3.3.2 2,3-Butanediol

2,3-Butanediol is also an industrially important chemical that can be produced from a variety of carbohydrates such as glucose, xylose and disaccharides present in the hydrolyzates of agro-residues. The bacterial species that have been used recently include *Klebsiella oxytoca, K. pneumoniae* and *Bacillus polymyxa.* In addition to 2,3-butanediol, most species also produce ethanol, acetate, acetoin, lactate, etc. The most important variable affecting the kinetics of its production appears to be oxygen transfer rate (Mas *et al.*, 1988; Nigam, 1989). A higher oxygen availability favors the formation of biomass at the expense of butanediol. On the other hand, decreased oxygen availability increases butanediol yield but results in decreased overall conversion rate due to lower cell concentration. Butanediol also strongly inhibits growth at higher concentrations but does not particularly affect the specific rate of butanediol production (Sablayrolles & Goma, 1984). The use of immobilized cells of *K. oxytoca* in a continuous reactor was found to result in low product concentration and productivity, primarily due to limited availability of oxygen to the cells (Nigam, 1989). Best results were obtained in a fed-batch culture system by maintaining the glucose substrate concentration below the inhibition level of 50 g/liter and by regulating the oxygen supply rate at 11·2 mmol/liter h. Under these conditions, product concentration of approximately 80 g/liter and productivity of 2·08 g/liter h have been achieved (Nigam, 1989). On xylose, however, butanediol production in excess of 11 g/liter has been reported for *K. pneumoniae* (Yu & Saddler, 1983). When the water-soluble hemicellulose fraction of steam-treated aspen wood was used as a substrate, both growth and butanediol production were severely retarded in *K. pneumoniae* due to the presence of several inhibitors such as furfural (Nishikawa *et al.*, 1988). The toxic effects of possible chemicals present in pretreated agro-residues should, therefore, be considered before using them for butanediol production. The possibility of utilizing lignocellulosic residues for butanediol production has also been studied by Yu and Saddler (1987). Using xylan and Solka floc™ at 100 g/liter level, butanediol concentrations of 18·5 g/liter and 22·6 g/liter have been reported in a combined hydrolysis and fermentation approach (Yu & Saddler, 1987).

4 FUTURE PROSPECTS

Agro-residues are a potential resource for bioconversion into glucose and a number of chemicals. However, due to the crystalline nature of cellulose in the residues, they have to be made susceptible to enzymatic action by subjecting them to some form of pretreatment. At present, organosolv, steam treatment and dilute acid prehydrolysis pretreatment processes appear to be promising for increasing the susceptibility. Despite the potential of these methods, there is rather a limited understanding as to how these pretreatments result in enhanced enzymatic hydrolysis. A more in-depth understanding of the phenomena is needed to devise technically feasible and cost-effective pretreatment methods.

A number of mutants exist which secrete a fairly large amount of cellulase, but none of the them possesses the properties that are optimal under actual process conditions of cellulose hydrolysis. Isolation of cellulases from new sources or developing them through modern techniques of recombinant DNA and protein engineering that can function efficiently in hydrolyzing crystalline cellulose would be a worthwhile approach in this direction.

For efficient bioconversion of cellulose into ethanol, the simultaneous saccharification and fermentation process needs to be improved by isolation of thermotolerant yeasts or bacteria (along with ethanol tolerance), so that compatibility problems could be minimized. The potential of thermotolerant bacteria in lignocellulose-based bioconversion processes, in view of their ability to utilize both the cellulose and hemicellulose fractions, is large. More fundamental studies concerning their physiology are required so that their ethanol tolerance can be enhanced and stable high-ethanol-yielding mutants can be obtained. There is also a dearth of process-oriented studies which include development of different types of reactors and removal of ethanol from dilute fermentation broths of thermophilic bacteria.

Advances in these and related areas of both scientific and engineering aspects of lignocellulose bioconversion are expected to result in realization of the fuller potential of the microbes and the enzymes in various biotechnological applications.

REFERENCES

Allen, A. L. & Andreotti, R. E. (1982). In *Fourth Symp. on Biotechnology in Energy Production and Conservation, Biotechnol. Bioeng. Symp. No. 12,* ed. C. D. Scott. John Wiley, New York, p. 451.

Allen, A. L. & Roche, C. D. (1989). *Biotechnol. Bioeng.*, **33,** 650.

Andreotti, R. E., Medeiros, J. E., Roche, C. & Mandels, M. (1981). In *Proc. Symp. Bioconversion and Biochemical Engineering,* Vol. 1, ed. T. K. Ghose. Indian Institute of Technology, New Delhi, p. 353.

Avgerinos, G. C., Fang, H.-Y., Biocic, I. & Wang, D. I. C. (1981). In *Advances in Biotechnology,* Vol. 2, eds M. Moo-Young & C. W. Robinson, Pergamon Press, Toronto, p. 119.

Bailey, R. B., Benitez, T. & Woodward, A. (1982). *Appl. Environ. Microbiol.*, **44,** 631.

Beguin, P. *et al.* (1988). In *Proc. Symp. Biochemistry and Genetics of Cellulose Degradation,* eds J.-P. Aubert, P. Beguin & J. Millet. Academic Press, New York, p. 267.

Beja da Costa, M. & Van Uden, N. (1980). *Biotechnol. Bioeng.*, **22,** 2429.
Ben Ghedalia, D. & Miron, J. (1981). *Biotechnol. Bioeng.*, **23,** 823.
Berghem, L. E. R., Pettersson, L. G. & Axio-Fredriksson, U.-B. (1975). *Eur. J. Biochem.*, **53,** 55.
Biely, P., MacKenzie, C. R., Puls, J. & Schneider, H. (1986). *Bio/Technology*, **4,** 731.
Binder, A., Haltmeir, T. & Fiechter, A. (1980). In *Proc. Symp. Bioconversion and Biochemical Engineering*, Vol. 1, ed. T. K. Ghose. Indian Institute of Technology, New Delhi, p. 315.
Bisaria, V. S. & Ghose, T. K. (1978). In *Proc. Symp. Bioconversion of Cellulosic Substances into Energy, Chemicals and Microbial Protein*, ed. T. K. Ghose. Indian Institute of Technology, New Delhi, p. 155.
Bisaria, V. S. & Ghose, T. K. (1981). *Enzyme Microbial Technol.*, **3,** 90.
Bisaria, V. S. & Mishra, S. (1989). *Crit. Rev. Biotechnol.*, **9,** 61.
Bisaria, R., Madan, M. & Bisaria, V. S. (1987*a*). *Biol. Wastes*, **19,** 239.
Bisaria, R., Gujral, G. S. & Bisaria, V. S. (1987*b*). *Crit. Rev. Biotechnol.*, **7,** 17.
Blanco, S., Gamarra, A., Cuevas, C. & Ellenrieder, G. (1982). *Biotechnol. Lett.*, **4,** 661.
Brener, D. & Johnson, B. F. (1984). *Appl. Environ. Microbiol.*, **47,** 1126.
Brown, J. A., Falconer, D. J. & Wood, T. M. (1987). *Enzyme Microbial Technol.*, **9,** 169.
Brownell, H. H. & Saddler, J. N. (1986). *Biotechnol. Bioeng.*, **28,** 792.
Brownell, H. H. & Saddler, J. N. (1987). *Biotechnol. Bioeng.*, **29,** 228.
Bruchmann, E.-E., Schach, H. & Graf, H. (1987). *Biotechnol. Appl. Biochem.*, **9,** 146.
Buchanen, R. A., Otey, F. H. & Hamerstrand, G. E. (1980). *Ind. Engg Chem. Prod. Res. Div.*, **19,** 489.
Carreira, L. H. & Ljungdahl, L. G. (1983). In *Liquid Fuel Developments*, ed. D. L. Wise, CRC Press, Boca Raton, Florida, p. 1.
Castanon, M. & Wilke, C. R. (1980). *Biotechnol. Bioeng.*, **22,** 1037.
Chakrabarti, S. (1989). Continuous bioconversion of carbohydrate feedstocks to acetone-butanol using immobilized cells. PhD thesis, Indian Institute of Technology, New Delhi.
Chang, M. (1971). *J. Polym. Sci., Part C*, No. 36, 343.
Chanzy, H., Henrissat, B., Voung, R. & Schulein, M. (1983). *FEBS Lett.*, **153,** 113.
Chen, L.-F. & Gong, C.-S. (1982). In *Fourth Symp. on Biotechnology in Energy Production and Conservation, Biotechnol. Bioeng. Symp. No.* 12, ed. C. D. Scott. John Wiley, New York, p. 57.
Chippaux, M. (1988). In *Proc. Symp. Biochemistry and Genetics of Cellulose Degradation*, eds J.-P. Aubert, P. Beguin & J. Millet. Academic Press, New York, p. 219.
Chou, Y. C. (1986). In *Eighth Symp. on Biotechnology for Fuels and Chemicals, Biotechnol. Bioeng. Symp. No. 17*, ed. C. D. Scott. John Wiley, New York, p. 19.
Chum, H. L. *et al.* (1988). *Biotechnol. Bioeng.*, **31,** 643.
Clark, T. A. & Mackie, K. L. (1986). In *Biotechnology in the Pulp and Paper Industry, 3rd Int. Conf.*, Stockholm, 16–19 June, p. 101.
Coudray, M. R., Canavascini, G. & Meier, H. (1982). *Biochem. J.*, **203,** 227.
Coughlan, M. P. & Ljungdahl, L. G. (1988). In *Proc. Symp. Biochemistry and Genetics of Cellulose Degradation*, eds J.-P. Aubert, P. Beguin & J. Millet. Academic Press, London, p. 11.
Cowling, E. B. (1975). In *Biotechnology and Bioengineering Symp. No. 5, Cellulose as a Chemical and Energy Resource*, ed. C. R. Wilke. John Wiley, New York, p. 163.
Crabbe, P. G., Tse, C. W. & Munro, P. A. (1986). *Biotechnol. Bioeng.*, **28,** 939.
Crawford, R. L. (1981). *Lignin Biodegradation and Transformation.* Wiley Interscience, New York.
Dadgar, A. M. & Foutch, G. L. (1986). In *Seventh Symp. on Biotechnology for Fuels*

and Chemicals, Biotechnol. Bioeng. Symp. No. 15, ed. C. D. Scott. John Wiley, New York, p. 611.
Dale, B. E. (1987). *Trends in Biotechnology,* **5,** 287.
Dale, B. E. & Moreira, M. J. (1982). In *Fourth Symp. on Biotechnology in Energy Production and Conservation, Biotechnol. Bioeng. Symp. No. 12,* ed. C. D. Scott. John Wiley, New York, p. 31.
Deeble, M. F. & Lee, J. M. (1986). In *Seventh Symp. on Biotechnology for Fuels and Chemicals, Biotechnol. Bioeng. Symp. No. 15,* ed. C. D. Scott. John Wiley, New York, p. 277.
Dekker, R. F. H. (1980). *J. Gen. Microbiol.,* **120,** 309.
Dekker, R. F. H. (1985). In *Biosynthesis and Biodegradation of Wood Components,* ed. T. Higuchi. Academic Press, New York, p. 505.
Dekker, R. F. H. & Wallis, A. F. A. (1983). *Biotechnol. Bioeng.,* **25,** 3027.
Demain, A. L. & Wu, J. H. D. (1989). In *Bioprocess Engineering: the First Generation,* ed. T. K. Ghose. Ellis Horwood, Chichester, UK, p. 68.
Deshpande, M. V. & Eriksson, K. E. (1984). *Enzyme Microbial Technol.,* **6,** 338.
Doppelbauer, R., Esterbauer, H., Steiner, W., Lafferty, R. M., & Steinmuller, H., (1987). *Appl. Microbiol. Biotechnol.,* **26,** 485.
Duff, S. J. B., Cooper, D. G. & Fuller, O. M. (1987). *Enzyme Microbial Technol.,* **9,** 47.
Durand, H., Baron, M., Calmels, T. and Tiraby, G. (1988). In *Proc. FEMS Symp. Biochemistry and Genetics of Cellulose Degradation,* eds J.-P. Aubert, P. Beguin & J. Millet. Academic Press, London, p. 135.
Enari, T. N. & Niku-Paavola, M. L. (1987). *Crit. Rev. Biotechnol.,* **5,** 67.
Eriksson, K.-E. (1981). In *Trends in the Biology of Fermentations for Fuels and Chemicals,* ed. A. Hollaender. Plenum Press, New York, p. 19.
Eriksson, K.-E. & Wood, T. M. (1985). In *Biosynthesis and Biodegradation of Wood Components,* ed. T. Higuchi. Academic Press, New York, p. 469.
Eriksson, K.-E., Gruenewald, A. & Vallender, L. (1980). *Biotechnol. Bioeng.,* **22,** 363.
Esser, K. & Karsch, T. (1984). *Process Biochem.,* **19,** 116.
Eveleigh, D. E. (1981). In *Alcohol Fuels Process R/O Newsletter (Summer),* Solar Energy Research Institute, Colorado, p. 50.
Fadda, M. B., Dessi, M. R., Rinaldi, A. & Satta, G. (1989). *Biotechnol. Bioeng.,* **33,** 777.
Fan, L. T., Lee, Y.-H. & Beardmore, D. H. (1980). In *Adv. Biochem. Eng., vol. 14,* ed. A. Fiechter. Springer-Verlag, Berlin, p. 101.
Fan, L. T., Lee, Y.-H. & Beardmore, D. H. (1981*a*). In *Proc. Symp. Bioconversion & Biochemical Engineering,* Vol. 1, ed. T. K. Ghose, Indian Institute of Technology, New Delhi, p. 233.
Fan, L. T., Gharpuray, M. M. & Lee, Y.-H. (1981*b*). In *Third Symp. on Biotechnology in Energy Production and Conservation, Biotechnol. Bioeng. Symp. No. 11,* ed. C. D. Scott. John Wiley, New York, p. 29.
Fengel, D. (1971). *J. Polym. Sci. Part C,* No. 36, 383.
Fengel, D. & Wegner, G. (1984). *Wood: Chemistry, Ultrastructure and Reactions.* Walter de Gruyter, New York.
Gallo, B. J. (1982). Paper presented at National Meeting of American Institute of Chemical Engineers, Orlando, Florida, 28 February–3 March.
Ghangas, G. S. & Wilson, D. B. (1987). *Appl. Environ. Microbiol.,* **53,** 1470.
Ghose, T. K. (1969). *Biotechnol. Bioeng.,* **11,** 239.
Ghose, T. K. (1987). *Pure Appl. Chem.,* **59,** 257.
Ghose, T. K. & Bisaria, V. S. (1979). *Biotechnol. Bioeng.,* **21,** 131.
Ghose, T. K. & Kostick, J. A. (1970). *Biotechnol. Bioeng.,* **12,** 921.
Ghose, T. K. & Pathak, A. N. (1973). *Process Biochem.,* **8(5),** 20.
Ghose, T. K. & Tyagi, R. D. (1979). *Biotechnol. Bioeng.,* **21,** 1387.

Ghose, T. K., Roychoudhury, P. K. & Ghosh, P. (1984). *Biotechnol. Bioeng.*, **26,** 377.
Ghosh, V. K., Ghose, T. K. & Gopalkrishnan, K. S. (1982*a*). *Biotechnol. Bioeng.*, **24,** 241.
Ghosh, P., Pamment, N. B. & Martin, W. R. B. (1982*b*). *Enzyme Microbiol. Technol.*, **4,** 425.
Gilbert, I. G. & Tsao, G. T. (1983). In *Annual Reports on Fermentation Processes,* Vol. 6, ed. G. T. Tsao. Academic Press, New York, p. 323.
Giuliano, C. & Khan, A. W. (1984). *Appl. Environ. Microbiol.*, **48,** 446.
Gong, C. S., Ladisch, M. R. & Tsao, G. T. (1979). In *Hydrolysis of Cellulose: Mechanisms of Enzymatic and Acid Catalysis, Adv. Chem. Ser.* **181,** eds R. D. Brown, Jr & L. Jurasek. American Chemical Society, Washington, DC, p. 261.
Gonzalez, G., Caminal, G. & Lopez-Santin, J. (1986). In *Eighth Symp. on Biotechnology for Fuels and Chemicals, Biotechnol. Bioeng. Symp. No. 17,* ed. C. D. Scott. John Wiley, New York, p. 535.
Gould, J. M. (1984). *Biotechnol. Bioeng.*, **26,** 46.
Grethlein, H. E. (1978). *Biotechnol. Bioeng.*, **20,** 503.
Gritzali, M. & Brown, R. D. (1979). In *Hydrolysis of Cellulose: Mechanisms of Enzymatic and Acid Catalysis, Adv. Chem. Ser.* **181,** eds R. D. Brown, Jr & L. Jurasek, American Chemical Society, Washington, DC, p. 237.
Grohmann, K., Torget, R. & Himmel, M. (1986). In *Eighth Symp. on Biotechnology for Fuels and Chemicals, Biotechnol. Bioeng. Symp. No. 17,* ed. C. D. Scott. John Wiley, New York, p. 135.
Guidoboni, G. E. (1984). *Enzyme Microbial Technol.*, **6,** 194.
Gusakov, A. V., Sinitsyn, A. P. & Klyosov, A. A. (1987). *Biotechnol. Bioeng.*, **29,** 898.
Hamilton, T. J., Dale, B. E., Ladisch, M. R. & Tsao, G. T. (1984). *Biotechnol. Bioeng.*, **26,** 781.
Han, Y. W. & Ciegler, A. (1982). In *Fourth Symp. on Biotechnology in Energy Production and Conservation, Biotechnol. Symp. No.* 12, ed. C. D. Scott. John Wiley, New York, p. 73.
Han, Y. W. *et al. Biotechnol. Bioeng.*, **23,** 2525.
Hatakka, A. I. (1983). *Eur. J. Appl. Microbiol. Biotechnol.*, **18,** 350.
Hawley, M. C., Downey, K. W., Selke, S. M. & Lamport, D. T. A. (1986). In *Cellulose: Structure, Modification and Hydrolysis,* eds R. A. Young & R. M. Rowell. John Wiley, New York, p. 297.
Hendy, N., Wilke, C. R. & Blanch, H. (1984). *Enzyme Microbial Technol.*, **6,** 73.
Herrero, A. A. (1983). *Trends in Biotechnology,* **1,** 49.
Hiltner, P. & Dehority, B. A. (1983). *Appl. Environ. Microbiol.*, **46,** 642.
Holtzapple, M. T. & Humphrey, A. E. (1984). *Biotechnol. Bioeng.*, **26,** 670.
Isaacs, S. H. (1984). *Ethanol Production by Enzymatic Hydrolysis, Parametric Analysis of a Base Case Process.* SERI/TR-231/2093, Solar Energy Research Institute, Golden, Colorado.
Jurasek, L., Paice, M. G., Yaguchi, M. & O'Leary, S. (1987). In *Biomass Conversion Technology: Principles and Practice,* ed. M. Moo-Young. Pergamon Press, New York, p. 131.
Kanellaki, M. *et al.* (1989). *Biotechnol. Bioeng.*, **34,** 121.
Kaushik, V. & Bisaria, V. S. (1989). *J. Sci. Ind. Res.*, **48,** 276.
Khorakiwala, K. H., Cheryan, M. & Mehaia, M. A. (1986). In *Seventh Symp. on Biotechnology for Fuels and Chemicals, Biotechnol. Bioeng. Symp. No. 15,* ed. C. D. Scott. John Wiley, New York, p. 249.
Kirk, T. K. & Farrell, R. L. (1987). *Ann. Rev. Microbiol.*, **41,** 465.
Kirk, T. K., Higuchi, T. & Chang, H. M. (1980). *Lignin Biodegradation: Microbiology, Chemistry and Potential Applications,* Vols 1 and 2. CRC Press, Boca Raton, Florida.
Klyosov, A. A. (1986). *Appl. Biochem. Biotechnol.*, **12,** 249.

Klyosov, A. A. (1989). In *Bioprocess Engineering: the First Generation,* ed. T. K. Ghose, Ellis Horwood, Chichester, UK, p. 59.
Klyosov, A. A., Mitkevich, O. V. & Sinitsyn, A. P. (1986). *Biochemistry,* **25,** 540.
Knowles, J., Lehtovarra, P. & Terri, T. (1987). *Trends in Biotechnology,* **5,** 255.
Kolot, F. B. (1984). *Process Biochem.,* **19,** 7.
Kosaric, N., Wieczorek, A., Cosentino, G. P., Magee, R. J. & Prenosil, J. E. (1983). In *Biotechnology,* Vol. 3, ed. H. Dellweg. Verlag-Chemie, Weinheim, Germany, p. 257.
Kundu, S., Ghose, T. K. & Mukhopadhyay, S. N. (1983). *Biotechnol. Bioeng.,* **25,** 1109.
Kyriacou, A., Neufeld, R. & MacKenzie, C. R. (1989). *Biotechnol. Bioeng.,* **33,** 631.
Lachke, A. & Deshpande, M. V. (1988). *FEMS Microbiol. Rev.,* **54,** 177.
Ladisch, M. R., Lin, K. W., Voloch, M. & Tsao, G. T. (1983). *Enzyme Microbial Technol.,* **5,** 82.
Lamed, R. & Bayer, E. A. (1988). *Adv. Appl. Microbiol.,* **33,** 1.
Lee, D. S. & Pack, M. V. (1987). *Enzyme Microbial Technol.,* **9,** 594.
Leisola, M. S. A. & Fiechter, A. (1985). *Adv. Biotechnol. Processes,* **5,** 59.
Lindley, N. D. & Durand, G. (1989). In *Bioprocess Engineering: the First Generation,* ed. T. K. Ghose. Ellis Horwood, Chichester UK, p. 324.
L'Italien, Y., Thibault, J. & LeDuy, A. (1989). *Biotechnol. Bioeng.,* **33,** 471.
Ljungdahl, L. G. & Wiegel, J. K. W. (1981). Anaerobic thermophilic culture system. US Patent 4292406.
Ljungdahl, L. G., Carreira, L. & Weigel, J. (1981). In *Ekman-Days. Int. Symp. Wood Pulping Chem.,* **4,** 23.
Lynd, L. R. (1989). In *Advances in Biochemical Engineering/Biotechnology,* Vol. 38, ed. A. Fiechter. Springer-Verlag, Berlin, p. 1.
Lynd, L. R. & Grethlein, H. E. (1987). *Biotechnol. Bioeng.,* **29,** 92.
Macdonald, D. G. & Mathews, J. F. (1979). *Biotechnol. Bioeng.,* **21,** 1091.
MacKenzie, C. R. & Bilous, D. (1982). *Can. J. Microbiol.,* **28,** 1158.
MacKenzie, C. R., Bilous, D. & Johnson, K. G. (1984). *Can. J. Microbiol.,* **30,** 1171.
Mairoella, B. L., Blanch, H. W. & Wilke, C. R. (1984). *Biotechnol. Bioeng.,* **26,** 1003.
Mandels, M. (1975). In *Cellulose as a Chemical and Energy Resource, Biotechnol. Bioeng. Symp. No. 5,* ed. C. R. Wilke. John Wiley, New York, p. 81.
Mandels, M. (1981). In *Annual Reports in Fermentation Processes,* Vol. 5, ed. G. T. Tsao. Academic Press, New York, 35.
Mandels, M. (1985). *Biochem. Soc. Trans.,* **13,** 414.
Mandels, M., Hontz, L. & Nystrom, J. (1974). *Biotechnol. Bioeng.,* **16,** 1471.
Mandels, M., Medeiros, J. E., Andreotti, R. E. & Bissett, F. H. (1981). *Biotechnol. Bioeng.,* **23,** 2009.
Mas, C., Jansen, N. B. & Tsao, G. T. (1988). *Biotechnol. Bioeng.,* **31,** 366.
Miller Jr, R. C. *et al.* (1988). In *Proc. Symp. Biochemistry and Genetics of Cellulose Degradation,* eds J.-P. Aubert, P. Beguin & J. Millet. Academic Press, London, p. 235.
Millet, M. A., Baker, A. J. & Satter, L. D. (1976). In *Enzymatic Conversion of Cellulosic Materials: Technology and Applications, Biotechnol. Bioeng. Symp. No. 6,* eds E. L. Gaden, M. H. Mandels, E. T. Reese & L. A. Spano. John Wiley, New York, p. 125.
Millet, M. A., Effland, M. J. & Caulfield, D. F. (1979). In *Hydrolysis of Cellulose: Mechanisms of Enzymatic and Acid Catalysis, Adv. Chem. Ser. No. 181,* eds R. D. Brown, Jr. & L. Jurasek. American Chemical Society, Washington, DC, p. 71.
Moloney, A. P., McCrae, S. I., Wood, T. M. & Coughlan, M. P. (1985). *Biochem. J.,* **225,** 365.
Monot, F., Martin, J. R., Petitdemange, H. & Gay, R. (1982). *Appl. Environ. Microbiol.,* **44,** 1318.

Montenecourt, B. S. & Eveleigh, D. E. (1979). In *Hydrolysis of Cellulose: Mechanisms of Enzymatic and Acid Catalysis, Adv. Chem. Ser. No. 181,* eds R. D. Brown Jr & L. Jurasek. American Chemical Society, Washington, DC, p. 289.
Montenecourt, B. S. *et al.*, Paper presented at National Meeting of American Institute of Chemical Engineers, Orlando, Florida, 28 February–3 March.
Morpath, F. F. & Jones, G. D. (1986). *Biochem. J.*, **236,** 221.
Mulder, M. H. V. & Smolders, C. A. (1986). *Process Biochem.*, **21,** 35.
Nanda, M., Bisaria, V. S. & Ghose, T. K. (1986). *J. Gen. Microbiol.*, **132,** 2761.
Neely, W. C. (1984). *Biotechnol. Bioeng.*, **26,** 59.
Nesse, N., Wallick, J. & Harper, J. M. (1977). *Biotechnol. Bioeng.*, **19,** 323.
Nevalainen, K. M. H., Palva, E. T. & Bailey, M. J. B. (1980). *Enzyme Microbiol. Technol.*, **3,** 59.
Ng, T. K. & Zeikus, J. G. (1982). *J. Bacteriol.*, **150,** 1391.
Ng, T. K., Weimer, P. J. & Zeikus, J. G. (1977). *Arch. Microbiol.*, **114,** 1.
Ng, T. K., Ben-Bassat, A. & Zeikus, J. G. (1981). *Appl. Environ. Microbial,* **41,** 1337.
Nigam, M. (1989). Studies on the kinetics of 2,3-butanediol formation by *Klebsiella oxytoca.* PhD thesis, Indian Institute of Technology, New Delhi.
Nishikawa, N. K., Sutcliffe, R. & Saddler, J. N. (1988). *Biotechnol. Bioeng.*, **31,** 624.
Nummi, M., Niku-Paavola, M.-L., Lappalainen, A., Enari, T.-M. & Rounio, V. (1983). *Biochem. J.*, **215,** 677.
Panda, T., Bisaria, V. S. & Ghose, T. K. (1987). *Biotechnol. Bioeng.*, **30,** 868.
Parisi, F. (1989). In *Advances in Biochemical Engineering/Biotechnology,* Vol. 38, ed. A. Fiechter. Springer-Verlag, Berlin, p. 53.
Park, C.-H., Okos, M. R. & Wankat, P. C. (1989). *Biotechnol. Bioeng.*, **34,** 18.
Pearce, P. D. & Bauchop, T. D. (1985). *Appl. Environ. Microbiol.*, **49,** 1265.
Pemberton, M. S., Brown Jr, R. D. & Emert, G. H. (1980). *Can J. Chem. Eng.*, **58,** 723.
Puls, J., Poutanen, K., Kroner, H. U. & Viikari, L. (1985). *Appl. Microbial Biotechnol.*, **22,** 416.
Reese, E. T. & Ryu, D. Y. (1980). *Enzyme Microbial Technol.*, **2,** 239.
Robson, L. M. & Chambliss, G. H. (1989). *Enzyme Microbial. Technol.*, **11,** 626.
Rogers, P. L., Lee, K. J. & Tribe, D. E. (1980). *Process Biochem.*, **15,** 7.
Rowland, S. P. & Roberts, E. J. (1972). *J. Polym. Sci., Part A*-1, No. 10, 2447.
Ryu, D. D. Y. & Mandels, M. (1980). *Enzyme Microbial Technol.*, **2,** 91.
Ryu, D. D. Y., Andreotti, R., Mandels, M., Gallo, B. & Reese, E. T. (1979). *Biotechnol. Bioeng.*, **21,** 1887.
Sablayrolles, J. M. & Goma, G. (1984). *Biotechnol. Bioeng.*, **26,** 148.
Saddler, J. N. & Chan, M. K.-H. (1982). *Eur. J. Appl. Microbiol. Biotechnol.*, **16,** 99.
Saddler, J. N. & Chan, M. K.-H. (1984). *Can. J. Microbiol.*, **30,** 212.
Saddler, J. N., Chan, M. K.-H. & Louis-Seize, G. (1981). *Biotechnol. Lett.*, **3,** 321.
Saddler, J. N., Brownell, H. H., Clermont, L. P. & Levitin, N. (1982*a*). *Biotechnol. Bioeng.*, **24,** 1389.
Saddler, J. N., Hogan, C., Chan, M. K.-H. & Louis-Seize, G. (1982*b*). *Can J. Microbiol.*, **28,** 1311.
Saddler, J. N., Mes-Hartee, M., Yu, E. K. C. & Brownell, H. H. (1983). In *Fifth Symp. on Biotechnology for Fuels and Chemicals, Biotechnol. Bioeng. Symp. No. 13,* ed. C. D. Scott. John Wiley, New York, p. 225.
Sahm, H. (1989). In *Bioprocess Engineering: the First Generation,* ed. T. K. Ghose. Ellis Horwood, Chichester, UK, p. 303.
Sarkarsen, K. V. (1980). In *Progress in Biomass Conversion,* Vol. 2, eds K. V. Sarkarsen & D. A. Tillman. Academic Press, New York, p. 127.
Sattler, W., Esterbauer, H., Glatter, O. & Steiner, W. (1989). *Biotechnol. Bioeng.*, **33,** 1221.

Scallan, A. M. (1971). *Text. Res. J.*, **41,** 647.
Schultz, T. P., Rughani, J. & McGinnis, J. D. (1989). *Appl. Biochem. Biotechnol.*, **20/21,** 9.
Sheir-Neiss, G. & Montenecourt, B. S. (1984). *Appl. Microbiol. Biotechnol.*, **20,** 46.
Shoemaker, S. P., Raymond, J. C. & Burner, R. (1981). In *Trends in the Biology of Fermentations for Fuels and Chemicals*, ed. A. Hollaender. Plenum Press, New York, p. 89.
Shukla, R., Kang, W. & Sirkar, K. K. (1989). *Biotechnol. Bioeng.*, **34,** 1158.
Slapack, G. E., Russell, I. & Stewart, G. G. (1987). *Thermophilic Microbes in Ethanol Production*. CRC Press, Boca Raton, Florida.
Soni, B. K., Soucaille, P. & Goma, G. (1986*a*). In *Eighth Symp. on Biotechnology for Fuels and Chemicals, Biotechnol. Bioeng. Symp. No. 17*, ed. C. D. Scott. John Wiley, New York, p. 239.
Soni, B. K., Das, K., Soucaille, P. & Goma, G. (1986*b*). In *Eighth Symp. on Biotechnology for Fuels and Chemicals, Biotechnol. Bioeng. Symp. No.* 17, ed. C. D. Scott. John Wiley, New York, p. 591.
Spangler, D. J. & Emert, G. H. (1986). *Biotechnol. Bioeng.*, **28,** 115.
Spindler, D. D., Wyman, C. E. & Grohmann, K. (1989). *Biotechnol. Bioeng.*, **34,** 189.
Srivastava, S. K. (1985). Kinetics of β-glucosidase production and its role in the enzymatic hydrolysis of cellulose. PhD Thesis, Indian Institute of Technology, New Delhi.
Srivastava, A. K. (1989). Bioconversion of cellulosic residues into microbial biomass protein by lignocellulolytic fungus *Coriolus hirsutus*. PhD Thesis, Indian Institute of Technology, New Delhi.
Stark, J. R. & Yin, X. S. (1987). *Enzyme Microbial Technol.*, **9,** 156.
Sternberg, D. & Dorval, S. (1979). *Biotechnol. Bioeng.*, **21,** 181.
Sternberg, D. & Mandels, G. R. (1979). *J. Bacteriol*, **139,** 761.
Sternberg, D., Vijay Kumar, P. & Reese, E. T. (1977). *Can. J. Microbiol.*, **23,** 139.
Sundstrom, D. W., Klei, H. E., Coughlin, R. W., Biederman, G. J. & Brouwer, C. A. (1981). *Biotechnol. Bioeng.*, **23,** 473.
Szczodrak, J. (1988). *Biotechnol. Bioeng.*, **32,** 771.
Szczodrak, J. (1989). *Biotechnol. Bioeng.*, **33,** 1112.
Szczodrak, J. & Targonski, Z. (1988). *Biotechnol. Bioeng.*, **31,** 300.
Tailliez, P., Girard, H., Millet, J. & Beguin, P. (1989). *Appl. Environ. Microbiol.*, **55,** 207.
Takagi, M., Abe, S., Suzuki, S., Emert, G. H. & Yata, N. (1978). In *Proc. Symp. Bioconversion of Cellulosic Substances into Energy, Chemicals and Microbial Protein*, ed. T. K. Ghose. Indian Institute of Technology, New Delhi, p. 551.
Tan, L. U. L., Yu, E. K. C., Mayers, P. & Saddler, J. N. (1987). *Appl. Microbiol. Biotechnol.*, **26,** 21.
Tanaka, M., Ikesaka, M., Matsuno, R. & Converse, A. O. (1988*a*). *Biotechnol. Bioeng.*, **32,** 698.
Tanaka, M., Fukuii, M. & Matsuno, R. (1988*b*). *Biotechnol. Bioeng.*, **32,** 897.
Taniguchi, M., Kobayashi, M. & Fujii, M. (1989). *Biotechnol. Bioeng.*, **34,** 1092.
Tassinari, T. & Macy, C. (1977). *Biotechnol. Bioeng.*, **19,** 1321.
Tjerneld, F., Persson, I., Albertsson, P.-A. & Hahn-Haegerdal, B. (1986). In *Seventh Symp. on Biotechnology for Fuels and Chemicals, Biotechnol. Bioeng. Symp. No. 15*, ed. C. D. Scott. John Wiley, New York, p. 419.
Toyama, N. (1969). In *Cellulases and their Applications, Adv. Chem. Ser.*, Vol. 95, ed. R. F. Gould. American Chemical Society, Washington, DC p. 359.
Viikari, L., Nybergh, P. & Linko, M. (1981). In *Advances in Biotechnology*, vol. 2, eds M. Moo-Young & C. W. Robinson. Pergamon Press, New York, p. 137.
Viikari, L., Ranua, M., Kantelinen, A., Sandquist, J. & Linko, M. (1986). In *Biotechnology in the Pulp and Paper Industry, 3rd Int. Conf.*, Stockholm, 16–19 June, p. 67.

Votruba, J., Volesky, B. & Yerushalmi, L. (1986). *Biotechnol. Bioeng.*, **28,** 247.
Wang, D. I. C., Avgerinos, G. C., Biocic, I., Wang, S.-D. & Fang, H.-Y. (1983). *Phil. Trans. Roy. Soc. Lond., Ser. B,* **300,** 323.
Wiegel, J. (1980). *Experientia,* **36,** 1434.
Wilke, C. R. (1978). *Process Development of the Bioconversion of Cellulose and Production of Ethanol.* Lawrence Berkeley Laboratory, University of California, Berkeley, Report No. LBL 7880.
Wilke, C. R., Yang, R. D. & Von Stockar, U. (1976). In *Enzymatic Conversion of Cellulosic Materials: Technology and Applications, Biotechnol. Bioeng. Symp. No. 6,* eds E. L. Gaden, Jr, M. H. Mandels, E. T. Reese & L. A. Spano. John Wiley, New York, p. 155.
Wilke, C. R. *et al.* (1983). *Enzymatic Hydrolysis of Cellulose—Theory and Applications.* Noyes Data Corp., Park Ridge, New Jersey.
Wood, T. M. (1985). *Biochem. Soc. Trans.*, **13,** 407.
Wood, T. M. (1988). In *Biodeterioration 7,* eds D. R. Houghton, R. N. Smith & H. O. W. Eggins. Elsevier Applied Science, London, p. 333.
Wood, T. M. & McCrae, S. I. (1977). *Carbohydrate Research,* **57,** 117.
Wood, T. M. & McCrae, S. I. (1979). In *Hydrolysis of Cellulose: Mechanism of Enzymatic and Acid Catalysis, Adv. Chem. Ser. 181,* eds R. D. Brown, Jr & L. Jurasek. American Chemical Society, Washington, DC, p. 181.
Wood, T. M. & Saddler, J. N. (1988). *Methods in Enzymology,* **160**(Part A), 3.
Wood, T. M., McCrae, S. I., Wilson, C. A., Bhat, K. M. & Gow, L. A. (1988). In *Proc. Symp. Biochemistry and Genetics of Cellulose Degradation,* ed. J.-P. Aubert, P. Beguin & J. Millet. Academic Press, London, p. 31.
Woodward, J., Whaley, K. S., Zachry, G. S. & Wohlpart, D. L. (1981). In *Third Symp. on Biotechnology in Energy Production and Conservation, Biotechnol. Bioeng. Symp. No. 11,* ed. C. D. Scott. John Wiley, New York, p. 619.
Wright, J. D., Power, A. J. & Douglas, L. J. (1986). In *Eighth Symp. on Biotechnology for Fuels and Chemicals, Biotechnol. Bioeng. Symp. No. 17,* ed. C. D. Scott. John Wiley, New York, p. 285.
Yablonsky, M. D. *et al.* (1988). In *Proc. Symp. Biochemistry and Genetics of Cellulose Degradation,* eds J.-P. Aubert, P. Beguin & J. Millet. Academic Press, New York, p. 249.
Yu, E. K. C. & Saddler, J. N. (1983). *Appl. Environ. Microbiol.,* **46,** 630.
Yu, E. K. C. & Saddler, J. N. (1987). In *Biomass Conversion Technology: Principles and Practices,* ed. M. Moo-Young. Pergamon Press, New York, p. 103.
Zadrazil, F. & Grabbe, K. (1983). In *Biotechnology,* Vol. 3, ed. H. Dellweg. Verlag Chemie, Weinheim, Germany, p. 145.
Zeikus, J. G., Ben-Bassat, A., Ng, T. K. & Lamed, R. J. (1981). In *Trends in the Biology of Fermentations for Fuels and Chemicals,* ed. A. Hollaender. Plenum Press, New York, p. 441.

Chapter 7

USE OF PHOTOSYNTHETIC BACTERIA FOR THE PRODUCTION OF SCP AND CHEMICALS FROM AGROINDUSTRIAL WASTES

KEN SASAKI

Hiroshima-Denki Institute of Technology, Nakano Akiku, Hiroshima 739-03, Japan

NAPAVARN NOPARATNARAPORN

Department of Microbiology, Faculty of Science, Kasetsart University, Bangkok 10900, Thailand

&

SHIRO NAGAI

Department of Fermentation Technology, Faculty of Engineering, Hiroshima University, Kagami-yama, Higashi-Hiroshima 724, Japan

CONTENTS

1 INTRODUCTION

Photosynthetic bacteria which are anoxygenic phototrophs are distributed widely in the natural world in habitats such as fields, lakes, ponds, rivers and oceans. The bacteria have the capability to utilise sulfate, thiosulfate and hydrogen, as well as various organic matter, as electron donors and are able to fix carbon dioxide and molecular nitrogen under anaerobic light conditions. (Pfennig, 1978; Truper & Pfennig, 1978; Hoshino & Kitamura, 1984). Hence, these bacteria play an important role in sulfur and nitrogen circulation in aquatic environments (Kobayashi & Kurata, 1978).

Photosynthetic bacteria are inherently different from algae and plants with regard to their photosynthetic organ because these bacteria harvest the energy as adenosine triphosphate (ATP) only via photosystem I (Morita & Shimada, 1984) compared with the two electron harvesting systems of algae and plants, i.e. photosystems I and II.

Photosynthetic bacteria can grow under essentially anaerobic-light conditions, though some of them can grow heterotrophically under aerobic-dark conditions similar to other heterotrophic microorganisms. Also, some of them can reduce nitrate and sulfate to generate ATP (Truper & Pfennig, 1978).

GENERAL CHARACTERISTICS

The photosynthetic bacteria are classified as green sulfur, purple sulfur and purple nonsulfur bacteria (Table 1). Among these photosynthetic bacteria, several bacteria in Rhodospirillaceae such as *Rhodobacter* (*Rb.*), *Rhodospirillum* (*Rsp.*), *Rhodocyclus* (*Rc.*) and *Rhodopseudomonas* (*Rp.*) are appropriate for application to Single Cell Protein (SCP) and chemical production because they can grow aerobically and/or photosynthetically, utilizing organic substrates. *Rhodospirillum* is being studied for biofuel (hydrogen) production from carbonaceous wastes (Takahashi, 1984). In addition, as the bacterial cells consist of a large amount of protein with valuable vitamins and photopigments, the cells are promising as a potent SCP source of animal feedstocks.

Some of these photosynthetic bacteria, e.g. *Rhodocyclus* spp., capable of hydrolyzing starch and protein (Shipman *et al.*, 1975), are also useful in treating agroindustrial wastes as well as being sources of SCP.

As Chromatiaceae, Ectothiorhodospiraceae, Chlorobiaceae and Chloroflexaceae grow only photosynthetically or photoautotrophically under anaerobic-light conditions, they are not thought to be a potent source of SCP due to their slow growth under light illumination.

In this chapter, Rhodospirillaceae capable of growing under aerobic-dark and/or anaerobic-light conditions with relatively rapid growth rate are studied with a view to their use for the production of SCP and chemicals from agroindustrial wastes.

APPLICATION OF PHOTOSYNTHETIC BACTERIA TO SCP AND CHEMICALS PRODUCTION FROM AGROINDUSTRIAL WASTES

Very large quantities of agroindustrial wastes are discharged throughout the world, e.g. in Japan about 7 000 000 tons/year of agricultural waste and 3 000 000 tons/year of animal waste are discharged (Hayami, 1985). Environ-

Table 1
Characteristics and Classification of Photosynthetic Bacteria[a]

Bacteria	*Color*	*Bacterio-chlorophyll*	*Main carbon source*	*Main electron donor*	*Growth condition*	*Typical genera*
Rhodospirillaceae	Red Violet Brown	Bchl a or b	Organic matter CO_2	Organic matter, H_2 (H_2S)[b]	Anaerobic-light Aerobic-dark Aerobic-light	*Rhodobacter* *Rhodospirillum* *Rhodopseudomonas* *Rhodocyclus*
Chromatiaceae	Red Pink Violet	Bchl a or b	Organic matter CO_2	H_2S, $S_2O_3^{2-}$ (S, H_2)[b]	Anaerobic-light	*Chromatium* *Thiocyctis* *Thiocapsa*
Ectothiorhodospiraceae	Red Pink Green	Bchl a or b	Organic matter CO_2	Organic matter H_2S, H_2	Anaerobic-light	*Ectothiorhodospira*
Chlorobiaceae	Green Brown	Bchl a or c, d, e	CO_2		Anaerobic-light	*Chlorobium* *Pelodictyon* *Chlorochromatium*
Chloroflexaceae	Green Orange	Bchl a or c, d, e	Organic matter CO_2	Organic matter H_2S, $S_2O_3^{2-}$	Anaerobic-light	*Chloroflexus* *Chloronema* *Oscillochloris*

[a] Modified from Kitamura, 1986.
[b] Only a few bacteria can utilize.

mental pollution caused by piling up these wastes causes trouble in the local country districts.

As to the treatment or reuse of such agroindustrial waste, methane fermentation (biogas production) is widely applied to chemical oxygen demand (COD) reduction together with energy recovery from the waste. Compost production is also practiced using animal and agricultural solid waste (Hayami, 1985).

Recently, attention has been paid to the utilization of the photosynthetic bacteria from agroindustrial waste to produce SCP (Shipman *et al.*, 1975; Kobayashi & Kurata, 1978; Sasaki *et al.*, 1981; Noparatnaraporn *et al.*, 1986*a*). The advantages are that these bacteria can grow at high growth rate under aerobic and light conditions. For example, Rhodospirillaceae can grow in a similar way to SCP yeasts, however, the protein content of the bacterial cells, 60–70% is relatively high compared with that of the yeasts (50–60%). In addition, these bacteria contain various vitamins, especially a high content of vitamin B_{12}. As an example of using photosynthetic bacteria for animal feedstock, the addition of the cells to hen feed (0·01–0·04%) increased the egg-laying rate by up to 10% and improved the quality of the egg yolk (Kobayashi & Kurata, 1978). In some fish cultures in Japan, the photosynthetic bacterial cells have been widely fed as the source of vitamins and trace minerals to cultivate healthy fish.

Thus, it is expected that photosynthetic bacterial cells will be enriching feedstock with a supplementary protein source together with valuable vitamin sources.

2 SCP PRODUCTION FROM THE WASTES

2.1 Pineapple Waste

Pineapple canning industries, which are widely distributed in tropical countries like Thailand, produce a large quantity of solid and liquid waste. These wastes are high organic loading wastes (COD *c.* 10 000 mg/liter) which contain mainly carbohydrates with little protein. The liquid waste is usually treated by aerated lagoon or oxidation pond process whereas the semi-solid wastes are normally discarded without proper treatment. Therefore, the semi-solid wastes consisting of the peel and core parts are the major causes of a serious pollution problem.

Photosynthetic bacteria as a source of single cell protein were used to treat pineapple waste due to their advantages of direct growth in high organic loading wastewater and the nutritive quality of the cell mass. In this section, a newly isolated strain of *Rhodobacter sphaeroides* strain P_{47} was used for cell mass production from pineapple peel waste (Noparatnaraporn *et al.*, 1986*a*).

CHEMICAL COMPOSITION

The chemical compositions of pineapple peel waste are shown in Table 2. These compositions vary considerably depending on the season, area, and

Table 2
Chemical Composition of Pineapple Peel Waste

	Filtrate	*After sterilization*[a]
Chemical composition (g/liter)[b]		
COD	100·8	103·7
Reducing sugar	39·2	41·2
Total sugar	100·0	100·9
Dextran	1·5	1·5
Raffinose	2·6	1·5
Sucrose	40·1	40·1
Glucose	23·6	23·6
Galactose	1·7	2·1
Fructose	14·0	15·6
Soluble protein	0·9	n.a.[c]
Kjeldahl nitrogen	0·2	n.a.
Trace elements (mg/liter)		
Fe	5·43	n.a.
Ca	3·31	n.a.
Mn	13·97	n.a.
Mg	62·50	n.a.
Co	0·07	n.a.
Cu	2·02	n.a.
Cd	0·03	n.a.
Na	8·61	n.a.
SO_4^{-2}	169·7	n.a.
PO_4^{-3}	223·8	n.a.
pH	4·0	4·0

[a] Autoclaved at 15 lb/in^2 for 15 min.
[b] The variation of sugar content and fraction depended on the harvesting season, area and canning process of the industry.
[c] Not analyzed.

canning process of the industry. The waste contains mainly sucrose, glucose and fructose, while dextrin, raffinose and galactose are minor. Although the waste contains a little nitrogen and soluble protein, the necessary elements for growth of photosynthetic bacteria such as Fe, Ca, Mg, Mn and Co are present in adequate concentrations. From the chemical compositions, the pineapple peel waste appears to be a good carbohydrate source for cultivation of photosynthetic bacteria.

EXPERIMENTAL PROCEDURE

The pineapple peel juice was obtained by a press dehydration of the peel and cores followed by filtration before use.

The pineapple waste medium for preculture was composed of diluted peel filtrate (*c.* 20 g/liter total sugar) supplemented with $(NH_4)_2SO_4$ (2·6 g/liter), $(NH_4)_2HPO_4$ (0·6 g/liter), nicotinic acid ($0{\cdot}8 \times 10^{-3}$ g/liter), thiamine-HCl ($1{\cdot}6 \times 10^{-3}$ g/liter) and biotin ($1{\cdot}2 \times 10^{-5}$ g/liter). Preculture was carried out on a rotary shaker (200 rpm) under aerobic-dark conditions at 35°C.

The waste medium for the main cultivation was composed of peel filtrate without dilution (total sugar *c.* 100 g/liter) supplemented with $(NH_4)_2SO_4$ (13 g/liter), $(NH_4)_2HPO_4$ (3 g/liter), nicotinic acid ($4{\cdot}0 \times 10^{-3}$ g/liter), thiamine-HCl (8×10^{-3} g/liter) and biotin (6×10^{-5} g/liter). Cultivations (inoculum: 10%) were carried out at 37°C in a 2·5 liter fermentor (1·5 liter working volume, MD-250 Marubishi Bioeng.) under sufficiently aerobic conditions at pH 7·0.

RESULTS

Growth and sugar utilization by *Rb. sphaeroides* P_{47} on pineapple waste medium (Fig. 1) indicates that the growth pattern was diauxic. After 40 h of culture, when glucose and fructose has been completely utilized, the remaining sucrose (45 g/liter) was utilized rapidly and the second growth phase could be observed up to 56 h of cultivation. The phenomena observed for sugar utilization would suggest that sucrase synthesis might be repressed while the glucose remained in the culture medium.

The culture yielded 26·5 g dry cell/liter after 60 h of cultivation with the

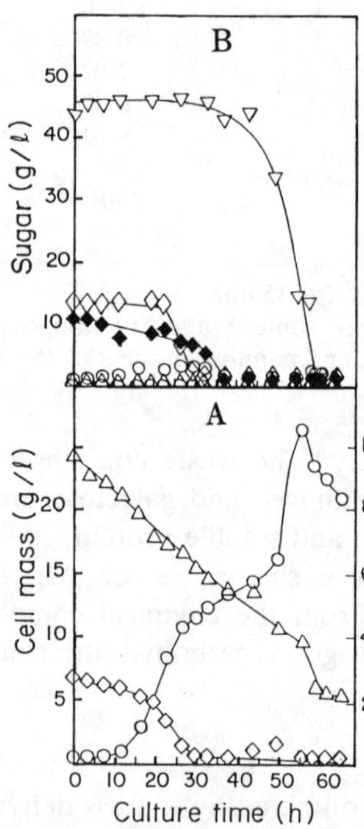

Fig. 1. Growth, sugar consumption and uptake of sugar components of *Rb. sphaeroides* P_{47} on pineapple peel waste medium (initial total sugar *c.* 100 g/liter). A, Growth and sugar consumption; B, uptake of sugar components. Symbols in A: ○: cell mass; △: total sugar; ◇: reducing sugar. Symbols in B: ▽: sucrose; ⌓: galactose; ◇: glucose; △: raffinose; ◆: fructose; ○: dextran.

maximum specific growth rate of 0·31/h and growth yield of 0·45 g cell/g sugar; COD removal was 85·3%.

A diagrammatic scheme for cell mass production of *Rb. sphaeroides* P_{47} on pineapple peel waste is proposed in Fig. 2 based on a medium-sized pineapple cannery which discharges daily 150 tons of pineapple solid waste. From 150 tons waste, 75 tons of juice (*c.* 100 g/liter of total sugar; pH 4·0) was obtained. Supplementation of nitrogen, phosphorus and essential vitamins (nicotinic acid, thiamine-HCl and biotin) together with pH adjustment to 6·8–7·0 was necessary before inoculation. After 56 h of cultivation, a cell mass of 2·5 ton dry cell/day will be recovered, and the remaining effluent after the cell recovery, containing approximately 23 g/liter of total sugar (COD 14·8 g/liter), should be further treated by ordinary treatment processes.

It would be concluded that *Rb. sphaeroides* P_{47} might be applicable to SCP cell mass production from pineapple waste, since the use of this bacterium is superior not only for recovering the nutritive cell mass but also for reducing the operation capacity of subsequent treatment.

2.2 Soybean Waste

In Japanese soybean curd and miso (fermented soybean) manufacture, a high COD wastewater (COD *c.* 20 000 mg/liter) which consists of mainly carbohydrates and protein is discharged from a steaming process of soaking soybean. Usually, the cooked soybean runoff has to be diluted 10–20 times with the process water before entering the activated aerobic sludge treatment. This greatly increases the total volume of wastewater which requires a large-scale plant with a high running cost.

An attempt was made to treat the cooked runoff directly by a photosynthetic bacterium, *Rhodocyclus gelatinosus,* which is capable of utilizing various carbohydrates and proteins so as to not only treat the COD but also harvest the bacterial cells for SCP (Sasaki *et al.*, 1981).

CHEMICAL COMPOSITION

The chemical composition of cooked soybean runoff (Table 3) shows that it contains a large amount of carbohydrates consisting mainly of stachyose, sucrose and raffinose, while galactose, fructose and glucose are minor components. Other than carbohydrate, the waste contains 10·4 g/liter of soluble protein. Considering these components, the use of *Rc. gelatinosus* appears to be suitable for the treatment and for SCP production.

EXPERIMENTAL PROCEDURE

The preculture of *Rc. gelatinosus* was carried out in a 500 ml shake-flask containing 150 ml of a medium MS (Lascelles, 1956) in which glutamate and malate were replaced to soluble starch. Preculture was conducted on a reciprocal shaker (115 rev., 7 cm amplitude) under dark conditions at 30°C.

The supernatant of cooked soybean runoff (centrifuged at 8000 g for 30 min,

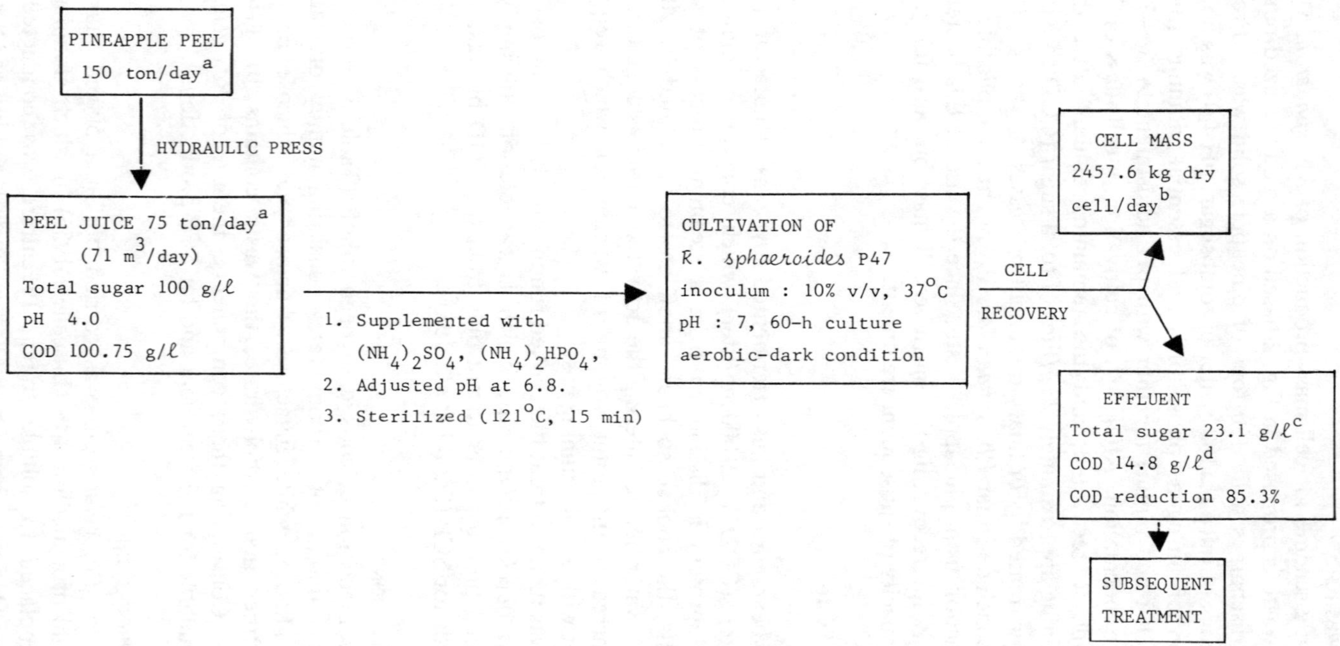

Fig. 2. Diagrammatic scheme of cell mass production from pineapple cannery waste by *Rb. sphaeroides* P_{47}. [a] Data obtained from a pineapple cannery in Cholburi area, Thailand. [b] Calculated from $Y_{x/s}$ of 0·45 g cell/g total sugar at 37°C. [c] Calculated from total sugar consumption of 76·9%. [d] Calculated from COD removal of 85·3% at 37°C.

Table 3
Chemical Composition of Cooked Soybean Runoff

		Cooked runoff	*Supernatant*[a]
pH		6·0	6·0
COD	(mg/liter)	$2·2 \times 10^4$	$1·8 \times 10^4$
Suspended solid	(mg/liter)	$2·24 \times 10^4$	0
Reducing sugar	(g glc./liter)	4·24	3·93
Total sugar[b]	(g glc./liter)	24·4	24·0
Stachyose[c]	(g/liter)	n.a.[d]	7·27
Raffinose[c]	(g/liter)	n.a.	3·42
Sucrose[c]	(g/liter)	n.a.	5·99
Galactose[c]	(g/liter)	n.a.	1·46
Fructose[c]	(g/liter)	n.a.	1·37
Glucose[c]	(g/liter)	n.a.	0·79
Mannose[e]		n.a.	trace
Arabinose[e]		n.a.	trace
Xylose[e]		n.a.	trace
Rhamnose[e]		n.a.	n.d.[f]
Uronic acid[e]		n.a.	n.d.
Cellobiose[e]		n.a.	n.d.
Total nitrogen[g]	(g/liter)	1·85	1·56
Soluble protein[h]	(g/liter)	10·4	10·4

[a] The cooked runoff was centrifuged at 8000*g* for 30 min. [b] Phenol-sulfuric acid method. [c] Quantitatively analyzed by high pressure liquid chromatography. [d] Not analyzed. [e] Qualitatively analyzed by thin layer chromatography. [f] Not detected. [g] Kjeldahl method. [h] Folin–Lowry method.

Table 3) was used for cultivation by supplementing the following inorganic materials (g/liter): $(NH_4)_2SO_4$, 5·7; K_2HPO_4, 3·6; KH_2PO_4 3·6; and $MgSO_4 \cdot 7H_2O$, 1·4. Cultivations (inoculum: 5%) using a 3-liter jar-fermentor (working volume 1·2 liter, Mitsuwa KMJ-2) were conducted under aerobic-dark conditions maintaining a sufficient level of dissolved oxygen (3–7 mg/liter) at pH of $7·0 \pm 0·1$.

RESULTS

Growth of *Rc. gelatinosus* on cooked soybean runoff medium (Fig. 3) shows that the COD was reduced from 20·5 to 3·5 g/liter in which *c.* 73% of total sugar and 59% of soluble protein were consumed for 40 h culture. An optimum temperature for the growth and COD removal was 35°C among the range of culture temperatures (30–45°C) tested.

During the cultivation (Fig. 3), sucrose, glucose, galactose, fructose and raffinose were almost completely consumed, although *c.* 50% of stachyose remained. This suggested that the organism could utilize raffinose and stachyose, for whose hydrolysis α-galactosidase might be induced to act. This finding encourages the use of *Rc. gelatinosus* since other photosynthetic

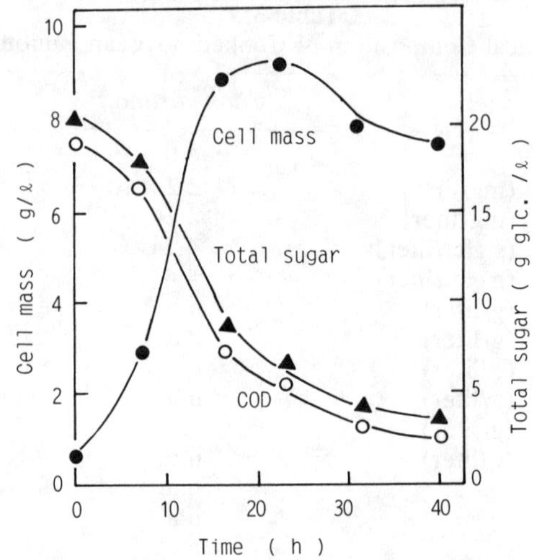

Fig. 3. Typical growth, total sugar and COD reductions of *Rc. gelatinosus* on soybean waste under aerobic-dark conditions (35°C).

bacteria usually do not utilize such sugars. The cell mass obtained contained 62–63% of protein and 33 μg/g dry cells of vitamin B_{12}, suggesting that the cell mass might be applicable as an SCP source.

Some advantages of using *Rc. gelatinosus* cells for soybean processing manufacture (COD 20 000 mg/liter) in place of activated sludge treatment are summarized in Fig. 4. In the figure, 22 m^3 of cooked soybean runoff is supposed to be discharged per day. In the activated sludge treatment, this has to be diluted *c.* 20 times; consequently about 440 m^3 of waste water have to be

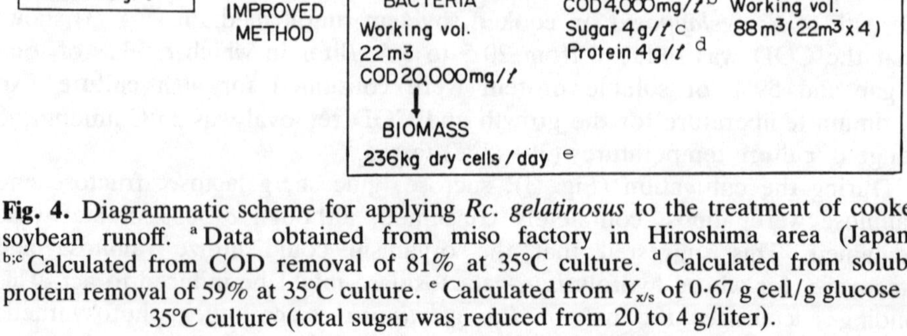

Fig. 4. Diagrammatic scheme for applying *Rc. gelatinosus* to the treatment of cooked soybean runoff. [a] Data obtained from miso factory in Hiroshima area (Japan). [b;c] Calculated from COD removal of 81% at 35°C culture. [d] Calculated from soluble protein removal of 59% at 35°C culture. [e] Calculated from $Y_{x/s}$ of 0·67 g cell/g glucose at 35°C culture (total sugar was reduced from 20 to 4 g/liter).

treated. In the case of *Rc. gelatinosus,* after 40 h of culture of intact cooked soybean runoff (22 m^3) under aerobic conditions, the initial COD might be reduced from 20 000 to 4000 mg/liter. After diluting this four times (COD *c.* 4000 mg/liter), the liquid, now 88 m^3, enters into an ordinary activated sludge treatment. Although an economic assessment is required for the optimization of the above process, a key point will be the chemical value of the cell mass produced with the appropriate amount of the cooked soybean runoff discharged.

2.3 Starch Waste

During the processing of cassava starch in Thailand, two kinds of wastes are produced: one is the residual pulp which is normally used as a carbohydrate source for animal feed ('cassava sundried solid waste') while the other is the 'high COD loading wastewater' discharged from starch processing. The latter has caused serious pollution problems in areas where starch factories are located.

Among purple nonsulfur photosynthetic bacteria (PSB), only *Rhodocyclus gelatinosus* is capable of utilizing cassava starch directly. Here, the cell mass productions of *Rc. gelatinosus* and a new isolate PSB strain VI from wastes are described (Noparatnaraporn, 1987).

2.3.1 Cassava Liquid Waste

Cassava starch processing either with or without a decantor system usually produces a large quantity of high organic loading wastewater. As a preliminary experiment, the separator wastewater was the most suitable portion for cultivation of photosynthetic bacteria (PSB).

Here, a new isolate (PSB VI) capable of growth in raw separator wastewater was directly used in an outdoor nonsterile culture system (Noparatnaraporn, N., unpublished).

CHEMICAL COMPOSITION

The waste contains cassava starch at a concentration of 8–22 g/liter with insufficient nitrogen and soluble protein (Table 4). Therefore, the necessary supplementation for PSB culture may be NH_4^+ and PO_4^{3-}.

EXPERIMENTAL PROCEDURE

The medium compositions for preculture and outdoor culture are as follows: 1 liter of separator wastewater (starch 8 g/liter) was supplemented with 8 g $(NH_4)_2HPO_4$, 4 g KH_2PO_4 and 4 g K_2HPO_4. Initial pH was adjusted to 6·8–7·0. The waste medium was used directly without sterilization.

Preculture was carried out at 35°C in a 1·5 liter Rhoux bottle (medium: 1·5 liter) incubated under microaerobic-light conditions which were achieved using a cotton plug with weak agitation. The light intensity was in the range 3–5 klux.

Table 4
Chemical Compositions of Separator Wastewater from Cassava Starch Processing

pH		6
COD	(g/liter)	5·2–5·3
Total nitrogen	(mg/liter)	72–73
Soluble protein	(mg/liter)	0–0·3
Starch	(g/liter)	8–22
CN^-	(mg/liter)	0–2
Ca^{2+}	(mg/liter)	0–450
Mg^{2+}	(mg/liter)	0–476
Mn^{2+}	(mg/liter)	0–11
Total iron	(mg/liter)	0–1

Outdoor nonsterile culture of the PSB strain VI was carried out in a 40 liter acrylic open chamber. The culture was a nonsterile open system without controlling temperature, pH and light. Culture broth was continuously circulated by a pump. A high inoculation concentration of 20% v/v is required to maintain the dominance of the PSB strain VI.

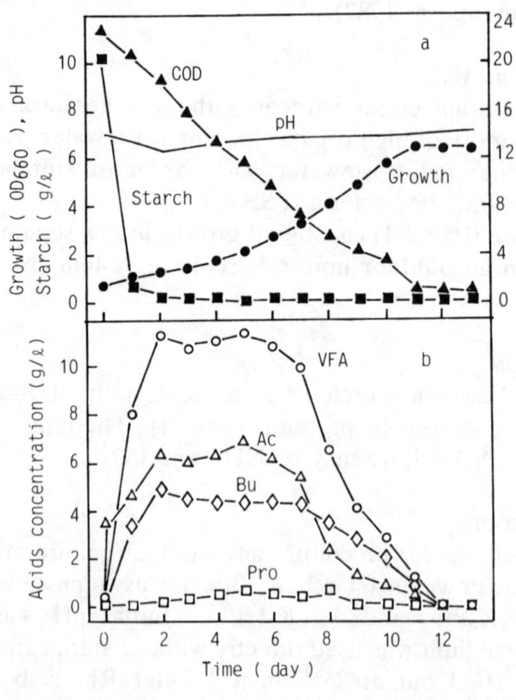

Fig. 5. Utilization of starch and profiles of VFA in cassava wastewater by PSB strain VI in 40 liters outdoor culture. VFA, Volatile fatty acids; Ac, acetic acid; Pro, propionic acid; Bu, butyric acid.

RESULTS

The results of outdoor culture (Fig. 5(a)) indicate that the maximum cell mass obtained was 4·7 g dry cells/liter after 11 days' culture. The COD removal was 95·6%. It is interesting to note that the pH was rather stable throughout the cultivation; even the rapid utilization of starch was observed. This suggested that starch might be converted quickly to other substances which might be useful substrates of PSB. Volatile fatty acids in the culture broth were detected (Fig. 5(b)), supporting the above hypothesis. In fact, three major volatile fatty acids, acetic, propionic and butyric acids, were produced in parallel with starch utilization and utilized completely after 12 days. The cell mass of the photosynthetic bacteria obtained from outdoor culture was of quite good nutritional quality in terms of amino acid and vitamin composition (see section 2.6).

Although the experiment showed the possibility of outdoor culture of PSB, this culture system has to be further improved in terms of growth rate and cell mass attained.

2.3.2 Cassava Solid Waste

Cassava solid waste is normally used as a carbohydrate source of animal feed due to its low cost and abundance. Since *Rhodocyclus gelatinosus* can utilize starch directly, the growth of these bacteria was tested (Noparatnaraporn *et al.*, 1987*a*).

CHEMICAL COMPOSITION

The chemical compositions of cassava waste extract (Table 5) indicate that the filtrate appears to be a good carbohydrate source for *Rc. gelatinosus* with the addition of nitrogen as well as other essential nutrients.

Table 5
Chemical Compositions of the Extract Solution[a] of Sun-dried Cassava Waste

Compositions (g/liter)	*Filtrate*	
	Before sterilization	*After sterilization*
Total sugar	11·3	12·0
Reducing sugar	0·2	0·2
Soluble protein	0·1	0·1
Sugar concentration		
Glucose	0·11	0·33
Maltose	0	0
Maltotriose	0·22	0·19
pH	6·0	5·5

[a] 25 g dried waste per liter were autoclaved (15 lb, 15 min) and filtered.

EXPERIMENTAL PROCEDURE

Cassava starch medium was composed of (g/liter extract, see Table 5): total sugar *c.* 11–12; $(NH_4)_2HPO_4$, 5·6; K_2HPO_4, 2·8; KH_2PO_4, 2·8; $CaCl_2{\cdot}2H_2O$, 0·3; $MgSO_4{\cdot}7H_2O$ 1·1; nicotinic acid, $11{\cdot}25 \times 10^{-3}$; biotin, $4{\cdot}5 \times 10^{-5}$; thiamine-HCl, $9{\cdot}0 \times 10^{-3}$; ferric citrate, $1{\cdot}4 \times 10^{-3}$; $MnSO_4{\cdot}5H_2O$, $6{\cdot}75 \times 10^{-3}$ and $CoCl_2{\cdot}6H_2O$, $5{\cdot}4 \times 10^{-3}$. All cultivations were carried out at 40°C by batch culture in a 2·5 liter fermentor (working volume 1·5 liter) under aerobic-dark conditions.

RESULTS

Growth and sugar utilization of *Rc. gelatinosus* on cassava waste medium are shown in Fig. 6. Sugar analyses revealed that the abundant reducing sugar at 20 h was mainly maltose and glucose and both sugars were rapidly utilized after 25 h. It was also shown that the total liquefying activity for starch tended to increase when the reducing sugar in the culture was reduced. These charac-

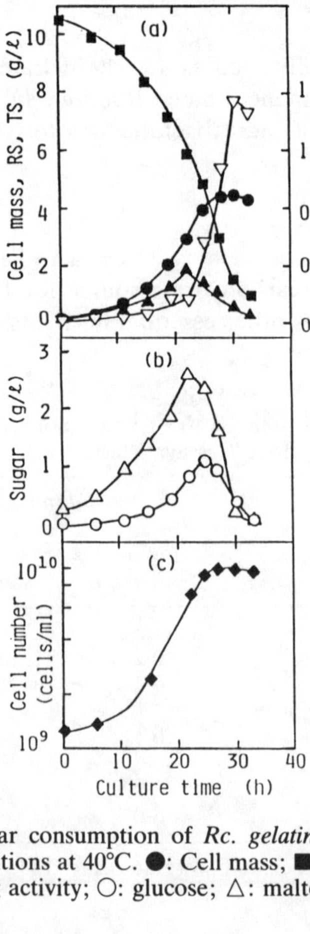

Fig. 6. Growth and sugar consumption of *Rc. gelatinosus* on cassava waste medium under aerobic-dark conditions at 40°C. ●: Cell mass; ■: total sugar; ▲: reducing sugar; ▽: liquefying activity; ○: glucose; △: maltose; ◆: cell number.

teristics suggested that α-amylase might be repressed during the culture of *Rc. gelatinosus* by the presence of large amounts of reducing sugars. Vitamin B_{12} content of the cells was 23 μg/g cell, and carotenoid content was 0·09 mg/g cell, suggesting the usefulness of *Rc. gelatinosus* cell for an SCP source.

2.4 Mandarin Orange Peel

Orange juice manufacture (juice yield: *c.* 50%) produces a large quantity of orange waste, *c.* 400 000 tons per year in Japan. Although part of the peel waste is converted to peel powder to use as an animal feed additive, most of these semi-solid wastes have to be treated by physical (heat) or chemical or biological means, causing a pollution problem. To solve the pollution problem, *Candida* sp. and *Aspergillus niger* as SCP and macerating enzyme sources, respectively, were cultivated directly on orange peel waste (Nishio & Nagai, 1979).

As an alternative means, effective acidogenic fermentation of orange waste has succeeded in eliminating methanogenesis by making the retention time of the culture broth shorter than the usual time required for methane fermentation (Nishio *et al.*, 1982). As a result, volatile fatty acids such as acetic and propionic, etc., have been accumulated as the end-products which are likely available substrates for photosynthetic bacteria. Hence, photosynthetic bacteria may use the culture medium obtained after acidogenic fermentation of orange waste. The feasibility of using photosynthetic bacteria is discussed.

CHEMICAL COMPOSITION

The chemical composition of mandarin orange peel is shown in Table 6. This waste consists mainly of total sugars including pectin, hemicellulose and

Table 6
Chemical Composition of Mandarin Orange Peel (Dry Basis)

	Orange peel (g/g peel)	*Orange peel[a] acid hydrolysis (g/g peel)*
Total sugar	0·37	0·57
Reducing sugar	0·19	0·39
Sucrose	0·18[b]	0·18[b]
Glucose	0·04	0·13
Total uronate	0·11	0·21
Crude protein	0·07	—
Crude fat	0·02	—
Crude fiber	0·12	—
Ash	0·02	—

[a] Treated at 1 kg/cm^2 for 15 min. [b] Total sugar minus reducing sugar.
Source: Nishio & Nagai, 1979.

cellulose with a remarkable amount of soluble sugars (sucrose, *c.* 0·2 g/g dry peel). The content of nitrogenous compounds such as crude protein is quite low compared with the total sugars (C/N ratio, 36·8). For the cultivation of photosynthetic bacteria, a nitrogen source such as $(NH_4)_2SO_4$ should be supplemented even if the acidogenesis of the waste was done.

EXPERIMENTAL PROCEDURE

Seed culture for acidogenesis: The seed sludge from an anaerobic digestor of an orange canning factory was inoculated into orange peel medium and anaerobic digestion was carried out at 35°C with pH maintained at 7·0. The orange peel medium contained (g/liter, tap water): dried orange peel powder 11·2 (200 mesh); NH_4Cl, 1·14; K_2HPO_4, 0·38; and $MgSO_4{\cdot}7H_2O$, 0·3. Anaerobic digestion was conducted using a 1·5 liter conical flask (seed sludge 300 ml and medium 700 ml). Repeated batch culture in which 100 ml of digested culture was replaced by fresh medium daily was continued for 30 days to obtain acclimatized sludge for acidogenic fermentation (Nishio *et al.*, 1982).

Acidogenic culture: A 300 ml sample of acclimatized sludge was inoculated into a 3-liter conical flask (medium: 700 ml) and anaerobic culture was conducted at 35°C for 5 days (Fig. 7). During fermentation pH was daily controlled at 7·0. Then the culture broth was centrifuged (12 000g, 30 min) to obtain a clarified supernatant. This supernatant was further supplemented by (g/liter) $(NH_4)_2HPO_4$, 2; KH_2PO_4, 0·5; K_2HPO_4, 0·5; and $MgSO_4{\cdot}7H_2O$, 0·2; the product was used for the cultivation of photosynthetic bacteria.

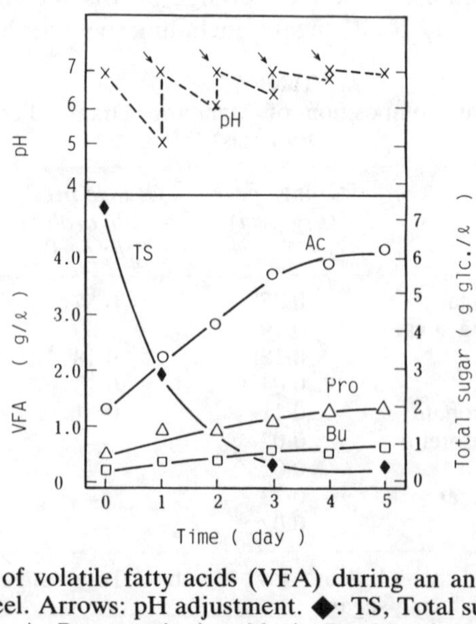

Fig. 7. Production of volatile fatty acids (VFA) during an anaerobic fermentation of mandarin orange peel. Arrows: pH adjustment. ◆: TS, Total sugar; ○: Ac, acetic acid; △: Pro, propionic acid; ▲: Bu, butyric acid.

Culture of photosynthetic bacteria: *Rhodobacter sphaeroides* S cells were cultivated aerobically in a modified medium MS (Lascelles, 1956) in which malate and glutamate were replaced by acetic and propionic acids. The preculture was inoculated into a 10-liter fermentor with 6 liters of medium (Marubishi MD-10). Three culture conditions were selected: (1) aerobic-dark (aeration rate 0·5 vvm (3 liters air/6 liters medium/min) and agitation rate 400 rpm to maintain concentration of dissolved oxygen (DO) $>$ 5 mg/liter); (2) microaerobic-light (aeration rate 0·05 vvm and agitation rate 250 rpm to maintain nil DO under illumination of 3 klux); and (3) anaerobic-light (N_2 gas was charged into the fermentor's head space and agitation rate was 150 rpm under 3 klux illumination). The fermentor was surrounded by six fluorescent lamps; the light intensity on the fermentor's surface was 3 klux. The inoculation size was 5–17% v/v. Supernatant from anaerobic acidogenesis (medium for photosynthetic bacteria) was not sterilized.

RESULTS

An acidogenic fermentation from mandarin orange peel medium (Fig. 7) indicates that during 5 days' digestion, total sugars were almost consumed up to the 3-day culture and acetic acid (4·5 g/liter) was mainly accumulated with 2 g/liter of propionic acid and 1 g/liter of butyric acids. The supernatant of the acidogenic fermentation was used for further cultivation of *Rb. sphaeroides* S (Sasaki, 1980).

Rhodobacter sphaeroides cells cultivated under aerobic-dark, microaerobic-light and anaerobic-light conditions are summarized in Fig. 8. In aerobic-dark and microaerobic-light cultures, growth and decrease of COD were remarkably different compared with results under anaerobic-light culture. In the fermentors, volatile fatty acids were also quickly consumed. In microaerobic-light culture, the maximum cell mass attained was relatively high at 3·7 g/liter compared with 2·0 g/liter observed in aerobic-dark culture, although the growth rate in microaerobic-light culture was relatively slow.

COD removal was *c.* 85% in both aerobic-dark and microaerobic-light cultures, indicating that the growth yield from COD consumed in the latter conditions was 0·83 g cell/g COD, which was 1·6 times higher than that in aerobic-dark conditions. This suggests that in microaerobic-light culture, photosynthetic carbon fixation and aerobic assimilation of organic matter were occurring simultaneously. In anaerobic-light culture, remarkably slow growth and COD reduction were observed compared with both aerobic cultures (Fig. 8) due to the limitation of light illumination.

Thus, from a practical point of view, a microaerobic-light culture of *Rb. sphaeroides* S may be promising for SCP production from volatile fatty acids rather than an aerobic culture, since the cell yield per COD consumed was 1·6 times higher than that from aerobic assimilation. Besides, as PSB contain appreciable amounts of chlorophyll, carotenoids and vitamins B_{12} (section 2.6), *Rb. sphaeroides* S may become a potent source of SCP.

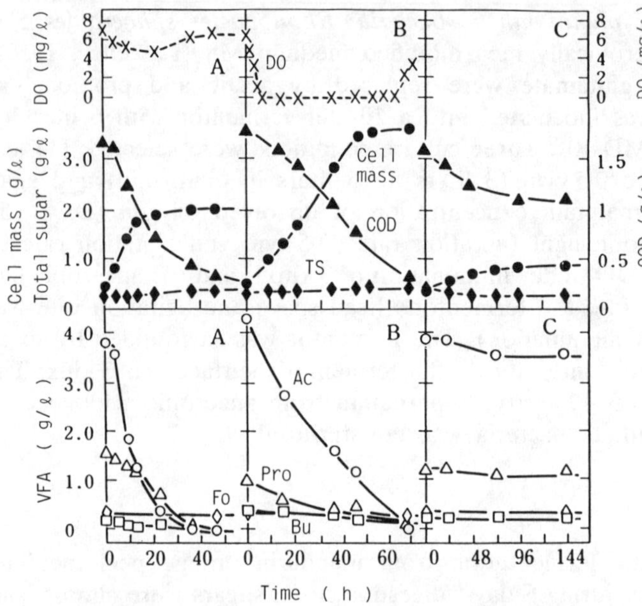

Fig. 8. Growth and volatile fatty acids consumption by *Rb. sphaeroides* S on the medium obtained from anaerobic fermentation of mandarin orange peel. A, aerobic-dark, B, microaerobic-light (3 klux), and C, anaerobic-light (3 klux). ●: Cell mass; ×: dissolved oxygen concentration (DO); ◆: total sugar (TS); ◇: Fo, formic acid; ○: Ac, acetic acid; △: Pro, propionic acid; □: Bu, butyric acid; ▲: COD.

2.5 Cow Dung and Swine Waste

The potential annual production of livestock discharges in Japan is *c.* 5 500 000 tons (Hayami, 1985). Part of the waste is being treated by methanogenic fermentation. However, the effluent from an anaerobic digestor still contains abundant organic matter, e.g. biochemical oxygen demand, BOD, 3000–4000 mg/liter, consisting mainly of volatile fatty acids. To treat the discharged effluent, aerobic degradation of activated sludge is commonly applied. Sometimes, dilution of the discharged effluent is required before treatment by aerobic degradation.

The application of photosynthetic bacteria to the secondary treatment of the effluent after anaerobic digestion of livestock discharges is described below.

CHEMICAL COMPOSITION

The chemical compositions of cow and swine discharges and of the effluent from their anaerobic digestion are summarized in Table 7. Swine discharges contain high amounts of organic matter in terms of COD or BOD with a high nitrogen source. After anaerobic digestion of the discharges, the COD decreased to 3·9 g/liter, one-tenth of its original value, consisting mainly of

Table 7
Compositions of Cow and Swine Discharges and the Effluent after Anaerobic Digestion of the Discharges

		Cow[a]		*Swine*[b,c]	
		Discharge	*Effluent after anaerobic digestion*	*Discharge*	*Effluent after anaerobic digestion*
pH		—	—	7·0–7·3	7·5–8·0
BOD	(g/liter)	—	—	62·7	—
COD	(g/liter)	—	—	35·0	3·9
Total nitrogen	(g/liter)	3·75	2·50	4·7	3·0
Protein	(g/liter)	3·00	2·00	0·8	0·8
NH_4^+-N	(g/liter)	—	—	2·8	2·5
Acetic acid	(g/liter)	0·50	0·39	—	0·48
Propionic acid	(g/liter)	0·50	0·32	—	1·17
Butyric acid	(g/liter)	0·10	0·20	—	0·28

[a] Vrati, 1984. [b] Sasaki *et al.*, 1987*a*. [c] Wagai, 1985.

volatile fatty acids. Since the volatile fatty acids are useful substrates for photosynthetic bacteria, instead of using the conventional activated sludge treatment, *Rb. capsulatus* and *Rb. sphaeroides* S were applied to SCP production from the effluent of anaerobic digestion of livestock discharges.

EXPERIMENTAL PROCEDURE

Anaerobic digestion: For the anaerobic digestion of cow dung, the discharges (10% dry weight/volume) were fermented anaerobically at 45°C in the presence of native microflora for 10 days (Vrati, 1984). When biogas production ceased, the culture broth was centrifuged to obtain a clarified supernatant, which was further sterilized, with pH adjusted to 7·0, and used as the medium for a photosynthetic bacterium, *Rb. capsulatus*.

For the anaerobic digestion of swine discharges, the discharge was diluted with urine and tap water in ratio dung/urine/water = 1/2/2. This slurry was anaerobically fermented for 7 days at 30°C in a 1·2 liter fermentor with medium of 1 liter (Fig. 9) (Sasaki *et al.*, 1987*a*). After the fermentation, the culture broth was centrifuged at 12 000 *g* for 30 min and the clarified supernatant was used for the medium of *Rb. sphaeroides* S without sterilization.

Cultivation of photosynthetic bacteria: *Rhodobacter capsulatus* was anaerobically cultivated at 37°C under light illumination (10 klux) on the medium prepared from the post-anaerobic digestion of cow dung (for details see Vrati, 1984).

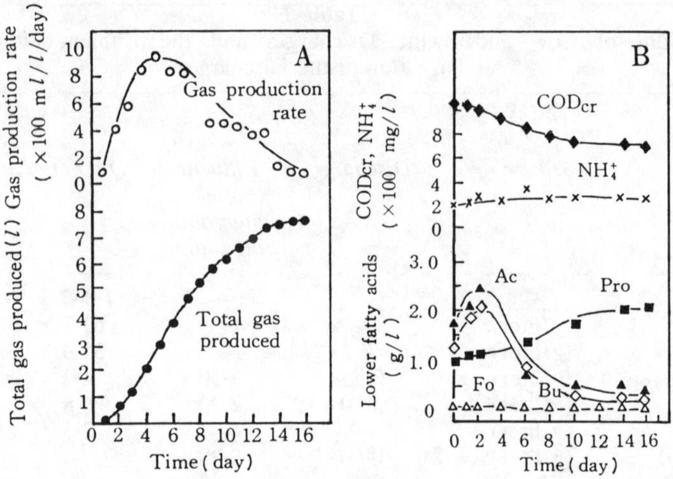

Fig. 9. Typical anaerobic digestion of swine discharges. (Sasaki *et al.*, 1987*a*) A, Biogas production rate (○) and total gas produced (●). B, Profiles of COD_{cr} (◆), NH_4^+ (×), acetic acid (▲, Ac), propionic acid (■, Pro), butyric acid (◇, Bu) and formic acid (△, Fo).

Rhodobacter sphaeroides S was aerobically cultured on the medium prepared from the swine discharges using a 1·5 liter fermentor with medium of 1 liter. Two hundred ml of *Rhodobacter sphaeroides* S cells grown on acetate–propionate medium (Sasaki *et al.*, 1978) were inoculated into the fermentor and cultivated at 30°C for 2 days. The dissolved oxygen concentration was kept above 2 mg/liter.

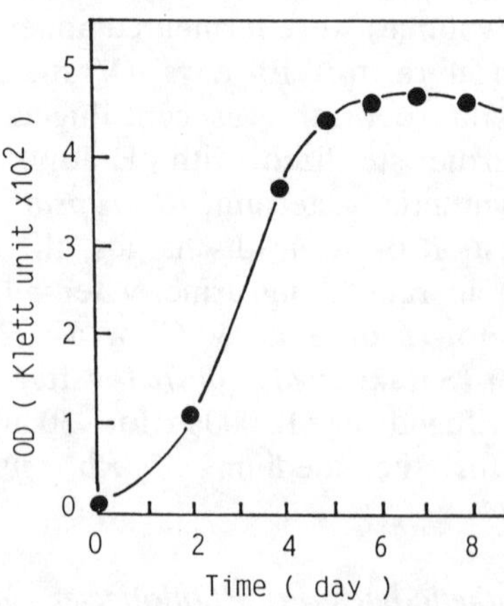

Fig. 10. Growth of *Rb. capsulatus* under anaerobic-light conditions on the medium prepared from the effluent of anaerobic digestion of cow dung (modified from Vrati, 1984).

RESULTS

Rhodobacter capsulatus cells grown on the medium prepared from cow dung (Fig. 10) indicate that, after 6 days' cultivation, maximum growth was attained at a value of 4·56 g wet cell/liter (0·91 g dry wt/liter). This suggests that *c.* 64 kg of *Rb. capsulatus* cells may be harvested from the effluent of anaerobic digestion of 10 tons of cow dung. Other photosynthetic bacteria such as *Rp. palustris, Rb. sphaeroides, Rsp. rubrum* and *Rc. tenuis* may also be available for SCP production from this process (Vrati, 1984).

An aerobic cultivation of *Rb. sphaeroides* S cells on the medium prepared from the post-anaerobic digestion of swine discharges (Fig. 11) indicates that for 20 h culture, the cell mass attained up to 1·8g dry cell/liter and COD was reduced from 4·4 g/liter to 0·9 g/liter (removal 80%) which corresponds to the utilization of volatile fatty acids. It is interesting to note that *Rb. sphaeroides* S can grow faster on unsterilized medium than the originally existing microorganisms from methane fermentation. Predominance of *Rb. sphaeroides* S cells after 20 h culture (Fig. 11) was *c.* 90% under microscopic observation. This suggests that this organism may be suitable for this process. Growth yield from COD removal was 0·52 g/g COD. Continuous cultivations of *Rb. sphaeroides* S under aerobic conditions and using the same medium, indicated that when the dilution rate of continuous culture was less than 0·125/h, almost one month of stable operation was continuing, with COD reduction of 70-80% and the predominance of *Rb. sphaeroides* was 80–90%. If 20 ton/day of swine discharges are produced, 42 kg of dry cells can be produced per day from this process.

Thus, SCP production using photosynthetic bacteria from cow dung and

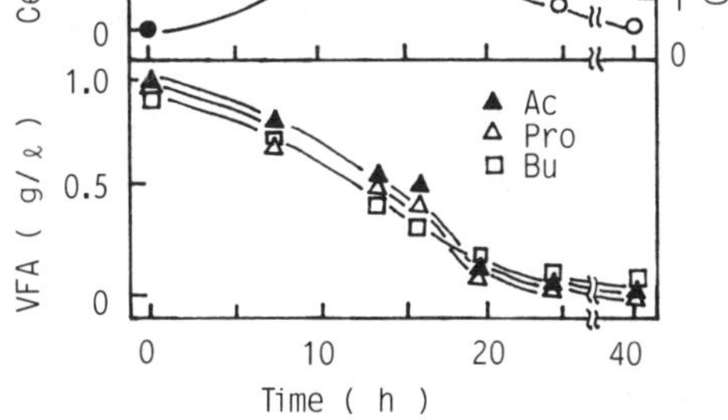

Fig. 11. Time course of cell growth, COD and volatile fatty acids (VFA) disappearance by aerobic culture of *Rb. sphaeroides* S in the medium prepared from the effluent of anaerobic digestion of swine discharges. Ac, acetic acid; Pro, propionic acid; Bu, butyric acid.

swine waste seems to be possible, although these discharges must be treated by anaerobic digestion before applying this process, followed by the separation of the solid to obtain the medium required.

2.6 Cell Yields and Compositions of PSB

To be suitable for SCP production, the microorganisms should grow fast with efficient growth yield from diverse carbon substrates and the cell mass should also contain balanced nutrients. As to these, purple nonsulfur photosynthetic bacteria (PSB) have been proposed by various researchers (Shipman *et al.*, 1975; Kobayashi & Kurata, 1978). This section deals with the growth characteristics of PSB for various waste materials together with their nutritional quality as useful animal feeds.

RESULTS

The growth characteristics of PSB in terms of growth rate, cell mass yields from carbon substrates, maximum cell mass attained and COD removal from several agroindustrial wastes are summarized in Table 8 (Sasaki, 1980; Noparatnaraporn *et al.*, 1986*a*).

From the table, the maximum cell mass ever obtained was 26·5 g/liter after 56 h aerobic culture in pineapple peel waste. Cell mass yields ranged from 0·4 to 0·67, all being acceptable for cell mass production. COD removal ranged from 79 to 95·6%, indicating the advantages of PSB for subsequent treatment. Table 8 suggests that photosynthetic bacteria, particularly *Rb.*

Table 8

Growth Yields and Growth Characteristics of the Photosynthetic Bacteria from the Various Wastes

	Pineapple	*Soybean*	*Mandarin orange peel*[a]	*Cassava*	*Swine*
Strain & culture conditions[b]	P_{47} AD	R.g. AD	S MAL	PSBVI MAD + MAL	S AD
X_{max} (g/liter)[c]	26·5	9·3	3·7	4·7	1·8
T_m (h)[d]	56	22	50	264	19
$Y_{X/S}$ (g/g)[e]	0·45 (g/g TS)	0·67 (g/g TS)	2·48 (g/g COD)	0·40 (g/g TS)	0·53 (g/g COD)
COD removal (%)	85	81	84	95·6	79

[a] Post-anaerobic digestion liquor (see sections 2.4 and 2.5).
[b] P_{47}: *Rb. sphaeroides* P_{47}; R.g.: *Rc. gelatinosus*; S: *Rb. sphaeroides* S; AD: aerobic-dark; MAL: microaerobic-light; MAD: microaerobic-dark.
[c] Maximum cell mass attained.
[d] Time required until maximum cell mass attained.
[e] Growth yield from total sugar (TS) or COD.

Table 9
Cell Compositions of the Photosynthetic Bacteria Compared with Other SCP Sources

	Rc. gelatinosus[a]	*Rb. sphaeroides* P_{47}[b]	*Rb. capsulatus*[c]	*Esso–Nestle (Yeast)*[d]	*Chlorella vulgaris*[c]	*Soybean*[d]	*FAO/WHO scoring pattern*[e]
Crude protein	56	66·6	66·0	54·0	55·5		
Crude fat	2·45	1·88	7·0	10·0	8·07		
Carbohydrate	26·4	24·9	23·0	26·0	21·0		
Crude fiber	na	2·95	na	na	12·1		
Ash	3·21	3·62	4·0	7·0	3·28		
Lysine	3·04	2·57	2·86	3·76	2·71	2·58	5·5
Histidine	1·15	0·96	1·25	0·90	1·06	na	na
Threonine	2·04	2·87	2·70	2·63	2·28	1·62	4·0
Valine	3·44	2·68	3·51	3·20	3·02	1·86	5·0
Methionine	1·92	1·47	1·58	0·51	0·27	0·43	1·9
Isoleucine	2·73	1·78	2·64	2·63	2·44	1·80	4·0
Leucine	5·84	3·90	4·50	3·54	4·46	2·70	7·0
Phenylalanine	3·08	2·36	2·60	2·20	2·65	1·98	3·0
Arginine	3·42	3·55	3·34		3·24		
Aspartic acid	4·92	5·18	4·56		4·74		
Serine	2·10	2·33	1·68		2·12		
Glutamic acid	5·84	6·22	5·34		4·62		
Proline	2·40	2·02	2·80		2·12		
Glycine	2·58	3·18	2·41		2·28		
Alanine	4·45	5·06	4·65		2·98		
Tyrosine	1·50	1·70	1·71		0·96		

[a] Growth on cassava starch medium (aerobic-dark culture, see section 2.3). [b] Growth on pineapple waste (aerobic-dark culture, see section 2·1). [c] Kobayashi & Kurata, 1978; [d] Kanamori, 1984; [e] Noparatnaraporn, 1987; na: not analyzed.

Table 10
Vitamin Content of the Photosynthetic Bacterial Cells[a]

	Rc. gelatinosus	*Rb. sphaeroides* P_{47}	*Rb. capsulatus*[b]	*Esso–Nestle (Yeast)*[c]
B_1	na[d]	na	12	11–13
B_2	33·2	13·0	50	110–130
B_6	na	na	5	4·8–7·6
B_{12}	33	78	21	trace
E	51	210	na	na
Carotenoid	90	800	na	na
Nicotinic acid	136	58	125	165–200
Folic acid	7·2	1·0	60	1·8–2·4
Pantothenic acid	na	na	30	14–23
Biotin	8·3	6·3	65	110–130

[a] μg Vitamins/g dry cells. [b] Kobayashi & Kurata, 1978; [c] Kanamori, 1984. [d] Not analyzed.

sphaeroides P_{47}, might be useful microbial resources for SCP production on a commercial scale, since the cell mass attained in commercial production should reach at least 10–20 g dry cell/liter.

The nutritional quality of the PSB cells from various wastes is shown in Tables 9 and 10. The proximate analyses indicate that the cell mass contained crude protein ranging from 56 to 67%, being comparable to yeast and algae, and that the amino acid patterns were also comparable to those of yeast and algae. In particular, the contents of lysine, methionine, leucine and phenylalanine, four essential amino acids, were appreciable in comparison to other SCP cells. Among them, the methionine content of PSB cells was much higher than that of other SCP and plant protein, e.g. soybean (0·43%). Since methionine is one of the limiting essential amino acids in animal feedstuff, PSB cells may be a useful source of this material in feedstocks and should thus be recommended as a supplement to the basic diet rather than as the sole protein source.

The PSB cells also contain considerable amounts of several vitamins necessary for feeding animals (see Table 10). *Rhodocyclus gelatinosus* cells from cassava waste were rich in vitamin B_2 (33·2 mg/kg cell) and niacin (135·8 mg/kg cell) while those of *Rb. sphaeroides* P_{47} from pineapple waste were rich in vitamin B_{12} (78·9 mg/kg cell) and vitamin E (210 mg/kg cell). It is noticeable that all four vitamins essential for animal feedstuff were contained in PSB cells in appreciable values.

In addition, the PSB cells contained carotenoid pigment to the extent of 0·09–0·68 mg/g cell. The biological functions of carotenoids were recently reported to be useful in intensifying the color of egg yolk, chicken flesh and aquarium-fish skin, and in increasing the viability and decreasing the mortality of teleost eggs—a noteworthy advantage from the use of PSB cells for animal feeds (Noparatnaraporn, 1987).

Therefore, based on the nutritional quality of PSB cells, these bacteria cultivated on various agroindustrial wastes are proposed as a good potential source of a multipurpose feed supplement for animal feedstuffs.

2.7 Problems

As described in sections 2.1–2.6, application of photosynthetic bacteria to SCP production from agroindustrial wastes seems to be promising. However, there are some practical problems, as follows:

(1) Wastes should be discharged in sufficient quantities and concentrations for processing to be economically viable.
(2) Sterilization of the waste before cultivation is no longer considered desirable prior to inoculation with the photosynthetic bacteria. It is recommended that the waste water discharged (e.g. in soybean curd manufacture) from heat treatment processes is used for this purpose. Otherwise, to maintain the predominance of photosynthetic bacteria throughout the culture, heavy inoculation is required to achieve the cultivation within a short period.
(3) Bacterial cell harvest is generally costly compared with the SCP yeast cells. Research and development of a novel separation technique are required to solve the problems.
(4) In some cases, research and development of photobioreactor construction are required, in which the efficiencies of light illumination to the cells, maintenance of homogeneous mixing and optimum temperature have to be considered.

Considering these problems, economic assessment should be made when plant-scale production is planned. With reference to (1) and (2) above, the manufacture of soybean curd (tofu) and fermented soybean mash (miso) seem to offer suitable opportunities for the use of photosynthetic bacteria because the wastes discharged after autoclaving contain abundant organic matter. In this case, the cultivation of photosynthetic bacteria under aerobic conditions is practicable, as with baker's yeast cultivation in a normal fermentor. Also, the use of photosynthetic bacteria for distilled waste from alcohol distilleries may be possible.

In situations such as using post-anaerobic digestion liquor from orange peel waste (section 2.4) and swine discharges (2.5), the necessity or otherwise of heavy inoculation may be a key factor, especially for the production of ALA (4·3), ubiquinone (4·2) and vitamin B_{12} (4·1).

3 APPLICATION TO FEEDSTOCK

It has been evident from this chapter as well as the reports of other workers (Shipman *et al.*, 1975; Kobayashi & Kurata, 1978) that photosynthetic bacterial cells could be enriched with high quality protein and other physiologically active substances for inclusion in animal feedstocks. However, research data on the use of PSB cells in feeding animals have been rather few, although application of PSB to fish farms is quickly taking hold in Japan. This particular application is therefore described below.

EXPERIMENTAL PROCEDURE

Fresh PSB cells of *Rb. sphaeroides* P_{47} cultivated on pineapple waste were used in the fish cultures of red *Tilapia* (Noparatnaraporn, 1987) and fancy carp whereas *Rc. gelatinosus* cells from cassava solid waste were used for goldfish (Noparatnaraporn, 1988, Noparatnaraporn *et al.*, 1987*b*).

Diet composition and preparations: The PSB diets (Table 11) were prepared as follows: all ingredients were homogeneously mixed with the addition of deionized water to adjust moisture content to *c.* 10% and then pelletized by wet extrusion; this was followed by air-drying for 6–8 h and preparation in convenient quantities.

RESULTS

Effects on growth of goldfish: Table 12 indicates that there was no difference in mortality between control and PSB-supplemented diets; the latter were readily accepted and utilized, and no toxicity was found in the test fish. Similar results were also obtained in red *Tilapia* and carp. Table 12 shows that growth was significantly superior in terms of growth rate (9·9–11·69%/day) and weight gain (99·1–116·97 mg/day, 18·03%).

Effects on gonad development and fecundity: Effects on gonad development of red *Tilapia* with the supplement of *Rb. sphaeroides* P_{47} cells to the control

Table 11

Compositions of Control and PSB Cells-Supplemented Diets[a]

Compositions (1%)	*Red Tilapia* *CD*	*PSBD*	*Goldfish* *CD*	*PSBD*	*Fancy carp* *CD*	*PSBD*
Fish meal	12	6	30	15	16	8
Peanut meal	23	11·5	24	24	24	24
Rice bran	40	20	45	45	30	30
Soybean meal	—	—	—	—	14	14
Sponge tree leaf	4	1	—	—	—	—
Broken rice	—	—	—	—	15	15
Rice husk	—	—	—	—	—	1·2
Alpha starch	20	10	—	—	—	7·5
Premix	1	0·5	1	1	1	1
PSB cells	0	50	0	15	0	6·8
Crude protein	24·3	44·7	27·0	27·1	28·6	28·9

[a] CD: control diet; PSBD: PSB supplemented diet (for *Tilapia* and carp, *Rb. sphaeroides* P_{47} was used, and for goldfish, *Rc. gelatinosus*).

Table 12
Effects of Supplementation of *Rc. gelatinosus* Cells to Control Diet on the Growth of Goldfish *Carassius auratus* (2-*month old*)[a]

Experimental analytical term	*Control diet*	*PSBD*[b]
Total number of fish	60	60
Initial body weight (g)	3·40	3·40
Initial total length (cm)	3·58	3·58
Final body weight (g)	15·49	17·67
Final total length (cm)	5·30	5·43
Weight gain (%)	355·59	419·71
Weight gain (mg/day)	99·10	116·97
Feed consumption (mg/day)[c]	364·75	395·88
Feed conversion ratio (mg/mg)[d]	3·68	3·38
Specific growth rate (%/day)[e]	9·91	11·69
Survival (%)	100	100

$$\text{Increase in weight gain} = \left(\frac{116{\cdot}97 - 99{\cdot}10}{99{\cdot}10}\right) \times 100\% = 18{\cdot}03\%$$

[a] Growth was determined monthly for 122 days (from 1 March to 30 June), with fish 2 months old at the start. [b] PSBD: PSB supplemented diet (see Table 11). [c] Averaged from monthly feed intake based on 5% of body weight. [d] Calculated from feed consumption/weight gain. [e] Calculated from (weight gained, g/day) × 100.

diet (data not shown) suggested that the vitamin E of PSB cells might stimulate the development of the reproductive system. In fact, vitamin E was an essential vitamin for the reproduction of feeding animals (Noparatnaraporn, 1987).

The fecundity development of 6-month old goldfish (see Fig. 12) supported the results obtained with red *Tilapia*. Figure 12 shows significant enhancement in fecundity in terms of ovary weight (increases of 8·6% and 41·2%) and total numbers of eggs (increases of 34·8% and 68·5%).

Effects on skin pigmentation: In fancy carp, the effects of *Rb. sphaeroides* P_{47} cell supplementation on skin pigmentation were tested with supplementations of 3·4, 6·5, 10·2 and 13·6% to the control diet (data not shown). It was found that the feed pellet substituted with more than 6·8% resulted in increased skin pigmentation, particularly in shades of deep orange and red (Noparatnaraporn, 1988), while the effect of carotenoid incorporation into muscle pigment could not be observed in the carp.

Summarizing, PSB cells might be useful as a multipurpose fish-feed supplement, and have been used successfully in this way at various fish farms in Japan (from a trade paper on fish farms, Minato Shinbun, Shimonoseki, Japan, 1989).

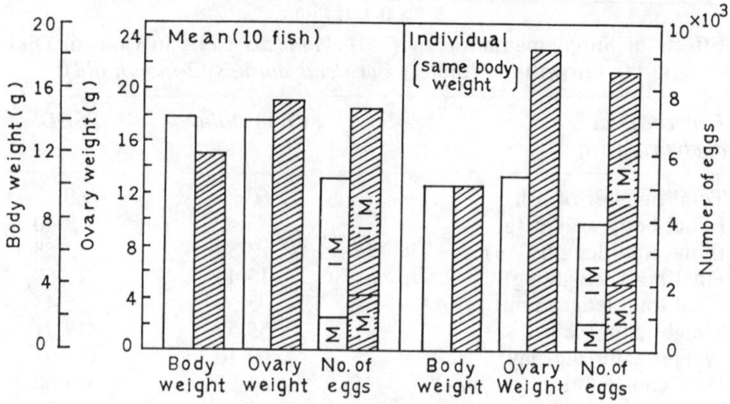

Fig. 12. Effect of the supplement of *Rc. gelatinosus* cells to the control diet (see Table 11) on the fecundity development of 6-month-old female goldfish (Noparatnaraporn *et al.*, 1987*b*). □: Control diet; ▨: PSB supplemented diet; M: mature; IM: immature.

4 CHEMICAL PRODUCTION

4.1 Vitamin B_{12} (Corrinoids)

Vitamin B_{12} is characterized as a carrier for methyl transfer and coenzymes for such enzymes as glutamate mutase, methylmalonyl-CoA mutase and so on (Fukui, 1980). Vitamin B_{12} is not only widely used for medicines for pernicious anemia and other disorders, but also applied as a supplement to animal feedstocks to enhance their nutritional value. Vitamin B_{12} is produced through the fermentation of sugars by, e.g. propionic acid bacteria.

As shown in Fig. 13, vitamin B_{12} synthesis is closely correlated with the biosynthesis of other tetrapyrrols (porphyrin, heme and chlorophyll). In the photosynthetic bacteria such as *Rhodobacter* spp., tetrapyrrols are mainly synthesized from glycine and succinate via 5-aminolevulinic acid (ALA) and Urogen III as the common intermediate (Shemin pathway) (Lascelles, 1978). However, in the algae, higher plants and methanogenic bacteria, tetrapyrrols are synthesized from glutamate via ALA (C-5 pathway) (Jaenchen *et al.*, 1981; Klein & Porra, 1982). In *Rb. sphaeroides,* the Shemin pathway of total tetrapyrrols synthesis seems to account for *c.* 90% of the synthesis compared with that of the C-5 pathway (Klein & Porra, 1982). In the Shemin pathway, ALA synthetase seems to be a key enzyme, playing an important role in tetrapyrrol synthesis. This enzyme is regulated by heme compound (see Fig. 13) by a feedback inhibition (Lascelles, 1978).

The dissolved oxygen concentration and light illumination during cultivation of photosynthetic bacteria were altered on the formation of ALA synthetase and consequently, a remarkable difference in tetrapyrrol accumulations was

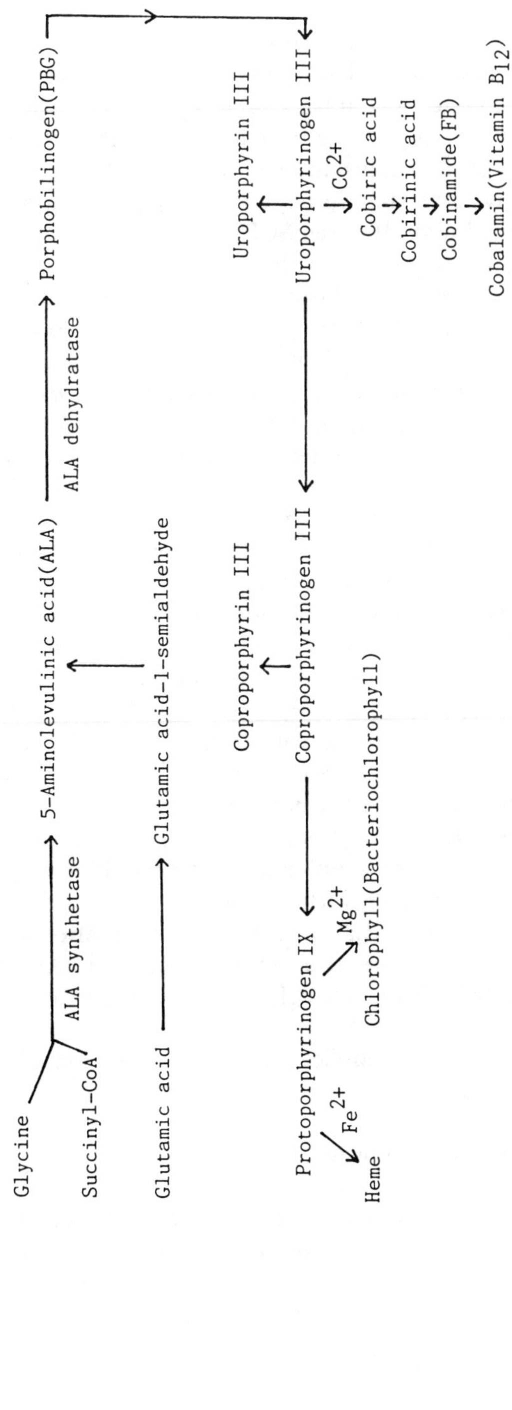

Fig. 13. Vitamin B_{12} biosynthesis pathway in photosynthetic bacteria (modified from Satoh & Shimizu, 1988).

observed (Noparatnaraporn *et al.*, 1986*b*). For example, under a low dissolved oxygen concentration and low light illumination, ALA synthetase activity was enhanced with increase in bacteriochlorophylls of *Rb. sphaeroides* (Sasaki & Nagai, 1979*a*).

In corrinoid formation by *Rb. sphaeroides,* cobalamin is the main product (Sasaki *et al.*, 1978) and incomplete corrinoids such as cobirinic acid and cobinamide (FB) are undetectable intracellularly.

In the application of photosynthetic bacteria to the production of SCP from agroindustrial wastes, it is important whether or not vitamin B_{12} can be produced intracellularly since vitamin B_{12} containing SCP is more valuable for animal feedstocks than vitamin B_{12}-free SCP.

In Table 13, vitamin B_{12} production from agroindustrial wastes by photosynthetic bacteria is summarized. Usually, the vitamin B_{12} content in the photosynthetic bacterial cells is 23–87 μg/g dry cells irrespective of culture conditions. These values are appreciable compared with a commercial SCP, e.g. ESSO–Nestle yeast (0·11–0·17 μg/g cell) (Kanamori, 1984).

The intracellular content of vitamin B_{12} produced from the orange peel waste (Table 13) increased under either microaerobic-light or anaerobic-light conditions. However, in anaerobic-light culture, growth rate and the cell mass attained were relatively lower than those in aerobic-dark cultivation. To overcome this drawback, it is possible to enhance the low vitamin B_{12} content of the aerobically grown cells by controlling the dissolved oxygen concentration at an extremely low level (Fig. 14) (Noparatnaraporn *et al.*, 1986*b*). In the aerobic-dark culture of *Rc. gelatinosus* on cassava starch medium, the first stage of the cultivation was conducted under sufficiently aerobic conditions (DO > 4 mg/liter) to support a high growth rate with high substrate uptake rate; then after the cell mass attained its maximum level, the aeration and agitation were drastically reduced (see the second stage in Fig. 14) to establish a microaerobic condition where redox potential in the broth was *c.* −200 mV. As shown in the figure, vitamin B_{12} content increased from 22 to 39 μg/g cell after 6 h incubation without light illumination. Bacteriochlorophyll and carotenoid contents increased together with vitamin B_{12} under the microaerobic conditions, indicating that the ALA synthetase activity also increased under these conditions (Noparatnaraporn *et al.*, 1986*b*). Protein content in the cells was kept unchanged (62–63%) throughout the aerobic and microaerobic conditions.

This simple technique may be applicable to the production of enriched SCP from the photosynthetic bacteria.

4.2 Ubiquinone from Starchy Waste

Ubiquinone consists of the structure 2,3-dimethoxy-5-methylbenzoquinone with a polyisoprenoid at the sixth position of the benzene ring. The natural ubiquinones are Q-6, Q-7, Q-8, Q-9 and Q-10, where the number refers to the carbon number of the isoprene unit of the side chain. These compounds play

Table 13
Vitamin B_{12} Production by Photosynthetic Bacteria from Agroindustrial Wastes

Source of waste	*Organism*	*Culture conditions*[a]	*Vitamin B_{12} produced*		*Ref.*
			μg per g dry cell	*μg per liter broth*	
Pineapple	*Rb. sphaeroides* P_{47}	AD	75	1 578	Noparatnaraporn *et al.*, 1986*a*
Soybean	*Rc. gelatinosus*	AD	33	307	Sasaki,*et al.*, 1981
Cassava starch	*Rc. gelatinosus*	AD	23	104	Noparatnaraporn *et al.*, 1987*a*
	(Mixed culture *Rc. gelatinosus*: *Rb. sphaeroides* P_{47}, 1:1)	AD	44	274	Noparatnaraporn *et al.*, 1987*a*
Mandarin orange peel[b]	*Rb. sphaeroides* S	AD (DO > 4 mg/liter)	37	74	Sasaki, 1980
		MAL (DO ≑ 0, 3 klux)	79	260	Sasaki, 1980
		ANL (3 klux)	87	67	Sasaki, 1980

[a] AD: Aerobic-dark (DO > 4 mg/liter); MAL: Microaerobic-light; ANL: Anaerobic-light.
[b] Supernatant after anaerobic digestion of orange waste.

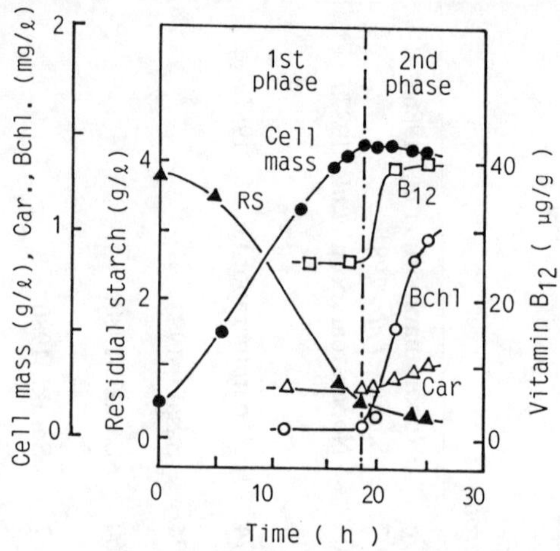

Fig. 14. Enhancement of photopigments and vitamin B_{12} formations of *Rc. gelatinosus* growing cassava starch medium after the culture condition was suddenly changed from aerobic to microaerobic (Noparatnaraporn *et al.*, 1986*b*); pH = 7, temperature = 40°C. 1st phase: aerobic-dark, DO > 4 mg/liter (ORP > +110 mV), aeration 1 vvm, agitation 500 rpm. 2nd phase: microaerobic-dark, DO = 0 (ORP = −200 ± 10 mV), aeration 0·1 vvm, agitation 200 rpm. B_{12}: Vitamin B_{12}, Car: carotenoid, Bchl: bacteriochlorophyll, RS: residual starch.

an important role in electron transfer in living systems (Morimoto, 1971). Q-11 and Q-12 are observed in higher plants (Morimoto, 1971).

Among the ubiquinones in nature, Q-10 is widely used as a medicine, being effective for anemia, hyperpiesia, pyorrhea and dystrophy. Photosynthetic bacteria are potent producers of ubiquinone which plays an essential role in the electron transport of both respiration and bacterial photosynthesis (Morita & Shimada, 1984).

Industrial production of ubiquinone Q-10 using photosynthetic bacteria has been carried out in Japan. Ubiquinone synthesis of photosynthetic bacteria is closely related to the carotenoid synthesis via the common pathway of terpenoid synthesis (Fig. 15). Benzoquinone moiety of ubiquinone is synthesized from phosphoenolpyruvate via shikimate and *p*-hydroxybenzoate. Intracellular accumulation of ubiquinone was frequently regulated by culture conditions such as dissolved oxygen concentration, light illumination and medium composition in the bacteriochlorophyll synthesis of photosynthetic bacteria.

Table 14 shows ubiquinone production using a starch medium by *Rc. gelatinosus* at different culture temperatures. At the optimal temperature of 35°C, ubiquinone (Q-8) content attained 1·63 mg/g cell (1·25 mg/liter broth) which was comparable with other photosynthetic bacteria such as *Rb. capsulatus* and *Rb. sphaeroides*. It is advantageous that *Rc gelatinosus* can utilize starchy waste directly as observed in section 2.3, but Q-8 produced

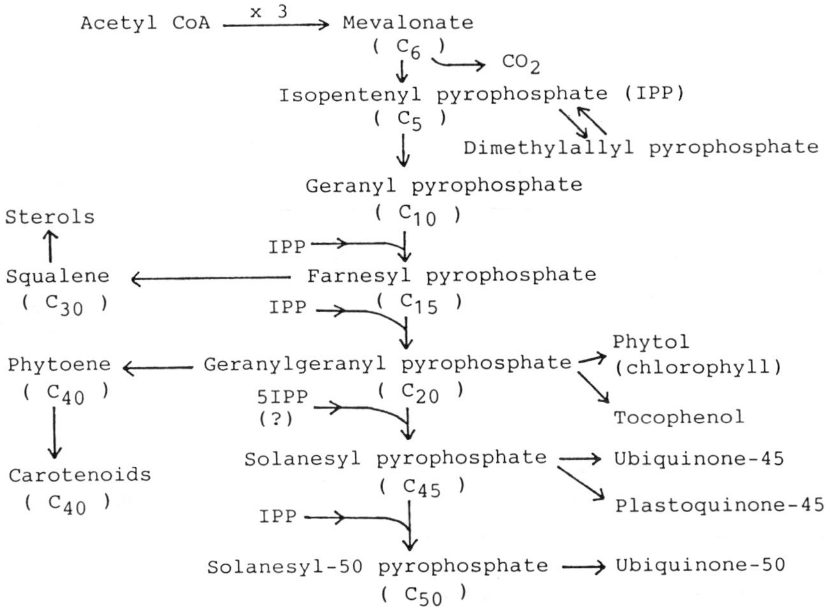

Fig. 15. Carotenoid and ubiquinone biosynthesis (Britton, 1976).

intracellularly is less valuable compared with Q-10 as a medicine (Sasaki & Nagai, 1979*b*).

In Table 15, ubiquinone, carotenoid and bacteriochlorophyll formations of *Rb. sphaeroides* P_{47} grown under anaerobic-light conditions indicate that intracellular ubiquinone (Q-10) content was *c.* 3 mg/g cell, which was twice that of aerobically grown cells. The different carbon source did not affect ubiquinone content (Table 15), suggesting that in ubiquinone production it might be possible to use the culture medium of acidogenic fermentation which contains volatile fatty acids (sections 2.4 and 2.5). Carotenoid contents in

Table 14
Ubiquinone Production of *Rc. gelatinosus* Aerobically Grown on a Starch Medium (pH = 7·0) as a Function of Culture Temperature

Temperature (°C)	*Cell mass (g/liter)*	*Ubiquinone (Q-8)*	
		mg/g cell	*mg/liter*
25	0·48	0·21	0·10
30	0·45	1·04	0·47
33	0·63	1·32	0·83
38	0·57	1·54	0·88
41	0·44	1·40	0·62
45	0·37	0·90	0·33

Source: Sasaki & Nagai, 1979*b*.

Table 15
Ubiquinone and Photopigment Formations of *Rb. sphaeroides* P_{47} Grown Under Anaerobic-light Conditions (5 klux, 35°C)

Carbon substrate	*Ubiquinone (Q-10)*		*Carotenoid*		*Bacterio-chlorophyll*
	mg/g cell	*mg/liter broth*	*Spheroiden*	*Spheroidenon*	*mg/g cell*
			mg/g cell		
Glutamate	2·7	2·2	3·8	1·0	33
Acetate	2·8	2·7	3·0	2·8	32
Propionate	3·0	1·7	3·0	4·7	30

terms of spheroiden and spheroidenone of *Rb. sphaeroides* P_{47} (Table 15) were rather high.

It may be possible to use photosynthetic bacteria for the production of ubiquinone and carotenoids simultaneously from agroindustrial waste, not only for producing SCP but also useful chemicals.

4.3 5-Aminolevulinic Acid from Swine Waste

5-Aminolevulinic acid (ALA), an intermediate of tetrapyrrol biosynthesis, is a new herbicide which damages weeds but does not harm crops, humans or other animals (Rebeiz *et al.*, 1984). However, due to the difficulties of chemical synthesis of ALA, biological processes of ALA production become attractive (Sasaki *et al.*, 1990).

With respect to ALA formation by microorganisms, it has been found that a bacterium, *Methanobacterium thermoautotrophicum* (Jaenchen *et al.*, 1981), and algae such as *Agnemellum quadruplicum, Anacystis marina* and *Chlorella vulgaris* (Beale, 1970) can excrete ALA extracellularly when levulinic acid (LA), a competitive inhibitor of ALA dehydratase in tetrapyrrol biosynthesis (C-5 pathway), was added to the medium. Although the amount of ALA excreted was quite low, the usefulness of LA for the extracellular formation of ALA may be applicable to industrial ALA production.

Rhodobacter sphaeroides, having the tetrapyrrol synthesis pathway (see Fig. 13) seems to be a potent producer of ALA in the presence of LA; in fact, ALA excretion from the growing cells was observed (Sasaki *et al.*, 1987*b*). If ALA can be produced by *Rb. sphaeroides* from waste such as anaerobic digestion liquor from swine waste containing volatile fatty acids, this kind of use of digestion liquor will be more advantageous because after cultivation by *Rb. sphaeroides* this liquor can be used directly for fertilizer having herbicide activity. In a usual pig house in Japan, a relatively high running cost is required for the treatment of swine waste by anaerobic digestion followed by the conventional activated sludge treatment.

In this section, ALA production by *Rb. sphaeroides* using the medium prepared from post-anaerobic digestion of animal discharges is described

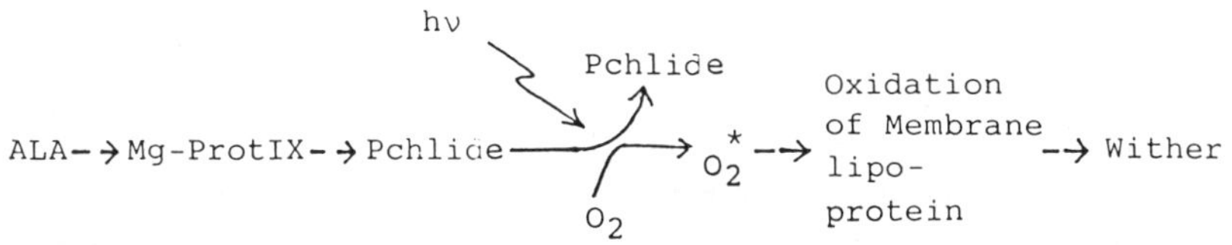

Fig. 16. Herbicide mechanism of ALA to weeds (based on information from Rebeiz *et al.*, 1984). ALA: 5-Aminolevulinic acid, Mg-ProtIX: Mg-protoporphyrin IX, Pchlide: protochlorophyllide.

together with a herbicide effect upon weeds using ALA containing culture broth.

In Fig. 16, the herbicide mechanism of ALA is summarized (Rebeiz *et al.*, 1984). The ALA sprayed on a leaf is converted to protochlorophyllides ('pchlides') under dark conditions. The six pchlides so far known are Mg-mono- and divinylprotoporphyrin IX, their monesters, and mono- and divinylprotochlorophyllide. Under sun irradiation, the overaccumulated pchlides act as photosynthesizers, i.e. convert triplet oxygen to the potently oxydizing singlet oxygen. The singlet oxygen superoxidizes the phospholipids of the cell membrane of the plant leaf. Damaged membranes become leaky and consequently weeds die. This herbicide action is sensitive for dicotyledons but less sensitive for monocotyledons.

ALA production by *Rb. sphaeroides* from the post-anaerobic digestion liquor is shown in Fig. 17. In the figure, LA addition alone (a) did not produce

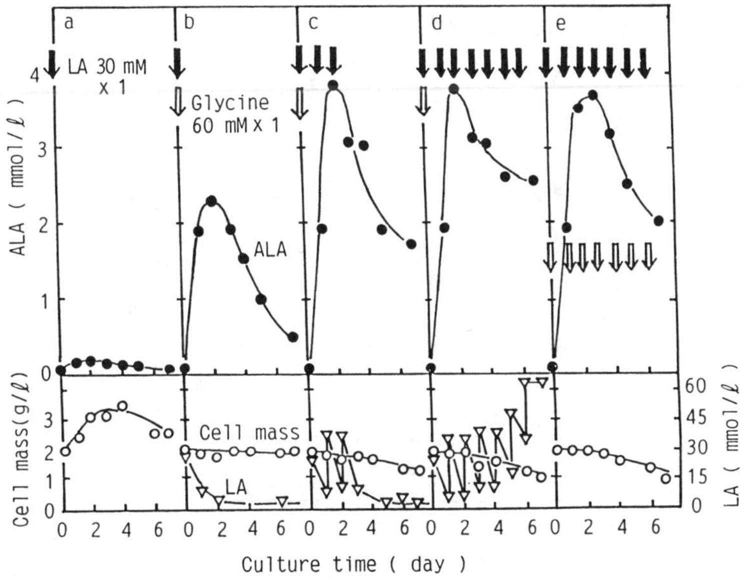

Fig. 17. ALA production of *Rb. sphaeroides* (IFO 12203) from the supernatant of post-anaerobic digestion of swine discharges (Sasaki *et al.*, 1990). Culture conditions: anaerobic-light illumination (5 klux), initial pH = 6·5, temperature = 30°C. Dense inoculation of preculture was done. (a)–(e): See text.

ALA; however, with LA and glycine, ALA was markedly excreted from the cells. Repeated addition of LA three times (c) accelerated ALA accumulation up to 3·9mM. It was suggested that the repeated addition of LA might maintain the ALA dehydratase activity at a low level. However, when LA was added seven times (d), ALA accumulation was no more enhanced compared with (c). The repeated addition of LA together with glycine seven times (e) did not enhance ALA compared with (c). The accumulated ALA was reutilized when LA had disappeared.

Profiles of ALA, cell mass, LA and glycine during culture of *Rb. sphaeroides* are shown in Fig. 18 which corresponds to Fig. 17(c). ALA accumulation of *c.* 4mM was twofold compared with another ALA producer, *Chlorella vulgaris* (Beale, 1970). Propionic acid seemed to be mainly utilized compared to acetic or butyric acid, although *Rb. sphaeroides* utilizes acetic, propionic and butyric acids simultaneously under normal growth conditions. It was suggested that propionic acid might play an important role in ALA formation as a source of succinyl-CoA supply via the methylmalonyl–CoA pathway (Sasaki *et al.*, 1978; Kitamura, 1988).

A practical use of this culture broth (containing ALA at 4mM) as a herbicide

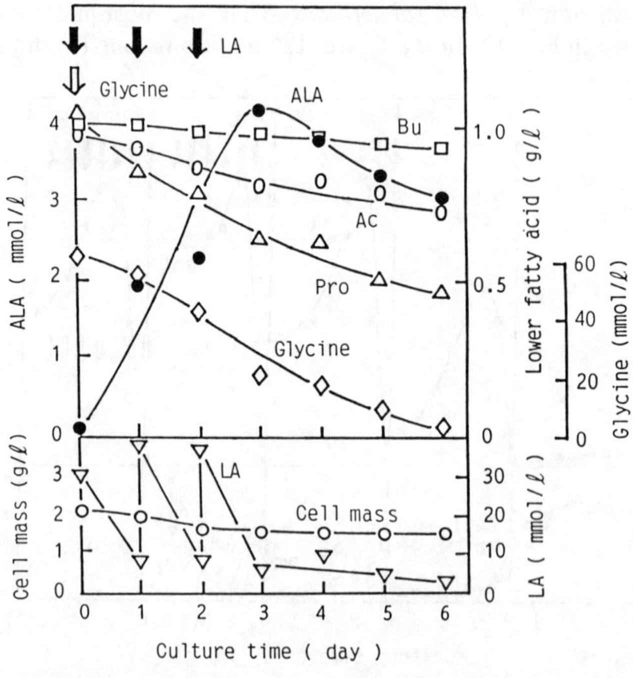

Fig. 18. Profiles of ALA (●), LA (▽), cell mass (○), glycine (◇) and volatile fatty acids (substrates) during culture of *Rb. sphaeroides* IFO 12203 on the medium prepared from post-anaerobic digestion of swine discharges (corresponding to Fig. 17(c)). ○: Ac, acetic acid; △: Pro, propionic acid; □: Bu, butyric acid.

Table 16
Herbicide Effect of Culture Broth (ALA 4mM) on the Leaves of Clover (*Trifolium repens* L.)[a]

	Herbicide activity (%)[b]		
	Spray after (days)		
	1	*2*	*3*
Control[c]	0	0	0
Culture broth[d]	83	90	95
Culture broth plus α,α'-dipilidyl[e]			
1mM	91	98	95
3mM	96	100	100
5mM	100	100	100

[a] Triton X-100 was added as the developing reagent (1% v/v).
[b] Area of damaged leaves ÷ area of healthy leaves × 100.
[c] Broth of 0-h culture (30mM LA plus 60mM glycine, containing no ALA). Sprayed 10 ml/150 cm^2.
[d] Broth of 4-day culture (ALA = 4mM). Sprayed 10 ml/150 cm^2.
[e] α,α'-Dipilidyl was added to enhance the herbicide effect.
Source: Sasaki, K., unpublished.

was tested in the same manner as reported by Rebeiz *et al.* (1984). As shown in Table 16, relatively strong herbicide activity was indicated within 3 days of spraying the culture broth directly on the leaves and stems of *Trifolium repens* (a clover) which was one of the most abundant plants in the fields. When α,α'-dipilidyl was added, the herbicide efficiency approached 100%. In addition, almost the same herbicide effects could be observed for wormwood, day flower and creeping woodsorrel which were common weeds in the fields, but a low effect was observed for monocot weeds such as crabgrass and goosegrass.

Thus, it is possible to utilize the post-anaerobic digestion liquor from swine waste for herbicide (ALA) production by *Rb. sphaeroides*. In addition, the culture broth can be utilized directly as a herbicide without any ALA concentration and purification treatment.

5 FUTURE PROSPECTS

Photosynthetic bacteria are potent producers of valuable SCP and physiologically active chemicals. In fact, these bacterial cells have been used on a practical scale as a nutritional ingredient for yellowtail and sea bream fish cultures in Japan. Stimulative effects are observed in fish cultures by enhancement of growth rate, improvement of the quality of fish meat, and prevention of sudden death of baby fish. In addition, ubiquinone (Q-10) production by a photosynthetic bacterium has been industrialized. ALA production for use as a biodegradable herbicide seems to be quite promising in the near future, since the synthetic chemicals widely used as herbicides have become serious environmental problems in farms and fields. In addition, production of antibiotics and antiviral substances by photosynthetic bacteria has been reported (Kobayashi, 1984). Thus, the range of application of the photosynthetic bacteria seems to be increasing.

With regards to future prospects of developing the utilization of photosynthetic bacteria for conversion of wastes to useful materials, the following subjects are considered.

Screening superior strains: Photosynthetic bacteria are widely distributed in the natural environment. Since the bacteria isolated so far are limited in number, a new strain may indicate a new ability to produce a physiologically active substance with a high growth rate. Recently, a potent producer of cellulase by a photosynthetic bacterium has been reported (Noparatnaraporn, not published).

Improvement of strains: Photosynthetic bacterial cells can be improved by means of mutation and genetic engineering techniques. For example, the nitrogenase unit of photosynthetic bacteria may transfer to other bacteria so that rapid N_2 fixation may become possible without light illumination. Mutation technologies have been widely applied to photosynthetic bacteria in which ubiquinone (Q-10) accumulation has increased more than 10 times (Asahikasei Co. Ltd, Japan, private communication), and these technologies may apply to the production of other physiologically active substances.

Improvement of culture systems: In SCP and chemicals production, a high cell density after cultivation is required from a practical point of view. In aerobic culture, this can be achieved by such means as fed batch culture, but the achievement of a dense culture is very difficult in anaerobic-light culture due to the limitation of light energy supply. In fact, some valuable cell components, e.g. vitamin B_{12}, increase in anaerobic-light culture compared with aerobic-dark culture. Also, for CO_2 and N_2 fixation by photosynthetic bacteria, light irradiation is essential. Therefore, for commercial application of photosynthetic bacteria, research and development into photobioreactors attaining a high cell density is very important. Recently, a special photobioreactor system using an optical fiber to provide light energy to the inside of the reactor has been reported to aid the development of new applications of photosynthetic organisms (Mori *et al.*, 1987). By using this system, fixation of CO_2 to materials on the metabolic pathway may become possible at a relatively high rate compared with algae and other plants. The mass-production of useful chemicals originating from photosynthetic bacteria may be realized in the near future.

REFERENCES

Beale, S. I. (1970). *Plant Physiol.*, **45,** 504.

Britton, G. (1976). In *Chemistry and Biochemistry of Plant Pigments,* Vol. I, ed. T. W. Goodwin. Academic Press, London, New York & San Francisco, 1976, p. 262.

Fukui, S. (1980). In *Vitaminology II,* ed. The Vitamin Society in Japan. Tokyo Kagaku Dozin, Tokyo, p. 477.

Hayami, A. (1985). In *Biomass,* ed. Nippon Nogei Kagakukai. Asakura Shoten, Tokyo, p. 1.

Hoshino, Y. & Kitamura, H. (1984). In *Kougousei-Saikin* (*The Photosynthetic Bacteria*), eds H. Kitamura, S. Morita & J. Yamashita. Gakkai Shuppan Center, Tokyo, p. 3.

Jaenchen, R., Harald, H. G. & Thauer, R. K. (1981). *FEMS Microbiol. Lett.*, **12,** 167.

Kanamori, M. (1984). In *Kougousei-Saikin* (*The Photosynthetic Bacteria*), eds H. Kitamura, S. Morita & J. Yamashita. Gakkai Shuppan Center, Tokyo, p. 329.

Kitamura, H. (1986). *Biseibuts* (*Japan*), **2,** 526.

Klein, O. & Porra, R. J. (1982) *Hoppe-Seyler's Z. Physiol. Chem.*, **363,** 551.

Kobayashi, M. (1984). In *Kougosei-Saikin* (*The Photosynthetic Bacteria*), eds H. Kitamura, S. Morita & J. Yamashita. Gakkai Shuppan Center, Tokyo, p. 339.

Kobayashi, M. & Kurata, S. (1978). *Process Biochem.*, **13,** 27.

Lascelles, J. (1956). *Biochem. J.*, **62,** 78.

Lascelles, J. (1978). In *The Photosynthetic Bacteria,* eds R. K. Clayton & W. R. Sistrom, Plenum Press, New York & London, p. 795.

Mori, K., Ohya, H., Matsumoto, K. & Furuune, H. (1987). *Adv. Space Res.*, **7,** 447.

Morimoto, H. (1971). *Tanpakushitau Kakusan Kouso,* **16,** 183.

Morita, S. & Shimada, K. (1984). In *Kougousei-Saikin* (*The Photosynthetic Bacteria*), eds H. Kitamura, S. Morita & J. Yamashita. Gakkai Shuppan Center, Tokyo, p. 293.

Nishio, N. & Nagai, S. (1979). *Eur. J. Appl. Microbiol. Biotechnol.*, **6,** 371.

Nishio, N., Kitaura, S. & Nagai, S. (1982). *J. Ferment. Technol.*, **60,** 423.

Noparatnaraporn, N. (1987). Studies on cell mass production by photosynthetic bacteria and cell utilization. PhD thesis, Faculty of Engineering of Hiroshima University, Hiroshima, Japan, p. 92.

Noparatnaraporn, N., (1988). In *Proceedings of the International Symposium on Applications of Biotechnology for Small Industries Development in Developing Countries,* Bangkok, Thailand, p. 25.

Noparatnaraporn, N., Wongkornchawalit, W., Kanchote, D. & Nagai, S. (1986*a*). *J. Fermet. Technol.*, **64,** 137.

Noparatnaraporn, N., Sasaki, K., Nishizawa, Y. & Nagai, S. (1986*b*). *Biotechnol. Lett.*, **8,** 491.

Noparatnaraporn, N., Trakulnaleumai, S., Silveira, R. G., Nishizawa, Y. & Nagai, S. (1987*a*). *J. Fermet. Technol.*, **66,** 11.

Noparatnaraporn, N., Trakulnaleumsai, S. & Duangsawat, S. (1987*b*). *J. Sci. Soc. Thailand,* **13,** 15.

Pfennig, N. (1978). In *The Photosynthetic Bacteria,* eds R. K. Clayton & W. R. Sistrom, Plenum Press, New York & London, p. 3.

Rebeiz, C. A., Montazer-Zouhoor, A., Hoppen, H. & Wu, S. M. (1984). *Enzyme Microbiol. Technol.*, **6,** 390.

Sasaki, K. (1980). Growth characteristics and their applications of a photosynthetic bacterium, *Rhodopseudomonas sphaeroides* S. PhD thesis, Faculty of Engineering of Hiroshima University, Hiroshima, Japan.

Sasaki, K. & Nagai, S. (1979*a*). *Eur. J. Appl. Microbiol. Biotechnol.*, **7,** 201.

Sasaki, K. & Nagai, S. (1979*b*). *J. Ferment. Technol.*, **57,** 383.

Sasaki, K., Hayashi, M. & Nagai, S. (1978). *J. Ferment. Technol.*, **56,** 200.

Sasaki, K., Noparatnaraporn, N., Hayashi, M., Nishizawa, Y. & Nagai, S. (1981). *J. Ferment. Technol.*, **59,** 471.

Sasaki, K., Kidani, T., Emoto, Y. & Hamaoka, T. (1987*a*). *J. Soc. Agric. Struct. (Japan)*, **18,** 38.

Sasaki, K., Ikeda, S., Nishizawa, Y. & Hayashi, M. (1987*b*). *J. Ferment. Technol.*, **65,** 511.

Sasaki, K., Tanaka, T., Nishizawa, Y. & Hayashi, M. (1990). *Appl. Microbiol. Biotechnol.*, **32,** 727.

Satoh, K. & Shimizu, S. (1988). *Vitamins (Japan)*, **62,** 69.

Shipman, R. H., Kao, I. C. & Fan, L. T. (1975). *Biotechnol. Bioeng.*, **17,** 1561.

Takahashi, H. (1984). In *Kougousei-Saikin (The Photosynthetic Bacteria)*, eds H. Kitamura, S. Morita & J. Yamashita. Gakkai Shuppan Center, Tokyo, p. 196.

Truper, H. G. & Pfennig, N. (1978). In *The Photosynthetic Bacteria*, eds R. K. Clayton & W. R. Sistrom, Plenum Press, New York & London, p. 19.

Vrati, S. (1984). *Appl. Microbiol. Biotechnol.*, **19,** 199.

Wagai, B. (1985). *Treatment and Utilization of Animal Discharges*. Youkendo, Tokyo, p. 1.

Chapter 8

BIOCONVERSION OF STARCH WASTES

Byong H. Lee

Agriculture Canada, Food Research & Development Centre, 3600 Casavant Blvd West, St Hyacinthe, Québec, Canada J2S 8E3

and

Department of Food Science, Macdonald College, McGill University, Ste-Anne-de-Bellevue, Québec, Canada, H9X 1C0

CONTENTS

1 INTRODUCTION

All our petroleum, natural gas and coal were once biomass. Since these fossil resources are no longer being created, they will eventually be exhausted or become too expensive to recover. Plant material represents an annually renewable resource which, if we learn to use it efficiently, should be able to supply an increasing portion of our fuel/chemical and ingredient needs. This chapter focuses on bioutilization of plant material, particularly the starches.

Starch is a carbohydrate occurring in granular form in the organs of plants; after cellulosics resource, starch is the next most abundant compound synthesized by plant cells (Knight, 1969). Starch can be extracted commercially from many raw materials including corn, cassava root, potatoes and wheat. World production of starch is based largely on corn because this cereal is widely available as an economic source of raw material. The total world starch production is estimated to have been at least $1{\cdot}1 \times 10^9$ ton in 1978 (Van Beynum & Roels, 1985). However, only 1% of the total starch production is further converted. The overwhelming majority of the agricultural production of starch feedstocks is used as a feed or a food. Approximately $1{\cdot}3 \times 10^9$ kg of potato processing wastes, containing about $2{\cdot}36 \times 10^8$ kg of starch, is generated annually in the United States, while $2{\cdot}0 \times 10^8$ kg of potato processing wastes is generated in Canada (Stevens & Gregory, 1987). In the last 10–20 years, important new starch conversion processes and their technologies have been developed which added new starch-based products such as High Fructose Corn Syrups (HFCS), glucose or dextrose syrups, modified starches, dextrins, and ethanol. Biochemical conversions of starch to other organic acids, alcohols and polyols do exist (Van Beynum & Roels, 1985), but unfortunately, the processes result in dilute aqueous solutions and an economic separation technology must still be developed.

Most current treatment processes are designed simply to get rid of the nuisance waste materials. However, quite often, wastes contain valuable raw materials that potentially could be converted by microorganisms to useful products. The emerging field of biotechnology offers promise for developing new microbial processes for handling wastes, which currently are not amenable to conventional biological treatment.

In this chapter, the current technology and our results on the use of microbial enzymes and microorganisms to convert starch to sugar, alcohol and Single Cell Protein (or Microbial Biomass Protein) will be reviewed. A major feature of the industry is its ability to divert production from starch to sweeteners, alcohol, SCP and vice versa, depending on market conditions (Fig. 1).

2 SOURCES, NATURE, PRODUCTS OF STARCH

Starch is a glucose polymer and is one of the most widely available plant polysaccharides. In the biosphere, carbohydrate matter including starch and cellulose exceeds the combined amount of all other organic compounds. Continued polymerization of glucose can occur by formation of new

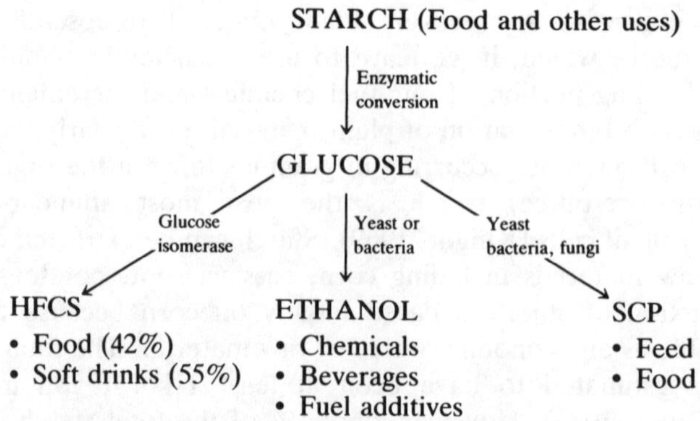

Fig. 1. Production of value-added products from starch bioconversion.

Table 1
Typical Analysis (%) of Selected Starch Containing Raw Materials[a]

Composition	*Corn*	*Wheat*	*Potato*
Starch	60·0	55·3	15·0
Protein	9·0	13·2	2·0
Fat and other	15·0	18·7	4·0
Moisture	16·0	12·0	79·0

[a] Starch raw materials produced in Canada (McNicol *et al.*, 1972).

1,4-glycosidic bonds. Amylose is a straight-chain polymer of glucose subunits whose molecular weight may vary from several thousand to half a million. Amylose typically constitutes about 20–27% of starch. While the amylose fraction of starch consists of straight-chain, water-insoluble polymers, the bulk of starch is amylopectin, a D-glucose polymer (molecular weight up to 1–2 million) which is distinguished by a substantial amount of branching 1,6-glycosidic linkages.

The commercial sources of starch are the seeds of cereal grains (corn, wheat, sorghum, rice), tubers (potato), roots (tapioca, sweet potato, arrowroot) and Sago palm (Swinkels, 1985). However, the major sources of starch produced in North America are corn, wheat and potatoes. The processes by which starch is obtained from these products have the commonality of separating starch from the other components present in the raw material.

Manufacture of starches from corn (maize), wheat and potato has been comprehensively reviewed elsewhere (McNicol *et al.*, 1972). The proportions of these ingredients vary greatly between raw materials and even between different grades of raw material. A typical analysis of three starch materials is shown in Table 1. The starch which is obtained from any one of the three raw materials may be (a) used in the form in which it is dried (unmodified starch, known as pearl or powdered starch), (b) further processed to produce many different types of starch or dextrin, or (c) converted to sweeteners (glucose/dextrose or HFCS) and other products. It is beyond the scope of this chapter to attempt to describe each of the several hundred product-grades produced by the industry. Most of this unmodified starch goes into the paper trade for paper coating and sizing and to the corrugated box industry as an adhesive. Unmodified starch also goes into fermentation for beer and chemical items.

3 STARCH-BASED PRODUCTS

In addition to modified starches produced by chemical reactions (hydrolysis, oxidation, cross-linking, substitution), starch may also be converted into products whose characteristics are distinctively different. Of these, the most important are the starch-based sweeteners, alcohol and Single Cell Protein (SCP) or Microbial Biomass Protein (MBP). Starch-containing raw materials require different pretreatments for processing into sweeteners or a fermentable mash depending on the particular raw materials used. While grains can be produced without pretreatment, tubes or roots are disintegrated mechanically (milling), thermally (boiling, steaming) or in a combined process. Native starch is hardly soluble and enzymatic hydrolysis must be preceded by the gelatinization (hydration) of the starch at around 65–80°C. The preparation of mashes by the classical process is carried out discontinuously in a steam cooker (disintegration of the raw material and gelatinization of the starch) and in a mash turn. In the mash turn, the starch is hydrolyzed with α-amylase (liquification) at about 90°C and with glucoamylase (saccharification at 60°C to fermentable sugars). In one example of a continuous process, the coarse, preliminary comminution of the raw material is followed by extremely fine milling (less than 100 μm) and the gelatinization of the starch by indirect (thermalizer–heat exchanger) or direct (steam injection) heat exchange. The feed of glucoamylase for saccharification is carried out continuously and heat can be recovered by countercurrent heat exchange (Drawert *et al.*, 1987).

3.1 Sweeteners

The conversion of starch to sugars is one of the most important biotechnological processes with an annual output of over 8 million tons. Sales of High Fructose Corn Syrup (HFCS) have exceeded 4 million tons, accounting for over 30% of the US sweetener market (Coker & Venkatasubramanian, 1986). Early discovery of the derivation of dextrose from starch by acid hydrolysis by Kirchhoff in 1811 and de Saussure in 1815 established the basis for the commercial production of syrups and crude sugars. Although acid-converted corn syrup was useful in many applications, the lack of sweetness relative to that of sucrose and reversion products limited the use of glucose syrups.

The introduction of fungal glucoamylase (amyloglucosidase) in the late 1940s made possible an acid–enzyme process for starch hydrolysis. This process involved a mild treatment with acid to give a Dextrose Equivalent (DE) of about 12; the resulting liquors were neutralized to about pH 6·0 and hydrolyzed with glucoamylase of *Aspergillus niger* to obtain a maximum DE of about 93. In the early 1960s, the all-enzyme process was developed using bacterial α-amylase from *Bacillus subtilis* for liquifaction–dextrinization and fungal glucoamylase for saccharification of starch to produce approximately 95% DE. By 1962, the large scale production of all enzyme glucose syrup was established and the significant features of this new technology were patented (Rentschler *et al.*, 1962). The next major development was the introduction of thermostable α-amylase from *B. licheniformis* and *B. amyloliquefaciens*, capable of operating at above 100°C after glucoamylase treatment. If the starch slurry contains high protein above 0·3% total protein, the syrup produced has a greater potential to develop color as a result of the Maillard reaction even with the carbon and ion exchange refining systems. Such a development was a necessary precursor to the introduction of High Fructose Corn Syrup (HFCS) or isoglucose (glucose–fructose mixtures).

The commercial production of glucose isomerase from *Arthrobacter cloacae* or *Streptomyces flavovirens* was a major milestone in the corn sweetener industry (Casey, 1977). The manufacturing process (isomerization) involved the direct addition of an isomerase enzyme, which converts glucose substrate in a batch system. The 42% fructose:58% glucose syrup is regarded as being as sweet as sucrose and thus the term 'isosweet' was used for the first commercial 42 DI HFCS material. Three commercial HFCS products contain 42, 55, and 90% fructose. Japanese developments of *Streptomyces* glucose isomerase moved the technology to the first commercial production of HFCS in 1967. However, a number of problems remained before the product could become a serious competitor to sucrose and invert sugar. Of these, the most important were (1) expensive enzyme cost in the batch systems, (2) the disadvantages of high-temperature operation due to the formation of colored products, (3) the high pH optimum of glucose isomerase under which conditions the isomeric mixture produces non-metabolizable sugar (D-psicose), and (4) inhibition of the enzyme by minerals, especially calcium ions. The major technological development occurred in 1972 when batch production was replaced by a continuous process utilizing enzyme immobilization. This continuous process for HFCS production using glucose isomerase was probably the first example of a large-scale industrial use of immobilized enzymes (Lee *et al.*, 1976). This immobilized enzyme process resulted in a significant reduction of production costs to levels competitive with the production of sucrose. The early process involved the immobilization of heat-treated whole cells or cell-free enzymes on various supports such as absorption on insoluble matrices (ion-exchange celluloses) or entrapment in insoluble matrices (cellulose, collagen, polyacrylamide). However, many of the larger US producers have licensed either the Novo Industry system of covalent bonding or the Corning Glass Company system of covalent bonding to glass. The design and operation of continuous immobilized isomerase for HFCS are described in detail elsewhere (Venkatasubramanian & Harrow, 1979). A schematic diagram of the production of a conventional high DE glucose syrup and of the High Fructose Corn Syrup (HFCS) process is shown in Fig. 2.

The Clinton Corn Processing Company (CPC) marketed the first syrup, called Isomerase 30 (first 30% fructose, then later 42% fructose). This was followed in 1972 by the introduction of the continuous process using immobilized whole cells or cell-free enzymes. Finally, in 1976 appeared the second generation (55%) HFCS which aims at direct replacement of the liquid invert sugar used in the food and beverage industry. The enriched product, containing approximately 90% fructose, is referred to as Ultra-High Fructose Syrups (UHFS) or Very Enriched Fructose Corn Syrup (VEFCS). This VEFCS fraction can be blended back with 42% HFCS feed material to obtain

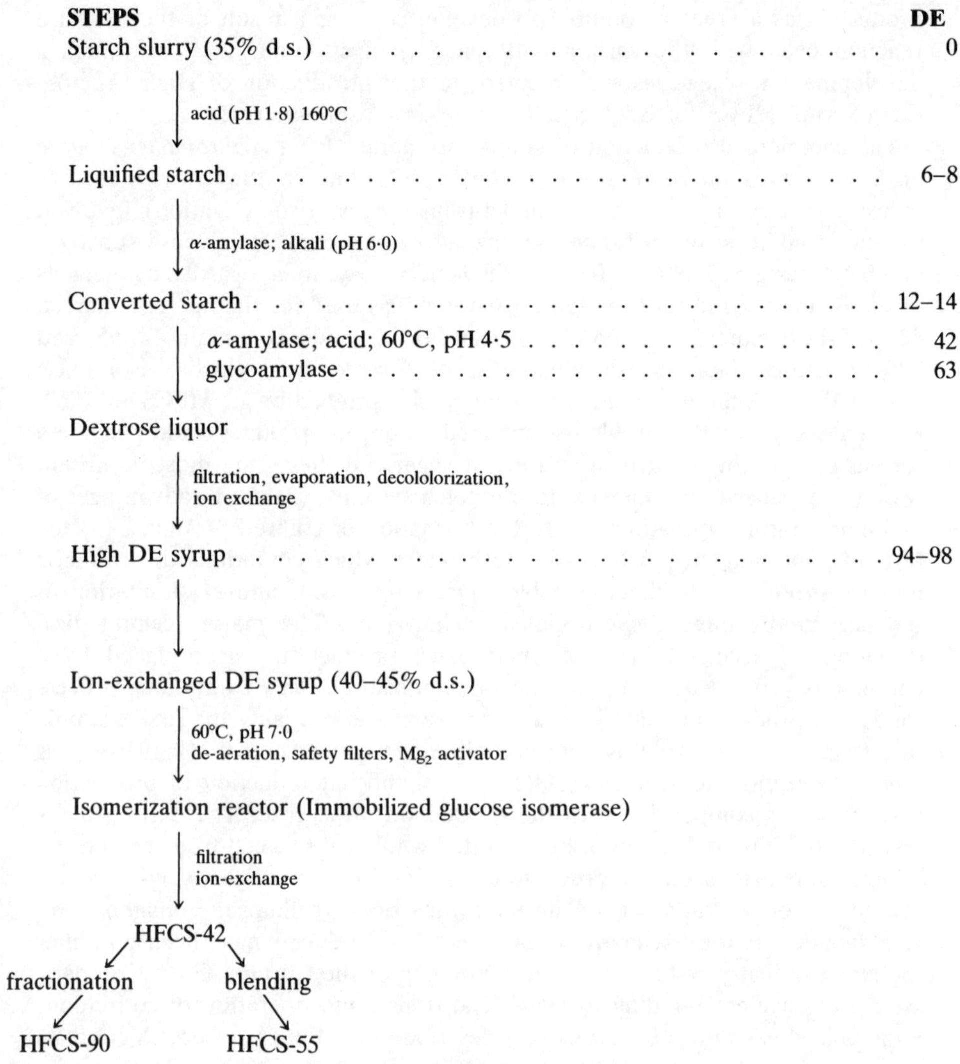

Fig. 2. Production of High DE syrup and HFCS.

products having a fructose content of 55%. The 55% HFCS can replace up to 100% sucrose, especially in the soft-drinks industry, while the 90% HFCS is limited to low calorie foods because of its high sweetness. The marketing and economic considerations of the sugar market are explained in detail by Crott (1986). Table 2 summarizes some of the commercially available glucose isomerases, and a more complete description of recent advances has been given by Barker and Petch (1985) and Van Tilburg (1985).

The worldwide economic role of the cassava plant is not well known, though cassava (tapioca, manioc) is the sixth most important nutrient plant on earth

Table 2
Commercially Available Glucose Isomerases (Data Compiled by B. H. Lee)

Company (trade name)	*Enzyme source*	*Immobilization*	*Reference*
Takasaki	*Streptomyces* spp.	Whole cells by heat treatment	Takasaki *et al.* (1969)
Clinton Corn Processing Co.	*Streptomyces* spp.	Cell-free enzyme or whole cells absorbed to ion-exchange resins	Wingard *et al.* (1976)
Reynolds Tobacco Co.	*Arthrobacter* spp.	Whole cells, flocculated, filtrated, dried	Vellenga (1978)
Gist-Brocades (Maxazyme)	*Actinoplanes missouriensis*	Whole cells in gelatin, cross-linked with glutaraldehyde	Shieh *et al.* (1974).
Novo Industry (Sweetzyme Q)	*Bacillus coagulans*	Whole cells, cross-linked with glutaraldehyde	Outtrup (1976)
Miles Kali-Chem (Optisweet 22)	*Streptomyces* spp.	Cross-linked whole cells (granulated) or purified enzyme bound on silica	
Corning Glass Works	*Streptomyces* spp.	Enzyme absorbed to inorganic matter	

(Berghofer & Sarhaddar, 1988). Only a few references are available for the industrial processing of cassava to glucose or high-fructose syrups. However, this field of sugar production is expected to become increasingly important in developing countries in addition to the contribution made by sugar cane. Tapioca starch from cassava root is a highly refined product and its composition differs only slightly from starch of other origin, such as corn or potatoes. Thus, liquifaction and saccharification are usually carried out in a similar way with the same good results (Aschengreen *et al.*, 1979). Berghofer and Sarhaddar (1988) produced less expensive syrups (glucose and fructose) from fresh and dried cassava roots by means of direct hydrolysis.

In the author's laboratory (Davids *et al.*, 1987), the complete hydrolysis of whole potatoes to glucose, maltose and soluble oligosaccharides by the multiple enzyme treatment was achieved (160 g of reducing sugar per kg of potato). Prolonged α-amylase treatment alone liquifies but does not sufficiently solubilize the starch to permit its total physical separation from the non-starch residues. Total solubilization was achieved with glucoamylase along with pectinase and cellulase: total recovery of the extract was hindered somewhat by the solid residues (protein, fibre) but a high recovery (98%) was obtained leaving only fiber. The course of isomerization of syrups from potato in our studies was almost identical with the procedure for syrups made from pure glucose (unpublished data, Lee, B. H., *et al.*, 1988).

Because HFCS costs 15–20% less than equivalent sucrose syrup, there has been widespread switching to its use in soft drink beverages. However, to capture the beverage market, it becomes necessary to refine further the 42% syrup to yield a 90% syrup by ion-exchange chromatography for blending back into the normal HFCS. Furthermore, high quality standards must be met for low color levels, low titratable acidity and pH of not less than 3·5. High saccharides and acetaldehyde contents should not be high. Acetaldehyde forms molecular complexes with fructose, and upon storage of cola drinks such complexes break down to release the acetaldehyde which produces off-flavors (Finn, 1987).

3.2 Ethanol

Basically, all raw materials that contain either fermentable sugars or constituents (starch, cellulose) can be converted into sugars and alcohol by suitable treatments. However, based on modern commercial and industrial energy demand, the two most important end products of carbohydrate conversion are ethanol as a liquid fuel and methane. It is still uneconomic to produce methane from pure starch, but when the starch wastes derived from potato-processing industries are considered, its conversion into methane is quite often justified (Venkatasubramanian & Keim, 1985). However, ethanol production is a far more attractive proposition and the economics of ethanol fermentation are particularly important due to its potential fuel and chemical uses.

The production of ethanol from corn dextrose by yeast fermentation is an old technology and the alcoholic beverage industry today is well established in almost every part of the world. In attempting to discuss this topic, the commercial aspects of fuel-grade ethanol from starch materials rather than the finer aspects of beverage use are focused. For many years, all ethanol, except for beverage use, was obtained by chemical catalytic hydration of ethylene. In the light of present-day energy shortages, ethanol from fermentation has gained renewed importance in the 1980s. In countries with a large agricultural surplus, such as Brazil, South Africa, Canada and the United States, intensive studies are being conducted on the production of ethanol from carbohydrate materials such as sucrose, starch and plant biomass, etc. (Dale, 1984).

At the present time, processes are in operation which convert starch enzymatically into sugar (mono- and di-saccharides) which are then fermented by yeast or *Zymomonas* bacteria to alcohol. The microorganisms involved and the costs of fermentation for ethanol from starch feedstocks are summarized in Table 3. Several microorganisms can convert sugars to ethanol. However, only some particular strains of yeast such as the genera *Saccharomyces* and *Kluyveromyces* lead to a rapid and efficient conversion. *Zymomonas mobilis* and *Clostridium thermosaccharolyticum* (Wang *et al.*, 1978) are also under development for production of fuel ethanol using fermentable carbon substrates. As is the case with yeast success on the industrial scale, conversion appears to be strain-dependent, as some strains of *Zymomonas* are susceptible to glucose or product inhibition or lactic bacterial contamination (Monica *et al.*, 1989).

Table 3
Fermentation Ethanol from Starch Feedstocks

Starch feedstocks	*Microorganisms*	*Net feedstock cost ($/gal)*	*Reference*
Maltose (starch)	*Saccharomyces cerevisiae*	—	Peppler (1979)
Wheat (Europe)	*Saccharomyces cerevisiae*	1·53	Martin (1982)
Wheat	*Saccharomyces cerevisiae*	1·23	Tyner (1980)
Corn (wet-milling)	*Saccharomyces cerevisiae*	0·38	Hertzmark *et al.* (1980)
Corn (wet-milling) (without by-products)	*Saccharomyces cerevisiae*	1·00	Hertzmark *et al.* (1980)
Cassava	*Saccharomyces cerevisiae*	0·69	DeCarvalho *et al.* (1977)
Inulin (Jerusalem artichokes) (without by-products)	*Kluyveromyces marxianus, Torulopsis colliculosa*	0·64	Guiraud *et al.* (1982)
	Saccharomyces cerevisiae S. diastaticus, K. cicerisporus	—	Duvnjak *et al.* (1982)
Sorghum	*Zymomonas mobilis*	—	Millichip & Doelle (1989)

Source: author's information.

A genetic modification of *Z. mobilis* to give a high product concentration without inhibition is currently being investigated in many parts of the world (Conway *et al.*, 1987).

The developments in the United States and Brazil particularly have sparked worldwide interest in fermentation alcohol as a full (Brazil) or partial replacement of liquid fuels and as a base for producing many organic chemicals (Venkatasubramanian & Keim, 1985). In Brazil in the 1980s, 20% of petroleum imports have been replaced with ethanol produced from sugar cane. In the United States, two-thirds of the starch comes from corn (maize), and of that fraction about 80% is converted to either dextrose or High Fructose Corn Syrup (HFCS). Currently, about 80% of fuel-grade alcohol in the USA comes from fermentation and the current ethanol production level is the equivalent of about 65 000 barrels per day of imported oil.

Because of pollution control plans to phase out the use of lead in gasoline, the present level of gasoline consumption (approximately 100×10^9 gallons annually) in the USA is expected to boost use of fuel-grade ethanol significantly. Ethanol in a 10% blend is known to increase the octane rating by about 3 points, or the ethanol can be used at the 2·5% level as a co-solvent with methanol.

Many other economic and market aspects of alcohol usage as an octane extender have been reviewed elsewhere (Venkatasubramanian & Keim, 1985; Sercovich, 1986).

Since sugar makes up the largest portion of the fermentation medium, the substrate is expected to account for a significant fraction of the overall

Table 4
Yields of Ethanol Production from Different Raw Materials

Crops	*Biomass (ton/ha)*	*Ethanol (liter/ha)*	*Ethanol (liter/ton biomass)*
Molasses	—	—	270
Sugar cane	9	3 600	70
Sugar beet	6	3 000	—
Sorghum	5	1 500	—
Potatoes	5	1 800	—
Corn	3	1 200	370
Cassava	7	1 800	180

Sources: Venkatasubramanian & Keim, 1985; Sercovich, 1986; and author's information.

fermentation costs. Thus, one of the most important commercial issues relating to fuel-grade alcohol production is the availability and cost of the feedstocks. For ethanol approximately 50–70% of total production costs are likely to be due to the cost of carbohydrate stock. Table 4 shows the potential alcohol yield from some raw materials. From the sugar source, only 51% by weight is usually transformed to ethanol and the rest is used as a carbon source. The conversion ratio of sugars to ethanol is commonly 90–95%. Therefore, 1 g of sugar converted with 90% efficiency will give 0·46 g of alcohol.

Other potential raw materials for alcohol production—sulfite waste, whey, the new crops and cellulosic materials—are of relatively little interest at the present time. Sulfite wastes and whey are produced in small quantities and are too dilute to transport. On a worldwide scale, lignocellulosic materials are the most abundant source, having been estimated at 150×10^9 tons per year (Prendergast *et al.*, 1983). However, one of the key problems in lignocellulosic conversion is the difficulty of hydrolyzing cellulose to glucose. The main disadvantage of these wastes is that the costs of collecting and transporting sufficient quantities to operate a processing facility large enough to be economically viable can be prohibitive, even in less-developed countries.

Three-quarters of the ethanol production uses the batch method, which is a slow process, followed by an efficient multistep distillation. If a whole-grain mash is used, recycling of the yeast is not possible and the by-product animal feed is sold as distillers' dried grains instead of as corn gluten. Recently, continuous processes have been developed in which the raw mash is fed in at the same rate as spent mash is removed, some yeast in the spent mash being fed back into the tank. There are three types of continuous process: the conventional continuous stirred tank reactor (CSTR), the CSTR coupled with a cell recycle, and the Immobilized Cell Reactor (ICR). The last two processes show improvement over the conventional CSTR due to the use of high cell densities and, hence, a faster rate of fermentation. There is also a potential for cross-flow or tangential-flow membrane separations. The Kyowa Co. in Japan used immobilized cell reactors (i.e. packed beds) on a commercial scale. Some companies use a semi-continuous cascade of fermentors rather than batch sequencing. Many advances in ethanol production by using immobilized technology have been discussed by Gôdia *et al.* (1987) and Sharma (1988). The solid-substrate fermentations of sugars and sorghum (Gibbons *et al.*, 1984) and the co-immobilization of glucoamylase and *S. cerevisiae* (Chithra & Baradarajan, 1989) have also been studied.

Most recently, the production of alcohols, primarily ethanol, using recombinant DNA technology has been a subject of considerable interest. Improvement of yeasts could overcome low tolerance to ethanol and high substrate concentrations. The process could be simplified if yeast strains were available to convert starch directly. Innis *et al.* (1985) were able to construct a yeast strain containing an *Aspergillus* glucoamylase gene and integrate it into the chromosome. Such recombinant yeast was superior to the natural starch fermenter, *S. cerevisiae. Aspergillus* glucoamylase contained isoenzymes I and II which are encoded by a single gene. By comparison, the glucoamylase genes (STA 1, 2, 3) of *Saccharomyces diastaticus* were highly homologous, each encoding a glucoamylase activity. The three unlinked STA genes were cloned into *S. cerevisiae* by functional complementation, using plasmid YEp13, (Yamashita *et al.*, 1985; Errat & Nasim, 1986; Lee, 1987).

Another area for strain improvement to reduce ethanol production costs is that of increasing the variety of abundant carbohydrates that yeast will economically ferment. Starch, cellulose and xylose are abundant carbohydrates found in plant materials. Some fungi as well as bacteria have been identified as being able to degrade these complex carbohydrates. One should be able to isolate the genes for these enzymes and create a unique yeast capable of synthesizing and secreting them extracellularly.

Pentoses constitute the second major compound of wood and a waste product in the paper industry. The utilization of these feedstocks requires either genetic modification of *S. cerevisiae* or the use of alternative ethanologens able to ferment this sugar (Alexander, 1986; Hollenberg & Wilhelm, 1987).

3.3 Single Cell Protein (SCP)

3.3.1 General

Unconventional SCP comprises yeast, bacteria, molds and higher fungi as well as algae. The use of microbial protein to contribute to the world's protein supply has been a subject for discussion and experimentation for several decades. The route for its use may be either indirect, as a protein component of animal feedstuffs, reducing the requirements for such materials as soybean meal and fish meal, or as a material for direct inclusion in the human diet.

Growing microorganisms for feed or food has two main attractions: first, the growth rate of microorganisms is very much faster than those of animals or plants, thus potentially shortening the time needed to produce a given mass of food; second, a range of materials including low-grade waste carbohydrates can be used to produce SCP, depending on the microorganisms chosen.

Modern, large-scale SCP production is about 30 years old but in most instances, the economic and technological constraints have limited their industrial application. The SCP processes are capital- and energy-intensive. The production cost of SCP is dependent upon substrate cost, operating expenses, and capital investment for fermentation equipment. The nature of SCP products also makes their production cost susceptible to influences other than scientific and economic. Marketing considerations are important aspects for the overall economics, and commercialization of SCP production has been strongly affected by sociological, psychological and political parameters. Another important factor for SCP production is product quality and safety.

A major factor in the cost is the raw materials and the efficiency of their conversion into SCP biomass (Faust, 1987). The yield coefficient (Y_s = g cell biomass produced/g carbon source utilized) is the single most important parameter influencing the economics of SCP production. The choice of substrate is usually dependent upon local availability, since the transportation of substrates over long distances is not cost-effective. It is usually advantageous to couple SCP production to another existing process whose waste by-products can be utilized as substrates for SCP.

Since the 1950s, there has been considerable interest in the large-scale cultivation of algae from the renewable resources, CO_2 and sunlight. Algae can also be grown either using artificial lighting or heterotrophically in the dark with organic carbon and energy sources. Algae production from wastewaters and animal slurries has also been tried since the 1950s but economics are still the bottleneck for commercial operation, due to the limited sunlight illumination intensity and the costs of harvesting algae due to the low cell density.

The main raw materials that have been considered for SCP are oil fractions (hydrocarbons, *n*-paraffins, *n*-alkanes, natural gas, methanol), and ethanol, carbohydrates, and cellulosic materials from various sources.

The licensing problems in the production of SCP from oil fractions, including that of toxicity, together with the increase in oil prices, stimulated efforts towards SCP production from methanol. By 1980, when Imperial Chemical Industries (ICI) and Phillips started marketing 'Pruteen' and 'Provesteen' (high density yeast cells: 120–160 g/liter) respectively (Hitzman, 1986), all the other companies around the world stopped producing SCP from methanol because of economic difficulties. Economic analyses have shown that the process is highly sensitive to changes in the market price, the substrate price, investment costs, and the market price of the traditional protein sources such as soybean meal and fish meal (Ericsson *et al.*, 1981).

Ethanol can be assimilated by a large number of different microorganisms and technically represents the simplest and neatest variant for the process for SCP production. However, the raw materials are obtained either from carbohydrate by fermentation or from ethylene via petrochemical precursors.

The market price of ethylene-derived ethanol, coupled with declining soybean prices, prevents large-scale commercialization at present.

Current economic trends favor substrates derived from carbohydrate wastes, either from tropical crops (sugar cane, cassava) or from food processing and agricultural wastes. The production of SCP from various carbohydrate-containing substrates is an old tradition when one thinks of the yeast fermentation of molasses, whey, and sulfite waste liquor (from the production of cellulose). Agricultural wastes which represent an inexpensive source of substrates include cellulose, hemicellulose, lactose and starches. For the utilization of these waste substrates by SCP microorganisms, saccharification and fermentation are necessary.

Lignocellulosic materials, widely available in agricultural or forestry-product residues, are the most abundant source. However, pretreatment prior to bioconversion is essential for the realization of economic yields. The most common physical methods, grinding and steam explosion, require the investment of more energy than that available from the raw material. Moreover, they do not render the raw material completely digestible and therefore SCP grown on such wastes will be consumed with the substrates, which limit their use as feed for ruminant animals.

Also many of these substrates are available only seasonally which would make it likely that SCP production facilities would be operating for only a part of the year. In general, carbon and energy source costs may range from 14% of total manufacturing costs for wastes to greater than 50% for highly purified substrates. Costs ranging from \$40 to \$50/ton (1985 prices) have been estimated for the collection of agricultural residues. Effective and economical pretreatments are thus crucial to the success of lignocellulosic conversion. Although the main chemical methods, such as acid, alkali, SO_2 at high temperatures, and ozonization, are available for the solubilization of these substrates, many problems such as growth inhibition, corrosion and pollution must be overcome and these are all high energy-consuming procedures. The production of SCP directly on cellulosic material without pretreatment has been very rare.

Large-scale SCP production facilities are being built elsewhere in countries where self-sufficiency in food production is important (e.g. in the USSR and Eastern Europe) and where other non-economic factors make it advantageous for these countries to produce their own animal feed supplements (Carrantino, 1983).

The safety of SCP products for human consumption is more critical than that of feed. High nucleic acid contents of microorganisms can result in the formation of kidney stones or gout if consumed as a significant portion of protein intake. Another important product for human consumption is the good functional proteins which are free from nucleic acids and toxicants. The food-grade yeasts such as *Saccharomyces cerevisiae* and *Kluyveromyces fragilis* have a good position in the food ingredients market but yeast cannot utilize lactose in whey or agricultural biomass (xylose) which are two promising substrates. Through using genetic engineering the required characteristic enzyme systems can be introduced into yeasts, consequently enabling the use of less expensive, more abundant substrates which will lower the costs (Pellon & Sinsky, 1983; Lee, 1990).

3.3.2 Starch Substrates for Single Cell Protein

The technology of yeast production has provided one of the most efficient methods of converting inexpensive substrate into food or feed, dating back to about 2300 BC (Dziezak, 1987). Traditionally, yeasts have been used by the food industry principally for their production of ethanol and carbon dioxide which are important to the brewing, wine, distilling and baking industries. Among the substrates that could be utilized as raw materials for the production of SCP, starches are especially attractive since they are easily metabolized by many yeasts. There is also a growing interest in the use of potato wastes because of the increase in costs of high-protein feeds. Starches are relatively poor in protein (approximately 2%) while being extremely rich in complex carbohydrates. This product of low economic value and low protein content can be enriched in protein by direct fermentation with an edible yeast. Since Wickerman and Kuehner (1953) first described a mixed culture of amylolytic *Saccharomyces* (*Saccharomycopsis*) *fibuligera* and *Candida utilis,* the following important processes have been developed using starch:

(1) the Symba process for potato wastes by *S. fibuligera* and *C. utilis* (Jarl, 1969; Stogman, 1976);
(2) the manioc process for cassava wastes by *C. lipolytica* (Revuz & Voisin, 1980);
(3) the cassava starch process by *C. utilis* using fed-batch culture (Ghoul & Engasser, 1983);
(4) the solid state fermentation process (Senez, 1983);
(5) the whole potato process (Lee & Simard, 1984; Kombila *et al.*, 1985; Davids *et al.*, 1987; Kombila *et al.*, 1988);
(6) the Provesteen process (Hitzman, 1986).

These processes are discussed in detail later in this section.

The direct fermentation of unhydrolyzed potato starch (4%) by a combination of yeasts (*S. fibuligera, C. utilis*) and *Aspergillus awamori* (Cheng & Pereira, 1970) and the assimilation of corn starch pastas (4%) by a mixed culture of *S. fibuligera* and *C. utilis* also have been studied (Lachance & Blackwood, 1972). The slow process by direct starch fermentation by yeast was the prime rate-limiting factor. This inefficient process can be improved by simultaneous enzymatic hydrolysis and fermentation by an industrially important yeast.

Tong *et al.* (1973) studied the growth of *Rhodotorula rubra* on acid-hydrolyzed potato starch. Gregory *et al.* (1977) also studied fungal growth on acid-hydrolyzed cassava starch but this process needed an expensive energy-requiring fermentor. An enzyme hydrolysis process has advantages over acid hydrolysis in that a high conversion (low reversion) of starch (95% yield in 30% dry solid) and a low concentration of degradation products (e.g. furfurals, which are inhibitory to microbial growth) can be obtained.

Many authors attempted to select yeast strains for the production of SCP from starch materials. Hattori (1961) selected an *Endomyces* sp among 27 strains. Spencer-Martins and Van Uden (1977) compared the growth of several yeasts in 0·2% of soluble starch and recommended the genus *Lipomyces*. The yields in terms of g biomass/g starch supplied were 0·489, 0·585, 0·384 and 0·442 for *L. starkeyi, L. konenenkoae, L. lipofer* and *L. tetrasporus,* whereas the yields of *Candida tropicalis* (0·234), *Hansenula anomala* (0·357), *Schwaniomyces alluvius* (0·376), *Pichia polymorpha* (0·293), and *Trichosporon pullulans* (0·392) were much lower than those of *Lipomyces* spp. *Lipomyces konenenkoae* (IGC-4052) produced an exocellular amylase system (capable of total starch hydrolysis) consisting of α-amylase, glucoamylase and debranching transferase that were subject to catabolite repression (Spencer-Martins & Van Uden, 1979).

For most amylolytic strains, the secretion of amylases depended on the composition of their growth medium (Oteng-Gyang *et al.*, 1980); a reduction in growth rate and biomass yield was observed by using the yeast nitrogen-base (YNB, Difco) starch medium containing potassium (0·2M). The potassium ion appears to inhibit the biosynthesis and excretion of the amylolytic enzymes. Other authors reported the excretion of amylase by *Schwaniomyces alluvius* (Gallejà *et al.*, 1981), *Saccharomyces fibuligera* (Clement *et al.*, 1980), and *Lipomyces kononenkoae* (Spencer-Martins & Van Uden, 1979). Azoulay *et al.* (1980) isolated a mutant from *Candida tropicalis* which utilizes quite effectively the soluble starch; its glucoamylase was inducible and intracellular. Continuous culture (35°C, pH 4·0, $D = 0{\cdot}26$/h) of thermally pretreated cassava suspensions gave a productivity of 0·038 g/liter h and a yield of 0·55 g biomass/g starch consumed. Lemmel *et al.* (1979) and Admasson *et al.* (1981) optimized the continuous production of *Saccharomyces fibruligera* on potato processing wastewater. An extracellular glucoamylase was produced at a maximum rate at $D = 0{\cdot}22$/h and cell productivities and yields were 0·43–0·52 g/liter h at $D = 0{\cdot}25$/h and 0·84–0·95, respectively. The production of biomass and amylases in batch and continuous culture of *L. konenenkoae* were also reported (Correira & Van Uden, 1981). Biomass productivities were 0·086 and 0·137 g/liter h for yeasts grown on 0·5% soluble starch medium. The depression of the enzymes permitted the use of dilution rates near μ_{max}(0·14/h).

Wilson *et al.* (1982) studied the growth of *Schwaniomyces alluvius* on 0·2% soluble starch and obtained growth rates of 0·21–0·25/h and yields of 0·59–0·62. Similar growth rates were obtained with glucose and the extracellular amylase system was not limiting growth. Touzi *et al.* (1982) studied 15 yeast strains on five different starch media and found *Schwaniomyces alluvius* best in terms of biomass yield (0·45) and batch growth rate (0·28/h). Moreton (1978) grew the *Candida utilis* strain in chemically or enzymatically hydrolyzed potato starch and obtained very high productivity (0·49–0·53 g/liter h) and yield (0·51–0·58). However, the cost of the pre-hydrolysis step of the process was an important drawback.

There are two different processes of protein enrichment, (a) by solid-state fermentation, or (b) by liquid-state (submerged) fermentation of starch materials.

SOLID-STATE FERMENTATION

Solid-state fermentations are distinguished from submerged (liquid) cultures by the fact that microbial growth and product formation occur at or on the surfaces of solid materials with low moisture contents. Commonly known in the East as koji and in the West as solid-substrate fermentations, such processes are well established in the Orient for the production of traditional food fermentations (soy sauce, miso, mushrooms) and of rice wine, enzymes and organic acids. Although the solid surface growth makes the process more difficult to control, this process may have significant advantages over submerged-culture methods for some of the food, pharmaceutical and chemical industries. Advantages of the solid-state fermentation were cited by Hasseltine (1972). Recently, research data have been published on fungal or mold biomass products obtained from starch or lignocellulosic materials as animal feeds.

ORSTOM/IRCHA process: Raimbault and co-workers at ORSTOM and IRCHA (Paris) developed a solid-state fermentation where filamentous fungi grow in a starch biomass (Senez *et al.*, 1980; Raimbault, 1981; Senez, 1983). The originality and economic interest of this process was that it was workable at the rural level for animal feeding. After glutinization of dried starch flour (75°C, 10 min), *Aspergillus hennebergii* spores (2×10^7) and nutrients (40% urea, 60% ammonium sulfate, salts) were mixed in a simple fermentor by gentle mechanical stirring. The substrate was incubated (37°C, pH 4·5) for 24–26 h under intermittent aeration. The products obtained with solid-state culture of different starchy materials (cassava, banana, potato, potato wastes) contained about 20% protein and 20–25% residual sugars. The amino acid profile was well balanced, with high contents of lysine and sulfur amino acids. The nutritional and toxicological assays on rats and chickens proved satisfactory.

Similar results were obtained with a number of other amylolytic fungi (*A. oryzae, A. awamori*) and various *Rhizopus* or *Penicillium* spp. isolated from traditional fermented foods in the Orient. Thus this process suggested the possibility of utilization of protein-enriched food for direct human consumption. This process has been run experimentally on a large pilot scale producing 700 kg dry weight in collaboration with a French bioengineering firm (SPEICHIM).

Azoulay et al. process: The food value of starch powders such as cassava and corn had been enhanced by increasing their protein content through solid-state fermentation using fungi (Brook *et al.*, 1969; Reade & Gregory, 1975). The growth of a yeast such as *Candida tropicalis* on corn powder could also enrich the protein content of this product by the presence of the synthesized yeast biomass (Gregory *et al.*, 1976).

Azoulay *et al.* (1980) developed a process whereby a strain of *C. tropicalis* grows directly on soluble starch powder, resulting in the improvement of corn or cassava for its direct utilization as animal feed. The growth of *C. tropicalis* in batch cultures was possible on cassava concentrations ranging from 2 to 100 g/liter. The maximum specific growth rate and productivity were about 0·4/h and 0·038 g cell/liter h respectively. The yield coefficient was 0·55 g cells/g metabolized starch. This compared favorably with the yield coefficient of 0·5 g cells/g starch found with several microbial systems (Spencer-Martins & Van Uden, 1977). In continuous cultures, the cassava powder was suspended in water and sterilized at 110°C for 15 min. The maximum biomass productivity obtained was 3·6 g/liter h at a dilution rate of 0·25/h. The resultant mixture of biomass and residual corn or cassava contained about 20% protein, which represents a balanced diet for either animal fodder or human food. The increase in the protein content paralleled an appreciable enrichment in lysine and cysteine.

Ramos-Validivia process: Among the great variety of amylolytic fungi, *Rhizopus oligosporus* is harmless and exhibited a good protein productivity on cassava in solid-substrate fermentaion (Sanchez, 1981; Ramos-Validivia, 1982). The fermentations of cassava meal were carried out in submerged and solid-state (packed columns or trays) cultures under non-aseptic conditions. The final product of the submerged fermentation contained 26% true protein, while those of solid-state fermentations contained 22·8% (packed columns) and 16·8% (trays). The specific growth rates of the fungus in the submerged fermentation and solid-state fermentations were 0·22 and 0·05, respectively. The main problem in solid-state fermentations was the limitation of oxygen for the fungal growth. However, Raimbault and Alazard (1980) found similar specific growth rates of *Aspergillus niger* in cassava, regardless of solid-state fermentation or submerged culture. Nevertheless the volumetric protein productivities of the solid cultures were higher than those of the submerged ones.

SUBMERGED CULTURE

Solid-state cultures are of minor importance today, although a number of enzymes are still produced in this way from *Aspergillus, Mucor,* or *Rhizopus.* In practice, shaken culture represents an important microbiological technique for carrying out strain selection and examining parameters (nutrients, aeration, agitation, or temperature) in submerged culture. While submerged cultures may be aimed at cell production, the objective is usually the formation of a product obtained by the microbial growth. The economy of the process is considerably affected both by the energy demand and by mass transport. Provided that the biochemical course of the process is known, almost complete automation of the total process is possible today. Although submerged cultures have related mainly to the use of filamentous microorganisms, the submerged culture method is often applied to other types of fermentations.

Filamentous fungal processes: Several fungal processes in submerged culture for treatment of starch wastes have been described. Balagopal and Maini (1976, 1977) grew fungi in a suspension culture containing 25 g/liter cassava starch wastes and found *Aspergillus niger* (NRRL 330) and *Rhizopus* sp. to be superior in terms of mycelial weight and protein production. Muindi and Hanssen (1981) grew *Trichoderma harzianum* in suspensions of cassava meal (4%) and obtained an enriched product with 38% protein (dry weight) from original cassava containing 2·4% protein.

Aspergillus niger mutants have been applied for increasing the protein content of cassava starch wastes up to 20% (Nga *et al.*, 1977). *Penicillium notatum* and *P. digitatum* also grew well on potato processing wastes, resulting in a biomass of 9–24 g/liter (Rolz, 1984).

Gregory *et al.* (1977) grew *Cephalosporium eichhornial* ATCC 3825 in acid-heat gelatinized cassava starch (40 g/liter) and obtained biomass productivities of 0·58–0·98 g/liter h, yields of 42% and carbohydrate consumption of 95%.

Symba process: The Symba process was initially invented by Tveit (1963) and Jarl (1969). In this process, waste starches from a potato processing plant were fermented with the two yeasts *Saccharomyces fibuligera* (amylolytic) and *Candida utilis* (non-amylolytic). The growth of the latter was dependent upon the release of glucose from starch by amylases (α and β) produced by the former. The small-celled *S. fibuligera* grew more slowly than *C. utilis* and thus it contributed only 1–2% of the solids in the final product. Organic material not hydrolyzed by *S. fibuligera* will be present in the final product, thus limiting its use to animal feeds. Although full details of the Symba process have not been published, data obtained from Jarl (1969) from the growth of *Saccharomyces* in shake flask experiments containing starch medium gave a specific growth rate of 0·21 (38% of μ_{max}). The slow process by direct fermentation by *Saccharomyces* and *Candida* was the prime rate-limiting factor.

Moreton process: Several processes for treatment of potato wastes have been described with the object of reducing the Biochemical Oxygen Demand (BOD) of potato effluents.

Since Rieser (1954) recovered 50% yeast biomass (*Candida utilis*) from potato starch wastes and Bloch *et al.* (1973) grew *Aspergillus foeticus* on alkaline potato peel wastes, Moreton (1978) grew mixed cultures of *Saccharomyces fibuligera* and *C. utilis* on enzymatically hydrolyzed potato wastes. Poor amylase production by *Saccharomyces* resulted in slow growth of *C. utilis*. A significant improvement was achieved by subjecting the slurry to a high-speed disintegrator or homogenizer and to enzymatic hydrolysis (α-amylase and glucoamylase). Shearing with a Silverson homogenizer followed by thinning (liquifying) with a thermostable α-amylase (thermamyl, NOVO) from *Bacillus licheniformis* at 95–100°C was by far the most effective method. The yields and growth rates in the separated liquid phase of the hydrolyzate derived from α-amylase and glucoamylase hydrolysis were similar to those obtained with conventional substrates ($Y_s = 0{\cdot}56$; $\mu = 0{\cdot}50$).

As *C. utilis* is a Crabtree-negative yeast (De Deken, 1966), much higher sugar concentrations can be used in batch culture without adversely affecting cell yield.

Lee and Simard process: The slow process by direct starch fermentation by yeasts is the prime rate-limiting factor. This inefficient process was significantly improved by simultaneous enzymatic hydrolysis and fermentation of whole potatoes by important food yeasts.

Lee and Simard (1984) reported the growth response of *Saccharomyces fibuligera* on α-amylase hydrolyzed potato starch in batch and continuous cultures. The maximum productivities of 0·155 g/liter h in batch and 0·29 g/liter h in continuous culture with *S. fibuligera* were obtained at 30 g/liter of substrate.

A comparative study was further carried out on the growth and yield of *S. fibuligera, S. diastaticus, Schwaniomyces alluvius* and *Candida utilis* in different concentrations (2–20%) of α-amylase hydrolyzed potato starch with and without control of pH (Kombila *et al.*, 1985). The pH-controlled fermentation did not significantly ($p < 0{\cdot}05$) increase cell yields but the best results in batch culture were obtained with substrate levels between 4 and 8%, regardless of strain. The highest biomass (15 g/liter) and yield (31%) were obtained at 4 and 6% with *S. diastaticus,* but the yields were generally lower than 26% with the other yeasts tested (Table 5).

The insufficient fermentation occurred not because of the unavailability of nitrogen source but because of repression of glucoamylase by metabolites in the medium. The fed-batch culture did not improve the biomass yields of *S. diastaticus* from α-amylase hydrolyzed potato starches. This was believed to be due to the inhibitory effects of ethanol on yeast growth and of various metabolites including glucose and dextrin on α-amylase and glucoamylase.

Besides α-amylase and glucoamylase, a debranching enzyme such as pullulanase is essential for the complete hydrolysis of starch (Sills & Stewart, 1982). *Saccharomyces diastaticus* is a typical brewing yeast, leaving the large dextrins unfermented, and three genes of glucoamylase appeared to control the fermentation of dextrin (Errat & Stewart, 1981).

Further comparative studies on the growth and yield of *S. diastaticus* in semi-continuous (fed-batch) and continuous cultures showed that the fed-batch culture of *S. diastaticus* on saccharified (glucoamylase) potato starch gave high

Table 5
pH Controlled Fermentation of Potato Starch Hydrolyzate by Yeasts

Yeast	*Cells* ($N \times 10^8/ml$)	S_0 (*g/liter*)	S (*g/liter*)	$S_0 - S$ (*g/liter*)	$X - X_0$ (*g/liter*)	Y (%)
Saccharomyces fibuligera	5·8	60	30·42	29·58	7·71 ± 0·27	26·10
	3·0	80	31·70	48·30	5·94 ± 0·45	12·30
Candida utilis	17·0	60	30·46	29·54	7·49 ± 0·19	25·36
	16·0	80	28·56	51·44	10·16 ± 1·60	19·75
S. diastaticus	8·0	60	10·60	49·40	15·29 ± 2·20	30·95
	4·2	80	12·42	67·58	7·74 ± 0·57	11·45
Schwaniomyces alluvius	6·3	60	13·36	46·64	12·43 ± 1·30	26·65
	6·5	80	12·12	67·88	10·73 ± 3·20	15·81

S_0 = Original starch hydrolyzate; S = residual starch; $S_0 - S$ = assimilated starch; $X - X_0$ = biomass; Y = yield, $(X - X_0)/(S_0 - S)$.

yields up to the theoretical value ($Y = 0{\cdot}50$), while yields obtained with liquified (α-amylase) potato starch were only 0·31 (Kombila *et al.*, 1988). The continuous culture of this yeast permitted an increase in the productivity from 0·38 to over 3 g/liter h (Fig. 3). Ethanol produced from sugar accumulation in the medium during yeast growth eventually decreased the conversion efficiency. The biomass obtained showed a high protein content and a good essential amino acid profile compared with FAO standards.

In most cases, the substrate is diluted and/or not clarified and the starch is the only fraction receiving an enzymatic pretreatment. The efficiency of the processes for utilization of potato wastes was enhanced by applying pectinase and cellulase, simultaneously reducing the water content of the residue and the quantity of non-fermentable colloidal impurities in the extracts (Davids *et al.*,

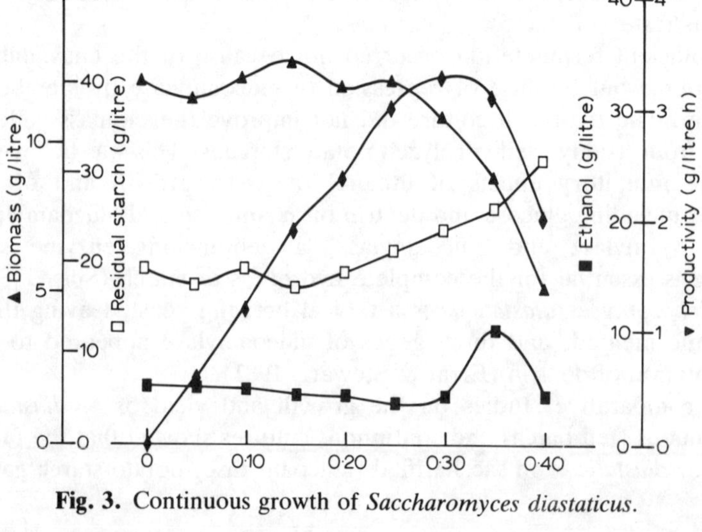

Fig. 3. Continuous growth of *Saccharomyces diastaticus*.

1987). Such an extract contains mostly glucose and mineral salts and non-protein nitrogen and may be used instead of mollases for the growth of baker's yeast, with an advantage that the final broth is filterable. Industrial amylases, pectinase and cellulase were used in conjunction with pressureless cooking to prepare from whole potato tubers a clear liquid extract containing *c.* 25% reducing sugar on which yeasts (*C. utilis, S. cerevisiae*) were grown in semi-continuous culture under non-sterile conditions. Nitrogen as either HNO_3 (56 g), $OC(NH_2)_2$ (18·8 g) or NH_4OH (73 g) and phosphorus as H_3PO_4 (8·5 g) per liter of extract were added to either the extract or pre-culture water. The culture was maintained at pH 4·0–4·2 by adding the acidified or alkalinized extract to the fermentor in response to the pH fluctuation due to the yeast metabolism. Conversion yields and specific growth rates were calculated for *C. utilis* (49%, 0·34/h) and *S. cerevisiae* (50%, 0·18/h) with increases in biomass (54 or 34·5 g/liter respectively) after 12–18 h of growth using urea (Tables 6 and 7). *Candida utilis* performed similarly using nitrate. Bacterial contamination was practically non-existent.

Ghoul and Engasser process: Ghoul and Engasser (1983) increased the yield of *C. utilis* by fed-batch culture in α-amylase and glucoamylase-hydrolyzed cassava starch. They obtained 50% biomass conversion after 20 h of fermentation which resulted in 70 g of yeast per liter at near-theoretical and near maximum growth rate. It was possible to control the feeding with gas-analysis equipment capable of detecting very fine changes in the respiratory quotient. One difficulty of this process is that some pre-culture growth (2 g yeast/liter) is required before the continuous feeding of the substrate to the fermentor.

Provesteen process: Phillips Petroleum Company and its wholly-owned subsidiary, Provesta Corporation, have developed an ultra-high density fermentation for the product on single-cell protein (Hitzman, 1986). The products are derived from the yeast cells *Pichia, Torula* and *Kluyveromyces* grown in continuous cultures on a variety of feedstocks such as methanol, ethanol, sugars, cassava, sorghum, bananas, sweet potatoes, whey and corn. The fermentor effluent usually contained 120–150 g of dry cells per liter that is dried directly after heat-killing the yeasts. Thus, filtration and centrifugation are eliminated in the process which simplifies plant design, engineering and operations. The versatility of this process technology offers many opportunities for the practical production of biomass and products which previously were uneconomic or difficult to achieve.

The yeast products produced by this process are known as Provesteen P, T, and K, derived from *Pichia, Torula* and *Kluyveromyces* cultures, respectively.

4 CONCLUSIONS

After lignocellulosic materials, starch feedstocks are the next most abundant plant material. The majority of the agricultural production of starch materials

Table 6
Regimes for Semi-continuous (Fed-batch) Cultures

	Regime[a]				
	A	*B*	*C*	*D*	*E*
Pre-culture[b]	Water, $CaCl_2$, $NaNO_3$, $MgSO_4$ 100 ml extract	Water, $CaCl_2$, $MgSO_4$ NH_4Cl 100 ml extract	Water, $CaCl_2$, $MgSO_4$ $CO(NH_2)_2$, Citrate	Potato juice, H_3PO_4, $CaCl_2$, $MgSO_4$ $CO(NH_2)_2$	Potato juice, $CaCl_2$, $MgSO_4$, NH_4Cl H_3PO_4
Feed	Whole extract HNO_3, H_3PO_4, (pH 1·5)	Whole extract NH_4OH, NH_4Cl (pH 9·5)	Whole extract, H_3PO_4, HCl (pH 3·5)	Whole extract, citrate, HCl, $CO(NH_2)_2$, (pH 3·0)	Whole extract, NH_4OH NH_4Cl, (pH 9·5)

[a] Nitrogen as either $HNO_3 + NaNO_3$, $CO(NH_2)_2$ or $NH_4OH + NH_4Cl$ and phosphorus as H_3PO_4 were added to either the extract or pre-culture water with the intention of producing yeasts containing 8·5% N and 2·5% P at 50% biomass yield (dry weight on the available reducing sugar). [b] $CaCl_2$, $2H_2O$ (0·4 g/liter), Mg $SO_4.7H_2O$ (1 g/liter) and citric acid H_2O (1.6 g/liter).

Table 7
Fed-batch Culture of *C. utilis* and *S. cerevisiae* in Various Media Regimes

		Regime				
		A	*B*	*C*	*D*	*E*
C. utilis		0·28	0·30	0·26	0·34	0·19
	Y^x	0·44	0·40	0·41	0·49	0·40
	X^s	43·8	39·6	39·0	54·0	44·6
S. cerevisiae			0·19	0·17	0·18	0·17
	Y^x		0·42	0·38	0·50	0·26
	X^s		22·0	40·0	34·5	20·5

Specific growth rate (average for the entire run) calculated as $\frac{0{\cdot}693}{t} \times \log_2 \frac{(X_h - X_i)}{X_i}$ units of h^{-1}. t: elapsed time (h). X_h: total harvested biomass, dry weight. X_i: inoculum, dry weight. Y^x: yield of biomass, gram dry weight per gram of glucose consumed. X^s: final biomass concentration g/liter.

is used as a feed or food and only 1% of the total starch production is further converted to HFCS or alcohol. Starch materials offer the advantages of availability and consistent quality, being a readily fermentable substrate at a more stable and competitive cost than other sugars. On the basis of starch (actually glucose) as a relatively inexpensive starting material, a wide variety of products can be prepared. Starch can be hydrolyzed to glucose or HFCS syrups or to ethanol or other chemicals and SCP.

The key to growth in fermentation-produced commodity chemicals will be the relative price increases over the next 10–20 years in agricultural commodities such as molasses, and particularly in starch-derived carbohydrate relative to petrochemical raw materials. Although the exact market potential for ethanol would depend on a number of competing factors, alcohol produced from starch or renewable resources (lignocellulosics) could make an important contribution as an octane enhancer in unleaded gasoline in the near future.

Many efforts should also be directed towards increasing the fraction of raw material such as starch wastes that are assimilated into cell mass. When evaluating SCP production from wastes as part of a pollution control scheme, different aspects should be considered since in most cases it is available in very dilute solutions. The availability of zero cost raw material and a simple nonsterile operation could improve the economics of SCP production and have a beneficial effect on the waste-treatment process. The market price of SCP will be determined by the local price of soybean meal and fishmeal. An important factor is the superiority of most SCP products over soya in terms of protein content and amino acid profile. Although large-scale SCP production has been subjected to many influences and difficulties, development of efficient technology will continue to be of primary importance to solve feed and food problems that exist in large areas and for any country to control its own protein supply sources.

The last 10 years or so have seen dramatic advances in recombinant DNA technology and the development of immobilization technology. It would be possible to construct a microorganism containing the genes for starch-hydrolyzing enzymes, thereby enabling it to utilize starch directly as a substrate or utilize a particular carbon substrate.

REFERENCES

Admasson, W., Dorus, R. A., Heimsch, R. C. & Lemmel, S. A. (1981) *Biotechnol. Bioeng.*, **23,** 2361.
Alexander, N. J. (1986). *Food Technol.*, **40,** 99.
Aschengreen, P. H., Helwüg, B., Rosendal, P. & Østergaard, J. (1979). *Starch*, **64,** 66.
Azoulay, E. *et al.* (1980). *Appl. Environ. Microbiol.*, **39,** 41.
Balagopal, C. & Maini, S. B. (1976). *J. Root Crops*, **2,** 49.
Balagopal, C. & Maini, S. B. (1977). *J. Root Crops*, **3,** 33.
Barker, S. A. & Petch, G. S. (1985). In *Enzymes and Immobilized Cells in Biotechnology*, ed. A. I. Laskin. Benjamin Publ. Co., Menlo Park, California, p. 93.
Berghofer, E. & Sarhaddar, S. (1988). *Process Biochem.*, **23,** 188.
Bloch, F., Brown, G. E. & Farkas, D. F. (1973). *Amer. Potato J.*, **50,** 357.
Brook, E. J., Stanton, W. R. & Wallbridge, A. (1969). *Biotechnol. Bioeng.*, **11,** 1271.
Carrantino, S. (1983). *Biofutur*, January, 17.
Casey, S. P. (1977). *Stärke*, **29,** 196.
Cheng, W. S. & Pereira, R. R. (1970). Paper presented at 13th Annual Conf. Can. Inst. Food Sci. & Technol, Windsor, Ontario, 8–10 June.
Chithra, N. & Baradarajan, A. (1989). *Process Biochem.*, **24,** 208.
Clement, F., Rossi, J., Costamagna, L. & Rosi, J. (1980). *Antonie van Leeuvenhoek*, **46,** 399.
Coker, L. E. & Venkatasubramanian, K. (1986). In *Food Biotechnology*, ed. D. Knorr. Marcel Dekker, New York, p. 443.
Conway, T., Byun, M. O. K. & Ingram, L. O. (1987). *Appl. Environ. Microbiol.*, **53,** 235.
Correira, S. A. & Van Uden, N. (1981). *Eur. J. Appl. Microbiol. Biotechnol.*, **13,** 24.
Crott, R. (1986). In *The Biotechnological Challenge*, eds S. Jacobsson, A. Jamison & A. Rothman. Cambridge University Press, Cambridge, UK, p. 96.
Dale, B. E. (1984). In *Annual Reports on Fermentation Processes*, ed. G. T. Tsao. Academic Press, New York, p. 107.
Davids, S. J., Lee, B. H. & Simard, R. E. (1987). *Microbiologie—Aliments—Nutrition*, **5,** 197.
DeCarvalho, A. V., Milfont Jr, W. N., Yang, V. & Trinidade, S. C. (1977). In *Int. Symp. on Alcohol Fuel Technology—Methanol and Ethanol*, Wolfsburg, Germany, 21–23 November.
De Deken, R. H. (1966). *J. Gen. Microbiol.*, **44,** 149.
Drawert, F., Klisch, W. & Sommer, G. (1987). In *Fundamentals of Biotechnology*, eds P. Präve, U. Faust, W. Sittig & D. A. Sukatsch. VCH, Weinheim, Germany, p. 382.
Duvnjak, Z., Kosaric, N. & Kliza, S. (1982). *Biotechnol. Bioeng.*, **24,** 2297.
Dziezak, J. D. (1987). *Food Technol.*, **41,** 104.

Ericsson, M., Ebbinghaus, L. & Lindblom, M. (1981). *J. Chem. Technol. Biotechnol.*, **31,** 33.
Errat, J. A. & Nasim, A. (1986). *J. Bacteriol.*, **166,** 484.
Errat, J. A. & Stewart, G. G. (1981). *Dev. Ind. Microbiol.*, **22,** 557.
Faust, U. (1987). In *Fundamentals of Biotechnology,* eds P. Präve, U. Faust, W. Sittig & D. A. Sukatsch. VCH, Weinheim, Germany, p. 601.
Finn, R. K. (1987). In *Biotech.* 1, *Microbial Genetic Engineering and Enzyme Technology,* eds C. P. Hollenberg & H. Sahm. VCH, New York, p. 101.
Gallejà, G. B., Moranelli, F., Nasim, A., Duck, P. D. & Levy, S. (1981). Paper presented at VIIth Int. Specialized Symp. on Yeast, Valencia, Spain, 1–3 September.
Ghoul, M. & Engasser, J. M. (1983). *Microbiologie—Aliments—Nutrition,* **1,** 277.
Gibbons, W. R., Westby, C. A. & Dobbs, T. L. (1984). *Biotechnol. Bioeng.*, **26,** 1098.
Gôdia, F., Casas, C. & Sola, C. (1987). *Process Biochem.* **22,** 43.
Gregory, K. F. *et al.* (1976). *Food Technol.*, **3,** 30.
Gregory, K. F. *et al.* (1977). *Anim. Feed Sci. Technol.*, **2,** 7.
Guiraud, J. P., Caillaud, J. M. & Galzy, P. (1982). *Eur. J. Appl. Microbiol. Biotechnol.*, **14,** 81.
Hasseltine, C. W. (1972). *Biotechnol. Bioeng.*, **14,** 517.
Hattori, Y. (1961). *Agric. Biol. Chem.*, **25,** 737.
Hertzmark, D., Flaim, S., Ray, D. & Parvin, G. (1980). *Amer. J. Agr. Econ.*, **62**(5), 965.
Hitzman, D. O. (1986). In *Microbial Biomass Proteins,* eds M. Moo-Young & K. F. Gregory. Elsevier Applied Science, London, p. 27.
Hollenberg, C. P. & Wilhelm, M. (1987). In *Biotech. 1: Microbial Genetic Engineering and Enzyme Technology,* eds C. P. Hollenberg & H. Sahm. VCH, New York, p. 21.
Innis, M. A. *et al.* (1985). *Science,* **228,** 21.
Jarl, K. (1969). *Food Technol.* (Champaign), **23,** 1009.
Knight, J. W. (1969). *The Starch Industry.* Pergamon Press, Oxford, UK.
Kombila, M. E., Lee, B. H. & Simard, R. E. (1985). *Dev. Ind. Microbiol.*, **26,** 781.
Kombila, E. M., Lee, B. H. & Simard, R. E. (1988). *Microbiologie—Aliments—Nutrition,* **6,** 239.
Lachance, M. A. & Blackwood, A. C. (1972). Paper presented at Congrès ACFAS, Ottawa, Canada, 2–5 October.
Lee, B. H. (1987). In *Proc. 7th Congr. Food Science and Technology,* Singapore, p. 60.
Lee, B. H. (1990). Genetic Engineering Part II: enzyme cloning. In *Encyclopedia of Food Science and Technology,* John Wiley, New York.
Lee, B. H. & Simard, R. E. (1984). *Dev. Ind. Microbiol.*, **25,** 459.
Lee, D. D., Lee, Y. Y., Reilly, P. J., Collins Jr, E. V. & Tsao, G. T. (1976). *Biotechnol. Bioeng.*, **18,** 253.
Lemmel, S. A., Heimsch, R. C. & Edwards, L. L. (1979). *Appl. Environ. Microbiol.*, **37,** 227.
Martin, S. R. (1982). *Chemical Engineer* (London), **50,** February.
McNicol, D. G. *et al.* (1972) *Grain starch utilization study.* Report of a study group appointed by the Canadian Wheat Board.
Millichip, R. J. & Doelle, H. W. (1989). *Process Biochem.*, **24,** 141.
Monica, B. D., Millichip, R. J. & Doelle, H. W. (1989). *Process Biochem.*, **24,** 137.
Moreton, R. S. (1978). *J. Appl. Bacteriol.*, **44,** 373.
Muindi, P. J. & Hanssen, J. F. (1981). *J. Sci. Food Agric.*, **32,** 647.
Nga, B. H., Eswaran, S., Tan, E. L. & Lee, T. K. (1977). *Malys. Appl. Biol.*, **6,** 123.
Oteng-Gyang, K., Moulin, G. & Galzy, P. (1980). *Acta Mikrobiol. Hung.*, **27,** 155.
Outtrup, H. (1976). (To Novo-Industri A/S) British Patent 1455993.
Pellon, J. R. & Sinsky, A. J. (1983). In *Food Microbiology: Advances and Prospects,* eds T. A. Roberts & F. A. Skinner. Academic Press, New York, London, p. 287.

Peppler, H. J. (1979). In *Microbiolial Technology 1*, 2nd edition, eds H. P. Peppler & D. Perlman. Academic Press, New York.

Prendergast, P., Booth, A. & Colleran, E. (1983). In *Production and Feeding of Single Cell Protein,* eds M. P. Ferranti & A. Fiechter. Applied Science, London & New York, p. 96.

Raimbault, M. (1981). Fermentation en milieu solide. ORSTOM, p. 291.

Raimbault, M. & Alazard, D. (1980). *Eur. J. Appl. Microbiol. Biotechnol.*, **9,** 199.

Ramos-Validivia, A. (1982). Enriquecimento proteico de la yuca (Manihot esculenta crantz). MS Thesis, Centro de Investigacion y de Estudios Avanzados del IPN, Mexico.

Reade, A. E. & Gregory, K. F. (1975). *Appl. Microbiol.*, **30,** 897.

Rentschler, D. F., Langlois, D. P., Larson, R. F., Alverson, L. M. & Liggett, R. W. (1962). US Patent 3039935.

Revuz, B. & Voisin, D. (1980). *Ind. Alim. Agric.*, **97**(10), 1079.

Rieser, C. O. (1954). *Agric. Food Chem.*, **2,** 70.

Rolz, C. (1984). In *Annual Reports on Fermentation Processes,* ed. G. T. Tsao. Academic Press, New York, p. 213.

Sanchez, M. R. (1981). Enriquecimento de la harina del tuberculo de yuca mediante fermentacion aerobica con hongos mucorales. Thesis, Universidad Autonoma del Edo. de Mexico, Mexico.

Senez, J. C. (1983). In *Production and Feeding of Single Cell Protein, eds* M. P. Ferranti & A. Fiechter. Applied Science, New York, p. 101.

Senez, J. C., Raimbault, M. & Deschamps, F. (1980). *FAO World Animal Rev.*, **35,** 36.

Sercovich, F. C. (1986). In *The Biotechnological Challenge,* eds S. Jacobsson, A. Jamison & H. Rothman. Cambridge University Press Cambridge, UK, p. 148.

Sharma, G. (1988). *Process Biochem.*, **23,** 139.

Shieh, K. K., Lee, H. A. & Donelley, B. J. (1974). (To Auheuser Busch) US Patent 3834988.

Sills, A. M. & Stewart, G. G. (1982). *J. Inst. Brew.*, **88,** 313.

Spencer-Martins, I. & Van Uden, N. (1977). *Eur. J. Appl. Microbiol. Biotechnol.*, **4,** 29.

Spencer-Martins, I. & Van Uden, N. (1979). *Eur. J. Appl. Microbiol. Biotechnol.*, **6,** 241.

Stevens, C. A. & Gregory, K. F. (1987). *Appl. Environ. Microbiol.*, **53,** 284.

Stogman, H. (1976). In *Food from Waste,* eds G. C. Birch, K. J. Parker & J. T. Worgan. Applied Science, London, p. 167.

Swinkels, J. J. M. (1985). In *Starch Conversion Technology,* eds G. M. A. Van Beynum & J. A. Roels. Marcel Dekker, New York, p. 15.

Takasaki, Y., Kosugi, Y. & Kanbayashi, A. (1969). In *Fermentation Advances,* ed. D. Perlman. Academic Press, New York.

Tong, P. Q., Riel, R. R. & Simard, R. E. (1973). *J. Inst. Can. Sci. Technol. Aliment.*, **6,** 212.

Touzi, A., Prebois, J. P., Moulin, G., Deschanps, F. & Glazy, P. (1982). *Eur. J. Appl. Microbiol. Biotechnol.*, **15,** 232.

Tveit, M. T. (1963). Conversion of starches and starch like substances into yeast cells. US Patent 3105799.

Tyner, W. E. (1980). *Agricultural Energy Production Potential,* 15. Dept of Agricultural Economics, Purdue University, Lafayette, IN.

Van Beynum, G. M. A. & Roels, J. A. (1985). *Starch Conversion Technology.* Marcel Dekker, New York.

Van Tilburg, R. (1985). In *Starch Conversion Technology,* ed. G. M. A. Van Beynum & J. A. Roels. Marcel Dekker, New York, p. 175.

Vellenga, K. (1978). The isomerization of D-glucose into D-fructose. PhD Thesis, State University of Groningen, The Netherlands.
Venkatasubramanian, K. & Harrow, L. S. (1979). *Ann. New York Acad. Sci.*, **326,** 141.
Venkatasubramanian, K. & Keim, C. R. (1985). In *Starch Conversion Technology,* eds G. M. A. Van Beynum & J. A. Roels, Marcel Dekker, New York, p. 143.
Wang, D. I. C., Cooney, C. L., Wand, S. D., Gordon, J. & Wang, H. Y. (1978). *Second Biomass Energy Systems Conference,* Troy, New York, p. 537.
Wickerman, L. J. & Kuehner, C. C. (1953). US Patent 2764487.
Wilson, J. J., Khachatourians, G. G. & Ingledew, W. M. (1982). *Biotechnol. Lett.,* **4,** 333.
Wingard, L. B., Katchalski-Katzir, E. & Goldstein, L. (1976). *Immobilized Enzyme Principles.* Academic Press, New York.
Yamashita, I., Suzuki, K. & Fukui, S. (1985). *J. Bacteriol.,* **161,** 567.

Chapter 9

FOOD PROCESSING WASTES

G. Th. Kroyer

Institute of Food Chemistry and Technology, Technical University Vienna, Getreidemarkt 9, A-1060 Vienna, Austria

CONTENTS

1 INTRODUCTION

The food industry produces large volumes of wastes, both solids and liquids, resulting from the production, preparation, and consumption of food. These wastes pose increasing disposal and potential severe pollution problems and represent a loss of valuable biomass and nutrients. In the past they have often been dumped or used without treatment for animal feed or as fertilizers. In the last few years, however, due to the increasing necessity to take into consideration aspects aimed at preventing pollution of the environment as well as for economic motives and the need to conserve energy and raw materials, new methods and policies towards waste handling and treatment have found their way in the recovery, bioconversion, and utilization of valuable constituents from food processing wastes. Beside their pollution and hazard aspects, in many cases, food processing wastes might have a potential for recycling raw materials, or for conversion into useful products of higher value as by-product, or even as raw material for other industries, or for use as food or feed after biological treatment. Particularly, bioconversion of food processing wastes is receiving increased attention in view of the fact that these wastes represent a possible and utilizable resource for conversion to useful products.

This chapter investigates the biological and biotechnical approaches to the feasibility of bioprocessing food processing waste materials for the recovery of bioresources and for the production of useful products of higher added value with emphasis on the recycling objective of the bioconversion.

2 CHARACTERISTICS OF FOOD PROCESSING WASTES

The composition of wastes emerging from food processing factories is extremely varied and depends on both the nature of the product and the production technique employed. There are, however, certain characteristics of the particular industries which may be useful in the selection of suitable methods for waste treatment. For example, wastes from meat processing plants will contain a high fat and protein content, whilst waste from the canning industry will contain high concentrations of sugars and starches (Isaac & Anderson, 1974). Thus, they are rich in easily degradable organic matter, but generally they are free from metals, pesticides, and other toxic substances and low in inorganic solids (James & Addyman, 1974). Also the waste may not only differ from site to site but also vary from one time of the year to another. Furthermore, the volume and concentration of the waste material will not be constant. This may cause some problems in managing a consistent working process due to fluctuations in the nature, composition, and quantity of raw materials (Kirsop & Hilton, 1981).

In general, wastes from the food processing industry have the following characteristics (Litchfield, 1987):

—large amounts of organic materials such as proteins, carbohydrates, and lipids;
—varying amounts of suspended solids depending on the source;
—high biochemical oxygen demand (BOD) or chemical oxygen demand (COD).

Vast quantities of water are used as a whole in the food processing industry, of which a small proportion is for consumptive use. Furthermore, the rapid growth in the production of factory-processed food (canned, bottled, frozen, dehydrated food) and the many other developments of modern food technology have led to fast growth in the production of liquid food wastes, which are characterized as dilute streams containing proteins, sugars, starches, and fats (Isaac & Anderson, 1974). However, regarding the utilization of food processing wastes, solid as well as liquid wastes are applicable to bioconversion to useful products.

The appraisal of food processing wastes as animal feed or human foodstuff necessitates information on their nutritional value and toxicological safety.

Suggestions are made regarding protein and non-protein content, oil and fatty acid composition, crude fibre and vitamin content, microbial content, and toxic organic compounds (Tacon, 1980).

In general, the major types of food processing industries whose wastes could be converted to useful products by biotechnological processes may be regarded as:

—the meat and fish processing industry;
—the fruit and vegetable industry;
—the dairy industry;
—the fermentation industry.

3 BIOTECHNOLOGICAL PROCESSES IN FOOD PROCESSING WASTE TREATMENT

The food industry may be recognized in many aspects as the largest, oldest and most potentially valuable form of biotechnology. The application of biotechnology to the modification of food components, to new methods for the production of food and food components, as well as to modern methods of food analysis is becoming more and more important in the food industry. Therefore, it is not surprising that biotechnology has also found its application in the treatment of wastes arising from the food processing industry. In this way food wastes could be utilized by modifying them to produce an upgraded material of higher value than the waste itself or to produce entirely new products, by either microbe-aided or enzyme-aided processes. In particular, the production of complete different products from waste, usually by applying biotechnological methods, has gained great attention.

The simplest biological treatment of waste is the use of aerobic filters, trickle beds or activated sludge plants. Thereby only detoxification of the waste is achieved and usually no by-products are obtained, at least on an economic scale (Hacking, 1988).

Under anaerobic conditions, food processing wastes containing carbohydrates, lipids, and proteins can be digested to yield biogas. Anaerobic digestion combines detoxification of waste materials by upgrading them to their use as animal feed or fertilizer and the production of biogas. Biogas varies in composition according to the composition of the waste treated. The content of methane generally lies between 60% and 80% with carbon dioxide and hydrogen sulphide as major remaining substances (Hacking, 1988). Biogas is used mainly as energy source. The anaerobic treatment processes can be applied to all segments of the food processing industries with particular reference to concentrated waste streams. Most of the full scale installations are within the sugar processing, meat rendering, brewing and distilling, bottling and soft drink, wheat starch and potato processing industries (Oleszkiewicz & Olthof, 1982). Also ethanol or organic acids can be produced from carbohydrate-containing waste materials by anaerobic microbiological processes (Litchfield, 1987).

Much research work has been focused on upgrading waste material to more valuable products using appropriate biotechnological processes. Thereby, the production of biomass, especially single cell protein (SCP), and the production of ethanol seem to be the most interesting and promising developments (Hacking, 1988).

Consideration of the substantial need for energy in the world as well as the increasing costs thereof makes the production of energy from waste in the form of ethanol very attractive. Biotechnological ethanol production on the industrial scale is already well known and large quantities are being produced from carbohydrate sources in many countries.

Single-cell protein, or biomass, produced from food waste, can be used either as animal feed or as human food. This imposes particular regulations regarding the consumption as a diet component. Furthermore, the product must not be contaminated with other micro-organisms, otherwise sterilization procedures of the waste are inevitable. Rigorous tests are usually necessary to confirm the purity of the biomass. All these steps are expensive, and this might be a limiting factor in practice for general application as a widespread method for utilizing food industry wastes for the production of biomass.

Production of solvents, chemical feedstocks, or liquid fuels by microbial digestion requires either aerobic or anaerobic digestion to be incomplete. Such incomplete digestions for production of ethanol, acetone, butanol, or butanediol are limited to fermentation of specific substrates such as carbohydrates and require selective conditions, and recovery is costly (Pirt, 1978).

4 PRODUCTION OF BIOMASS FROM FOOD PROCESSING WASTES

Due to a renewed awareness of the increasing world population, the increasing costs of energy and raw materials as well as the increased concern about environmental pollution, intensive research is underway into the use of fermentation processes to convert waste materials into useful and profitable products. In particular, the direct conversion of waste materials into microbial biomass suitable for animal feed and human consumption is one of the main targets of study well documented in the literature, and various waste materials from the food processing industry have been evaluated for use as substrates in biomass production (Mateles & Tannenbaum, 1968; Tannenbaum & Wang, 1975; Birch *et al.*, 1976; Gray & Berry, 1980; Herzka & Booth, 1981; Litchfield, 1983; Solomons, 1983; Goldberg, 1985).

In general, the composition of food wastes is such that they are suitable as growth media for micro-organisms. This can be exploited either by conventional waste treatment using a heterogeneous population of bacteria, fungi, and protozoa to convert the organic matter, or by growing food yeasts or other edible micro-organissm on the waste to get more or less pure cultures of microbial mass (James & Addyman, 1974).

Biomass, or single-cell protein (SCP), is the name given to the total dry matter of bacteria, yeasts, moulds, and higher fungi when used for animal feed or human food (Kirk-Othmer, 1980). The use in feed is related primarily to considerations of nutritional value and balance of the ingredients, whilst the level of utilization in food is certainly limited by the concentration of nucleic acid in the final product (Cooney *et al.*, 1980).

Regarding liquid wastes of an appropriate chemical composition, large quantities are produced in the food processing industry to be used as a substrate for microbial fermentations. It has been estimated that in general for most of the food industries, for every tonne of food produced an equal amount of carbohydrate waste and effluent is produced (Imrie & Righelato, 1976). This waste has often caused pollution resulting from its disposal into rivers or the sea or has increased the costs of the process by treatment in an effluent plant. For example, effluents which contain simple sugars (beet processing wastes) are easier to handle than those containing high molecular weight polysaccharides such as starch processing wastes, where it is necessary to break the wastes down to assimilable mono- and disaccharides before they can be metabolized by the micro-organisms (Anderson *et al.*, 1975).

A wide variety of micro-organisms has been studied as possible organisms for biomass production, including several bacteria and algae or filamentous fungi and yeasts (Gray & Berry, 1980). In general, bacteria are not as useful as yeasts and moulds, particularly where separation of the cell product from the residual waste is necessary, and cell recovery by centrifugation turns out to be uneconomic (Litchfield, 1987).

For the establishment of the microbial cultures different designs of fermentors are used, e.g. stirred tank reactors, tubular or tower fermentors, or multistage reactors, resulting in different microbial growth kinetics and rates in the fermentor, and characterized by different temperature effects, oxygen requirements, and substrate conversion efficiencies (Gray & Berry, 1980).

Regarding biomass production, it is important that the yields of cell mass and protein per unit of growth-limiting substrate consumed are maximized. It is necessary for growth to be carbon-limited, to prevent accumulation of carbon which would result in conditions of nitrogen limitation (Sucuru *et al.*, 1975). Furthermore the initiation of secondary metabolism which may lead to toxin production is largely avoided under conditions of carbon-limited growth (Litchfield, 1977).

To obtain an optimal yield of biomass and protein it is essential that all major nutrients which are required for growth are present together in an adequate proportion. However, generally, the effects of the various nutritional and physiological parameters of the whole process on the behaviour of the micro-organisms are very complex, and for each combination of organism and substrate the different combinations of parameters should be tested at experimental and production scale level in order to determine the optimum conditions. Values of substrate conversion efficiencies and yields of biomass of different organisms growing on different food processing wastes, obtained from several authors, are documented in the literature (Gray & Berry, 1980).

Above all, microbial biomass should be acceptable as a food for the animal or human consumption for which it has been produced. Animals normally accept microbial biomass in their diet; however, further treatment of the biomass is usually necessary if it is to be used for human consumption (Gromov, 1967). For the production of animal feed-grade single-cell protein from waste, a clean but non-sterile mode of operation is sufficient; for food-grade products, however, it may be necessary to apply sterile conditions to the substrate and fermentor (Litchfield, 1987).

The nutritional quality of microbial biomass depends not only on the high level of protein, but also on a good amino acid profile, that is, a high content of nutritionally valuable essential amino acids. The levels of essential amino acids in protein from a range of filamentous fungi have been detailed in the literature (Anderson *et al.*, 1975). According to this, fungal protein compares well with the more conventional foods, although most fungal protein is deficient in the sulphur amino acids.

A high content of nucleic acids, which are present in all microbial biomass, can present a nutritional problem, since this can possibly cause gout-like conditions in man (Kihlberg, 1972). Values of ribonucleic acid (RNA) between 3·2 and 4·7% w/w up to 10% w/w have been reported in filamentous fungi (Worgan, 1976). Techniques have been investigated to reduce the level of RNA in biomass by controlling the fermentation process or after it has been harvested (Sinskey & Tannenbaum, 1975).

For safety reasons harmful substances should not be present in the substrate, which can become incorporated into the final product (e.g. heavy metals, herbicides, toxins, etc.). Furthermore, care must be taken to prevent contamination by toxic materials (mycotoxins) produced by the organism during the fermentation.

Therefore, procedures for carrying out preclinical tests on novel foods for man have been recommended by the Protein Advisory Group of the United Nations, which includes aspects of safety, nutritional value, sanitation acceptability and technological properties (Tannenbaum & Wang, 1975).

The possibility of using activated sludge for the production of biomass as animal feed is described in recent literature (Vriens *et al.*, 1989). The activated sludge process is a process in which wastewater containing biogradable organic compounds is brought into contact with a fluidized mixed-culture of microorganisms in an aerobic environment. In an aeration tank an assimilative process removes from the water a portion of organic carbon, chemically bound nitrogen, phosphorus, and other micronutrients. Biomass is formed and simultaneously other organic and inorganic compounds are adsorbed by various mechanisms. Biomass is then separated from the liquid phase supernatant in a settling tank.

The activated sludge process for the treatment of waste water is in fact analogous to the production of SCP, with the exception that a mixed culture of micro-organisms is used instead of pure cultures. Activated sludge grown on food processing wastewater may be used successfully as a feed material because normally significant concentrations of toxic heavy metals as in sewage sludge would not be expected to be present, and contamination by faecal wastes or pathogenic bacteria could be prevented. Effluents from many food processing industries contain mainly simple sugars readily assimilable by micro-organisms, and activated sludge treatment is considered to require less capital investment, with lower operating costs, than the conventional SCP production. Extensive studies on different animals proved the suitability of activated sludge from the brewing, dairy, citrus fruit and fruit canning, and slaughterhouse industries. There may exist some problems concerning toxicological aspects and safety tests, but these could probably be solved on the basis of present technology. The feasibility of this process seems to be a matter of economics. The use of activated sludge from food processing wastewater treatment might therefore be expected to contribute indirectly to human nutrition in the near future (Vriens *et al.*, 1989).

5 PRODUCTION OF BIOGAS FROM FOOD PROCESSING WASTES

Food industry effluents of all types present a potentially large source of methane production. The biogradable content of food effluents is converted by anaerobic digestion to methane and carbon dioxide with minor amounts of hydrogen sulphide and ammonium salts or nitrate depending upon the ratio of C, H, O, N and S in the substrate. Methane is liberated as a gas and there is substantial experience of its utilization as a fuel. The advantages of utilization of the gas as an energy source together with the reduction of COD and thus the costs of effluent disposal would seem likely to render the production of methane from food processing wastes as an attractive process.

Methane production through anaerobic digestion is a three-step process. The complex biopolymers in the waste material are hydrolysed enzymatically to their monomeric units, which are transformed into low molecular weight organic acids, other organic compounds, and gases, mainly carbon dioxide and hydrogen, from which finally methane is generated (Rolz, 1978). Some of the processes that have been used for biogas production from different types of food processing wastes, such as sugar refining, potato processing, dairy, brewery, distillery, fruit and vegetable processing wastes, are cited in the literature (Litchfield, 1987).

Although anaerobic digestion is used mainly for treatment of liquid food industry effluents, preliminary laboratory work has been carried out on the anaerobic digestion of solid wastes from biscuit and chocolate production, which were collected from the mixing and packaging units and which were rich in easily biogradable carbohydrates and fats. The data collected showed that the waste is amenable to anaerobic digestion, and the biogas produced could be used in the factory (Ranade *et al.*, 1989).

Anaerobic digestion of cheese whey to yield biogas is reported in the literature (Clanton *et al.*, 1985; Yan *et al.*, 1988).

6 MEAT AND FISH PROCESSING WASTES

Due to an increasing demand for protein, there is a great need to develop processes to upgrade high-protein meat and fish processing wastes which currently may be used as animal feed or get lost from the food chain altogether. In the literature several studies have been reported on the utilization of meat and fish processing wastes which generally have a high pollution liability. Different micro-organisms, including yeasts, filamentous fungi, bacteria and algae, can be used to produce single-cell proteins (Litchfield, 1977; Hang, 1979). These processes are very energy- and capital-intensive, and it may be necessary to supplement the feedstock nutritionally (Litchfield, 1977). Regarding meat processing wastes the utilization of collagen for the production of single-cell protein was studied on a laboratory scale (Bough *et al.*, 1972). Further experimental studies were performed on the production of single-cell protein from different fish processing wastes (Reva-Moissev & Carroad, 1981; Cosio *et al.*, 1982; Welsh & Zall, 1984). However, none of these processes have been practiced on a larger scale, and further scale-up studies would be required (Litchfield, 1987).

Meat packing plants may also consider anaerobic digestion processes for the production of biogas, especially methane, from their waste materials (Hansen, 1983). Slaughterhouse blood, which is a protein-rich residue of meat production, can be utilized for the recovery of applicable protein preparates. About 90% of the thick blood protein can be recovered, and the iron-containing haem fraction could be used as a dietetic ingredient for specialized products and as a feed additive (Kobald & Holley, 1989). For the recovery of applicable protein preparates decolourization of the thick blood could be performed by enzymatic hydrolysis and separation of the coloured haem from the colourless globin protein fraction (Drepper & Drepper, 1981).

The production of biogas, and the recovery of protein and fat from meat, fish, and poultry wastes by using biological and biotechnological treatment methods, are reviewed in recent literature (McComis & Litchfield, 1989).

7 FRUIT AND VEGETABLE PROCESSING WASTES

Large amounts of fruit and vegetable processing wastes are produced from packing plants, canneries, etc., which may be disposed in several ways including immediate use for landfill or as an animal feed, or which, alternatively, may be processed biotechnologically in order to produce a more valuable product. In recent times, industry continues to make progress in solving the waste problem through recovery of by-products and waste materials such as peel, pulp, or molasses by employment of fermentation or other related biological processes.

The protein content of fruit and vegetable processing wastes with an adequate level of fermentable carbohydrates can be increased to 20–30% by using a solid substrate fermentation process. The production of single-cell protein from these wastes is supposed to be economically attractive if there exists a local market for the product and fuel costs could be minimized (Davy, 1981).

Solid-state fermentation, in which a moist water-insoluble solid substrate is utilized by micro-organisms in the absence of free water, is applied for the enrichment of the protein content of orange peel, banana and carrot wastes (Davy, 1981), coffee pulp (Penaloza *et al.*, 1985), and citrus peel (Rodriguez *et al.*, 1985).

Developments in the citrus fruit processing industry brought about handling problems of wastes remaining after extraction of juice or after making other products from the fruits which are rich in carbohydrates and account for about 45–60% of the weight of processed fruits. Studies on the utilization of orange waste as a substrate for microbial protein production by the fungus *Sporotrichum pulverulentum* were performed, and biomass containing about 32% protein was obtained (Karapinar & Okuyan, 1982). Further investigations into the utilization of orange peel by the fungal species *Memnoniella echinata* and *Fusarium roseum* emphasized the technical feasibility of microbial protein production from untreated citrus waste, possibly achieving crude protein yields greater than 16% and protein enrichment of the final biomass ranging from 40% to 46% (Clementi *et al.*, 1985). Citrus fruit pulp from the fruit canning industry was used as a substrate for single-cell production, showing that the fungus *Trichoderma koningii* yielded the maximum total protein productivity (Ghai *et al.*, 1979).

In the cane and sugar beet processing industries molasses left after sugar extraction can be utilized for growing yeasts. The molasses supply carbohydrates, mineral elements, and amino nitrogen for the fermentation process, but it may be necessary to add nitrogen and phosphorus. An aerobic fermentation process has been applied for biosynthesis of yeast protein using *Saccharomyces cerevisiae*. For the production of 100 kg yeast dry matter containing 50% protein, 400 kg of molasses, 25 kg of aqueous ammonia, 15 kg of ammonium sulphate, 7 kg of mono-ammonium phosphate, and 4810 m^3 air were required (James & Addyman, 1974).

Starch-containing wastes such as effluents from the processing of potatoes, corn, and rice have been treated in recent years by combined enzymatic hydrolysis and yeast fermentation with *Candida utilis* (Tveit, 1967; Jarl, 1969).

Onion juice, obtained as a by-product or waste of the dehydrated onion industry, has been used after hydrolysation as a carbon and nitrogen source for the cultivation of *Saccharomyces cerevisiae* strains yielding up to 46% protein (Abou-Zeid *et al.*, 1979).

Palm oil processing effluents have been utilized as substrates for microbial protein production by the fungus *Aspergillus oryzae,* yielding biomass of about 500 g/kg organic matter, with BOD reduction of 85% and COD reduction of 75–80% (Barker & Worgan, 1981).

Wastes from potato and wheat starch processing factories can be fermented to ethanol. Potato pulp is, however, difficult to ferment to ethanol due to its content of gelling materials such as pectins and hemicelluloses, and hydrolysis processes have been developed to obtain hydrolysates high in dry matter content with respect to the subsequent fermentation (Marihart, 1982).

Anaerobic digestion can be used to convert solid wastes from fruit and vegetable processing into methane fuel (Lane, 1979). In the citrus processing industry large amounts of solid wastes are generated whose disposal has caused an enormous pollution problem. However, the presence of peel oil makes citrus peel toxic to anaerobic digestion, which makes reduction of the oil content prior to the fermentation process necessary (Lane, 1980). A practical means for reducing the peel oil content for the subsequent utilization of the peels as feedstock for anaerobic digestion was established, whereby yields of $0{\cdot}5\,m^3$ of biogas (50–55% methane) per kilogram of total solids with conversion rates of solids to gas up to 100% could be achieved (Lane, 1984).

Wastes from bean bleaching, and pear and potato peeling, have been used for the production of methane by an anaerobic contact process, and more recently by a UBF upflow blanket process, with high efficiencies in converting organic solids to methane (Van den Berg & Lentz, 1977; Van den Berg *et al.*, 1983). An upflow sludge blanket process has also been used for treating beet sugar refining wastes (Pette *et al.*, 1981).

Production of fat from sweet potato processing wastes from starch manufacture, and from molasses has been performed by fermentation processes with micro-organisms from the *Lipomyces* species. The sugar sources were treated with diluted sulphuric acid before fermentation, and ammonium sulphate or peptone was used as a nitrogen source, yielding about 20% fat based on consumption of the sources (Watanabe, 1974).

The possibility of the production of antibiotics from plant food processing wastes is described in the literature. Egyptian blackstrap molasses obtained from different sugar cane manufacturers, and freeze-dried olive water obtained from the pressing and separation of olive oil, have been used for the production of oxytetracycline by *Streptomyces rimosus* (Abou-Zeid *et al.*, 1980).

An alternative treatment of vegetable oil refinery wastewater might be the microbiological conversion of the unsaturated fatty acids in the waste vegetable oil into hydroxy- and keto-fatty acids. Biotransformation of oleic acid and linoleic acid by *Rhodococcus rhodochrous* has been achieved for the production of the corresponding hydroxy- and keto-stearic acids (Litchfield & Pierce, 1986).

Olive oil extraction effluents are major pollutants of the environment. Their utilization in producing useful by-products by fermentation could limit environmental pollution, since the effluents contain various amounts of carbohydrates depending on the variety of olives, growth conditions and extraction procedures. Some yeast strains of *Torulopsis* sp. MK-1, *Saccharomyces norbiensis* MC-1, *S. oleaceus* MC-2 and *S. oleaginosus* were grown well on the effluents, and fermented the sugar into alcohol, but relatively low final alcohol concentrations in the fermentation broth were achieved (Bambalov *et al.*, 1989).

An interesting approach to the utilization of food processing wastes has been proposed by studies performed with solid or liquid wastes, such as from molasses from alcohol fermentation, from the corn industry, from wastewater from coconut processing, but also from the malting and brewing industry, and from residues from the bakery industry, employing semi-solid or submerged fermentation processes to produce *Bacillus thuringiensis* endotoxins in order to get bacterial insecticides (Moraes *et al.*, 1989).

Almost any substrate which can be used to produce single-cell protein (SCP) could be used to produce single-cell oil (SCO). Suitable substrates are molasses (sugar beet and cane), whey, starch, and wastes such as date extracts and blanching liquors from vegetable processing units. After extraction of the oil from the micro-organism a protein-rich residue remains which is then saleable as single-cell protein. Thus, a micro-organism grown for oil is producing two valuable products (Ratledge, 1978). Factors affecting the production of single-cell oil from different micro-organisms (temperature, incubation period, C/N ratio, spore count) were studied using molasses and carrot wastes (Hassanien *et al.*, 1986).

8 DAIRY INDUSTRY WASTES

Wastes from the dairy industry include four major types (Jelen, 1983):

—milk from flushing and spills;
—dairy products from machinery malfunctions and retail returns;
—whey from cheese/casein production;
—ultrafiltration permeate from production of cheese and whey protein.

Research on the utilization of whey has attracted great attention. Whey is the residue which drains from the curd in cheese production and has a high organic strength and a high chemical oxygen demand (COD), and therefore often causes disposal problems (Moulin & Galzy, 1984). It is composed of approximately 95% water, 4–5% lactose, 0·7% protein, 0·5–0·6% salts and 0·3% fat, with slight variations depending on the source. About 50–60% is either used directly or converted into spray-dried whey powder, as a food ingredient and/or an additive to feedstuff. Lactose, the main constituent of whey, still causes some problems because it is metabolized only poorly—if at all—by most industrial micro-organisms, and it is only poorly soluble in water at ambient temperature (Hacking, 1988). But even 80% of lactose-containing whey residue is already used in technical and pharmaceutical products (Kobald & Holley, 1989).

Anaerobic digestion of raw cheese whey has been reported in the literature, and might help to solve the problem by reducing COD with the production of biogas, especially methane (Van den Berg, 1984; Clanton *et al.*, 1985; Yan *et al.*, 1988). Recent studies indicate that anaerobic digestion of cheese whey using an upflow anaerobic blanket reactor (UASB) would reduce pollution and produce energy. In general, over 98% removal of oxygen demand was achieved, the highest methane production rate of 9·57 litre CH_4/litre feed/day being obtained at a loading rate of 5·96/g COD/litre and an influent concentration of 28·8 kg COD/litre (Yan *et al.*, 1988). An anaerobic fixed-bed reactor was used to treat prefermented, neutralized, diluted cheese whey, with reduction of COD of 85–87% and production of 0·40–0·42 m^3 biogas/kg COD (De Haast *et al.*, 1983).

Whey can also be used as raw material for different fermentation processes. Yeasts from the *Kluyveromyces* strain (*K. fragilis*) are used for ethanol fermentation (Bernstein *et al.*, 1977; Maiorella & Castillo, 1984; Hacking, 1988). Thereby, BOD reduction achieved was about 90% yielding up to 9% ethanol. It was further concluded that, economically, the production of fuel ethanol from cheese whey or whey ultrafiltration permeate seems nowadays not to be attractive in comparison with petroleum prices, but premises may alter in times of petroleum shortage or restriction, or in view of environmental pollution aspects (Litchfield, 1987).

Whey is also proposed for single-cell protein production. In general, bacteria, yeasts, and fungi can be used for converting cheese whey to single-cell protein (Simpson & Young, 1957; Butany & Ingledew, 1973; Bernstein & Plantz, 1977; Kim & Lebeault, 1981; Moulin *et al.*, 1983). However, at present processes are still in marginal operation; only processes based on the yeast *K. fragilis* are practised on a commercial scale. *Fragilis* yeasts have been granted food additive approval by the US Food and Drug Administration (Litchfield, 1987).

In preparation the whey is heated to a sterilizing temperature by steam injection, and supplementation of whey is necessary for maximum yields (James & Addyman, 1974). When the yeast is separated most of the milk protein is carried with it. New technology has devised methods of ultrafiltration and gel filtration which can extract the milk protein in undenatured form from the whey streams to leave essentially a lactose solution (O'Sullivan, 1972). Although conventionally yeast grown on whey is generally produced in batch culture, experiments have been made with a continuous propagator (Amundson, 1967).

Studies have been carried out on the utilization of cheese whey both alone and mixed with bakery bread waste by *Saccharomyces cerevisiae* and *Kluyveromyces fragilis* for the production of protein and ethanol. The best ethanol yields were obtained by fermenting sweet whey containing 18% solids at pH 6·0 and acid whey containing 21% solids at pH 4·5 with *K. fragilis* alone, whilst the best protein yields were obtained by fermenting sweet whey alone or mixed with bakery waste at both pH 4·5 and 6·0 (Collins & Rajagopal, 1983).

Another waste material from the dairy industry whose utilization has received great attention in recent times is the ultrafiltration permeate generated from the production of cheese and whey protein by ultrafiltration, which creates an even worse problem than the whey (Jelen, 1983). The main problem from both the whey and the ultrafiltration permeate is the lactose content. The enzymatic hydrolysis of lactose by β-galactosidase (free or immobilized) seems to be the most promising method available for managing this problem; and the most economically feasible process is considered to be the application of immobilized enzymes (Coton, 1979; 1981). One of the most interesting uses for lactose is its hydrolysis into sweetening sugars, which has already become industrially applicable (Coton, 1981; Jelen, 1983). Hydrolysis can be accomplished in the deproteinated ultrafiltration permeate as well as in whole whey, though additional demineralization is required for conversion into sweetening syrups. Hydrolysed lactose syrups will have enhanced value based on their enhanced sweetness, and represent potential sweetening agents for use in the food industry such as in ice creams, confectionery and fruit products, for yeast production, and as an additive in beer production (Jelen, 1983).

A fluidized-bed, immobilized lactose process including ultrafiltration and demineralization operations is described for application to cheese whey treatment (Coughlin *et al.*, 1974, 1976/77).

Furthermore, whey permeate can also be used as a substrate for single-cell production; however, the cell product must be washed extensively because of the relative high ash content of the permeate, or the whey must be demineralized (Lane, 1977; Shay & Wegner, 1986).

The pretreatment of ultrafiltration whey permeate by an anaerobic process with 80% BOD reduction and with recovery of 0·3 m^3 methane/kg COD is reported in the literature (Li *et al.*, 1982).

Various lactic acid bacteria (*Lactobacillus casei, L. helveticus, L. bulgaricus*) can be utilized for the production of lactic acid from whey ultrafiltration permeate by application of different fermentation systems using batch fermentors, membrane recycle reactors, or immobilized cells (Tuli *et al.*, 1985; Mehaia & Cheryan, 1986; Roy *et al.*, 1986).

In general, by the utilization of whey using biotechnical processes the following products could be obtained: ethanol, butanol, methane, single-cell protein, enzymes (β-galactosidase), antibiotics, vitamins (L-ascorbic acid, B_{12}, B_6, B_2), organic acids (lactic acid, citric acid), and food gums (Kobald & Holley, 1989).

9 WASTES FROM THE FERMENTATION INDUSTRY

Recoverable by-products from the malting, brewing and distilling industries include spent grains, surplus yeasts, carbon dioxide, and fermentation residues from the distillery, which are mostly utilized directly for animal or human consumption, or recycled (Isaac & McFiggans, 1981).

Development work on the upgrading of these by-products has been carried out for the utilization of brewery wastes for the production of ethanol (Litchfield, 1987), as well as of distilling wastes for the production of biogas (Szendray, 1983; Henry, 1985).

Processing water from the brewery industry was submitted to an activated sludge treatment for the production of animal feed. The product was shown to have feed values comparable to soybean meal, and the high vitamin content led to defining the product as a vitamin B_{12} supplement (Vriens *et al.*, 1989).

10 CONCLUSION AND FUTURE OUTLOOK

Nowadays our society, in which there is great demand for appropriate nutritional standards, is beset by rising costs and often decreasing availability of raw materials together with much concern about environmental pollution, leading to the consequence that there is much occupation with recovery and recycling of wastes. This applies particularly to the food and food-processing industry where wastes, effluents and by-products could be recovered and often upgraded to useful products of higher value.

Many of the problems found in treating most food-processing wastes arise from the fact that they are usually diluted and available only seasonally, and often the treatment plants are restricted to scattered locations, thus incurring high transport costs. Furthermore, food-processing waste treatment processes are characterized by high investment costs in research and development and in the establishment of adequate process units with the appropriate technical equipment, which must be evaluated in relation to the economic value and benefit of the recovered and upgraded products. It should also be kept in mind that disposal or dumping of the waste also needs prior treatment, due to environmental constraints and to meet environmental legislation requirements which are becoming more and more important in our time. Wherever it is economically and technologically possible, emphasis should be laid on the utilization of waste materials for recycling, recovering and upgrading purposes, even if improved technologies of waste disposal are available.

Bioconversion of food processing wastes is receiving increased attention with the realization that the wastes contain valuable components which can be utilized for conversion into useful and high-value products. Current research into methods of food-processing waste treatment is focusing on marketable products. A number of appropriate biotechnological processes are available to produce food or feedstuffs, supplements or additives, chemicals, or products which are commercially exploited for their energy value. In a food process the recovery of waste components and bioconversion into higher added-value products can contribute significantly to improvements in overall profitability, and enables the food processor to reduce the unit costs of the production process. In its simplest form product recovery can be the regaining of materials for use as a low-grade product for food or feed purposes, but is more effective if the waste components are upgraded, usually by biotechnological processes.

It is reasonable to expect that single-cell protein will ultimately contribute a major part of mankind's future food needs. Worldwide protein deficit makes the production of single-cell protein from food wastes particularly attractive, since a much-needed food supplement would be produced from sources which otherwise would be disposed of and wasted.

The crux of the recent advances in biotechnological processes for bioconversion and upgrading waste materials to useful products is the scale-up of new biological techniques to commercial applications at industrial level in a cost-effective and convenient manner.

With regard to future work and trends in bioconversion of food processing wastes to upgraded and useful products, emphasis should be laid on the following aspects.

—Much greater efforts are required to recover the large and increasing amounts of energy present in food processing wastes.
—Improved techniques should be developed to solve the problems arising from the dwindling worldwide supply of protein by production of single-cell protein, with special emphasis on the development of effective low-technology and small-scale processes for the utilization of food processing wastes, particularly in developing countries.
—Further evaluation of the microbiological aspects of food industry waste treatment processes should be carried out with emphasis on genetically engineered micro-organisms. Genetic engineering can be used to improve the economics of fermentation processes by increasing productivity, final product concentrations and production rates, but also for correction of the deficiencies and problems in the utilization of single-cell protein, e.g. by applying mutants with a lower nucleic acid content.
—Further research work is needed on scale-up of laboratory-tested processes to commercial applications for the upgrading of food-processing waste materials to useful products based on improved understanding of the mechanisms of the fermentation process. In order to achieve commercial success, food-processing waste treatment must reliably fulfil the required technical role, be cost-effective, and be acceptable to regulatory authorities. The scale-up from laboratory investigations to a commercially viable level seems to be the most difficult step in the development of new biotechnologically based processes, together with the expensive and time-consuming safety testing procedures of the products of the new technologies.
—Applied research work should be aimed at the development of tailored waste treatment equipment suitable for the management of specific food process systems, and to determine which processes offer the most benefit and advantages to the different types of food processing industries.

—Strict legislative control over food-processing waste treatment has to cover both environmental aspects and evaluation of the novel products obtained.
—Increased efforts should be focused on marketing and propagating the upgraded waste-derived products. Marketing of some products and processes may be hindered by legislative regulations. Further research and development in these areas would provide regulation authorities with the evidence needed for demonstrating the safety of these products and processes.
—Altered policies should be initiated in favour of waste treatment, aiming at bioconversion of waste materials to useful products rather than at disposal, discarding, or simple downstream processes.
—Data on the chemical and physical properties of the waste-derived products and processing parameters should be collected to help to develop potential markets.
—Development of models is required for prediction of profitable and economic processes for bioconversion of food-processing waste materials.
—Additional research is needed for optimization of the processes which would require minimal, simple steps from the waste material to the useful end-product.
—Last but not least, increased efforts should be concentrated in the motivation of consumers, producers, and authorities to conserve energy, water and raw materials, and to minimize the generation of waste materials by re-utilization and upgrading techniques, and, in consequence, to prevent pollution of the environment.

REFERENCES

Abou-Zeid, A. Z. A., Abd El-Fattah, A. F. & Farid, M. A. (1979). *Agric. Biol. Chem.*, **43**(9), 1977.
Abou-Zeid, A. Z. A., El-Diwany, A. I., Shaker, H. E. D. & Salem, H. M. (1980). *Microbiol. Agric. Wastes*, **2**(4), 293.
Amundson, C. H. (1967). *Amer. Dairy Food Mfg Rev.*, **29**(7), 22, 23, 96, 99.
Anderson, C., Longton, J., Maddix, C., Scammell, G. W. & Solomons, G. L. (1975). In *Single Cell Protein II*, eds S. R. Tannenbaum & D. I. C. Wang. MIT Press, Cambridge, Massachusetts, p. 314.
Bambalov, G., Israilides, C. & Tanchev, S. (1989). *Biol. Wastes*, **27,** 71.
Barker, T. & Worgan, J. T. (1981). *Eur. J. Appl. Microbiol. Biotechnol.*, **11**(4), 234.
Bernstein, S. & Plantz, P. E. (1977). *Food Eng.*, **48**(11), 74.
Bernstein, S., Tzeng, C. H. & Sisson, D. (1977). *Biotechnol. Bioeng. Symp.*, **7**(7), 1.
Birch, G. G., Parker, K. J. & Worgan, J. T. (1976). *Food from Waste*. Applied Science, London.
Bough, W. A., Brown, W. L., Porsche, J. D. & Doty, D. M. (1972). *Appl. Microbiol.*, **24,** 226.
Butany, G. & Ingledew, W. M. (1973). *Can. Inst. Food Sci. Technol. J.*, **6,** 291.
Clanton, C. J., Goodrich, P. R., Morris, H. A. & Backus, B. D. (1985). In *Proc. 5th Int. Symp. on Agricultural Wastes, 1985*, p. 475.

Clementi, F., Moresi, M. & Rossi, J. (1985). *Appl. Microbiol. Biotechnol.*, **22,** 26.
Collins, J. A. & Rajagopal, S. N. (1983). *Cult. Dairy Prod. J.*, **18**(2), 22, 35.
Cooney, C. L., Rha, Ch. & Tannenbaum, R. (1980). *Adv. Food Res.*, **26,** 1.
Cosio, I. G., Fisher, R. A. & Carroad, P. A. (1982). *J. Food Sci.*, **47,** 901.
Coton, G. (1979). Paper presented at IDF Technical Session, Geneva, Switzerland.
Coton, G. (1981). Paper presented at Seminar on Dairy Ingredients in Food Industry, Luxembourg.
Coughlin, R. W., Charles, M., Allen, B. R., Paruchuri, E. K. & Hasselberger, F. X. (1974). *AIChE Symp. Ser. 144,* **70,** 199.
Coughlin, R. W., Charles, M. & Julkowski, K. (1976/77). *AIChE Symp. Ser. 172,* **74,** 40.
Davy, C. A. E. (1981). In *Food Industry Wastes: Disposal and Recovery,* eds A. Herzka & R. G. Booth. Applied Science, London & New Jersey, Chap. 16, p. 219.
De Haast, J. *et al.* (1983). *New Zealand J. Dairy Sci. Technol.*, **18,** 261.
Drepper, G. & Drepper, K. (1981). *Fleischwirtschaft,* **61**(9), 1393.
Ghai, S. K., Kahlon, S. S. & Sood, S. M. (1979). *J. Res. (Punjab Agric. Univ.),* **16**(1), 64.
Goldberg, I. (1985). *Single Cell Protein.* Springer Verlag, New York.
Gray, P. & Berry, D. R. (1980). In *Handbook of Organic Waste Conversion,* ed. M. W. M. Bewick. Van Nostrand Reinhold, New York, p. 330.
Gromov, B. V. (1967). In *Man, Algae and the Environment,* ed. D. F. Jackson. Syracuse University Press, Syracuse, New York, p. 249.
Hacking, A. J. (1988). In *Food Biotechnology 2,* eds R. D. King & P. S. J. Cheetham. Elsevier Applied Science, London & New York, Chap. 2, p. 25.
Hang, Y. D. (1979). In *Food Processing Waste Management,* eds J. H. Green & A. Kramer. AVI, Westport, Connecticut, p. 442.
Hansen, C. (1983). *Food Technol.*, **37**(2), 77.
Hassanien, F. R., Ragab, M. & El-Makhzangy, A. (1986). *Fette–Seifen–Anstrichmittel,* **88**(1), 33.
Henry, M. (1985). In *Energy from Biomass and Wastes IX,* Lake Buena Vista, Florida. Institute of Gas Technology, Chicago, Illinois.
Herzka, A. & Booth, R. G. (1981). *Food Industry Wastes: Disposal and Recovery.* Applied Science, London.
Imrie, F. K. E. & Righelato, R. C. (1976). In *Food from Waste,* eds G. G. Birch, K. J. Parker & J. T. Worgan. Applied Science, London, p. 79.
Isaac, P. C. G. & Anderson, G. K. (1974). In *Proc. Symp. Treatment of Wastes from the Food and Drink Industry,* Institute of Water Pollution Control and The University of Newcastle-upon-Tyne. Paper No. 1, p. 7.
Isaac, P. C. G. & McFiggans, A. (1981). In *Food Industry Wastes: Disposal and Recovery,* eds A. Herzka & R. G. Booth. Applied Science, London and New Jersey, Chap. 10, p. 144.
James, A. & Addyman, C. L. (1974). *Proc. Symp. Treatment of Wastes from the Food and Drink Industry,* Institute of Water Pollution Control and The University of Newcastle-upon-Tyne. Paper No. 18, p. 149.
Jarl, K. (1969). *Food Technol.*, **23,** 1009.
Jelen, P. (1983). *Food Technol.*, **37**(2), 81.
Karapinar, M. & Okuyan, M. (1982). *J. Chem. Technol. Biotechnol.*, **32,** 1055.
Kihlberg, R. (1972). *Ann. Rev. Microbiol.*, **26,** 427.
Kim, J. H. & Lebeault, J. M. (1981). *Eur. J. Appl. Microbiol. Biotechnol.*, **13,** 151.
Kirk-Othmer, R. E. & Othmer, D. F. (1980). Foods, non conventional. In *Encyclopedia of Chemical Technology,* 3rd edition, Vol. 11, John Wiley, New York–London.
Kirsop, B. H. & Hilton, M. G. (1981). In *Food Industry Wastes: Disposal and Recovery,* eds A. Herzka & R. G. Booth. Applied Science, London & New Jersey.

Kobald, M. & Holley, W. (1989). Paper presented at 5th Int. Congr. on Engineering and Food, Köln, Germany.
Lane, A. G. (1977). *J. Appl. Chem. Biotechnol.*, **27,** 165.
Lane, A. G. (1979). *Food Technol. Aust.*, **31,** 201.
Lane, A. G. (1980). *J. Chem. Technol. Biotechnol.*, **30,** 345.
Lane, A. G. (1984). *Food Technol. Aust.*, **36**(3), 125.
Li, A. *et al.* (1982). In *Proc. 1st Int. Conf. Fixed-film Biol. Proc. (GB)*, **3,** 1521.
Litchfield, J. H. (1977). *Adv. Appl. Microbiol.*, **22,** 267.
Lithfield, J. H. (1983). *Science*, **219,** 740.
Litchfield, J. H. (1987). *Food Biotechnol.*, **1**(1), 29.
Litchfield, J. H. & Pierce, G. E. (1986). US Patent 4582804.
Maiorella, B. L. & Castillo, F. J. (1984). *Process Biochem.*, **19**(4), 157.
Marihart, J. (1982). *Starch*, **34**(9), 290.
Mateles, R. J. & Tannenbaum, S. R. (1968). *Single Cell Protein*. MIT Press, Cambridge, Massachusetts.
McComis, W. T. & Litchfield, J. H. (1989). *Journal WPCF*, **61**(6), 855.
Mehaia, M. A. & Cheryan, M. (1986). *Enzyme Microbiol. Technol.*, **8,** 289.
Moraes, I. O., Capalbo, D. M. F. & Moraes, R. O. (1989). Paper presented at 5th Int. Congr. on Engineering and Food, Köln, Germany.
Moulin, G. & Galzy, P. (1984). *Biotechnol. & Genetic Eng. Rev.*, **1,** 347.
Moulin, G., Malige, B. & Galzy, P. (1983). *J. Dairy Sci.*, **66,** 21.
Oleszkiewicz, J. A. & Olthof, M. (1982). *Food Technol.*, **36**(6), 78.
O'Sullivan, A. C. (1972). *Dairy Ind.*, October.
Penaloza, W. M., Molina, R., Brenes, R. G. & Bressani, R. (1985). *Appl. Environ. Microbiol.*, **49,** 388.
Pette, K. C., de Vletter, R., Wind, E. & van Gils, W. (1981). *Proc. 35th Ind. Waste Conf.*, Purdue University. Ann Arbor Science Publishers, Ann Arbor, Michigan, pp. 635.
Pirt, S. J. (1978). *J. Appl. Chem. Biotechnol.*, **28,** 232.
Ranade, D. R., Yeole, T. Y., Meher, K. K., Gadre, R. V. & Godbole, S. H. (1989). *Biol. Wastes*, **28,** 157.
Ratledge, C. (1978). Paper presented at Symposium on Resources Conservation by Novel Processes, London, 25 May.
Reva-Moissev, S. & Carroad, P. A. (1981). *Biotechnol. Bioeng.*, **23,** 1067.
Rodriguez, J. A. *et al.* (1985). *Biotechnol. Lett.*, **7,** 577.
Rolz, C. (1978). *J. Appl. Chem. Biotechnol.*, **28,** 321.
Roy, D., Goulet, J. & LeDuy, A. (1986). *Appl. Microbiol. Biotechnol.*, **24,** 206.
Shay, L. K. & Wegner, G. H. (1986). *J. Dairy Sci.*, **69,** 676.
Simpson, E. G. & Young, H. (1957). US Patent 2809113.
Sinskey, A. J. & Tannenbaum, S. R. (1975). In *Single-cell Protein II*, eds S. R. Tannenbaum & D. I. C. Wang. MIT Press, Cambridge, Massachusetts, p. 158.
Solomons, G. L. (1983). *CRC Crit. Rev. Biotechnol.*, **1,** 21.
Sucuru, G. A., Engelbrecht, R. S. & Chian, E. S. K. (1975). *Biotechnol. Bioeng.*, **17,** 1639.
Szendray, L. M. (1983). In *Energy from Biomass and Wastes, VI Conf.*, Lake Buena Vista, Florida. Institute of Gas Technology, Chicago, Illinois.
Tacon, A. G. (1980). In *Effluent Treat. Biochem. Ind., Conf. Paper.*, 3rd Process Biochem. Int. Conf., 1979, Paper No. 13.
Tannenbaum, S. R. & Wang, D. I. C. (eds) (1975). *Single Cell Protein II*. MIT Press, Cambridge, Massachusetts.
Tuli, A., Sethi, R. P., Khanna, P. K. & Marwaha, S. S. (1985). *Enzyme Microbiol. Technol.*, **7,** 164.
Tveit, M. (1967). In *Biology and the Manufacturing Industry*. Academic Press, London, p. 3.

Van den Berg, L. (1984). *Can. J. Microbiol.*, **30,** 975.
Van den Berg, L. & Lentz, C. P. (1977). In *Food, Fert. Agric. Residues, Proc. 9th Cornell Agric. Waste Manage. Conf.*, ed. R. C. Loehr. Ann Arbor Science, Ann Arbor, Michigan, p. 381.
Van den Berg, L., Kennedy, K. J. & Hamoda, M. F. (1983). *Proc. 38th Ind. Waste Conf.*, Purdue University. Ann Arbor Science, Ann Arbor, Michigan, p. 686.
Vriens, L., Nihoul, R. & Verachtert, H. (1989). *Biol. Wastes,* **27,** 161.
Watanabe, D. (1974). *Hakko Kyokaishi,* **32**(2), 62.
Welsh, F. W. & Zall, R. R. (1984). *Process Biochem.*, **19**(3), 122.
Worgan, J. T. (1976). In *Food from Waste,* eds C. G. Birch, K. J. Parker & J. T. Worgan. Applied Science, London, p. 23.
Yan, J. Q., Liao, P. H. & Lo, K. V. (1988). *Biomass,* **17,** 185.

Chapter 10

BIOCONVERSION OF CHEESE WHEY TO ORGANIC ACIDS

R. D. Tyagi,[a] D. Kluepfel[b] & D. Couillard[a]

[a]*Institut National de la Recherche Scientifique, Université du Québec, INRS-Eau, 2700 rue Einstein, C.P. 7500, Sainte-Foy, Québec, Canada G1V 4C7*
[b]*Institut Armand-Frappier, 531,, boulevard des Prairies, C.P. 100, Succ. L-Q-R, Ville de Laval, Québec, Canada H7N 4Z3*

CONTENTS

1 INTRODUCTION

Whey is a byproduct of cheese production, when fat and casein are removed, generated at a rate of about 9 kg for every 1 kg of cheese or 6 kg for every 1 kg of cottage cheese (91% of the milk processed). One hundred kg of whey is equivalent to the sewage produced by 45 people (Jelen, 1979). At this rate a medium sized cheese plant discharging $50–150 \times 10^3$ kg of whey per day constitutes a polluting strength equal to the sewage from 22 500–67 500 people on a daily basis (Modler, 1982). In many countries, including Canada, most of the whey is discarded as waste, creating severe pollution problems due to its high biochemical oxygen demand (BOD) (35 000–40 000 ppm) (Kosikowski, 1976; Scott, 1981). This BOD is due mainly to the lactose which is present at concentrations between 4·5 and 5% (Kosikowski, 1976; Garoutte *et al.*, 1983).

Table 1
Composition of Cheese Whey (Nickerson, 1974)

	% Dry weight
Fat	0·3
Protein	0·9
Lactose	5·0
Ash	0·5
Nitrogenous compounds	1·0
Minerals and vitamins	0·8

Whey is a very rich source of minerals, calcium, phosphorus, potassium, sodium, copper and iron. These minerals have very good digestive qualities. It is also a very good source of vitamins of the B-complex group, riboflavin and pantothenic acid. But their availability varies considerably. A typical composition of cheese whey is shown in Table 1.

The food and dairy wastes in fact should be viewed as a rich, inexpensive potential source of raw material from which valuable products can be produced. An additional advantage of these wastes is that they are available all year round at a fairly constant rate.

2 PRODUCTION OF WHEY

Annual production of whey in the USA and Canada is 16×10^9 kg and $1{\cdot}5 \times 10^9$ kg respectively (Singh & Ghaly, 1984). The production of cheese in the world has been increasing steadily and in 1981 reached $11{\cdot}6 \times 10^6$ tons. The total amount of whey generated in this activity was nearly $10{\cdot}4 \times 10^7$ tons. In Québec the production of cheese has increased from 32 535 tons in 1965 to 76 181 tons in 1984 (Tyagi, 1986). In 1985 cheddar and speciality cheese production in Canada was 178 711 tons and that in the province of Québec 60 675 tons. This generated 5 746 422 tons and 1 697 755 tons of sweet whey in Canada and Québec respectively.

3 POLLUTION CONTROL

In the past, the application of pollution control technology to the process effluents from the cheese making industry has pursued three alternatives: direct discharge to the municipal system; pretreatment at the cheese plant followed by discharge to the municipal system; and full treatment at the cheese plant.

The high organic content in whey coupled with a relatively balanced nutrient composition indicates that some method of biological treatment (activated sludge with its many process variations: extended aeration, contact stabilization, the oxidation ditch and more recently the pure oxygen injection systems and deep shaft) is appropriate, but experience has shown that this is not entirely without problems. Even though whey is nutrient-balanced, the lactose fraction causes the bacterial population to swing towards filamentous organisms which are unable to use proteins present as a nitrogen source. Protein molecules, because of their large size, are unable to pass through the bacterial cell wall. The extracellular enzymes (proteases) break down the large protein molecules but the overall reaction rates are low (Riddle & Chandler, 1974). Moreover, high BOD of wastes makes conventional aerobic treatment very difficult and expensive. Anaerobic treatment of waste to generate methane is possible, but in many cases gives a poor return on investment.

A further consideration is the effect which whey discharge may have on an existing biological treatment plant. In many cases a cheese plant will contribute an organic load many times greater than that from the community. Also, because of the operational nature of a cheese plant, the loading rate to the treatment plant will not be constant unless load equalization is practiced. Experience has shown that for successful treatment of cheese plant waste, the influent five day BOD, (BOD_5) concentration to the biological system should not exceed 1000 mg/liter. At concentrations higher than this, severe bulking problems have been experienced. The impact on a community's waste treatment facility can be reduced by a pretreatment at the cheese plant (screening followed by flow equalization prior to continuous discharge to the municipal system is one possible solution).

It should also be noted that most municipalities have a sewer discharge by-law and impose a sewer surcharge to industry if certain maximum values are exceeded. While the economics of waste treatment are site specific, it is readily apparent that opportunities for significant cost saving through pretreatment can be realized.

The problem of whey treatment demands simple economical solutions, particularly in the case of small cheese plants in which oxidative disposal methods may be too costly to install (Groves, 1972; Ripley, 1979). The large amount of whey among fewer factories can have both advantages and disadvantages. If satisfactory methods of whey disposal are not available, additional supplies of whey can mean greater economic returns (economy of scale). The larger volumes can make it economically feasible to adopt processing techniques which are not practical for smaller quantities of whey. In other words, if disposal is a problem, additional volume can increase returns.

4 CURRENT DISPOSAL METHODS OF WHEY

The most common methods of disposal and recovery are drying to produce whey powder; returning whey to farmers for animal feeding; selling it to processors; paying processors to collect it; and dumping it as a waste or sewage.

Table 2
Distribution of Cheese Production Plants

Capacity (kg/day)	*Number of plants*
<11 000	90
11 001–24 000	28
24 001–45 000	38
45 001–91 000	14
91 001–227 000	7
>227 000	5

The most widely used method is evaporation and drying of whey to produce whey powder which is used in bakery and other food preparations. The increase in cheese manufacture has occurred mainly in big plants that are equipped with whey drying equipment. Although installations for drying and processing might be economical for large plants, their costs were reported to be prohibitive for medium and small plants (Muller, 1979). Also whey utilization requires adequate cooling and storage facilities. Thus only large facilities with an output exceeding 150×10^3 kg of whey per day are able to realize a return on investment from processed whey products (Modler, 1982). Almost 50% of Canada's cheese producing plants fall into the small scale category (Table 2). The main source of waste whey in North America is also due to smaller plants (Jelen, 1979).

In 1978, about 35·6% of whey was disposed into the environment, whereas in 1976 the volume of whey discharged in the environment was only 28% of the total production. This amounted to 507 000 tons of whey in Canada and 161 000 tons of whey in Québec. In 1976 only 1% of the total volume was used in liquid form for animal feeding. Muller (1979) reported that of the $1{\cdot}2 \times 10^9$ kg of liquid whey produced in 1979 in Canada, 49% was used for whey powder, 8% was fed to hogs, 17% was dumped into sewers, and 26% was disposed of on land. In 1984 about $7{\cdot}36 \times 10^6$ litres of cheese whey was produced in Québec; about 8% was discharged into the environment, about 16% was consumed by animals and about 76% was dried.

Cheese whey has been considered as a feed to animals. Provided there are enough hogs, the cheese whey can be substituted in the hogs' diet up to 25–30% of the total. It has been estimated that hogs can consume up to 1000 liters of whey per animal annually. But due to high trucking costs, liquid whey can only be used economically for hog feeding within a radius of 30–40 km. In many cases, however, the hog population is too small in relation to the number of cheese plants (Muller, 1979).

A recent approach for whey utilization is ultrafiltration to separate the proteins from the so-called permeate containing mostly lactose. The protein fraction is used for food and to replace egg white while the permeate is normally disposed of without treatment or recovery (see section 6). A few small cheese plants are selling their whey to industries within a radius of 80 km, for recovery of the proteins by ultrafiltration.

5 GLOBAL UTILIZATION OF WHEY

On a global basis, a divergence of whey disposition practices exist. In Denmark no whey is wasted and 90% of the whey is fed to animals. In Sweden it is illegal to discharge whey into the sewerage system whereas in Israel 84% of whey produced is wasted and not used for animal feed.

Byproduct recovery for proteins and lactose specifically is accomplished using concentration technologies ranging from spray drying to ultrafiltration coupled with gel filtration. Most countries practice some degree of whey conversion into useful products. The Netherlands lead the way with 93% of their whey being processed. In Canada 48% is processed. In New Zealand, casein whey was used mainly for spray irrigation and pig feeding, whereas cheese whey was used for lactose manufacture and pig feeding (Galpin, 1981). Forty-five of the plants dispose of effluents by spray irrigation, 38 by discharge into natural waterways and municipal sewers, and one by means of a biological treatment unit operated by the dairy. Food uses of whey are increasing and have encompassed pharmaceutical products, dairy products, bakery products, confections and coatings, frozen and canned foods as well as powder foods. For example, production of whey powder in Canada has increased from 41×10^6 kg in 1977 to 57×10^6 kg in 1981 (Dairy Market Review, 1982). The choice of recovery technology is a function of end-product use criteria.

6 MANAGEMENT STRATEGIES

A correct strategy for industrial pollution control consists of minimizing feedstock leakage, optimizing product conversion efficiency, maximizing recovery and recycle of process losses, and optimizing byproduct generation. The objective is to reduce the amount of waste material which results from the particular industry and thus to exert the minimum economic burden to that company by way of waste treatment requirements. There are many examples in industry where byproduct generation and recovery have led to economic spin-offs which more than paid for the cost of pollution control (Campbell & Glenn, 1982).

The cheese producer does have a number of options when addressing the whey problem and some of them are discussed in Section 4. Recovery of sugars, proteins and minerals in the waste effluents by crystallization, evaporation and spray drying are practiced in many food and dairy plants. However the increased energy cost, especially that of natural gas, has made many of the above operations economically unattractive.

One management approach applies the principles of pollution control and resource recovery and involves application of ultrafiltration or hyperfiltration and reverse osmosis for concentrating the whey solids (Kosaric & Asher, 1985).

Ultrafiltration has provided the cheese maker with a means of channeling the component of whey with the greatest value into the market place. Whey protein concentrate (WPC) is an item of worldwide commerce, with significant dollar value. That is the bright side. The dark side is that for each kilogram of WPC, 2·3 kg of permeate solids are obtained, with a lactose content of approximately 85% of the whole whey.

A number of options have been proposed to convert permeate to value-added products and other more profitable alternatives are constantly being evaluated. These alternatives include alcohols, organic acids, solvents and proteins, all of which have been discussed in a number of review articles (Kosaric & Wieczorek, 1984; Kennedy, 1985; Kosaric & Asher, 1985; Goursaud, 1986; Moebus & Teuber, 1986; Tyagi, 1986).

The absolute dependence of Canadian fermentation industries on sugar-cane molasses associated with problems like import, fluctuating prices, variations in chemical and microbiological qualities, clarification burden, storage, pumping, and high residual BOD after fermentation, has resulted in the search for an alternative raw material. Among a number of options proposed to convert the whey permeate to a value-added product, replacement of molasses for the production of a number of chemicals (alcohols, organic acids, solvents, proteins, vitamins, enzymes) seems to be most attractive. Utilization of whey or whey permeate as a raw material for fermentation industries is not only a waste treatment/recovery alternative, but also reduces those problems encountered in the use of molasses. Thus lactic acid produced from whey/permeate may serve as a raw material for the fermentation industry while reducing the BOD of the whey by 90–95%.

The following section is devoted to a review of the organic acid production from cheese whey.

7 LACTIC ACID

The art of lactic acid fermentation stems from antiquity, and the biotechnology now is well advanced. Lactic acid, a natural organic acid, has long been of use in the pharmaceutical, chemical and food industries, primarily as an acidulant and as a preservative. About half the total world production of lactic acid is for non-food uses. The fermented product containing lactic acid has potential commercial use as a nitrogen-enriched feed supplement for ruminant animals (Gerhardt & Reddy, 1978; Marriott, 1985) or as a substitute for molasses in bakers' yeast production. Homofermentative lactic acid bacteria produce no carbon dioxide from sugars, and thus the substrate carbon is more productively utilized (Tuli *et al.*, 1985). Yield of 0·9 g lactic acid/g lactose or better should be obtained, compared to typically half that with ethanol (0·45–0·5 g/g).

The lactic acid fermentation of non-concentrated whey and whey permeate has been studied in batch, continuous and immobilized cell processes using different bacterial strains.

7.1 Microorganisms Involved in Lactic Acid Fermentation

Lactic acid-producing microorganisms are found in four genera (Kandler, 1982): *Streptococcus, Pediococcus, Lactobacillus* and *Leuconostoc.* The biochemical characteristics of these organisms have been discussed by Buchta (1983). The most important producers of lactic acid belong to the family of Lactobacillaceae.

Lactobacillus helveticus is specially advantageous for the following reasons:

(1) it provides an alternative solution to the phase contamination in dairy industries (this problem is generally encountered using *L. bulgaricus*);
(2) it produces almost twice the amount of lactic acid in milk as compared to other common lactic acid bacteria, including *Streptococcus thermophilus* and *L. bulgaricus* (Roy *et al.*, 1986);
(3) it provides a racemic lactic acid (DL) as compared to only dextro-rotatory lactic acid (D) produced by *L. lactis, L. delbrueckii* and *L. bulgaricus*;
(4) it can ferment lactose and all its breakdown products (glucose and galactose) (Bergey's Manual, 1974);
(5) it is more advantageous for preparation of culture starters from milk base media;
(6) it produces only lactic acid (Roy *et al.*, 1987).

Lactobacillus helveticus cells do not have a preference for utilizing glucose or galactose in a free cell system; however, for an unknown reason, galactose is less metabolized when the same microorganism is entrapped in Ca-alginate gel, especially at low dilution rates. During batch fermentation, *L. helveticus* gave inferior productivity compared to *L. bulgaricus* (Roy *et al.*, 1986) but a higher conversion of substrate was observed using *L. helveticus* (Roy *et al.*, 1987).

Streptococcus thermophilus has two drawbacks: (1) only a few strains of *S. thermophilus* are able to ferment galactose, one of the two sugars of the lactose molecule (Somkuti & Steinberg, 1979), and (2) *S. thermophilus* requires some growth factors (peptide) for lactic acid production in a milk-based medium due to its limited amount of active cell-bound proteinase (aminopeptidases) (Roy *et al.*, 1986). Other homofermentative lactic acid bacteria (*L. delbrueckii, L. lactis, L. casei* and *S. lactis*) produce other organic compounds (acetic acid, formic acid, ethanol) depending on fermentation conditions.

By using fermentation temperatures of 43–45°C, pH 5·5 and lactic acid tolerant organisms such as *L. bulgaricus,* unsterilized and unpasteurized whey can be used (Keller & Gerhardt, 1975; Reddy *et al.*, 1976). The unsterilized system has been used for as long as 42 days by these authors without encountering putrefactive spoilage.

7.2 Batch process

Studies on lactic acid fermentation in a batch process have been conducted by a number of workers (Campbell, 1953; Hanson & Tsao, 1972; Marshall & Earle, 1975; Reddy *et al.*, 1976; Tewari *et al.*, 1985; Roy *et al.*, 1986).

Lactobacillus bulgaricus, L. casei, L. fermentii and *L. plantarum* were screened on paneer whey (a type of cheese whey) for their ability to produce lactic acid (Tewari *et al.*, 1985). The acid production was efficient with 10% inoculum of *L. bulgaricus,* 4·5% whey lactose, pH 5·5 and a fermentation period of 24 h at 30°C. Addition of peptide and mustard oilseed cake to whey improved acid production; $FeCl_3$ (5 ppm) increased acid fermentation. Lactic acid production was induced by glutamic acid.

Inoculum prepared in skim milk medium or in lactose synthetic medium resulted in the best performance in fermentation of cheese whey ultrafiltrate (in terms of growth, lactic acid production, lactic acid yield and maximum productivity of lactic acid), compared to that prepared in glucose synthetic medium (Roy *et al.*, 1986). The addition of yeast extract at a concentration of 1·5% (w/v) in the whey ultrafiltrate, as a source of nitrogen/growth factors, gave the highest maximum productivity of lactic acid of 2·7 g/liter h. Addition of corn steep liquor (CSL) or the tryptone gave inferior results. Lactic acid production, lactose utilization and maximum productivity of lactic acid increased linearly with pH (in the range 4·7–6·3). Specific activity of the CSL-grown cells was observed to be higher than that of skim milk-grown cells. However, cell yield and cell growth were lower in CSL medium.

Lactose in unpasteurized whey was fermented to lactic acid by *L. bulgaricus* at a temperature of 43°C and pH 5·5 controlled with NH_4OH (Reddy *et al.*, 1976). The fermented product when concentrated by evaporation yielded 70% of total solids and adjusted to pH 6·8 with addition of ammonia. The concentrated product contained about 5·5% crude protein (approximately 6–8% of the crude protein is derived from bacterial cells, 17% from whey proteins, and 75–77% from ammonium lactate). The efficiency of conversion of lactose to lactic acid exceeded 95%. The fermentation time was greatly reduced upon the addition of 0·2% yeast extract or 0·1% CSL as a source of growth factors. Whey containing lactose at concentrations up to 7% can be fermented efficiently, but higher concentrations of lactose are fermented incompletely. Reddy *et al.* (1976) have shown that a 2·5% inoculum produces just as rapid and contamination-free fermentation as that obtained with a 10% inoculum.

Batch method for the production of lactic acid by fermentation of cheese whey/permeate suffers from one or more of the following disadvantages: there is a long lag period prior to active lactic acid fermentation; there are unusually long fermentation times that required greater fermentor capacity and increased operational costs; and the manual procedures used for the addition of ammonium or calcium ion to neutralize the lactic acid produced during the fermentation are tedious and contribute to the increased product variability.

7.3 Continuous Process

The continuous bioconversion process has the advantages of high productivities achievable with the selected microbial strains. For example at a dilution rate of 0·2/h a single 6500 liter continuous bioreactor could replace two 30 000 liter bioreactors used in batch operated plants. The continuous process in a free cell system has been studied by several workers (Hanson & Tsao, 1972; Keller & Gerhardt, 1975; Cox & MacBean, 1977; Stieber & Gerhardt, 1979*a*; Major & Bull, 1985), and also as a dialysis continuous process (Friedman & Gaden, 1970; Stieber *et al.*, 1977; Stieber & Gerhardt, 1979*b*, 1981). *Lactobacillus helveticus* was used for continuous fermentation of cheese whey, yeast extract and permeate medium. A maximum productivity of 9·7 g/liter h at a dilution rate of 0·352/h was obtained with a 37·4 g/liter feed lactose. Under high dilution rates, the cells were elongated to several times their normal size, resulting in a wall growth (Roy *et al.*, 1987).

The productivity of the fermentation of whey permeate to lactic acid could be vastly improved using a membrane recycle bioreactor. Table 3 is a comparison of several types of bioreactors used for the production of lactic acid. The batch productivity values do not take into account the time needed between batches for cleaning, sterilization, etc. It is important that comparison be made considering the feed, microorganism, product concentration and substrate utilization. In this regard the superiority of the membrane recycle

Table 3
Comparison of Bioreactors for Lactic Acid Production[a]

Bioreactor	*Organism (Lactobacillus)*	*PD*	*D*	*LA*	*Reference*
Batch	*L. bulgaricus*	3–5	—	44	Mehaia & Cheryan (1986)
Continuous	*L. bulgaricus*	11	0·334	33	Cox & MacBean (1977)
Continuous	*L. bulgaricus*	1·77	0·032	55·3	Keller & Gerhardt (1975)
Continuous	*L. bulgaricus*	5·0	0·17	29·4	Stieber & Gerhardt (1979*b*)
Continuous	*L. helveticus*	9·7	0·352	27·5	Roy *et al.* (1987)
Dialysis	*L. bulgaricus*	10·7	0·136	78·6	Stieber & Gerhardt (1979*a*)
Immobilized	*L. casei*		0·015	33	Tuli *et al.* (1985)
Immobilized	*L. bulgaricus*	0·5	0·083	14·4–19·2	Stenroos *et al.* (1982)
Immobilized	*L. helveticus*	1·2–1·6	0·144	18·0	Roy *et al.* (1987)
	(milano)	2·6			
	L. bulgaricus		0·55	30·9	Authors' work
Immobilized	*L. bulgaricus*	17·0	1·6	10·5	Authors' work
Immobilized	*L. helveticus*	16·77	1·0	15	Authors' work
Immobilized	(ATCC-10797)	15·0			
Hollow fiber	*L. bulgaricus*	2	0·166	12	Mehaia & Cheryan (1986)
Membrane	*L. bulgaricus*	25	0·89	28	Mehaia & Cheryan (1986)
	L. bulgaricus	85	1·97	43	Mehaia & Cheryan (1986)

[a] *PD* = productivity (g/liter h), *D* = dilution rate (h^{-1}); *LA* = lactic acid (g/liter).

fermentor is remarkable. This is a direct result of much higher cell concentrations used in these bioreactors. According to Mehaia and Cheryan (1986), the membrane recycle configuration operated in the CSTR (continuous stirred tank reactor) mode has several advantages over immobilized cell and hollow-fibre systems that are operated in the plug-flow mode. Control of the improvement, especially the pH, is much easier and substrate–cell contact is much better due to the completely mixed conditions. With such vast improvements in productivity, the additional cost of the membrane module will quickly repay itself.

7.4 Product Inhibition in Lactic Acid Fermentation (Design Aspects)

In lactic acid fermentation, there is an inhibitory effect caused by the product lactic acid (Friedman & Gaden, 1970; Keller & Gerhardt, 1975; Reddy *et al.*, 1976; Hongo *et al.*, 1986) on the lactic acid productivity. The specific growth rate, μ, varies as a function of product concentration.

Keller and Gerhardt (1975) assumed that specific growth rate of *L. bulgaricus* as a function of product concentration exhibited the following linear relationship:

$$\mu = \mu_{max}(1 - P/P_m) \tag{1}$$

where μ is the specific growth rate at a given product concentration, μ_{max} is the maximum specific growth rate (when $P = 0$), P is the product concentration, and P_m is the product concentration at which μ first equals zero.

In the pH range 4·8–5·6, μ is approximately inversely proportional to product concentration. Luedeking and Piret's (1956) data on lactic acid inhibition at different values of pH used by Keller and Gerhardt (1975) (eqn (1)) were reassessed according to the following product inhibition equations proposed by different researchers:

$$\mu = \mu_{max}K_1/(K_1 + P) \tag{2}$$

$$\mu = \mu_{max}e^{-K_2P} \tag{3}$$

$$\mu = \mu_{max}[1 - (P/P_m)^n] \tag{4}$$

where K_1, K_2 = constants of inhibition.

The data were found to give a best fit using eqn (4) as shown in Fig. 1. The values of the constants thus obtained are presented in Table 4. The values of n (which represents the nature of inhibition) at different values of pH are close to unity (except at pH 4·8) which shows a linear relationship as represented by eqn (1). The value of n at pH 4·8 is 0·73 (<1) which represents a parabolic inhibition curve.

Due to product inhibition, many of the kinetic advantages of continuous fermentation are not fully realized in lactic acid fermentation. In continuous lactic acid fermentation, as D (dilution rate = reactor working volume/liquid inflow rate) decreases, the product concentration increases and causes μ to

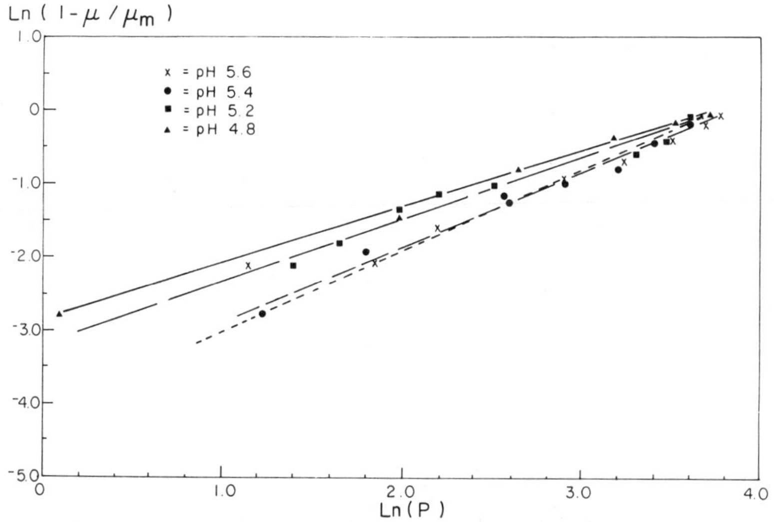

Fig. 1. Linear plot of the effect of product (lactate) concentration on the specific growth rate in batch fermentation.

decrease. Therefore, further increase in retention time is counteracted by reductions in μ. At pH 5·5, approximately the same time is required to process a given volume in a two-stage continuous fermentation as in batch fermentation (Keller & Gerhardt, 1975). However, the retention time in continuous fermentation is reduced by increasing pH to 5·8. Except at extremely low substrate concentrations, μ is more a function of product than substrate concentration (Keller & Gerhardt, 1975).

Lactic acid, calcium lactate, sodium lactate and ammonium lactate were added to the medium at different concentrations at the time of inoculation to find their inhibitory effects on the growth of *L. delbrueckii* (Hongo *et al.*, 1986). At a concentration of 0·5% lactic acid, cell growth was 20% compared to that in the absence of lactic acid. At a concentration of 1·0%, cell growth ceased entirely. With the addition of 1·0% of ammonium lactate, sodium lactate or calcium lactate, reductions of 50, 67 and 72% cell growth were

Table 4
Values of Constants

pH	P_m *(g/liter)*	*n*	r^2
5·6	47·2	0·997	0·991
5·4	46·7	0·97	0·90
5·2	43·8	0·99	0·88
4·8	41·8	0·73	0·995

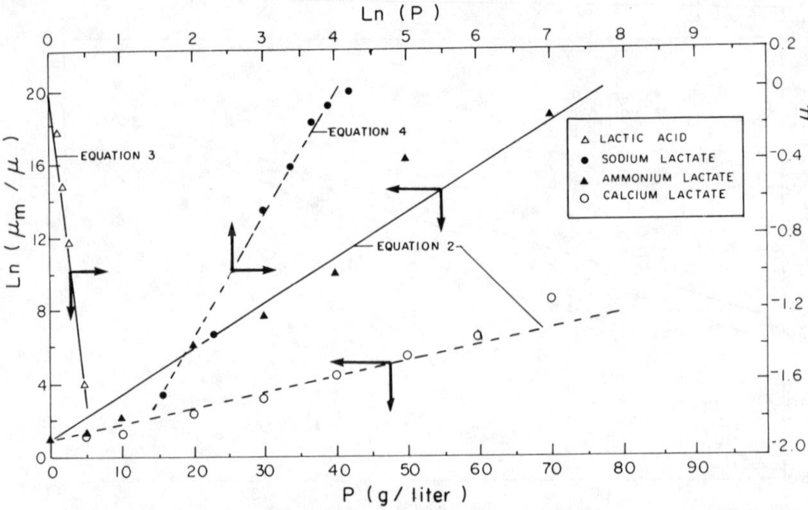

Fig. 2. Linear plot of the effect of lactic acid and various lactates on cell growth based on data from Hongo *et al.* (1986).

observed, respectively. Thus lactic acid and ammonium lactate showed the most inhibitory effect on cell growth, and even neutralized lactates clearly showed inhibitory effects. However, the faster rate of lactic acid production and the best cell growth among pH-controlled fermentations was obtained when the pH was controlled with ammonium hydroxide rather than NaOH or $CaCO_3$ (Hongo *et al.*, 1986). This observation is contradictory to the inhibition observed by lactate additions. This could be due to the fact that NH_4OH not only neutralized the pH but also served as a nitrogen source. At a given pH, greater inhibition of batch lactic acid fermentation was also observed with ammonium lactate than with calcium lactate by Keller and Gerhardt (1975).

The inhibition data of Hongo *et al.* (1986) on ammonium, calcium, sodium lactates and lactic acid gave different kinetic patterns. As shown in Fig. 2, ammonium lactate and calcium lactate followed eqn (2), whereas sodium lactate and lactic acid followed eqns (4) and (3) respectively. This behavior suggests different inhibition mechanisms by different salts of lactic acid. The values of constants for this analysis are given in Table 5.

It has been demonstrated that the concentration of undissociated lactic acid, rather than the actual pH of the medium, inhibits lactic acid bacteria (Friedman & Gaden, 1970; Keller & Gerhardt, 1975; Reddy *et al.*, 1976).

The inhibitory effects of lactic acid have been alleviated to a certain extent by conducting fermentation in a continuous dialysis process (Keller & Gerhardt, 1975; Coulman *et al.*, 1977; Stieber *et al.*, 1977; Stieber & Gerhardt, 1979*a*, 1981), in a hollow-fiber fermentor (Vick Roy *et al.*, 1982) and in an electrodialysis system (Hongo *et al.*, 1986) (discussed in Section 7.5).

Table 5
Inhibition Constants for Lactates and Lactic Acid

Inhibitor	*Equation no.*	*Constants*
Calcium lactate	2	$K = 10{\cdot}14$ g/liter
Ammonium lactate	2	$K_1 = 3{\cdot}84$ g/liter
Lactic acid	3	$K_2 = 0{\cdot}43$
Sodium lactate	4	$P_m = 57{\cdot}15$ g/liter
		$n = 1{\cdot}51$

7.5 Immobilized Cell Process

Increasing the cell mass concentration in the reactor is one way of increasing both the rate and extent of substrate conversion. Recycling (feedback) of cells in a continuous fermentation process is used to retain cells and thereby increase the biomass concentration in the fermentor, allowing increased throughput and decreased retention time. Recycling also reduces substrate concentration in the fermentor effluent. The cell concentration and reusability of cells in the bioreactor can also be increased by immobilizing cells to inert support and recently several such studies have appeared on lactic acid production from whey and whey permeate (Stenroos *et al.*, 1982; Hamilton & Howell, 1983; Tuli *et al.*, 1985).

Two matrices were assessed for their ability to immobilize *L. casei* cells for lactic acid fermentation in whey permeate medium by Tuli *et al.* (1985). Agar at 2% concentration was better than polyacrylamide gel in its effectiveness to entrap the bacterial cells to carry out batch fermentation for up to three repeat runs. Temperature and pH had no significant influence on the fermentative ability of the immobilized organism. Temperatures of 40–50°C and pH 4·5–6·0 were optimum when fermentation was carried out under stationary conditions. In batch fermentation approximately 90% conversion of the substrate (lactose) was achieved in 48 h using immobilized cell gel cubes of 4 × 2 × 2 mm size, containing 400 mg dry bacterial cells per flask and 4·5% (initial) whey lactose content as substrate. However, further increase in substrate levels tested (more than 4·5%) did not improve the process efficiency. Supplementation of Mg^{2+} (1 mM) and agricultural byproducts (mustard oil cake, 6%) to the whey permeate medium further improved lactate production (Tuli *et al.*, 1985).

A fixed film unit using gelatin-coated packing cross-section was prepared by Compere and Griffith (1975). It was seeded with a kefir culture (lactobacilli and lactose fermenting yeasts) and tested using cheese whey. This packed bed column reactor increased the lactic acid production of the whey from 1·4 to 2·1% in a single pass. Lactic acid bacteria have been immobilized in several bioreactor systems: *L. delbrueckii* and *L. bulgaricus* in a hollow-fiber fermentor (Vick Roy *et al.*, 1982; Mehaia & Cheryan, 1987) and dialysis culture (Friedman & Gaden, 1970); and *L. casei* and *L. bulgaricus* in a polyacrylamide gel lattice (Divies & Siess, 1976; Ohmiya *et al.*, 1977). Biochemical fuel cell systems based on immobilized lactic acid bacteria have also been developed (Matsunaga *et al.*, 1978; Kovach & Meyerhoff, 1982). *Lactobacillus delbrueckii, L. bulgaricus, or L. helveticus* was used in immobilized form for conversion of whole cheese whey or synthetic media to lactic acid by several workers (Stenroos *et al.*, 1982; Vick Roy *et al.*, 1982, 1983; Boyaval *et al.*, 1985; Champagne & Boyaval, 1986; Boyaval & Goulet, 1988). Morimura *et al.* (1986) used carrageenan gel to immobilize adapted sludge for organic acid production.

Microorganisms were immobilized on porous cellulose acetate particles which have higher mechanical strength than conventional carriers such as polyacrylamides, carrageenan, and poly(vinyl alcohol) by Nagai *et al.* (1986). *Sporolactobacillus inulins* TUA 343L was cultured in a medium containing glucose, yeast extract, polypeptone, $MgSO_4$, $MnSO_4$, $FeSO_4$, NaCl, and $CaCO_3$ at 37°C for 24 h, and to this were added sterilized porous cellulose acetate particles (particle size 5·0 mm). The culture was kept at 40 mm Hg for 2 h for immobilization. The particles were collected, washed and added to a medium containing glucose, yeast extract, $CaCO_3$, $MgSO_4 . 7H_2O$, $MnSO_4 . 4H_2O$, $FeSO_4 . 7H_2O$, and NaCl, which was incubated at 37°C for 48 h for lactic acid production (Nagai *et al.*, 1986). The lactic acid concentration reached a value of 35 g/liter.

Mehaia & Cheryan (1986) studied the membrane recycle bioreactor (immobilization without carrier) for the continuous production of lactic acid from cheese whey by *L. bulgaricus.* At a cell concentration of 10 g/liter, optimum productivity of lactic acid was 35 g/liter h. Increasing the cell concentration to 30 g/liter enabled the use of a dilution rate of 1/h with complete substrate utilization. At 60 g/liter, productivity was over 80 g/liter h with complete substrate utilization, which is vastly superior to conventional batch and continuous fermentations.

An electrodialysis fermentation method in which lactic acid is continuously removed from the fermentation broth resulted in a continuous fermentation with a productivity three times that obtained in non-pH-controlled culture. With this method, the amount of lactic acid produced (pure) was 82·2 g/liter, approximately 5·5 times that produced in non-pH-controlled fermentation. However, the fouling of anion exchange membranes by cells was observed (Hongo *et al.*, 1986). The products obtained from these processes have commercial potential, including use as a nitrogenous feed supplement for ruminants (Marriott, 1985).

A fermentation apparatus for the continuous production of lactic acid from whey has been described by Prigent (1983). The apparatus contains a fermentor equipped with a mechanical stirrer, a source of NH_3 or soda to a pH regulator, as well as a means for continuous delivery of whey and supplements to the fermentor. A pump inserted between the fermentor and an ultrafiltration apparatus allows for recycling the liquor, as well as processing the filtrate by electrodialysis for lactate recovery. The apparatus also allows efficient separation of lactose from lactic acid. Empirical exponential equations correlating the volume of NH_4OH solution added for pH control of the medium, with the number of base additions to a batch lactic acid fermentation has been described by Borzani and Baralle (1983).

Calcium-alginate-entrapped *L. helveticus* cells were characterized by higher fermentation rates and optimum pH than free cells (Boyaval & Goulet, 1988). After a heat treatment, cell activity was higher for free cells than for immobilized cells. The volumetric productivity in a continuous packed bed reactor was 8 g/liter h if the total volume was considered and 30·8 g/liter h if the free volume only was considered. The lactose conversion was 50%. Cell leakage was also observed by these authors. Treatment of Ca-alginate beads with polyethyleneimine to increase the stability of the gel did not reduce the cell leakage and caused severe shrinkage of the beads. Incubation in glutaraldehyde has been demonstrated to kill all the cells in the beads (Boyaval *et al.*, 1985). Fewer than 1% of free cells were observed in the bioreactor. After a week of operation a plugging in the packed bed occurred which the authors attributed to decalcification of the Ca-alginate by the lactic acid produced in the bioreactor. Stenroos *et al.* (1982) proposed introducing fresh medium at the top of the bioreactor as the best feeding mode to avoid plugging. This effect was very pronounced at low dilution rates and caused higher leakage of cells. Based on volumetric productivity (free vs immobilized cells), it was concluded that the Ca-alginate beads entrapment method is not suitable for industrial purposes (Boyaval *et al.*, 1985; Boyaval & Goulet, 1988). Higher lactic acid productivity was obtained with multistage packed bed columns (Roy *et al.*, 1987) than with the single-stage system. This is due to the fact that the maximum pH gradient is lower than in a single stage.

The immobilization of *L. bulgaricus* on polyacrylamide and alginate beads was investigated (Lee, 1981). The most active immobilized cells were obtained by entrapment in Ca-alginate beads, which is in contrast to the observations of Boyaval and Goulet (1985) and Roy *et al.* (1987). These immobilized microbial cells, when introduced into 4·5% lactose solutions and whey, showed an improved conversion to lactic acid by 28 and 18% respectively compared to free cell cultures.

In our laboratory, we studied the production of lactic acid in an immobilized system utilizing different strains of *Lactobacillus*. The productivity of lactic acid in comparison to that of other workers is shown in Table 3. *Lactobacillus bulgaricus* gave 2–12 times higher productivity whereas *L. helveticus* gave 2–6 times higher productivity than that obtained from the other types of reactors (batch, continuous, dialysis and fixed bed). This productivity was obtained at residence times varying between 1·8 h and 36 min (unpublished results). Higher productivities have only been achieved in membrane recycle reactors. The lactic acid fermentation in our bioreactor, run without pH control, is probably subject to product inhibition, as seen elsewhere. Thus pH stabilization in future experiments should lead to still higher yield and productivity.

8 ACETIC ACID AND PROPIONIC ACID

The annual demand for propionic and acetic acids is approximately $1{\cdot}3 \times 10^9$ kg (Tyagi, 1986). Acetic acid is used extensively in the manufacture of cellulose acetate, vinylacetate, acetic esters, etc. Propionic acid is used in a variety of industrial processes such as those making cellulose propionate, an important thermoplastic, and esters which are used in the perfume industry.

Acetic and propionic acids may be produced biologically by the fermentation of sugars using a species of *Propionibacterium* (Lee *et al.*, 1974). *Propionibacterium acidi-propionici* produces high concentrations of propionic and acetic acids at pH values of 4·1–4·9 (the natural pH of acid whey) by utilizing sugars. Approximately 2 moles of propionic acid and 1 mole of acetic acid are produced per 1·5 mole of glucose or galactose (both present in whey).

Whey conversion of high concentration of propionic acid is achieved by fermentation in two steps, first with lactic acid-producing and second with propionic acid-producing bacteria. The latter product is used for preserving bakery products, acting as a mycostatic agent (Ahern *et al.*, 1985). In this process fresh pasteurized whey (6·5% solids) was inoculated with *Lactobacillus bulgaricus* and *Streptococcus thermophilus* to carry out a lactic acid fermentation at pH 4·3–6·0. To obtain propionic acid, the broth was adjusted to pH 7·0, sterilized and inoculated with *P. shermanii.* Final yields of propionic acid were 1·6% (by weight) (Ahern *et al.*, 1985). Mixed culture studies for the production of propionic acid to preserve bakery products have also been carried out by Bodie *et al.* (1983) in three stages. *Propionibacterium shermanii* and *L. casei* grown together through three stages in the draw-and-fill mode produced about 4·5% propionic acid in 70 h and all the lactose was consumed.

Whey fermented with *P. acidi-propionici* resulted in higher propionic acid yields, and consequently greater mycostatic activity, than those produced using *P. shermanii* (Anderson, 1982). Twenty-five thousand gallons of sweet whey containing 7% solids and 0·5% yeast extract was sterilized and inoculated with 2500 gallons of *P. acidi-propionici* preculture and incubated for 66 h at 35°C and pH 7·0, maintained by periodic addition of NaOH. The fermented broth containing 0·96% propionic acid and 0·2% acetic acid (compared with 0·8 and 0·3%, respectively, for *P. shermanii* fermentation) was spray-dried and packaged for use as a mycostatic agent in the manufacture of pastry, bread and other bakery products (Anderson, 1982).

Propionibacterium acidi-propionici is a slow-growing microorganism, requiring fermentation times of 12–14 days to achieve a 56% conversion (Prescott & Dunn, 1949; Clausen *et al.*, 1982). A batch fermentation of 8 days resulted in a conversion of only 30%. In mixed culture, the time required is about 3 days.

Immobilized whole cell system fermentation processes have the benefits of a high productivity with reduced substrate and product inhibition (as is the case for organic solvents and organic acids). Using the immobilized cell reactor for the production of organic acids and solvents using cheese whey as substrate, the fermentation time will be shortened with consequent reduction of the size of bioreactor and costs of the overall process. In continuous fermentation, a 90% conversion of the glucose was achieved in the continuously stirred tank reactor (CSTR) with a 72 h retention time and a 30 g/liter sugar concentration. The fermentation in the CSTR was shown to be about four times faster than that in a batch reactor, while 90% conversion of glucose was obtained after 28 h retention time in an immobilized cell reactor (ICR) (Clausen & Gaddy, 1984).

Production of propionic acid by *P. shermanii* was studied in our laboratory in batch reactors. Cells were immobilized on an inert support. *Propionibacterium shermanii* was cultivated followed by centrifugation and resuspension of cells in 1/10 volume of supernatant and immobilization. The carrier with immobilized cells was transferred to a bioreactor of 750 ml capacity and 500 ml working volume to which whey permeate, fortified with 1% yeast extract, was added. After the fermentation was completed, the medium was drained and reactor was refilled with the fresh permeate and fermentation was continued. Typical results of this fermentation are shown in Fig. 3, comparing acid production and lactose utilization under free cells without the addition of $CaCO_3$, and immobilized cells with and without addition of $CaCO_3$ and increased lactose concentration.

A maximum concentration of 12·0 g/liter propionic acid was obtained in the immobilized system against 2·25 g/liter propionic acid in the free system in the same time. A higher concentration of acid was obtained when $CaCO_3$ was added to the fermentation medium to control pH (Table 6). Additions of $CaCO_3$ to the fermentation medium not only gave a higher final concentration of propionic acid (seven times that obtained in the free cell system) but also increased the productivity and the final yield of propionic acid per unit weight

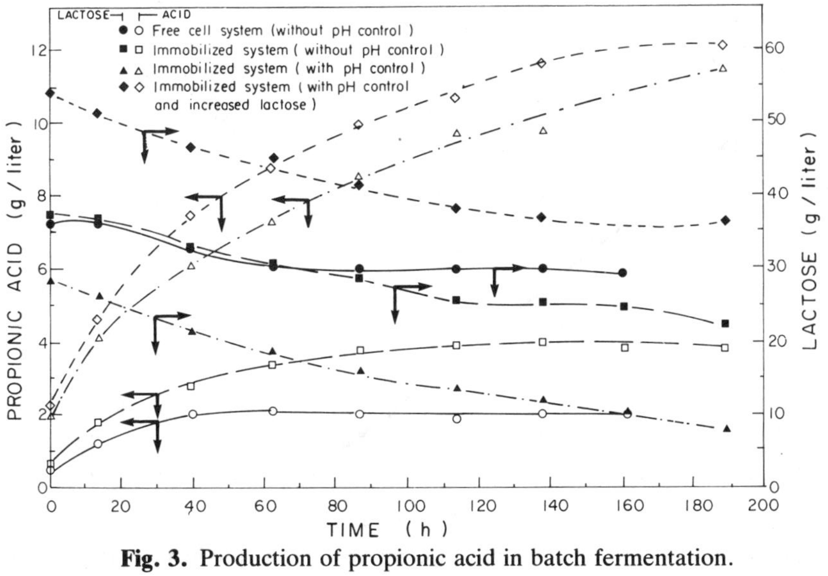

Fig. 3. Production of propionic acid in batch fermentation.

Table 6
Comparison of Free and Immobilized Cell Systems

System	*$CaCO_3$ added*	*pH Change*	*Propionic acid (g/liter)*	*Acetic acid (g/liter)*	*Yield*	
					Propionic acid	*Acetic acid*
					(moles/mole of lactose)	
Free	No	0·85	2·0	2·25	1·10	1·90
Immobilized	No	0·5	4·0	3·0	0·906	1·28
Immobilized	Yes	0·53	11·6	4·0	1·71	0·733
Immobilized	Yes	0·51	12·0	3·2	2·48	0·97

of substrate consumed. In immobilized cell systems the yield was also higher due to low growth of cells and therefore more and more energy being diverted to maintenance of cell function, resulting in higher yield of product. It was also noticed that if pH was not controlled, yield did not increase even in the immobilized system (in fact it was diminished slightly). This suggests that addition of $CaCO_3$ has a profound effect on the yield of product.

The highest yields of acetic acid with respect to lactose consumptions were obtained in the free cell system or in the immobilized cell system without pH control (Table 6). Addition of $CaCO_3$ resulted in decreased yields of acetic acid. This suggests that the addition of $CaCO_3$ to the fermentation medium distorts the metabolism of *P. shermanii* by reducing the yields of acetic acid (0·5 times) while increasing the yield of propionic acid (2·25 times).

Propionic acid appeared in the medium as a result of fermentation inhibiting the growth and acid production activity of *P. shermanii*. A linear correlation

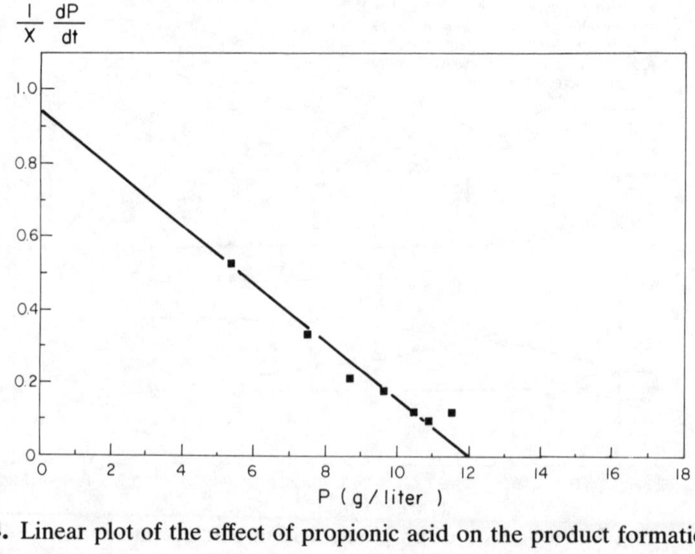

Fig. 4. Linear plot of the effect of propionic acid on the product formation rate.

was found between acid concentration and specific rate of acid production (v) which can be represented by the following equation (Fig. 4):

$$v = v_m(1 - P/P_m) \tag{5}$$

where v_m is the specific acid production at $P = 0$. This equation is similar to eqn (4) when μ is replaced by v and $n = 1$. This equation can be used as a design equation.

9 CONCLUSIONS

Cheese whey is available in large quantities and is presently not fully utilized for useful innovative products. This dairy byproduct is highly nutritious and various possibilities for its total utilization exist as a fermentation substrate. Production of organic acids from whey or permeate has high potential for several industrial uses.

Essential information on the feasibility of various microbial species for organic acid production has been reviewed. Batch and continuous organic acid fermentation processes are well documented and an up-to-date review of these methods has been provided. The rate of fermentation and microbial growth is inhibited by the product (organic acid) formed. Based on inhibitory effects, bioreactor design equations have been presented and discussed. Various information on immobilized cell supports and immobilization methods with respect to organic acid production has been reviewed and examined with special reference to product concentration and productivity. Decrease in pH in a packed bed reactor (as a result of organic acid production) decreases the product formation rate and hence demands high residence time. Further developments are required in this area.

ACKNOWLEDGEMENTS

Sincere thanks are due to Mr Ravishankar for reading the manuscript. This research was made possible through a grant from Université du Québec (Fonds de Développement académique du Réseau).

REFERENCES

Ahern, W. P., Skogerson, L. E. & Andrist, D. F. (1985). Eur. Patent Appl. EP 160417, November.

Anderson, T. M. (1982). Eur. Patent Appl. EP 95268 (cl. A2 3C21/02), 30 November 1983, US Appl. 379841, 20 May 1982.

Bergey's Manual of Determinative Bacteriology (1974), 8th edition, Williams & Wilkins, Baltimore, Maryland.

Bodie, E. A., Schwartz, R. D. & Anderson, T. M. (1983). Eur. Pat. Appl. EP 141642, 15 May 1985. US Appl. 548170, 2 November 1983.
Borzani, W. & Baralle, S. B. (1983). *Agric. Biol. Technol.*, **26**(4), 475.
Boyaval, P. & Goulet, J. (1988). *Enzyme Microbiol. Technol.*, **10,** 725.
Boyaval, P., Lebrun, A. & Goulet, J. (1985). *Lait,* **65,** 185.
Buchta, K. (1983). In *Biotechnology,* eds H. J. Rehm & G. Reed. Vol. 3: *Biomass, Microorganisms, Energy from Renewable Resources,* ed. H. Dellweg. Verlag Chemie, Weinheim, p. 409.
Campbell, L. A. (1953). *Can. Dairy & Ice Cream J.*, **32,** 29, 77.
Campbell, M. E. & Glenn, W. M. (1982). *Profit from Pollution Prevention—a guide to industrial waste reduction of recycling.* Pollution Probe Foundation Report, Toronto, Ontario, Canada.
Champagne, C. P. & Boyaval, P. (1986). *Technol. Lait. Marketing,* **1015,** 26.
Clausen, E. C. & Gaddy, J. L. (1984). *Chem. Eng. Progr.*, December, 59.
Clausen, E. C., Shah, R. B., Najafpour, G. & Gaddy, J. L. (1982). *Biotechnol. Bioeng. Symp.*, **12,** 67.
Compere, A. L. & Griffith, W. L. (1975). *Dev. Ind. Microbiol.*, **17,** 247.
Cox, G. C. & MacBean, R. D. (1977). *Aust. J. Dairy Technol.*, **32,** 19.
Coulman, R. A., Stieber, R. W. & Gerhardt, P. (1977). *Appl. Env. Microbiol.*, **34,** 725.
Dairy Market Review (1982). Marketing and Economics Branch, Agriculture Canada, Ottawa, August 1982.
Divies, C. & Siess, M. H. (1976). *Annales de Microbiologie* (Inst. Pasteur), **127B,** 525.
Friedman, M. R. & Gaden Jr, E. L. (1970). *Biotechnol. Bioeng.*, **12,** 961.
Galpin, D. B. (1981). *New Zealand J. Dairy Sci. Technol.*, **16**(3), 289.
Garoutte, C., Lim, J., Amundson, C. H. & Breslau, B. (1983). *Process Biochem.*, **18**(4), 2.
Gerhardt, P. & Reddy, C. A. (1978). *Dev. Ind. Microbiol.*, **19,** 71.
Goursaud, J. (1986). *Ind. Aliment. Agric.*, **103,** 349.
Groves, F. (1972). In *Proc. Whey Products Conf.*, p. 5 (Agric. Res. Service, US Dept of Agriculture, ERRL Publication No. 3779).
Hamilton, K. M. & Howell, J. A. (1983). In *Advances in Fermentation 83, Proc. Conf. Wheatland,* Rickmansworth, UK. Chelsea College, University of London, 21–23 September, p. 171.
Hanson, T. P. & Tsao, G. T. (1972). *Biotechnol. Bioeng.*, **14,** 233.
Hongo, M., Nomura, Y. & Iwahara, M. (1986). *Appl. Environ. Microbiol.*, **52,** 314.
Jelen, P. (1979). *J. Agric. Food Chem.*, **27**(4), 658.
Kandler, O. (1982). *Forum Mikrobiol.*, **5,** 16.
Keller, A. K. & Gerhardt, P. (1975). *Biotechnol. Bioeng.*, **17,** 997.
Kennedy, P. K. (1985). *Cult. Dairy Prod. J.*, **20**(1), 13.
Kosaric, N. & Asher, Y. J. (1985). *Adv. Biochem. Eng. Biotechnol.*, **32,** 25.
Kosaric, N. & Wieczorek, A. (1984). *Dev. Food. Sci.*, **9,** 229.
Kosikowski, F. V. (1976). *Cheese and Fermented Milk Foods.* Edward Brothers, Ann Arbor, Michigan.
Kovach, M. P. & Meyerhoff, M. E. (1982). *Anal. Chem.*, **54,** 217.
Lee, K. H. (1981). *Hanguk Nonghwa Hakhoe Chi* (Korea), **24,** 149.
Lee, I. H., Fredrickson, A. G. & Tsuchiya, M. H. (1974). *Appl. Microbiol.*, **28**(5), 831.
Luedeking, R. & Piret, E. L. (1956). *J. Biochem. Microbiol. Technol. Eng.*, **1,** 513.
Major, N. C. & Bull, A. T. (1985). *Biotechnol. Lett.*, **7,** 401.
Marriott, T. A. (1985). *J. Soc. Dairy Technol.*, **38,** 109.
Marshall, K. R. & Earle, R. L. (1975). *New Zealand J. Dairy Sci. Technol.*, **10,** 123.
Matsunaga, T., Karube, I. & Suzuki, S. (1978). *Anal. Chem. Acta,* **98,** 25.
Mehaia, M. A. & Cheryan, M. (1986). *Enzyme Microbiol. Technol.*, **8**(5), 289.
Mehaia, M. A. & Cheryan, M. (1987). *Appl. Biochem. Biotechnol.*, **14,** 21.

Modler, H. W. (1982). *Cult. Dairy Prod. J.*, **17**(2), 11.

Moebus, O. & Teuber, M. (1986). *Kiel. Milchwirtsch. Forschungsber.*, **38**(2), 119.

Morimura, S., Kishimoto, M. & Kida, K. (1986). Japanese patent 6156087, 20 March.

Muller, P. G. (1979). *Economic evaluation of feeding liquid whey to livestock.* Technical Report, Food Research Institute, Research Branch, Agriculture Canada, Ottawa.

Nagi, S., Ozaki, M., Fukunishi, K. & Yamazaki, K. (1986). Japanese patent 6158588, 25 March.

Nickerson, T. K. (1974). In *Fundamentals of Dairy Chemistry*, 2nd edition, eds B. H. Webb, A. M. Johnson & J. A. Alford. AVI, Westport, Connecticut, p. 273.

Ohmiya, K., Ohashi, H., Kobayashi, T. & Shimizu, S. (1977). *Appl. Environ. Microbiol.*, **33,** 137.

Prescott, S. C. & Dunn, C. G. (1949). *Industrial Microbiology*, 2nd edition, Chapter 21. McGraw-Hill, New York.

Prigent, Y. (1983). French Patent Demande Fr 2555200, 24 May 1985. Appl. 83/18631, 23 November 1983.

Reddy, C. A., Henderson, H. E. & Erdman, M. D. (1976). *Appl. Environ. Microbiol.*, **32,** 776.

Riddle, M. J. & Chandler, W. D. (1974). *Proc. Whey Utilization Symp.*, 27–28 June, Ottawa, Ontario, Canada.

Ripley, P. (1979). *Process Biochem.*, **14**(1), 8.

Roy, D., Goulet, J. & Leduy, A. (1986). *Appl. Microbiol. Biotechnol.*, **24,** 206.

Roy, D., Goulet, J. & Leduy, A. (1987). *J. Dairy Sci.*, **70,** 506.

Scott, R. (1981). *Cheesemaking Practice.* Applied Science, London.

Singh, R. K. & Ghaly, A. E. (1984). In *Agricultural Waste Utilization and Management, Proc. 5th Int. Symp. on Agricultural Wastes*, Chicago, Illinois. ASAE Publication 13-85, p. 546.

Somkuti, G. A. & Steinberg, D. H. (1979). J. Food Prot., **42,** 885.

Stenroos, S. L., Lindo, Y. Y. & Lenko, P. (1982). *Biotechnol. Lett.*, **4,** 159.

Stieber, R. W. & Gerhardt, P. (1979*a*). *Appl. Environ. Microbiol.*, **37,** 487.

Stieber, R. W. & Gerhardt, P. (1979*b*). *J. Dairy Sci.*, **62,** 1558.

Stieber, R. W. & Gerhardt, P. (1981). *Biotechnol. Bioeng.*, **23,** 535.

Stieber, R. W., Coulman, G. A. & Gerhardt, P. (1977). *Appl. Environ. Microbiol.*, **34,** 733.

Tewari, H. K., Seth, R. P., Sood, A. & Singh, L. (1985). *J. Res.* (*Punjab Agric. Univ.*), **22**(1), 89.

Tuli, A., Sethi, R. P., Khanna, P. K., Marwaha, S. S. & Kennedy, J. F. (1985). *Enzyme Microbiol. Technol.*, **7**(4), 164.

Tyagi, R. D. (1986). *Evaluation of Dairy Industry Waste* (*Cheese Whey*) *as Substrate for Bioconversion.* Scientific Report, INRS-Eau, Université du Québec.

Vick Roy, T. B., Blanch, H. V. & Wilke, C. R. (1982). *Biotechnol. Lett.*, **4,** 483.

Vick Roy, T. B., Mandal, D. K., Dea, D. K., Blanch, H. W. & Wilke, C. R. (1983). *Biotechnol. Lett.*, **5,** 665.

Mathieu, F. W. (1983). [illegible] Dairy Technol., 112, 126.
Mombas, C. & [illegible], M. (1980). [illegible] Milch. [illegible] Forschungsbericht, [illegible]
[illegible] & [illegible], M. [illegible] (1980). Japanese patent [illegible] 1605, [illegible]
Muller, [illegible] (1979). [illegible] evaluation of [illegible] liquid [illegible] Technical Report, [illegible] Research Institute, Research Branch, Agriculture Canada, Ottawa.
[illegible], [illegible], M., [illegible] & [illegible], K. (1979). Japanese patent [illegible] 5.
[illegible], [illegible] (1974). In [illegible] of Dairy [illegible], 2nd edition, ed. B. H. Webb, A. H. Johnson & J. A. Alford. AVI, Westport, Conn., [illegible] p. 2.
Ohniwa, K., Ohashi, [illegible] & [illegible], S. (1977). [illegible] Microbiol. 33, 137.
[illegible], S. C. (1979). [illegible] 2nd edition. Chapter 21. [illegible] New York.
Pagan, [illegible] Patent [illegible] 25, 322 [illegible] (1958). [illegible] 1963.
Reddy, C. V. [illegible] (1970). Appl. [illegible] 32, 70.
[illegible] (1977). [illegible] Research Branch, Agriculture Canada, Ottawa, Ontario, Canada.
[illegible] 1960.
[illegible] (1980). J. Appl. [illegible] 23.
[illegible] (1970). J. Dairy Sci., 53, 598.
[illegible] (1981). [illegible] London.
[illegible] (1984). [illegible] Illinois, [illegible] 8.
[illegible]
[illegible] Appl. Bacteriol., [illegible]
[illegible] (1958). Dairy Sci. [illegible] 1958.
[illegible] (1962). [illegible] 23.
[illegible] (1977). [illegible] 24.
Towers, [illegible] (1983). J. [illegible] 38.
[illegible] Kennedy, [illegible] (1964).
[illegible] (1982). [illegible] Milchwissenschaft [illegible]
[illegible] (1959). [illegible]
[illegible] (1980). [illegible] 3.

Chapter 11

PRODUCTION OF INDUSTRIAL ENZYMES FROM WASTE MATERIALS AND UTILIZATION OF WASTE ATP ENERGY FOR SYNTHESIS

TATSUROKURO TOCHIKURA

Department of Food Science and Technology, Faculty of Agriculture, Kyoto University, Sakyo-ku, Kyoto 606, Japan

CONTENTS

1 INTRODUCTION

The natural recycling of various substances, including carbon, sulfur, and nitrogen, is an essential part of ecological systems. It was stated (Forster & Jones, 1976) that the one exception to this general rule is post-Industrial Revolution man, who voraciously consumes natural products and prefers 'disposal by destruction' to any form of recycling. The finite nature of the chemical elements available on Earth together with the increasing demands of consumers and the slow rates of natural conversion mean that the natural cycle has to be bypassed and a recycling revolution must begin. In this context, organic waste material needs to be treated as a renewable reservoir of carbon–nitrogen complexes.

Certainly, public pressure to abate pollution is justified in the modern world, and organic wastes comprise a substantial part of total environmental pollution. The modern Earth is facing the following difficult problems: population, environmental pollution, consumption of natural resources, and food production.

Thirty years ago, it was stated (Beckhorn, 1960) that although enzymes from plants and animals were more important than those from microbial sources, for both technical and economic reasons microbial enzymes will play an increasingly important role in the future. The fermentation industry has now developed the bulk cultivation of microorganisms, which is the major source of many of the enzymes used commercially (Vincent & Priestley, 1975). The economic reasons are evident—bulk production of enzymes leads to ease of extraction, predictable output and enzyme activity, and independence from overseas suppliers in most cases. The technical reasons are even more convincing—the selection of suitable bacteria can be performed among the countless possibilities offered by the genetic flexibility and ease of manipulation of microorganisms. Undoubtedly, the range of economic possibilities provided by enzymes from plants and animals is much more limited.

All living cells are well known to be endowed with highly effective devices for converting energy (Lehninger, 1971). Adenosine triphosphate (ATP) acts as the carrier of chemical energy from the energy-generating oxidation of foodstuff molecules, whether by aerobic or anaerobic processes, to those reactions of cells that do not take place spontaneously and can proceed only if chemical energy is supplied. We have been confronted in our modern technology with the requirement for new modes of energy conversion, and microorganisms are today providing us with models and ideas for such new methods.

This chapter describes the production of industrial enzymes from waste materials and the utilization of the waste ATP energy of fermentation for synthesis.

2 PRODUCTION OF INDUSTRIAL ENZYMES FROM WASTE MATERIALS

2.1 Production of β-*N*-Acetylhexosaminidase from a Mold *Mucor* Grown in Bovine Blood

Animal blood produced by inspected slaughterhouses is wasted biomass in many countries. This unusable biomass causes an increase in the biochemical

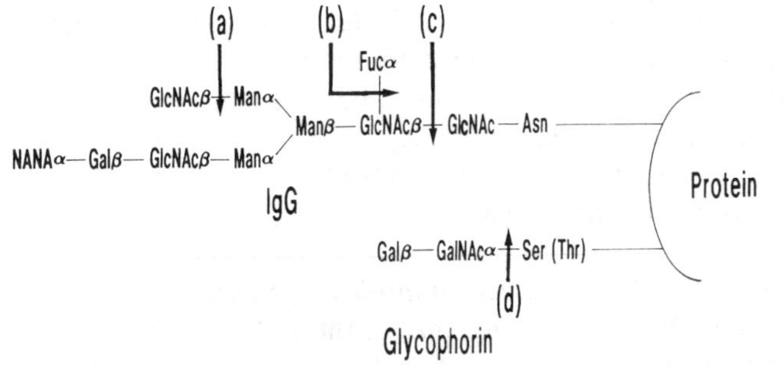

Fig. 1. Various glycosidases acting on the sugar chains of glycoconjugates: (a) exo-β-*N*-acetylhexosaminidase; (b) α-L-fucosidase; (c) endo-β-*N*-acetylglucosaminidase; (d) endo-α-*N*-acetylgalactosaminidase. GlcNAc, *N*-acetylglucosamine; GalNAc, *N*-acetylgalactosamine; Gal, galactose; Man, mannose; Fuc, fucose; NANA, *N*-acetylneuraminic acid (sialate).

oxygen demand (BOD) value in drainage. On the other hand, its profitable use could help decrease the pollution problems encountered in disposing of blood. In several European countries, blood is used as an ingredient of various foods, such as sausages and bread (Wismer-Pedersen, 1979).

We have investigated ways of reusing this biomass by using microorganisms. Blood is rich in glycoproteins and glycolipids. Therefore, we attempted to obtain useful glycosidases (Fig. 1) produced by microorganisms, which can grow on bovine blood as a medium. As a result, we found that some fungi grown on bovine blood as the sole carbon sources produced large amounts of β-*N*-acetylhexosaminidase. The enzyme activity of *Mucor fragilis* was several times higher on the blood than that of a culture grown on glucose medium.

β-*N*-Acetylhexosaminidase (EC 3.2.1.52) is widely distributed among animal tissues, higher plants, and microorganisms. The enzyme catalyzes the hydrolysis of the β-*N*-acetylglucosaminyl or β-*N*-acetylgalactosaminyl moiety of the nonreducing end of oligosaccharides and of the sugar chains of glycoconjugates. The enzymes of some fungi including *Aspergillus* (Bahl & Agrawal, 1969), *Sclerotinia* (Reyes & Byrde, 1973), *Tremella* (Sone & Misaki, 1978), *Paecilomyces* (Eriquez & Pisano, 1979) and *Pycnoporus* (Ohtakara *et al.*, 1981) were purified from cells or culture filtrates. We also purified an uncommon enzyme from a culture filtrate of *Penicillium oxalicum* which had higher β-*N*-acetylgalactosaminidase (β-GalNAc-ase) activity than β-*N*-acetylglucosaminidase (β-GlcNAc-ase) activity (Yamamoto *et al.*, 1985). Recently, β-*N*-acetylhexosaminidases have been recognized as a very useful tool for elucidation of the structures and functions of glycoproteins and glycolipids, since β-linked *N*-acetylhexosaminyl units are common constituents of biologically important substances such as blood group substances, immunoglobulins, cell walls of microorganisms and others.

Here, we describe the production, purification and properties of an extracellular β-*N*-acetylhexosaminidase by *M. fragilis* grown on bovine blood (Yamamoto *et al.*, 1986*a*).

Microorganisms and cultivation: Fungi from our collection including *M. fragilis* IFO 6449 were used. The medium was composed of 1% bovine blood lyophilized powder, 0·5% peptone, 0·5% yeast extract, and 0·5% NaCl, pH 6·5. Cultivation was carried out at 28°C on a reciprocal shaker. After cultivation, the culture broth was filtered, and the clarified supernatant was used as the source of the enzyme.

Production of β-N-acetylhexosaminidase: When *M. fragilis* was grown on bovine blood as the sole carbon source, the β-*N*-acetylhexosaminidase activity in the culture broth was several times higher than that in the supernatant of a culture grown on 1% glucose medium. The maximum enzyme activity was observed after 4–5 days, when fungal autolysis took place. On the other hand, little activity was found when *M. fragilis* was grown on glucose medium, although both the maximum mycelial yield and final pH were almost the same as for cells grown on bovine blood medium (Fig. 2). No glycosidase activities except β-*N*-acetylhexosaminidase were found in the culture broth of either glucose-grown or bovine blood-grown fungi. The enzyme activity depended on the initial pH of the bovine blood medium; the maximum activity was obtained with an initial pH of about 8.

Two different molecular weight forms of the enzyme: We found two β-*N*-acetylhexosaminidases having different molecular weights (MW) in the culture broth on Sephadex G-150 gel filtration only when the fungus was cultivated on bovine blood medium. The high-MW form of enzyme (named EI) seemed to be induced by bovine blood since it almost completely disappeared in the culture broth with glucose medium. It showed somewhat higher β-GalNAc-ase activity than the other enzyme, a low-MW form (named EII), which showed mainly β-GlcNAc-ase activity (Fig. 3). The activity of the low-MW form of enzyme was also induced by bovine blood.

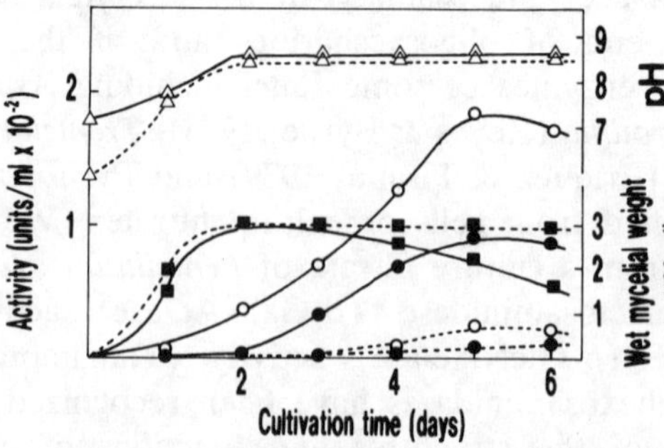

Fig. 2. Time-course of growth and enzyme production in cultures of *M. fragilis*. The cultures were carried out in 500-ml shake flasks containing 120 ml of 1% glucose medium (- - - - -) or 1% bovine blood medium (———) at 28°C on a reciprocal shaker. Mycelial wet weights were measured after culture fluids were filtered through filter paper. Symbols: ■, mycelia; ○, β-GlcNAc-ase; ●, β-GalNAc-ase; △, pH. Source: Modified from Yamamoto *et al.*, 1986*a*.

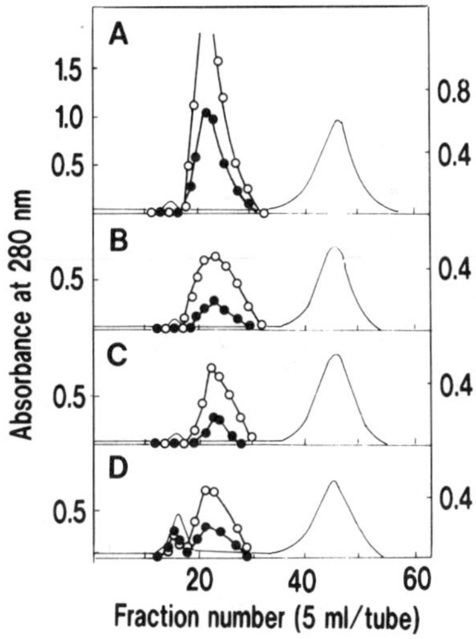

Fig. 3. Sephadex G-150 gel filtration of β-*N*-acetylhexosaminidases produced with various amino sugars: (A) 0·5% GalNAc medium; (B) 0·5% GlcNAc medium; (C) 0·5% *N*-acetylmannosamine medium; (D) 1% bovine blood medium. Fraction volume, 5 ml. Symbols: ○, β-GlcNAc-ase; ●, β-GalNAc-ase; ———, absorbance at 280 nm. Source: Modified from Yamamoto *et al.*, 1986*a*.

Influence of various carbon sources on enzyme production The fungus was cultivated on media in which various carbon sources replace the glucose, and the enzyme activity in each culture filtrate was investigated. β-*N*-Acetylhexosaminidase increased after addition of *N*-acetylaminosugars to the medium. Addition of GalNAc revealed a marked effect on the activity (about 17-fold higher than on glucose medium). Maximum activity was obtained on addition of 0·8% GalNAc to the medium. These amino sugars seemed to specifically induce the low-MW form of enzyme (EII) (Fig. 3).

Purification of β-N-acetylhexosaminidase We tried to purify the two enzymes, EI and EII, in the culture filtrate of bovine blood-grown fungi, but EI was produced in smaller amounts and was also rather unstable compared with EII. We did not succeed in purifying EI, but the purification of EII is summarized in Table 1. The overall purification was about 3500-fold, with a yield of 10%. At the DEAE-Sepharose CL-6B column chromatography step, most of EI was separated from EII: EII eluted first (0·2M NaCl), followed by EI (0·4–0·5M NaCl), on stepwise elution with an increasing NaCl concentration. After this step, the ratio of the β-GlcNAc-ase and β-GalNAc-ase activities of EII remained almost constant throughout the purification procedure (Table 1). The purified enzyme showed β-GalNAc-ase activity equivalent to about 40% of the β-GlcNAc-ase activity.

Purity and molecular weight: The purified β-*N*-acetylhexosaminidase pre-

Table 1
Purification of β-*N*-Acetylhexosaminidase from *M. fragilis*

	Total protein (mg)	β-GlcNAc-ase			β-GalNAc-ase			Ratio of total activity
		Total activity (units)	*Specific activity (units/ml)*	*Yield (%)*	*Total activity (units)*	*Specific activity (units/ml)*	*Yield (%)*	*β-GalNAc-ase/ β-GlcNAc-ase*
Culture filtrate	7 820	78·2	0·010	100	54·7	0·007	100	0·70
$(NH_4)_2SO_4$ precipitation	1 750	77·0	0·044	98	40·3	0·023	74	0·52
DEAE-Sepharose CL-6B column chromatography	251	42·7	0·17	55	17·1	0·068	31	0·40
Hydroxylapatite column chromatography	11·2	22·7	2·03	29	9·7	0·87	18	0·43
Sephadex G-150 column chromatography	3·1	18·7	6·07	24	7·3	2·35	13	0·39
Con A-Sepharose 4B column chromatography	0·22	7·7	35·2	10	3·1	14·2	6	0·40

Source: Modified from Yamamoto *et al.*, 1986*a*.

paration (EII) was found to be homogeneous by polyacrylamide gel electrophoresis. This enzyme seemed to be a glycoprotein. The MW of the enzyme was determined by gel filtration on a Sephadex G-200 column at approximately 125 000. Upon SDS gel electrophoresis, the enzyme exhibited a single band corresponding to a MW of about 70 000. Therefore, the enzyme consists of two subunits of identical MW.

Effect of pH on activity and stability: The optimum pH was 4·0–5·0 for β-GlcNAc-ase activity and 5·5–6·5 for β-GalNAc-ase activity (Fig. 4). The stable pH range of the enzyme was relatively wide, from pH 4·5 to 8·5 (Fig. 4). Both activities showed the same extent of impairment at the various values of pH.

Thermal stability: The purified enzyme (EII) was stable up to about 45°C after incubation for 10 min in 0·01M phosphate buffer (pH 7·0), and 50% of the original activity remained after incubation at 57·5°C for 10 min. The activities of both β-GlcNAc-ase and β-GalNAc-ase showed the same dependence on temperature.

K_m and substrate specificity: The apparent Michaelis constants of EII, calculated from Lineweaver–Burk plots, were 0·28mM for *p*-nitrophenyl-β-*N*-acetylglucosaminide and 0·46mM for *p*-nitrophenyl-β-*N*-acetylgalactosaminide.

The enzyme (EII) hydrolyzed di-*N*-acetylchitobiose and tri-*N*-acetylchitotriose and liberated GlcNAc residues. The hydrolysis rate for di-*N*-acetylchitobiose was faster than that for tri-*N*-acetylchitotriose. The

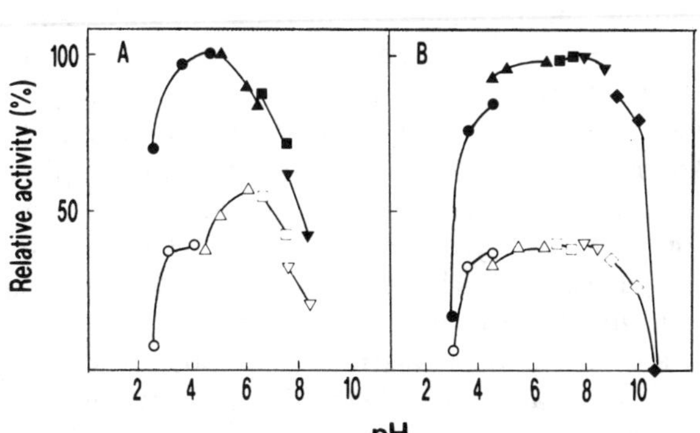

Fig. 4. Effects of pH on activity and stability of β-*N*-acetylhexosaminidase. (A) Enzyme activity was assayed in 0·05M buffers of various pH values under standard conditions, and β-GalNAc-ase activity was expressed relative to that of β-GlcNAc-ase. (B) Enzyme was kept at various pH values at 4°C for 46 h in various 0·05M buffers, followed by assay of the remaining activity under standard conditions. β-GalNAc-ase activity was expressed relative to that of β-GlcNAc-ase. Buffers: ○, ●, sodium citrate–HCl; △, ▲, sodium citrate–citric acid; □, ■, potassium phosphate; ▽, ▼, Tris–HCl; ◇, ◆, glycine–NaOH. Solid symbols, β-GlcNAc-ase; open symbols, β-GalNAc-ase. Source: Modified from Yamamoto *et al.*, 1986*a*.

enzyme was found to be capable of hydrolyzing β-galactosidase-treated desialized fetuin. The enzyme liberated GlcNAc from lysozyme-treated cell walls of *Micrococcus lysodeikticus*.

2.2 Production of α-L-Fucosidase from a Mold *Fusarium* Grown on Sludge

At present, increasing surplus active sludge which is produced in the treatment of city sewage and plant drainage is becoming a social problem, since it is one of the great unutilizable biomasses. We have investigated ways of reusing surplus sludge with microorganisms. We took advantage of the abundant glycoconjugates in the sludge and tried to obtain useful glycosidases produced by microorganisms able to assimilate the sludge as a culture medium.

During the screening of glycosidases capable of degrading complex carbohydrates, we found that a strain of fungus isolated from a soil sample produced a novel α-L-fucosidase in the culture broth. The fungus was identified as *Fusarium oxysporum*.

L-Fucose is a very important component of various glycoconjugates. It is usually located on the nonreducing termini of the carbohydrate moieties of several biologically active molecules such as serum glycoproteins, immunoglobulins, submaxillary mucins and blood group substances. It is also present in oligosaccharides of human milk, marine algae polysaccharides and plant gums. This L-fucose moiety of some complex carbohydrates is known to play an important role in cellular recognition, the immune response and the others.

α-L-Fucosidase (EC 3.2.1.51) has been found to release L-fucose from these molecules by splitting the terminal α-L-fucosidic linkage. All enzymes from various sources have been reported to generally show a narrow aglycon specificity. They seem to be classifiable into two types. One type comprises the microbial enzymes. The α-L-fucosidases from *Trichomonas* (Watkins, 1959), *Clostridium* (Aminoff & Furukawa, 1970), *Aspergillus* (Bahl, 1970) and *Bacillus* (Kochibe, 1973) were found to release fucose from blood group substances but did not hydrolyze simple nitrophenyl fucosides. They may be $\alpha(1\rightarrow 2)$-specific L-fucosidases. The enzyme from almond emulsin also showed this type of specificity. On the other hand, the α-L-fucosidases from mammalian tissues, abalone and a marine gastropod readily hydrolyzed *p*-nitrophenyl α-L-fucoside and also liberated fucose slowly from mucin, but seemed not to hydrolyze blood group substances.

Our fungal α-L-fucosidase showed a novel substrate specificity in that it could hydrolyze both *p*-nitrophenyl α-L-fucoside and blood group substances. Here, we describe the production, purification and properties of the unique α-L-fucosidase from *Fusarium oxysporum* isolated from soil which was grown on active sludge hydrolyzate as a medium (Yano *et al.*, 1985).

Preparation of the sludge hydrolyzate Active sludge was obtained after waste treatment from a Japanese sake company. Its water content was about

85%. The hydrolyzate of the active sludge as a specific medium for microorganisms was prepared according to the following method. Active sludge was added to an equal volume of 0·6N NaOH solution, and then the mixture was hydrolyzed by autoclaving it at 120°C for 30 min. After centrifugation, the supernatant was dialyzed against tap water overnight and then lyophilized. 0·3N HCl solution was added to the lyophilized powder, and then the mixture was hydrolyzed by autoclaving it for 30 min again. The hydrolyzate was neutralized by adding NaOH.

Production of α-L-fucosidase: After *Fusarium oxysporum* had been cultivated at 28°C for 2 days with shaking in a liquid medium (pH 7·0) composed of 2% sludge hydrolyzate, various glycosidase activities in the culture broth were investigated. We found high α-L-fucosidase activity and slight β-glucosidase and β-galactosidase activities, the latter being about one-eighth of the former. Then, we investigated the effect of the concentration of sludge hydrolyzate used as the sole nutrient source on α-L-fucosidase activity.

The enzyme activity on 2 days' cultivation increased with increasing sludge hydrolyzate concentration in the medium (Fig. 5). On the other hand, no activity was found when the concentration of glucose was increased in a medium containing 0·5% peptone, 0·5% yeast extract and 0·5% NaCl although the growth of the fungus was very good. We also could not detect any activity when the fungus was cultivated in the active sludge medium without the hydrolysis treatment.

For maximum production of the enzyme in the case of cultivation on 2% sludge hydrolyzate medium, we determined the enzyme activity in the culture broth at different stages of growth of the fungus. As shown in Fig. 6, the enzyme activity was highest at 3 days, when the mycelial weight and pH of the medium became constant.

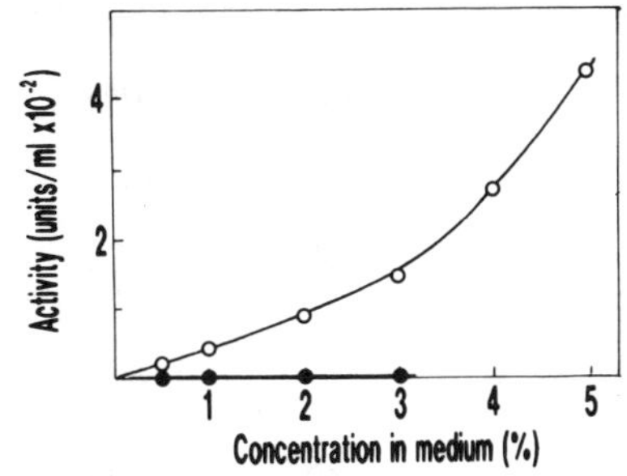

Fig. 5. Effect of concentration of sludge hydrolyzate as culture medium on enzyme activity. Enzyme activities were assayed in culture filtrates after 2 day cultivations with various concentrations of sludge hydrolyzate (○) and various concentrations of glucose in medium containing 0·5% peptone, 0·5% yeast extract and 0·5% NaCl (●). Source: Modified from Yano *et al.*, 1985.

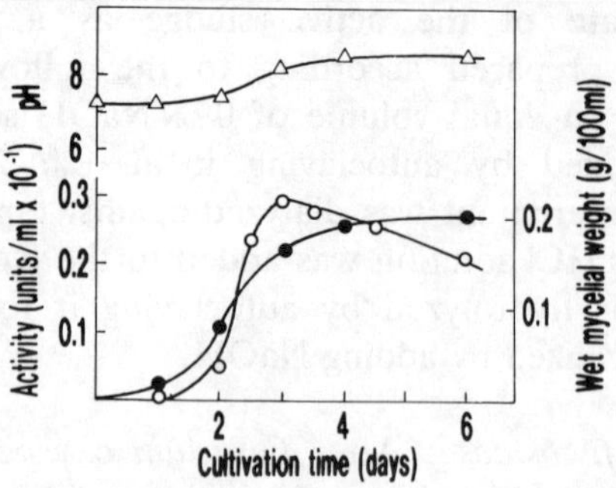

Fig. 6. Time-course of growth and enzyme production in 2% sludge hydrolyzate cultures of *F. oxysporum*. Symbols: ●, wet mycelial weight; ○, α-L-fucosidase activity; △, pH. Source: Modified from Yano *et al.*, 1985.

Purification of α-L-fucosidase from Fusarium oxysporum: The enzyme from culture broth of the fungus was purified, about 360-fold with a recovery of 20%, to homogeneity by polyacrylamide gel electrophoresis after ammonium sulfate fractionation and successive column chromatographies on DEAE-Sephadex A-50, hydroxylapatite, Sephadex G-150 and Con A-Sepharose 4B (Table 2).

Molecular weight: The molecular weight of the enzyme was estimated by gel filtration on Sephadex G-200 to be about 80 000.

Effect of pH on the activity and stability: The enzyme showed maximum reactivity in the pH range of 4·5–5·5. The enzyme was stable over the wide pH range of 4–8.

Table 2
Purification of α-L-Fucosidase from *F. oxysporum*

	Total protein (mg)	*Total activity (units)*	*Specific activity (units/mg)*	*Yield (%)*
Culture filtrate	538	3·82	0·007 1	100
Ammonium sulfate precipitation	376	3·68	0·009 8	96
DEAE-Sephadex A-50 column chromatography	51	2·19	0·043	57
Hydroxylapatite column chromatography	26	2·02	0·078	52
Sephadex G-150 column chromatography	5·4	1·56	0·29	40
Con A-Sepharose 4B column chromatography	0·3	0·77	2·58	20

Source: Yano *et al.*, 1985.

Effect of temperature on the activity and stability: The optimum temperature for the enzyme reaction was about 50°C under the standard assay conditions (pH 4·5). The enzyme was found to be stable up to 45°C on 10 min incubation.

Effect of the substrate concentration: The K_m value for *p*-nitrophenyl α-L-fucoside, calculated from a Lineweaver–Burk plot, was 0·87 mM.

Substrate specificity: The enzyme showed a novel substrate specificity in that it could hydrolyze porcine mucin and blood group substances of human saliva besides nitrophenyl compounds. Such a specificity has not been found for any other α-L-fucosidase from various sources.

Effects of carbon sources on α-L-fucosidase production (*Yamamoto et al.*, 1986*b*): In order to investigate the effects of different carbon sources on α-L-fucosidase production, the fungus was cultivated in liquid medium containing 0·5% various carbon sources, 0·5% peptone, 0·5% yeast extract and 0·5% NaCl (pH 7·0). After 4 days' cultivation, the culture broths were filtered and then the enzyme activity in the filtrates was determined. Among the various carbon sources, only L-fucose and D-arabinose caused α-L-fucosidase production. An especially significant level of enzyme production was seen in the culture filtrate of the L-fucose-grown fungus. When the fungus was grown with D-arabinose, which has a very similar structure to that of L-fucose, the level of enzyme production was only about one-eighth of that in the culture filtrate with the L-fucose medium. The following carbon sources tested showed no induction of α-L-fucosidase: D-fucose, L-arabinose, D-glucose, D-galactose, D-fructose, D-mannose, D-xylose, L-rhamnose, D-glucosamine, D-galactosamine, *N*-acetyl-D-glucosamine, *N*-acetyl-D-galactosamine, maltose and sucrose, although good growth of the fungus on these carbon sources was observed except on D-fucose, which was only slowly utilized by the fungus.

2.3 Production of α-L-Fucosidase from a *Bacillus* Grown on Porcine Gastric Mucin

As stated in the previous section, a *Fusarium* α-L-fucosidase can easily hydrolyze *p*-nitrophenyl α-L-fucoside, but it releases only a small amount of L-fucose from natural substrates. α-L-Fucosidase, which has a broad aglycon specificity and acts on natural macromolecules, is required to investigate the structure and function of glycoconjugate. We looked for such an α-L-fucosidase in the culture fluids of various microorganisms using a fucose-containing glycoprotein, porcine gastric mucin (PGM), as a substrate to assay the enzyme activity and also as the sole carbon source for bacterial cultivation. A bacterial strain, *Bacillus circulans* M28, isolated from a soil sample was recently found to produce a high level of a unique α-L-fucosidase in culture fluid. The enzyme seems to have broad aglycon specificity, unlike other microbial α-L-fucosidase, though it can not hydrolyze *p*-nitrophenyl α-L-fucoside (Tsuji *et al.*, 1990).

Microorganism and cultivation: *Bacillus circulans* was inoculated to a 500-ml Sakaguchi flask containing 100 ml of a medium consisting of 0·5% glucose, 0·5% peptone, 0·5% yeast extract, and 0·5% NaCl (pH 6·5). The culture was carried out overnight at 30°C with shaking and then its broth was transferred to a 10-liter Mitsuwa jar fermentor containing 8 liters of a medium containing 2% PGM, 0·5% peptone, 0·5% yeast extract, and 0·5% NaCl. The cultivation was done for 3 days at 30°C with aeration (3·5 liter/min) and agitation (350 rpm).

Effect of carbon source on α-L-fucosidase production: In order to investigate the effects of various carbon sources on α-L-fucosidase production, the bacterium was cultured in liquid medium containing 1% of various carbon sources. After 3 days' cultivation, the culture broths were centrifuged and the supernatants obtained were dialyzed overnight against 10 mM potassium phosphate buffer (pH 7·0) at 4°C, and then the enzyme activities were assayed by using PGM and *p*-nitrophenyl α-L-fucoside as substrates. Table 3 shows that enzyme activity toward *p*-nitrophenyl α-L-fucoside was found at relatively high levels in culture fluids from each medium with glucose, galactose,

Table 3
Effects of Different Carbon Sources on Enzyme Production[a]

Carbon source	*Hydrolysis activity (munits/ml)* *Substrates*	
	p-Nitrophenyl α-L-fucoside	*Porcine gastric mucin*
None	56	0·2
Glucose	94	0·7
Galactose	59	3
Mannose	81	3
Fructose	72	2
Maltose	3	0·4
Sucrose	8	0
Lactose	8	3
L-Fucose	82	0
Arabinose	13	0
Xylose	2	0
Glucosamine	1	0
Galactosamine	0	0
Porcine gastric mucin	58	107
Human milk oligosaccharides	65	22
Human red blood cells	0	0

[a] *Bacillus circulans* was grown in a medium containing 1% carbon sources, 0·5% peptone, 0·5% yeast extract, and 0·5% NaCl. After 3 days of cultivation, the α-L-fucosidase activity in the culture fluid was determined.
Source: Tsuji *et al.*, 1990.

mannose, fructose, L-fucose, PGM, and human milk oligosaccharides. On the other hand, a remarkable level of α-L-fucosidase activity toward PGM was found only in culture fluids of media with PGM and human milk oligosaccharides. In particular, when the bacterium was grown in PGM medium, the activity toward PGM was about 150 times higher than that in the culture broth obtained with glucose medium. The addition of human O(H) group red blood cells caused no enzyme production because these were hardly utilized by the bacterium.

Production of two α-L-fucosidases in culture medium containing PGM: The time courses of α-L-fucosidase production and cell growth were investigated when the bacterium was grown in medium containing 1·5% PGM as a carbon source. Figure 7 shows that the enzyme activity toward PGM in the culture broth was first detected at 12 h and attained a maximum after 2 days.

On the other hand, the activity toward *p*-nitrophenyl α-L-fucoside was first observed after a 24-h cultivation when cell autolysis took place, and became constant at 3 days at a low level. These data suggested there were two different α-L-fucosidases in the culture broth: one was able to act on PGM and the other was able to act not on PGM but on *p*-nitrophenyl α-L-fucoside.

Purification of α-L-fucosidase: The α-L-fucosidase acting on PGM was purified from the culture fluid obtained with PGM medium by ammonium sulfate fractionation and subsequent column chromatographies. The results of the purification are summarized in Table 4. The enzyme was purified about 750-fold from the culture fluid.

Enzyme purity and molecular weight: The purified enzyme migrated as a single protein band on polyacrylamide disc gel electrophoresis. Of the glycosidases detected in the crude preparation, i.e. α-L-fucosidases, β-galactosidase, and β-*N*-acetylhexosaminidase, only the α-L-fucosidase acting

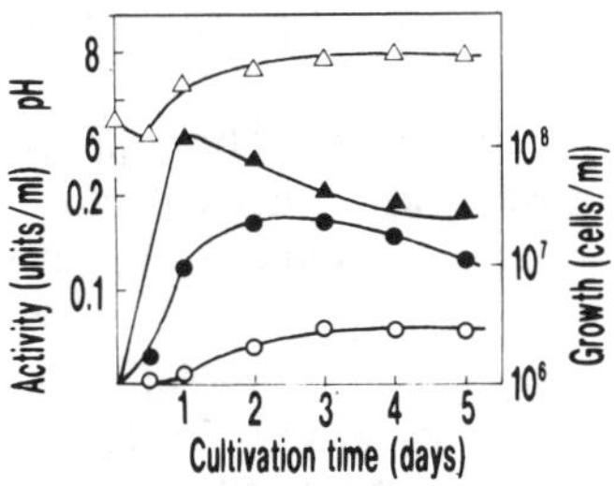

Fig. 7. Time-course of growth and enzyme production in 1·5% porcine gastric mucin culture of *B. circulans* M28. The culture was carried out in a 500-ml shake flask containing 100 ml of medium at 30°C on a reciprocal shaker. Symbols: ▲, cell growth; ●, α-L-fucosidase activity toward porcine gastric mucin; ○, α-L-fucosidase activity toward *p*-nitrophenyl α-L-fucoside, △, pH. Source: Modified from Tsuji *et al.*, 1990.

Table 4
Purification of α-L-Fucosidase Acting on Porcine Gastric Mucin

	Total protein (mg)	*Total activity (units)*	*Specific activity (units/ml)*	*Yield (%)*
Culture fluid	55 000	1 510	0·027 5	100
Ammonium sulfate precipitation	14 100	1 510	0·107	100
DEAE-Toyopearl 650M column chromatography	7 660	1 280	0·167	85
TSK-gel HW 65C column chromatography	342	922	2·70	61
Hydroxylapatite column chromatography	46·2	156	3·38	10
TSK-gel HW 65F column chromatography	18·1	80·3	4·44	5·3
Butyl–Toyopearl 650M column chromatography	0·70	14·0	20·0	0·9

Source: Tsuji *et al.*, 1990.

on PGM appeared in the purified preparation. No protease or neuraminidase activities were observed.

The molecular weight of the enzyme was estimated by gel filtration to be about 285 000. SDS-PAGE gave a single protein band corresponding to a molecular weight of about 135 000. These results suggest that the enzyme consists of two identical subunits.

Effect of pH and temperature on the activity and stability: Maximal activity was obtained at pH 5·5–6·5. The enzyme was stable in the pH range of 4·5–9·0 when kept at 4°C for 63 h. The enzyme was found to be stable up to about 45°C and could be stored frozen for over a year without any loss of activity.

Effect of the substrate concentration: The apparent Michaelis constants toward PGM, 2′-fucosyllactose and 3-fucosyllactose were calculated from Lineweaver–Burk plots to be 0·88, 0·63 and 1·25 mM, respectively.

Action of α-L-fucosidase on blood group O(H) active substances: The enzyme decomposed various blood group O(H) active substances such as PGM, human milk and human saliva. Moreover, the H activity of O-erythrocytes treated with the enzyme (hemagglutination titer, 1:16) was remarkably reduced in comparison with that of untreated O-erythrocytes (1:128), and the H activities of A- and B-erythrocytes treated with the enzyme were also reduced. However, neither the A activity of A-erythrocytes nor the B activity of B-erythrocytes was affected by treatment with the enzyme.

Substrate specificity: In order to investigate the substrate specificity of the enzyme, purified α-L-fucosidase (10 munits) was incubated with various synthetic and natural substances (corresponding to 0·1 μmol of bound fucose) in 50 mM potassium phosphate buffer (pH 6·0) at 37°C. After incubation for 30 and 120 min, the amount of L-fucose released was assayed. A summary of the substrate specificity studies is given in Table 5. The enzyme was found to have relatively broad aglycon specificity. The enzyme readily hydrolyzed L-fucose residues bound by α-(1→2)-linkage to D-galactose in 2′-fucosyllactose,

Table 5
Substrate Specificity of α-L-Fucosidase from *B. circulans* M28[a]

Substrate	*Structure*[b]	*Hydrolysis (%)*	
		30 min	*120 min*
p-Nitrophenyl α-L-fucoside		0	0
p-Nitrophenyl β-L-fucoside		0	0
Methyl α-L-fucoside		0	0
Fucan sulfate		0	0
Porcine gastric mucin (H)		85	93
Porcine submaxillary mucin (A⁻ type)	Fucα1→2Galβ1→3GalNAc 6 ↑ NGNAα2	86	88
Desialylated porcine submaxillary mucin		94	100
Human milk (H)		54	68
Human saliva (H)		65	84
2′-Fucosyllactose	Fucα1→2Galβ1→4Glc	94	96
3-Fucosyllactose	Galβ1→4Glc 3 ↑ Fucα1	34	74
Orosomucoid		8	17
Desialylated orosomucoid		48	85
Lacto-*N*-fucopentaose I	Galβ1→3GlcNAcβ1→3Galβ1→4Glc 2 ↑ Fucα1	82	97
Lacto-*N*-fucopentaose II	Galβ1→3GlcNAcβ1→3Galβ1→4Glc 4 ↑ Fucα1	75	93
Lacto-*N*-fucopentaose III	Galβ1→4GlcNAcβ1→3Galβ1→4Glc 3 ↑ Fucα1	65	92
Lactoferrin		17	31
6-O-α-L-Fucopyranosyl-*N*-acetylglucosamine	Fucα1→6GlcNAc	0	0
Thyroglobulin (bovine)		0	0

[a] The conditions used for enzymatic hydrolysis are given in the text.
[b] Fuc, L-fucose; Gal, D-galactose; GalNAc, *N*-acetylgalactosamine; Glc, D-glucose; GlcNAc, N-acetylglucosamine; NGNA, *N*-glycolylneuraminic acid.
Source: Tsuji *et al.*, 1990

lacto-*N*-fucopentaose I, PGM, and both intact and desialylated porcine submaxillary mucins. 2′-Fucosyllactose and desialylated porcine submaxillary mucin were the most effectively cleaved of all the substrates tested. The enzyme was also able to cleave α-(1→3)-L-fucosidic linkage to D-glucose in 3-fucosyllactose and to *N*-acetylglucosamine in orosomucoid (α_1-acid glycoprotein) and lacto-*N*-fucopentaose III, though native orosomucoid was hydrolyzed very slowly compared with the desialylated form. Similarly, the enzyme hydrolyzed α-(1→4)-L-fucosidic linkage to *N*-acetylglucosamine in lacto-*N*-fucopentaose II. The enzyme hydrolyzed the fucosidic linkages in lacto-*N*-fucopentaose I, II, and III in the following order of velocity: lacto-*N*-fucopentaose I > lacto-*N*-fucopentaose II > lacto-*N*-fucopentaose III. The enzyme failed to hydrolyze α-(1→6)-L-fucosidic linkage in 6-O-α-L-fucopyranosyl-*N*-acetylglucosamine or bovine thyroglobulin. The enzyme liberated a little L-fucose from lactoferrin, in which fucose is linked by α-(1→3)- and (1→6)-linkage to *N*-acetylglucosamine. It seemed that only α-(1→3)-L-fucosidic linkage in this substrate was cleaved by the enzyme. *p*-Nitrophenyl α-L- and β-L-fucosides, methyl α-L-fucoside and fucan sulfate were not hydrolyzed by the enzyme.

Thus, α-L-fucosidase from *B. circulans* grown on PGM medium had broad aglycon specificity and was able to act on native erythrocytes and macromolecular glycoproteins including porcine submaxillary and gastric mucins, etc. The enzyme is an extracellular one, unlike *Bacillus fulminans* (Kochibe, 1973) and *Trichomonas foetus* (Watkins, 1959) enzymes, and a large amount of the enzyme was obtained. This enzyme may be a useful analytical tool for elucidating the functions of fucose-containing substances.

2.4 Production of Endo-α-*N*-Acetylgalactosaminidase from a Bacterium *Alcaligenes* Grown on Porcine Gastric Mucin

Two types of carbohydrate chains are known to exist in glycoproteins. One is attached to an asparagine residue of a protein through an N-glycosidic linkage and the other to the hydroxyl groups of serine or threonine residues through an O-glycosidic linkage (Sharon & Lis, 1982). O-Glycosides are widely found as components of cell surface glycoproteins in erythrocytes and various animal tissues. The O-linked oligosaccharides seem to be involved in a variety of biological phenomena such as blood group specificity, cell recognition, embryonic development and differentiation. Recently, such oligosaccharides were found in hepatoma ascites cell membranes and human gastric cancer cells, and some relationships between O-glycoside carbohydrate chains and cancerous cells were proposed.

Endo-α-*N*-acetylgalactosaminidase (EC 3.2.1.97) hydrolyzes the O-glycosidic linkages between α-*N*-acetyl-D-galactosamine and serine or threonine residues in glycoproteins, and releases a disaccharide, galactosyl-*N*-acetylgalactosamine.

Galβ1$\rightarrow$3GalNAcα1$\rightarrow$Ser or Thr-Protein $\longrightarrow$

Galβ1$\rightarrow$3GalNAc + Ser or Thr-Protein

Since the disaccharide Galβ1$\rightarrow$3GalNAc exists in many O-glycosides as a core structure, the endo-glycosidase would be one of the most useful tools for structural studies on O-glycosidic glycoproteins, and for investigating the nature and functions of cell-surface complex carbohydrates.

However, the endo-glycosidase has been found only in the culture broth of *Clostridium perfringens* (Huang & Aminoff, 1972) and *Diplococcus pneumoniae* (Endo & Kobata, 1976; Umemoto *et al.*, 1977) which are both anaerobic and pathogenic bacteria. Although the enzymes have been effectively used in studies on glycoproteins, it is difficult to obtain the pure enzyme in large amounts and to treat these bacteria because of their anaerobic character and pathogenicity. In order to obtain a better source of endo-α-*N*-acetylgalactosaminidase, we looked for the enzyme in the culture broths of various bacteria isolated from soil.

We found the enzyme activity able to release a disaccharide, galactosyl-*N*-acetylgalactosamine, from an O-glycoside linked glycoprotein (asialo-fetuin), in the culture fluid of an aerobic and non-pathogenic *Alcaligenes* sp. (No. F-1906) isolated from a soil sample. We investigated the culture conditions for the bacterial strain and found that porcine gastric mucin induced high production of the enzyme (Fan *et al.*, 1988). This material is a sort of wasted and unusable biomass.

Cultivation of Alcaligenes sp. (*F*-1906 *strain*): The bacterium was cultured at 28°C on a reciprocal shaker in a liquid medium composed of 2% porcine gastric mucin, which was adjusted to pH 7·0.

Production of endo-α-N-acetylgalactosaminidase: Enzyme production in the culture fluid at different stages of cell growth was investigated. As shown in Fig. 8, the bacterial growth gradually reached a maximum level in 2 days,

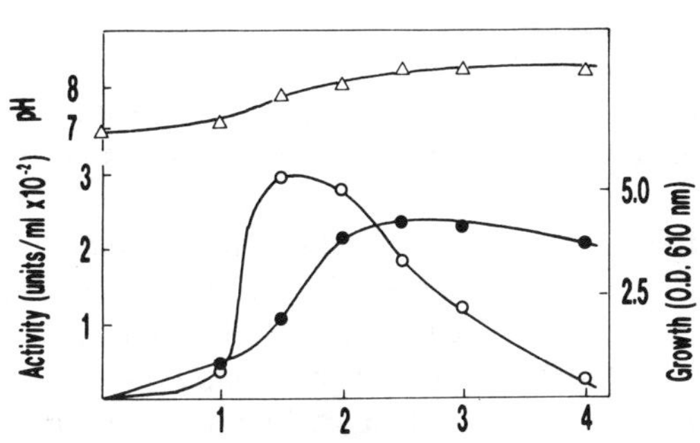

Fig. 8. Time-course of growth and enzyme production in a culture of *Alcaligenes* sp. The culture was carried out at 28°C on a reciprocal shaker . Symbols: ●, cell growth; ○, endo-α-*N*-acetylgalactosaminidase; △, pH. Source: Modified from Fan *et al.*, 1988.

Table 6
Purification of Endo-α-*N*-Acetylgalactosaminidase from *Alcaligenes* sp.

	Total protein (mg)	*Total activity (units)*	*Specific activity (units/mg)*	*Yield (%)*
Culture filtrate	60 400	233	0·003 9	100
Ammonium sulfate precipitation (80% saturation)	27 200	180	0·006 6	77
DEAE-Sephadex A-50 column chromatography	1 970	163	0·083	70
Hydroxylapatite column chromatography	9·8	21	2·18	9

Source: Fan *et al.*, 1988.

whereas the enzymatic activity rapidly increased after one day of cultivation and was maximal after 1·5–2 days. However, the activity gradually decreased after 2–3 days.

Purification of endo-α-N-acetylgalactosaminidase: The enzyme was purified, about 560-fold with a recovery of 9%, from the culture fluid, by fractionation with ammonium sulfate, and column chromatographies on DEAE-Sephadex A-50 and hydroxylapatite (Table 6).

Enzyme purity and molecular weight: The purified enzyme migrated as a single protein band on polyacrylamide disc gel electrophoresis. SDS-polyacrylamide gel electrophoresis also gave a single band. The molecular weight of the enzyme was estimated to be 160 000 by gel filtration and SDS-polyacrylamide gel electrophoresis. These results indicate that the enzyme has a monomeric structure. The isoelectric point of the enzyme was found to be 5·0 on chromatofocusing.

Amino acid composition: The amino acid composition of the enzyme is shown in Table 7. It seemed that the enzyme contained no methionine.

Effect of pH and temperature on the activity and stability: The optimum pH of the enzyme was found to be between 4·5 and 5·0. The enzyme was stable in the pH range of 4·5–6·5 when incubated at 30°C for 1 h. The enzyme showed the highest activity at 40–45°C under the standard assay conditions. Its thermal stability was examined by heating it at various temperatures for 10 min. The enzyme was stable below 35°C and completely inactivated at 55°C.

Effect of the substrate concentration: The K_m values were 3·7mM and 3·2mM for asialofetuin and asialo κ-casein glycopeptide as substrates, respectively.

Table 7
Amino Acid Composition of the Endo-α-*N*-Acetylgalactosaminidase[a]

Amino acid	*Residues/mol of protein*
Aspartic acid	167
Threonine	118
Serine	110
Glutamic acid	123
Proline	64
Glycine	134
Alanine	160
Valine	113
Methionine	ND[b]
Isoleucine	55
Leucine	74
Tyrosine	26
Phenylalanine	59
Lysine	54
Histidine	55
Arginine	88
Tryptophan	47
Half cystine	26

[a] The molar contents were calculated on the basis of a molecular weight of 160 000.
[b] ND = not detected.
Source: Fan *et al.*, 1988.

Substrate specificity: In order to investigate the specificity of the enzyme, the purified enzyme (2 munits) was incubated with various natural substrates for 24 h. The enzymatic products were detected by means of the color formation on the Morgan–Elson reaction (Reissig *et al.*, 1955). Besides asialofetuin and asialo κ-casein glycopeptides, the enzyme also acted on asialoglycophorin of human red cell membranes, that contains Galβ1$\rightarrow$3GalNAc linked to serine or threonine (Thomas & Winzler, 1969). As shown in Fig. 9, the enzyme released about 40% of the sugar chains of total disaccharides from asialoglycophorin within 2 h. However, the enzyme was not able to hydrolyze sugar chains containing sialate from fetuin and κ-casein glycopeptide. The enzyme had no ability to act on sugar chains of either sialo or asialo bovine submaxillary mucin. These results are summarized in Table 8.

Induction of endo-α-N-acetylgalactosaminidase: The enzyme production was induced by some potent materials in porcine gastric mucin. As shown in Fig. 10, when glucose-grown cells of *Alcaligenes* sp. were transferred to inducing media containing various concentrations of mucin, the enzyme activity increased with the amount of mucin, although cell growth was not significantly affected.

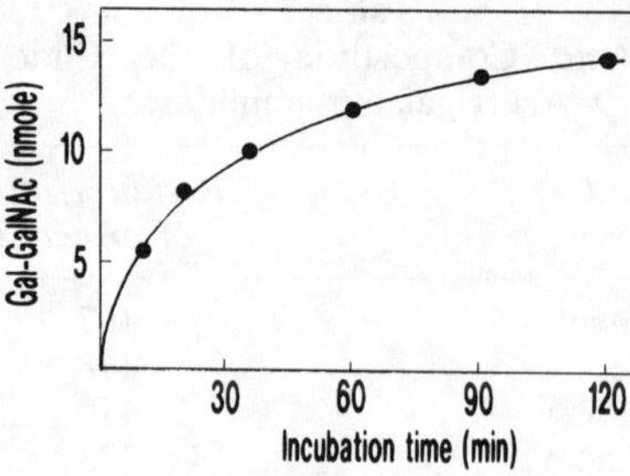

Fig. 9. The release of the disaccharide from asialoglycophorin by endo-α-*N*-acetylgalactosaminidase. 2·4 nmol asialoglycophorin was used as a substrate. Source: Modified from Fan *et al.*, 1988.

Using the inducing medium containing 2% mucin, which was prepared by protease (pronase) treatment, in 50mM phosphate buffer (pH 6·0), enzyme production was induced greatly and the enzyme could be easily purified from the inducing culture medium by ammonium sulfate fractionation, ion exchange chromatography on DEAE-Sephadex A-50 and gel filtration on Sephadex G-200, with a high recovery of 50%.

Table 8
Substrate Specificity of the Endo-α-*N*-Acetylgalactosaminidase

Substrate	*Structure*[a]	*Hydro-lyzed*
Asialofetuin	Galβ1→3GalNAcα1→Ser(Thr)—Protein	+
Asialo κ-casein glycopeptide	Galβ1→3GalNAcα1→Thr—Peptide	+
Asialo glycophorin	Galβ1→3GalNAcα1→Ser(Thr)—Protein	+
Fetuin	Galβ1→3GalNAcα1→Ser(Thr)—Protein 3 6 ↑ ↑ α2 α2 NANA NANA	–
κ-Casein glycopeptide	Galβ1→3GalNAcα1→Thr—Peptide 3 ↑ α2 NANA	–
Asialo bovine submaxillary mucin	GalNAcα1→Ser(Thr)—Protein	–
Bovine submaxillary mucin	GalNAcα1→Ser(Thr)—Protein 6 ↑ α2 NANA	–

[a] Gal, D-galactose; GalNAc, *N*-acetylgalactosamine; NANA, neuraminic acid.
Source: Fan *et al.*, 1988.

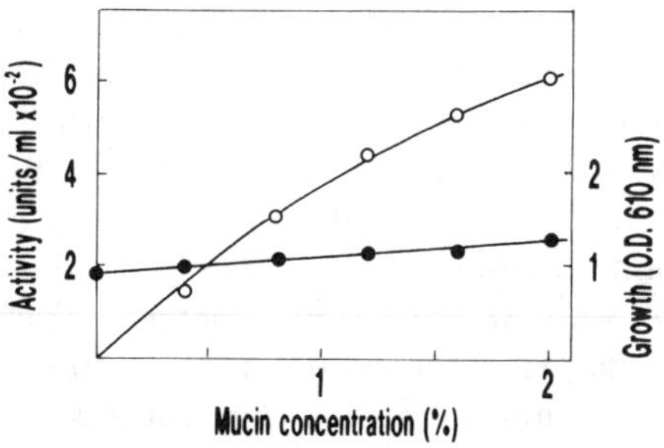

Fig. 10. Effect of the mucin concentrations on the enzyme production in an inducing culture of *Alcaligenes* sp. After growth in a medium (pH 7·0) containing 0·5% glucose, 0·5% peptone, 0·5% yeast extract and 0·5% NaCl for 18 h, the cells were collected by centrifugation and then suspended in inducing media (pH 7·0) containing various concentrations of the mucin supernatant. The inducing cultures were carried out for 10 h. Symbols: ●, cell growth; ○, endo-α-*N*-acetylgalactosaminidase. Source: Modified from Fan *et al.*, 1988.

3 MICROBIAL CONVERSION OF ANIMAL WASTE TO ANIMAL FEED

It has become increasingly difficult to dispose of animal excreta by traditional methods such as land spreading, due to increased government controls on environmental pollution, public pressure to limit pollution, and the dangers of disease transmission, especially near urban areas. Because of the potential nutritional value of animal wastes, we investigated recycling wastes into animal feed through microbial processes (Tochikura *et al.*, 1978). Prior to the cultivation of potent microorganisms, we applied a heat sterilization process to the animal wastes, including cattle, swine, and poultry excreta. This heating process has the following advantages: (1) the process can be carried out as a means of sanitation; (2) the sterilized wastes can be used as a culture medium in which the selective cultivation of potent microbes can be effectively accomplished; (3) the cultured waste containing useful microbes can be used as fodder.

We investigated various types of microbial conversion after separating animal feces and urine wastes.

(1) *Liquid culture with animal feces fraction* (Tochikura *et al.*, 1978). As the first step in conversion, requisite amounts of water and sugar or starch (final concentration of 1–4%) were added to the fecal waste (final concentration of 2–6% when dried) of cattle, swine or poultry. After a heat sterilization process, the fecal medium was inoculated with various yeasts or bacteria and incubated aerobically at 28°C for 1–2 days. The upper layer of the culture broth contains single-cell protein and can be utilized as a liquid fodder. When *Saccharomyces cerevisiae, Hansenula*

anomala and *H. miso* were used, cultivation resulted in a remarkable decrease in the offensive odor of the feces. As the second step, the fecal residue, after separation of the liquid fodder, was further inoculated with *Aspergillus, Neurospora* or *Rhodotorula* and incubated at 28°C for several days under static conditions. This solid culture, enriched with microbial carotinoid and ergosterol, can be used as fodder.

(2) *Liquid culture with animal urine fraction* (Tochikura *et al.*, 1978). Adjustment of the pH of the animal urine fraction after heat sterilization was found to be important for the successful commencement and progress of cultivation. After heating, the urine waste was adjusted to a pH of 6–5 by the addition of hydrochloric acid. The waste was then diluted to a 25–50% concentration with water and supplemented with glucose or starch in concentrations of 1–3%. The medium was incubated with potent yeasts, molds and bacteria aerobically at 28°C for 2–3 days. Good growth rates of the following organisms were observed: *Candida parapsilosis, C. tropicalis, Torulopsis candida, Pichia polymorpha, R. rubra, H. anomala, H. miso, Debaryomyces hansenii, S. lactis, Saccharocodes ludowigii, A. oryzae, N. crassa, Penicillium oxalicum, Bacterium gracile* and *Micrococcus glutamicus*. The cells produced were separated from the waste broth and utilized as fodder.

(3) *Culture of animal feces in a solid state with* A. oryzae *group* (Tochikura *et al.*, 1981*a*, 1986). The method for preparing solid culture (so-called koji culture) is summarized as follows: Fresh animal feces 4·5 kg (as wet matter) plus wheat bran 450 g→mixed→autoclaved→inoculated with *A. oryzae*→cultivated for 2-3 days at 28°C→autoclaved to kill the mold→storage. The koji culture resulted in a remarkable decrease of the fecal odor.
The stored culture could be used as fodder. In another case, the koji culture was digested with 2 volumes of water for 24 h at 45–55°C. The resulting culture could be used as the nutrient in a growth media of microbes.

(4) *Production of* Bifidobacterium *concentrates from koji cultures of animal feces* (Tochikura *et al.*, 1981*a*, 1986). *Bifidobacterium* species (Fig. 11) (Honma & Mitsuoka, 1978) are the normal flora in the lower intestines of animals. They have the ability to maintain a normal microbial balance in the intestine, in that they suppress many undesirable organisms and promote beneficial metabolism. We demonstrated that pure concentrates of these bacteria, including *B. pseudolongum, B. breve, B. bifidum, B. longum* and *B. thermophilum,* and *Lactobacillus acidophilus,* were produced in the digested koji cultures of animal feces supplemented with 1–3% glucose. For these pasty solid cultures, it was much easier to maintain a strictly anaerobic condition than for the liquid cultures. These pure cultures could be used as fodder additives.

(5) *Production of bacterial proteases from animal feces.* Following sterilization by heating, animal feces were efficiently converted to a culture

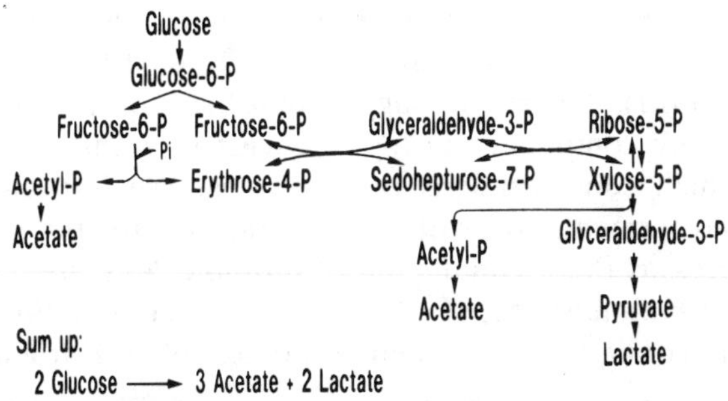

Fig. 11. Fermentation pathway of *Bifidobacterium* species.

medium for potent bacteria. Further, we investigated protease production by bacteria, including *Bacillus subtilis* and *B. natto*. We found that a remarkable production of neutral and alkaline proteases took place when the bacteria were aerobically cultivated at 28°C for 3 days in a medium containing 30–40% cattle feces (as fresh wet matter) as the sole nutrient. The proteases produced could be used as fodder additives and digestive chemicals for waste water treatment.

4 UTILIZATION OF WASTE ATP ENERGY FOR SYNTHESIS

4.1 Development of Coupled Fermentation with Energy Transfer

In alcoholic brewing, the chemical energy of ATP is liberated during the fermentation of glucose by yeast according to the following equation:

$$\text{Glucose} \xrightarrow[\text{ADP} \rightarrow \text{ATP}]{} 2C_2H_5OH + 2CO_2 \quad \Delta G^{0'} = -40{\cdot}0\ \text{kcal/mole}$$

The symbol $\Delta G^{0'}$ designates the standard free energy change.

However, no utilization of ATP energy itself for synthesis is known in the fermentation industry, where ATP was recognized as a form of unusable waste energy. In other words, with regard to alcoholic fermentation, people and yeast have quite opposite purposes: the former are able to obtain alcohol of industrial value through the non-productive discharge of ATP energy during yeast fermentation, while the latter is able to obtain ATP energy indispensable for anaerobic life where alcohol is the major waste which accumulates.

We have established processes for the production of various substances (Tochikura *et al.*, 1967, 1974*a*; Tochikura, 1976), employing dried yeast cells as a means to increase cell membrane permeability and as an enzyme source. The processes are driven by the ATP energy released by the alcoholic fermentation of sugar by the yeast. We have also designed a new process in which the fermentation of glucose by yeast is combined (coupled) with an endergonic reaction catalyzed by an enzyme or an enzyme system involving different microorganisms (Tochikura *et al.*, 1974*b*, 1981*b*; Tochikura, 1978; Tachiki *et al.*, 1981*a*). The results suggest that the process offers the possibility of producing many compounds in higher yields by using various biosynthetic reactions and high concentrations of substrates. We introduced the general idea of 'coupled fermentation with energy transfer' for the process. The first step in the process is the accumulation of unstable ATP energy in the form of stable fructose 1,6-diphosphate (FDP) using yeast fermentation of sugar, based on the Harden–Young effect (Harden & Young, 1906; Harden, 1923):

$$C_6H_{12}O_6 + HPO_4^{2-} \rightarrow C_2H_5OH + CO_2 + H_2O + \tfrac{1}{2}C_6H_{10}O_4(PO_4^{2-})_2$$

The second step is the coupling of the fermentation of FDP with endergonic reactions (energy-utilizing system) through catalytic ATP–ADP (adenosine diphosphate) system. The energy-utilizing system can involve the enzymes of yeast itself or those of other organisms. According to the coupled fermentation with energy transfer, alcoholic fermentation may be estimated as a renewable reservoir of ATP energy for biosynthesis. The principle of coupled fermentation is shown in Fig. 12 and it should be noted that a high yield of alcohol can be obtained as a byproduct. Finally, the relationship between the standard free energy change ($\Delta G^{0\prime}$) and ATP formation in alcoholic fermentation may be summarized as follows (Lehninger, 1971):

$$\text{Glucose} + 2\text{ADP} + 2\text{P}_i \rightarrow 2\text{ATP} + 2C_2H_5OH + 2CO_2 + 2H_2O$$

$$\Delta G^{0\prime} = -25{\cdot}4 \text{ kcal/mole}$$

$$[\text{ADP} + \text{Pi} \rightarrow \text{ATP} + H_2O;\ \Delta G^{0\prime} = +7{\cdot}3 \text{ kcal/mole}]$$

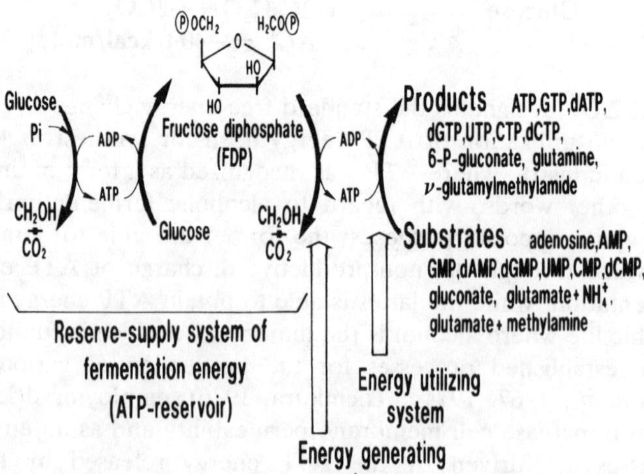

Fig. 12. Principle of coupled fermentations with energy transfer.

4.2 Synthesis of Nucleotides Using Waste ATP Energy Liberated by Yeast Fermentation

4.2.1 Production of Nucleoside Polyphosphate

PHOSPHORYLATION OF RIBONUCLEOSIDE MONOPHOSPHATES (Tochikura *et al.*, 1974*a*; Tochikura, 1978)
Phosphorylation of ribonucleoside monophosphates, including AMP, GMP, UMP or CMP (adenosine monophosphate, guanosine monophosphate, uridine monophosphate and cytidine monophosphate, respectively), was investigated by means of coupled fermentation (Fig. 12). The essence of the phosphorylation linked with glucose fermentation is as follows: (1) production of nucleoside polyphosphates (ATP, GTP, UTP, CTP) does not commence immediately after incubation, that is, there exists a time lag before phosphorylation of the nucleoside monophosphates, because of the accumulation of FDP; (2) accumulation of FDP tends to slow as both the inorganic phosphate (P_i) and glucose concentrations increase; and (3) high yields of the polyphosphates are obtained with a high level of P_i against the low yields obtained at a low level of P_i. The maximum yield of UTP or CTP was about 90mM per 8 h at 28°C with a medium containing P_i 800mM, glucose 600mM, $MgSO_4$ 30mM, UMP or CMP 100mM, catalytic AMP or ATP 5mM and dried baker's yeast 100 mg/ml. In the case of AMP phosphorylation with glucose, ATP yields were remarkably lower than those with FDP. Addition of high concentrations of AMP effected substantial production of IMP and inosine.

PHOSPHORYLATION OF ADENOSINE TO ATP (Tochikura *et al.*, 1974*a*; Tochikura, 1978)
An attempt was made to produce large amounts of ATP through the coupling of phosphorylation of adenosine with the fermentation of glucose using either dried baker's or brewer's yeast (Fig. 13). Although a large amount of adenosine is minimally soluble in fermentation media, ATP production can take place when adenosine crystals are in a suspended state. Typical results are as follows: (a) 400 mg/ml dried baker's yeast produced about 100mM ATP

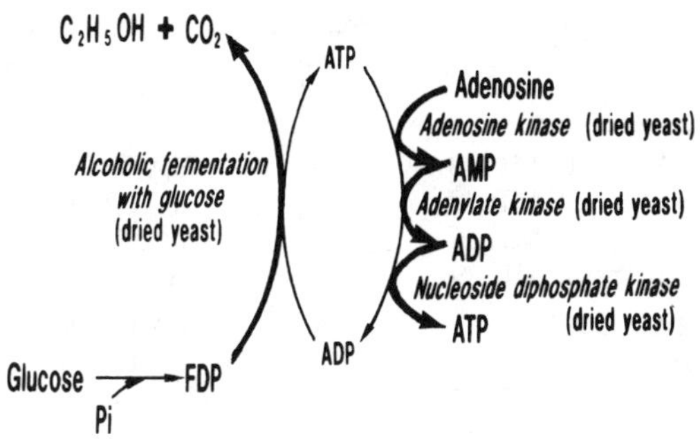

Fig. 13. Phosphorylation of adenosine to ATP through the coupled fermentation of yeast with energy transfer.

(50 g/liter) and 40mM ADP (16 g/liter) after a 4 h incubation period at 28°C in a medium containing P_i (pH 7·0) 1000mM, glucose 800mM, $MgSO_4$ 30mM and adenosine 200mM; (b) 400 mg/ml dried brewer's yeast produced 190mM ATP (about 96 g/liter) and 20mM ADP (9 g/liter) after 8 h incubation in a medium of P_i 1200mM, glucose 1600mM, $MgSO_4$ 30mM and adenosine 300mM. Based on the following fermentation equation,

$$\text{glucose} + \text{adenosine (MW 267)} + P_i \rightarrow \text{ATP (MW 507)} + C_2H_5OH + CO_2$$

ATP production mentioned here may be termed 'phosphate-fixing fermentation'.

PHOSPHORYLATION OF DEOXYRIBONUCLEOSIDE MONOPHOSPHATE (Kawaguchi *et al.*, 1970; Yano *et al.*, 1989)
We have tried to produce deoxyribonucleoside triphosphates (dNTPs) via a coupled fermentation with energy transfer process (Fig. 12), in order to maintain a sufficient supply of these compounds, for which demand has been increasing with the recent progress in the field of applied genetics.

Reaction conditions: The standard reaction mixture for dNTP production contained 100mM deoxyribonucleoside monophosphate (dNMP), 400mM glucose, 400mM potassium phosphate buffer (K-PB, pH 7·0), 10mM $MgSO_4$ and 100 mg/ml dried baker's yeast. Incubation was carried out at 28°C with shaking.

Preparation of hydrolyzed DNA: The hydrolysis of DNA was carried out by the method of Fujimoto *et al.* (1974) with a slight modification. Commercial DNA (100 mg), dissolved in 10 ml of water, was heated in boiling water for 10 min and then quickly cooled with ice. To the solution were added 1 ml of 0·02M acetate buffer (pH 5·5) and 0·001 mg of Nuclease P_1. The mixture was

Table 9
Production of Deoxyribonucleoside Triphosphates from Wasted Fish DNA

Substrates (mM)		*Reaction time (h)*	*Products (mM)*	
Hydrolyzed DNA	dAMP 40	6	dATP 27	dADP 5
(Salmon testes)	dGMP 32		dGTP 16	dGDP 5
	dCMP 36		dCTP 6	dCDP 6
	dTMP 29		dTTP 0	dTDP 0
Hydrolyzed DNA	dAMP 32	8	dATP 19	dADP 4
(Herring sperm)	dGMP 28		dGTP 9	dGDP 3
	dCMP 37		dCTP 4	dCDP 3
	dTMP 24		dTTP 0	dTDP 0

Source: Yano *et al.*, 1989.

incubated at 37°C for 15 h. The lyophilized mixture was used as the mixture of dNMPs.

Production of deoxyribonucleoside triphosphates from waste fish DNA: DNA is a polymer of dNMP which is cheaper than dNMPs, and Nuclease P_1 is able to attack all kinds of RNA (DNA) to produce 5′-NMPs (5′-dNMPs). Therefore, the production of dNTPs was investigated using DNA hydrolyzed with Nuclease P_1 in place of dNMPs (Table 9). When purified DNA (100 mg) from salmon testes was used, 49mM dNTPs was produced after 6 h incubation. On the other hand, 32mM dNTPs was produced after 8 h incubation with crude herring sperm. The rate of sugar fermentation by yeast in the latter mixture was lower than that in the former. Accordingly, it was thought that dNTPs can be produced cheaply by this method without dTTP.

4.2.2 Production of Sugar Nucleotides and CDP-Choline

SYNTHESIS OF SUGAR NUCLEOTIDES AND COUPLED FERMENTATION WITH DOUBLE OR MULTIPLE ENERGY TRANSFER (Tochikura *et al.*, 1972; Tochikura, 1976; Kawaguchi *et al.*, 1980)
The synthesis of sugar nucleotides (Figs 14–16) by fermentation appears to be relatively complex and refined in its energy requirement for this process. The following points are noteworthy.

In the first step of synthesis, glycosyl donors such as glucose, galactose and glucosamine added to the media are phosphorylated to sugar-6-phosphate (glucose-6-P, mannose-6-P) or sugar-1-phosphate (galactose-1-P) with the consumption of ATP, which is liberated by glycolysis in yeasts. Subsequently, sugar-6-phosphates are converted to sugar-1-phosphates. In the synthesis of UDP-*N*-acetylglucosamine (UDP-GlcNAc), glucosamine is converted to *N*-acetylglucosamine-1-P with the consumption of ATP and acetyl-CoA. On the other hand, either UMP or GMP, which are used as nucleotide donors for the synthesis of sugar nucleotides, is phosphorylated to UTP or GTP by coupling with glycolysis. In the next step, the sugar-1-phosphate reacts with UTP or GTP to form sugar nucleotide and pyrophosphate. Finally, the pyrophosphate

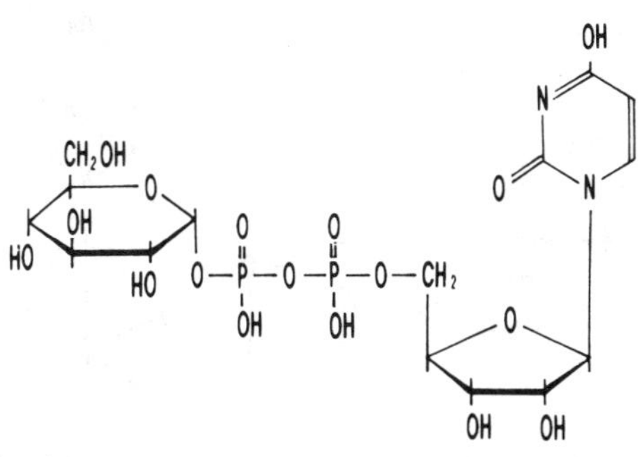

Fig. 14. Uridine diphosphate glucose (UDPG).

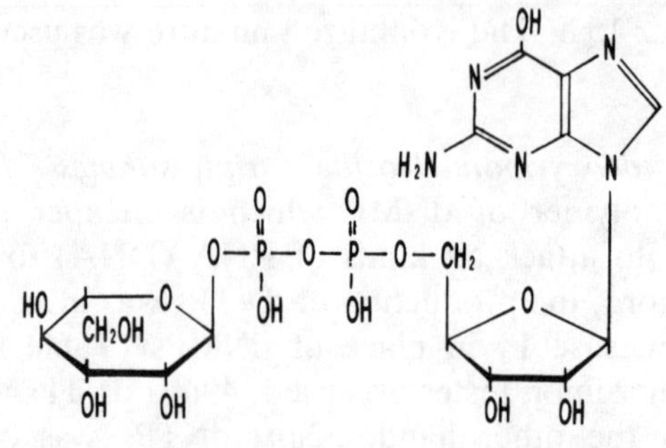

Fig. 15. Guanosine diphosphate mannose (GDP-Man).

is broken down by pyrophosphatase, and the enzymatic synthesis of the sugar nucleotide proceeds to completion. Therefore, the fermentative synthesis of sugar nucleotides can be accomplished by coupling endergonic reactions in the synthetic process with glycolysis. This type of fermentative synthesis is a practical application of coupled reactions involving double or multiple energy transfer systems. One energy transfer system is an ATP–ADP reaction responsible for the phosphorylation of the glycosyl donors in the synthesis of sugar nucleotides, and another is an ATP–ADP reaction responsible for the phosphorylation of the nucleotide donors, i.e. for the conversion of UMP or GMP to UDP or GDP and then to UTP or GTP. Thus, this method of microbial synthesis may be termed coupled fermentation with double or multiple energy transfer. The mechanism of fermentative synthesis of several sugar nucleotides is outlined in Fig. 17, according to the principle of coupling with energy transfer. Table 10 shows the reactions in the enzymatic synthesis of UDPG, which illustrate the pattern involved in the synthesis of most sugar nucleotides. For each mole of UDPG synthesized, three moles of ATP are consumed. One mole of ATP is consumed in the phosphorylation of glucose to glucose-6-P and two moles of ATP in the phosphorylation of UMP to UTP.

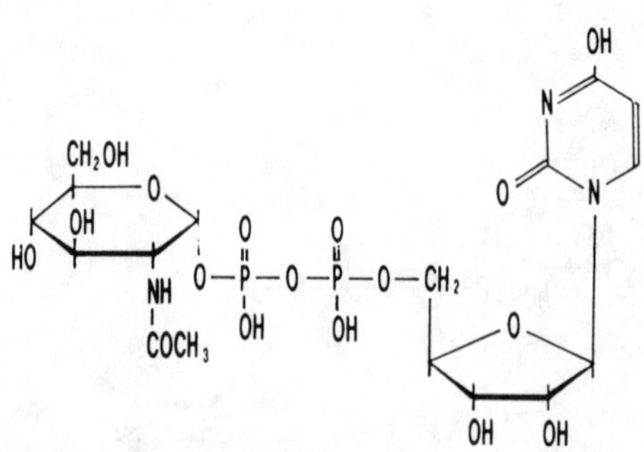

Fig. 16. Uridine diphosphate *N*-acetylglucosamine (UDP-GlcNAc).

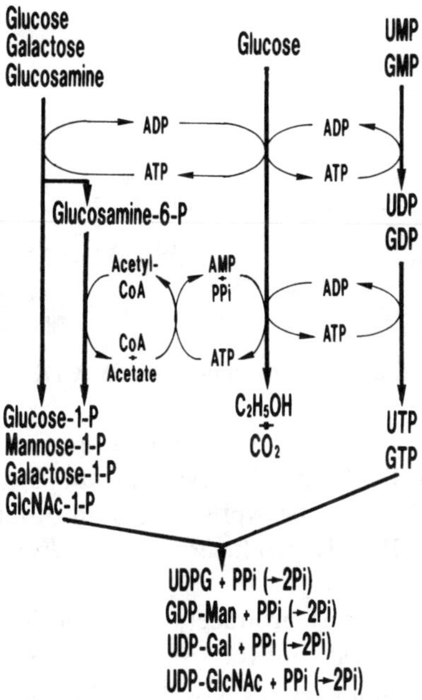

Fig. 17. Fermentative synthesis of sugar nucleotides and coupled fermentation with multiple energy transfer.

Table 10
Reactions Involved in the Enzymatic Synthesis of UDPG

ATP generating system:

$$1{\cdot}5\ \text{Glucose} + 3\ \text{ADP} + 3\ \text{P}_i \rightarrow 3\ \text{C}_2\text{H}_5\text{OH} + 3\ \text{CO}_2 + 3\ \text{ATP}$$

ATP utilizing system:

$$\text{Glucose} + \text{ATP} \xrightarrow[\text{hexokinase}]{} \text{Glucose-6-P} + \text{ADP}$$

$$\text{Glucose-6-P} \xrightarrow[\text{phosphoglucomutase}]{} \text{Glucose-1-P}$$

$$\text{UMP} + \text{ATP} \xrightarrow[\text{UMP kinase}]{} \text{UDP} + \text{ADP}$$

$$\text{UDP} + \text{ATP} \xrightarrow[\text{UDP kinase}]{} \text{UTP} + \text{ADP}$$

$$\text{Glucose-1-P} + \text{UTP} \xrightarrow[\text{UDPG pyrophosphorylase}]{} \text{UDPG} + \text{PP}_i$$

$$\text{PP}_i \xrightarrow[\text{inorganic pyrophosphatase}]{} 2\ \text{P}_i$$

Overall:

$$2{\cdot}5\ \text{Glucose} + \text{UMP} + \text{P}_i \rightarrow \text{UDPG} + 3\ \text{C}_2\text{H}_5\text{OH} + 3\ \text{CO}_2$$

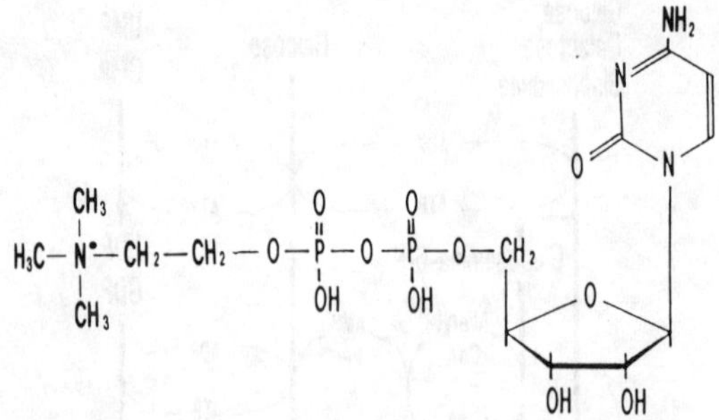

Fig. 18. Cytidine diphosphate choline (CDP-choline).

DISTRIBUTION OF SUGAR NUCLEOTIDES AND CDP-CHOLINE PRODUCERS IN YEASTS

—UDP-glucose (UDPG) producers: *Torulopsis candida, Candida parapsilosus*, baker's yeast, brewer's yeast and others.
—UDP-galactose (UDP-Gal) producers: *T. candida, T. sphaerica* and *C. intermedia.*
—GDP-mannose (GDP-Man) producers: *Hansenula jadinii, Debaryomyces subglobosus*, baker's yeast, brewer's yeast and others.
—UDP-*N*-acetylglucosamine (UDP-GlcNAc) producers: baker's yeast, *D. subglobosus, D. cantubellii.*
—CDP-choline (Fig. 18) producers: baker's yeast, brewer's yeast, *Saccharomyces rouxii, Brettanomyces petrophilum.*

Table 11 shows suitable yeast strains and substrates for sugar nucleotides and CDP-choline production, together with the fermentation yields.

THE FACTORS AFFECTING SUGAR NUCLEOTIDES AND CDP-CHOLINE PRODUCTION

Suitable conditions for the production of nucleotide derivatives are summarized as follows:

(1) *Reaction mixtures for UDPG.* UMP 20 μmol/ml, glucose 400 μmol/ml, potassium phosphate (P_i, pH 7·0) 180 μmol/ml, $MgSO_4$ 12 μmol/ml, dried baker's yeast 60 mg/ml; 28°C, 8–12 h with shaking.
(2) *Reaction mixtures for UDP-Gal.* UMP 20–60 μmol/ml, galactose 200–400 μmol/ml, P_i (pH 7·0) 200 μmol/ml, $MgSO_4$ 12 μmol/ml, dried *T. candida* (lactose grown cells) 100 mg/ml; 28°C, 8–24 h with shaking.
(3) *Reaction mixtures for GDP-Man.* GMP 20 μmol/ml, glucose 500–800 μmol/ml, P_i (pH 7·5–8·5) 360 μmol/ml, $MgSO_4$ 20 μmol/ml, dried baker's yeast 100 mg/ml; 28°C, 8–12 h with shaking.
(4) *Reaction mixtures for UDP-GlcNAc.* UMP 20 μmol/ml, glucose (fructose) 70 μmol/ml, P_i (pH 7·5–8·5) 170 μmol/ml, $MgSO_4$ 5 μmol/ml, glucosamine 20 μmol/ml, dried baker's yeast 100 mg/ml; 28°C, 8–12 h with shaking.
(5) *Reaction mixtures for CDP-choline* (Tochikura *et al.*, 1970; Shibukawa, 1976). CMP 20–30 μmol/ml, glucose 200–800 μmol/ml, Pi (pH 7·0)

Table 11
Sugar Nucleotide and CDP-Choline Production by Yeasts

Nucleotides produced	*Substrates*	*Yeast*[a]	*Yields of nucleotides (mg/ml)*
UDPG	UMP + glucose + P_i	Baker's yeast	6·7
		T. candida	10·0
UDP-Gal	UMP + galactose (lactose) + P_i	*T. candida*	51·5
UDP-GlcNAc	UMP + glucose (fructose) + glucosamine + P_i	Baker's yeast	4·1
		Debaryomyces sp.	6·0
GDP-Man	GMP + glucose + P_i	Baker's yeast	5·4
		Hansenula jadinii	10·8
CDP-Choline	CMP + (P)-choline + P_i	Brewer's yeast	2·7
		Hansenula jadinii	18·0

[a] Pressed baker's yeast was purchased from Oriental Yeast Co. Ltd. Some yeasts were grown on a medium containing 50 g of glucose or lactose, 5 g of peptone, 2 g of yeast extract, 2 g of KH_2PO_4, 2 g of $(NH_4)_2HPO_4$ and 1 g of $MgSO_4{\cdot}7H_2O$ in 1000 ml of tap water. Dried cells were prepared as follows. After incubation at 28°C for 48 h on a reciprocal shaker, the cells were harvested and washed three times with tap water. The washed cells were dried with an electric fan at room temperature for 12–24 h. The air-dried cells were further dried over P_2O_5 *in vacuo*. These cells, containing 4–8% water, were used throughout this work.

200 μmol/ml, P-choline 40–100 μmol/ml, $MgSO_4$ 10–15 μmol/ml, dried brewers' yeast 60–100 mg/ml; 28°C, 10 h with shaking.

4.3 Synthetic Processes by Coupling Bacterial Endergonic Reactions with the Waste ATP Energy of Yeast Fermentation

4.3.1 Production of 6-Phosphogluconate

Production of 6-phosphogluconate (Tochikura *et al.*, 1974*b*; Tochikura, 1978) consists of two reactions (Fig. 12), one being the phosphorylation of ADP to produce ATP through glucose fermentation by either bakers' or brewers' yeast, and the other being phosphorylation of gluconate to produce 6-phosphogluconate by the gluconokinase of *coli-aerogenes* bacteria grown in a gluconate–peptone medium. Although the ATP energy for the coupled reaction was supplied by glucose fermentation by yeast, the addition of large amounts of the yeast preparation to the gluconokinase brought about a remarkable decrease in the yield of 6-phosphogluconate. This means that the ratio of sugar-fermenting activity of the yeast to gluconokinase activity of the bacteria is a very important factor in 6-phosphogluconate production. Production of 6-phosphogluconate was demonstrated using various yeast preparations (cell-free extract, dried cells or solvent-treated cells) and the bacteria (cell-free preparations, dried cells or surface active agent-treated cells). Good yields of

6-phosphogluconate were obtained under the following conditions: (a) with P_i (pH 7·4) 800mM, adenosine 5mM in place of the catalyst ATP, glucose 800mM, sodium gluconate 300mM, cell-free preparation (crude gluconokinase) of *A. aerogenes* 9·6 mg/ml and dried bakers' yeast 100 mg/ml, 6-phosphogluconate was produced in a yield of 240mM (66 mg as free acid/ml) after 11 h at 37°C; (b) with P_i 600mM, ATP 5mM, glucose 600mM, gluconate 200 mM, *A. aerogenes* 20–40 mg dried cells/ml and dried yeast 50–100 mg/ml, 6-phosphogluconate was produced in yields of 100–160mM after 3–5 h at 37°C.

4.3.2 Production of Glutamine

Tachiki *et al.* (1981*a,b*) developed a method for glutamine production with bacterial glutamine synthetase in which ATP was regenerated with the energy of alcoholic fermentation (Fig. 12). This was an application of coupled fermentation with energy transfer that was established for the production of various substances. After a preliminary demonstration of the fermentation principle, they produced a high concentration of glutamine using glutamine synthetase from glutamate-producing bacteria (Tachiki *et al.*, 1983*b*; Tachiki & Tochikura, 1986). The organism produced a large amount of glutamine synthetase when glutamate was used as a nitrogen source, and a high yield of the enzyme can be obtained by a simple procedure (Tachiki *et al.*, 1981*c*). Cell-free extracts or toluol-treated cells were usable as a glutamine synthetase source.

As shown in Fig. 19, the amount (80mM) of glutamine produced was much more than that equivalent to ATP (2·5mM) added, indicating that the reaction generating glutamine synthesis was coupled with the ATP-regenerating system of sugar fermentation by yeast. The glutamine produced was relatively stable

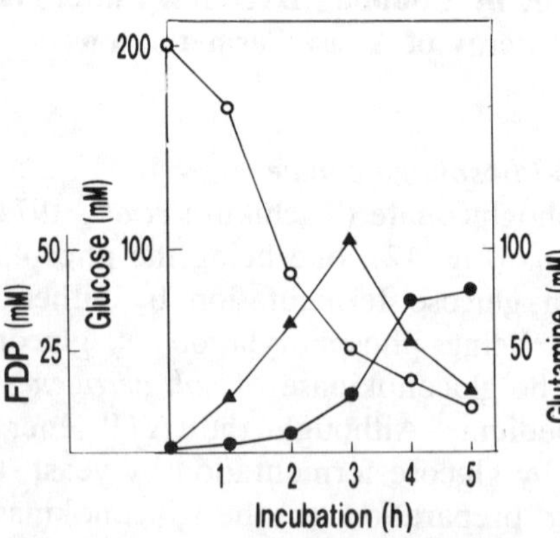

Fig. 19. Production of glutamine by combination of yeast cells and *Corynebacterium* glutamine synthetase. The reaction mixture contained 200mM glucose, 200mM potassium phosphate buffer (pH 7·0), 150mM glutamate, 150mM ammonium chloride, 15mM $MgCl_2$ 2·5mM ATP, 54 units/ml partially purified glutamine synthetase and 20 mg/ml dried yeast cells. Incubation was carried out at 28°C with shaking. Symbols: ●, glutamine; ○, glucose; ▲, FDP.

in the fermentation mixture after complete consumption of the energy source. The results also indicate that the first step in glutamine production was the accumulation of FDP. This means that sugar fermentation by the yeast in this system proceeded according to the Harden–Young equation and that the ATP energy of fermentation was reserved in the low-energy form of FDP in this step. The second step, after the consumption of glucose, is the coupling of the breakdown of FDP to ethanol with the reaction of glutamine synthetase through ATP–ADP. The activity ratio of sugar fermentation (energy generating system) and glutamine synthetase (energy utilizing system) is very important for the satisfactory production of glutamine with a limited energy source. In reaction mixtures containing more than the requisite amount of yeast cells, glutamine production decreased because glucose was rapidly consumed and most of the ATP energy supplied by sugar fermentation was lost without being utilized in glutamine synthetase.

The addition of a trace of ATP, the catalyst for energy transfer, was highly effective in increasing glutamine production with the higher concentrations of glutamate, ammonium chloride and P_i but less effective with larger amounts of yeast cells. When more than 10mM ATP was added to the mixture containing 200mM glucose, 200mM glutamate, 200mM ammonium chloride, 200mM potassium phosphate buffer and 20 mg/ml yeast cells, glutamine production was delayed due to the inhibition of glutamine synthetase by AMP and ADP formed in the early stage of the fermentation.

Consumption of sugar by yeast was slightly stimulated by less than 2mM Mg^{2+} but further addition had little effect (Table 12). On the contrary, metal ions greatly influenced glutamine production: the glutamine synthesis required

Table 12
Effect of Mg^{2+} and Co^{2+} on Glutamine Production by Coupled Fermentation[a]

Metal ion(s) added, with concentration (mM)			*Glutamine produced (mM)*		*Fermentation rate (glucose consumed mM/initial 60 min)*
			3 h	*10 h*	
(A) Mg^{2+}	0		2	7	80
	2		4	18	90
	15		10	30	105
	30		11	34	106
	50		10	35	106
(B) Mg^{2+}	0	Co^{2+} 5	16	55	85
	2	5	22	57	95
	15	5	25	62	100
	30	5	28	65	105
	50	5	29	73	105

[a] The mixture contained 300 mM glucose, 300 mM potassium phosphate buffer, 150 mM glutamate, ammonium chloride, 25 units/ml partially purified glutamine synthetase and 40 mg/ml yeast cells. The concentrations of Mg^{2+} and Co^{2+} are shown in the table.

higher concentrations of Mg^{2+} and was strikingly enhanced by the simultaneous addition of 1mM Co^{2+} or more. The purified glutamine synthetase required Mg^{2+} as a cofactor and its saturation curves at fixed levels of ATP were sigmoidal with the maximum activities at the Mg^{2+}–ATP ratio of 2:1 (Tachiki *et al.*, 1981*c*). Addition of a small amount of Co^{2+} to the mixture containing suboptimal levels of Mg^{2+} greatly stimulated the synthesizing reaction (Tachiki *et al.*, 1983*a*). Stimulation of glutamine production by Co^{2+} (Table 12) may be a reflection of this property and may be of importance for its industrial production.

The yield of glutamine based on fermentation energy in this process almost reached the theoretical value (90–100%) when FDP was substituted for glucose as the energy source (Table 13). Moreover, production of highly concentrated glutamine with high yield based on a substrate glutamate was also achieved in the FDP system. These findings suggested that the yield of glutamine in the glucose system might be improved under conditions where glucose is efficiently converted to FDP according to the Harden–Young effect of alcoholic fermentation.

Table 13
Glutamine Production by Coupled Fermentation

Energy source, consumed[a]	*Fermentation period*	*Glutamine formed*	*Yield based on fermentation energy (ATP)*[b]	*Yield based on substrate, glutamate*
(1) Glucose, 200 mM	3–4 h	120 mM, 18 g/liter	30%	100%
(2) Glucose, 200 mM	4–5 h	170 mM, 25 g/liter	43%	49%
(3) Glucose, 450 mM	24 h	280 mM, 41 g/liter	31%	70%
(4) FDP, 30 mM	4 h	120 mM, 18 g/liter	100%	80%
(5) FDP, 80 mM	9 h	280 mM, 41 g/liter	88%	70%
(6) FDP, 80 mM	12 h	315 mM, 46 g/liter	98%	79%

[a] Conditions:
(1) 120 mM glutamate, 120 mM NH_4Cl, 200 mM glucose, 200 mM phosphate buffer, 15 mM Mg^{2+}, 40 mg/ml yeast cells, 429 units/ml glutamine synthetase.
(2) 350 mM glutamate, 350 mM NH_4Cl, 200 mM glucose, 200 mM phosphate buffer, 15 mM Mg^{2+}, 40 mg/ml yeast cells, 429 units/ml glutamine synthetase.
(3) 400 mM glutamate, 450 mM NH_4Cl, 450 mM glucose, 160 mM phosphate buffer, 15 mM Mg^{2+}, 5 mM Co^{2+}, 3 mM adenosine, 50 mg/ml yeast cells, 566 units/ml glutamine synthetase.
(4) 150 mM glutamate, 200 mM NH_4Cl, 30 mM FDP, 100 mM phosphate buffer, 15 mM Mg^{2+}, 20 mg/ml yeast cells, 429 units/ml glutamine synthetase.
(5) 400 mM glutamate, 450 mM NH_4Cl, 80 mM FDP, 250 mM phosphate buffer, 15 mM Mg^{2+}, 15 mM Co^{2+}, 50 mg/ml yeast cells, 400 units/ml glutamine synthetase.
(6) 400 mM glutamate, 450 mM NH_4Cl, 80 mM FDP, 250 mM phosphate buffer, 15 mM Mg^{2+}, 15 mM Co^{2+}, 50 mg/ml yeast cells, 700 units/ml glutamine synthetase.

[b] Calculated as follows:

$$\frac{\text{glutamine formed (mM)}}{\text{glucose consumed (mM)} \times 2} \times 100\% \quad \text{or} \quad \frac{\text{glutamine formed (mM)}}{\text{FDP consumed (mM)} \times 4} \times 100\%$$

Table 14
Optimum Conditions for Glutamine Production with Glucose as Energy Source

Glucose, consumed[a]	*Fermentation period*	*Glutamine produced*	*Yield based on*	
			fermentation energy (ATP)	*substrate, glutamate*
(1) 80 mM	8 h	130 mM, 19 g/liter	81%	81%
(2) 180 mM	12 h	270 mM, 39 g/liter	75%	70%
(3) 300 mM	20 h	400 mM, 58 g/liter	67%	75%

[a] Conditions:
(1) 300 mM phosphate buffer, 30 mM Mg^{2+}, 5 mM Co^{2+}, 3 mM adenosine, 30 mg/ml yeast cells, 100 units/ml glutamine synthetase.
(2) 400 mM phosphate buffer, 30 mM Mg^{2+}, 5 mM Co^{2+}, 3 mM adenosine, 50 mg/ml yeast cells, 400 units/ml glutamine synthetase.
(3) 400 mM phosphate buffer, 30 mM Mg^{2+}, 5 mM Co^{2+}, 6 mM adenosine, 2 mM ATP, 40 mg/ml yeast cells, 400 units/ml glutamine synthetase.

Table 14 shows the typical results under the optimal conditions, using relatively higher concentrations of phosphate, and with the addition of Mg^{2+} and catalytic adenosine derivatives, determined by the potential FDP formation by yeast (75–110 g/liter of FDP was produced at a 60–80% yield based on glucose in other investigations).

Glutamine synthetase of glutamate-producing bacteria utilizes methylamine and ethylamine with 3 and 0·3%, respectively, of the reactivity of ammonia under the standard conditions and synthesizes the corresponding γ-glutamyl-derivatives (Tachiki *et al.*, 1986). When methylamine was added to the reaction mixture, 110mM (17 g/liter) γ-glutamyl methylamide was produced after 4 h cultivation.

5 CONCLUSION

There are many reports of attempts to design ways of recycling waste material. This is a very large subject, and it is not possible to deal here with all of the various arguments raised about waste recycling. Instead, the author's conclusions are summarized from the review of Birch *et al.* (1976).

(1) Reuse and recycling of wastes is made easier if the different types of wastes can be kept segregated. Segregation of the different types of municipal and industrial wastes after collection is expensive and is one of the disincentives to the greater utilization of the organic material they contain.
(2) Plaskett (1976) stated that organic wastes have to be treated on their merits according to composition. It is obvious that many organic wastes

are most unattractive and even offensive materials. Decaying plants, food processing wastes, animal excrement, and even human excrement are among the organic resources we cannot continue to ignore. Undoubtedly animal wastes contain fiber and significant amounts of nitrogenous compounds, especially protein of considerable value. Most people are untrained in chemistry and biology. Therefore, such unattractive wastes can be utilized in the community without this being drawn sharply to people's notice, and this way would appear to be the best solution.

(3) If hygienic standards can fully be met, there is going to be increasing economic pressure to utilize all unusable organic wastes. For example, partial replacement of animal feed by animal excrement is being explored and represents a valuable way of producing fodder from waste. Of course, suitable treatment to render such animal wastes free from pathogenic organisms and safe for use is essential.

REFERENCES

Aminoff, D. & Furukawa, K. (1970). *J. Biol. Chem.*, **245,** 1659.

Bahl, O. P. (1970). *J. Biol. Chem.*, **245,** 299.

Bahl, O. P. & Agrawal, K. M. L. (1969). *J. Biol. Chem.*, **244,** 2970.

Beckhorn, E. J. (1960). *Wallerstein Laboratories' Communication,* December, 201.

Birch, G. G., Parker, K. J. & Worgan, J. T. (1976). *Food from Waste.* Applied Science, London.

Endo, Y. & Kobata, A. (1976). *J. Biochem.*, **80,** 1.

Eriquez, L. A. & Pisano, M. A. (1979). *J. Bacteriol.*, **137,** 620.

Fan, J. Q., Kadowaki, S., Yamamoto, K., Kumagai, H. & Tochikura, T. (1988). *Agric. Biol. Chem.*, **52,** 1715.

Forster, C. F. & Jones, J. C. (1976). In *Food from Waste,* eds G. G. Birch, K. J. Parker & J. T. Worgan. Applied Science, London, p. 278.

Fujimoto, M., Fujiyama, K., Kuninaka, A. & Yoshino, H. (1974). *Agric. Biol. Chem.*, **38,** 2141.

Harden, A. (1923). *Alcoholic Fermentation.* Longmans, Green & Co., London.

Harden, A. & Young, W. J. (1906). *Proc. Roy. Soc. Lond., Ser. B,* **77,** 405.

Honma, S. & Mitsuoka, T. (1978). *Bifidobacteria* (in Japanese). Yaklut, Tokyo.

Huang, C. C. & Aminoff, D. (1972). *J. Biol. Chem.*, **247,** 6737.

Kawaguchi, K., Ogata, K. & Tochikura, T. (1970). *Agric. Biol. Chem.*, **34,** 908.

Kawaguchi, K., Kawai, H. & Tochikura, T. (1980). In *Methods in Carbohydrate Chemistry,* Vol. III, eds R. L. Whisler & J. N. BeMiller. Academic Press, New York, p. 261.

Kochibe, N. (1973). *J. Biochem.* (Tokyo), **74,** 1141.

Lehninger, A. L. (1971). *Bioenergetics,* 2nd ed. W. A. Benjamin, New York, p. 1.

Ohtakara, A., Yoshida, M., Murakami, M. & Izumi, T. (1981). *Agric. Biol. Chem.*, **45,** 239.

Plaskett, L. G. (1976). In *Food from Waste,* ed. G. G. Birch, K. J. Parker & J. T. Worgan. Applied Science, London, p. 264.

Reissig, J. L., Strominger, J. L. & Leloir, L. F. (1955). *J. Biol. Chem.*, **217,** 959.

Reyes, F. & Byrde, R. J. W. (1973). *Biochem. J.*, **131,** 381.

Sharon, N. & Lis, H. (1982). In *The Proteins,* Vol. 10, eds H. Neurath & R. L. Hill. Academic Press, New York, p. 1.

Shibukawa, M. (1976). In *Microbial Production of Nucleic Acid-related Substances,* eds K. Ogata, S. Kinoshita, T. Tsunoda & K. Aida. Kodansha, Tokyo, p. 289.

Sone, Y. & Misaki, A. (1978). *J. Biochem.* (Tokyo), **83,** 1135.

Tachiki, T. & Tochikura, T. (1986). In *Biotechnology of Amino Acid Production,* eds K. Aida, I. Chibata, K. Nakayama, K. Takinami & H. Yamada. Kodansha, Tokyo, & Elsevier, Amsterdam, 121.

Tachiki, T., Matsumoto, H., Yano, T. & Tochikura, T. (1981*a*). *Agric. Biol. Chem.,* **45,** 705.

Tachiki, T., Matsumoto, H., Yano, T. & Tochikura, T. (1981*b*). *Agric. Biol. Chem.,* **45,** 1237.

Tachiki, T., Wakisaka, S., Kumagai, H. & Tochikura, T. (1981*c*). *Agric. Biol. Chem.,* **45,** 1487.

Tachiki, T., Wakisaka, S., Suzuki, H., Kumagai, H. & Tochikura, T. (1983*a*). *Agric. Biol. Chem.,* **47,** 287.

Tachiki, T., Suzuki, H., Wakisaka, S., Yano, T. & Tochikura, T. (1983*b*). *J. Gen. Appl. Microbiol.,* **29,** 355.

Tachiki, T., Suzuki, H., Wakisaka, S., Yano, T. & Tochikura, T. (1986). *J. Gen. Appl. Microbiol.,* **32,** 545.

Thomas, D. B. & Winzler, R. J. (1969). *J. Biol. Chem.,* **244,** 5943.

Tochikura, T. (1976). In *Microbial Production of Nucleic Acid-related Substances,* eds K. Ogata, S. Kinoshita, T. Tsunoda & K. Aida. Kodansha, Tokyo, p. 264.

Tochikura, T. (1978). *Hakkokogaku Kaishi* (in Japanese), **56,** 508.

Tochikura, T. *et al.* (1967). *J. Ferment. Technol.,* **45,** 511.

Tochikura, T., Kimura, A., Kawai, H., Tachiki, T. & Gotan, T. (1970). *J. Ferment. Technol.,* **48,** 769.

Tochikura, T., Kawai, H., Kawaguchi, K., Mugibayashi, Y. & Ogata, K. (1972). *Proc. IV IFS, Fermentation Technology Today,* Kyoto, p. 463.

Tochikura, T., Kariya, Y., Yano, T., Tachiki, T. & Kimura, A. (1974*a*). *Fifth Int. Fermentation Symp.,* Berlin, p. 411.

Tochikura, T., Kariya, Y., Yano, T., Tachiki, T. & Kimura, A. (1974*b*). *Amino Acid and Nucleic Acid* (in Japanese), **29,** 59.

Tochikura, T., Tachiki, T., Yano, T., Moriguchi, M. & Kimura, A. (1978). In *Microbiology for Environment Cleaning,* ed. K. Arima. Published under the support of Ministry of Education, Grant No. 212204-1977, p. 612.

Tochikura, T. *et al.* (1981*a*). In *Annual Report of Environmental Science* (B72-S512-S513) (in Japanese). Published under the support of Ministry of Education, p. 26.

Tochikura, T., Tachiki, T., Yano, T. & Kadowaki, S. (1981*b*). *Yukigouseikagaku Kyokaishi* (in Japanese), **39,** 487.

Tochikura, T., Yano, T. & Hatanaka, M. (1986). In *Annual Report of Environmental Science* (B304-R30) (in Japanese). Published under the support of Ministry of Education, Vol. 4, p. 95.

Tsuji, Y. *et al.* (1990). *J. Biochem.,* **107,** 324.

Umemoto, J., Bhavanandan, V. P. & Davidson, E. A. (1977). *J. Biol. Chem.,* **252,** 8609.

Vincent, W. A. & Priestley, G. (1975). In *Handbook of Enzyme Biotechnology,* ed. A. Wiseman. Ellis Horwood, Chichester, UK, p. 27.

Watkins, W. M. (1959). *Biochem. J.,* **71,** 261.

Wismer-Pedersen, J. (1979). *Food Technol.,* **33,** 76.

Yamamoto, K., Lee, K. M., Kumagai, H. & Tochikura, T. (1985). *Agric. Biol. Chem.,* **49,** 611.

Yamamoto, K., Tsuji, Y., Matsushita, S., Kumagai, H. & Tochikura, T. (1986*a*). *Appl. Environ. Microbiol.,* **51,** 1019.

Yamamoto, K., Tsuji, Y., Kumagai, H. & Tochikura, T. (1986*b*). *Agric. Biol. Chem.,* **50,** 1689.

Yano, T. *et al.* (1985). *Agric. Biol. Chem.,* **49,** 3179.

Yano, T. *et al.* (1989). *Agric. Biol. Chem.,* **53,** 1935.

Chapter 12

LIGNOCELLULOSIC WASTES: BIOLOGICAL CONVERSION

D. S. CHAHAL

Institut Armand-Frappier, 531, boulevard des Prairies, C.P. 100, Succ. L-Q-R, Ville de Laval, Québec, Canada H7V 1B7

CONTENTS

1 INTRODUCTION

The lignocellulosic wastes are composed of mainly polysaccharides (cellulose and hemicelluloses) and lignin. After an appropriate pretreatment the polysaccharides can be biologically converted into single-cell protein (SCP) to be used as food for human consumption or as feed for animals, and also into fuel ethanol. Some processes for the production of SCP have been described. Lignocelluloses seem to be the best substrates for solid-state fermentation (SSF) for production of cellulase systems containing all the necessary enzymes (filter paper cellulase, β-glucosidase and xylanases) in the right proportion to hydrolyze cellulose into glucose, or polysaccharides (cellulose and hemicelluloses) of lignocelluloses into monomeric sugars for their further fermentation into various products. In another case of SSF high yields of gibberellic acid are achieved. The lignocelluloses can also be used as inert substrates when impregnated with suitable nutrients for production of spores of various microorganisms in SSF to be used as bioinsecticides and bioherbicides. Although voluminous literature on the degradation of lignin is available, there is no process at present by which the enormous amount of lignin, obtained from the various processes of biconversion of lignocelluloses, could be utilized for the synthesis of new products.

'Lignocelluloses' are usually defined as any of the several closely related substances composed of plant (wood) cell walls where cellulose is intimately associated with lignin. In addition to these compounds lignocelluloses also contain other polysaccharides commonly called 'hemicelluloses'. On the other hand it would be difficult to define 'lignocellulosic wastes' because every so-called waste has some uses. However, in the present context they could be defined broadly as comprising that portion of the entire Plant Kingdom which is not being properly utilized for the welfare of human beings. In this chapter, therefore, the term 'lignocellulosic wastes' would include mainly the crop residues, and wood and forestry wastes.

2 COMPOSITION AND STRUCTURE OF LIGNOCELLULOSES

The general chemical composition of lignocelluloses is given in Table 1. The data presented in this table indicate that crop residues have less cellulose and lignin than do wood residues. They also have lower density and thus are more voluminous than wood residues. On the other hand, the contents of hemicelluloses are generally higher in most crop residues than in wood residues. Some crop residues, such as rice straw and sugarcane bagasse, have a high content of silica. Very little is known about the role of silica in the bioconversion of lignocellulose into various products. Some research on this aspect has recently been taken up at our laboratory.

All the lignocelluloses are composed of plant (wood) cells with a thin primary wall that surrounds the relatively thick secondary wall (Fig. 1). The

Table 1
Chemical Composition of Lignocelluloses (% Dry Weight)

Constituents	*Crop residues*[a]	*Wood residues*[b]
Cellulose	30–45	45–56
Hemicelluloses	16–27	10–25
Lignin	3–13	18–30
Protein (N × 6·25)	3·6–7·2	NA[c]

[a] Source: Sloneker, 1976. [b] Source: Wenzel, 1970. [c] NA = Not available.

primary wall is only 0·1–0·2 μm in thickness and contains a randomly and loosely organized network of cellulose microfibrils. The outer layer of the secondary wall, S_1, has a crossed fibrillar structure. In the S_2-layer, the main portion of the secondary wall (1–5 μm thick), the microfibrils are oriented almost parallel to the lumen axis. In the thin S_3-layer the microfibrils form a flat helix. The innermost portion of the cell wall consists of the so-called warty layer, probably formed from protoplasmic debris. The central empty portion, formed after the disintegration of protoplasm at the time of aging, is called the lumen. The primary walls of two adjacent cells are cemented together with pectic compounds and lignin. The portion between the two adjacent cells is called the middle lamella.

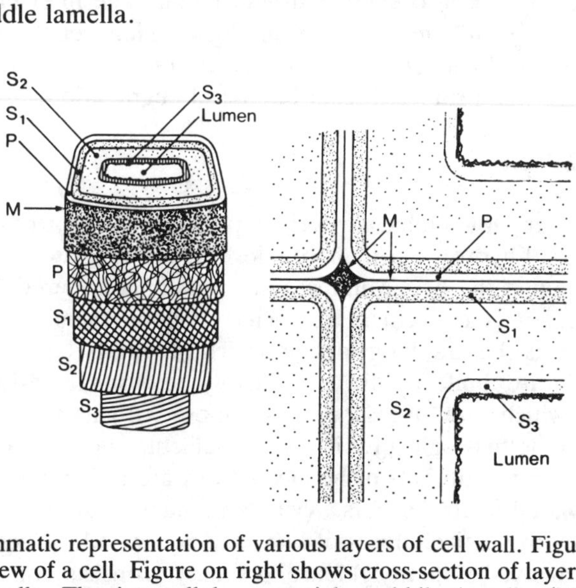

Fig. 1. Diagrammatic representation of various layers of cell wall. Figure on left shows front and top view of a cell. Figure on right shows cross-section of layers of cell walls of four adjacent cells. The intercellular materials—middle lamella (M) and adjacent primary walls (P) of contiguous cells—comprise the compound middle lamella. The secondary walls are composed of outer (S_1), middle (S_2) and inner (S_3) layers. Each layer has a different orientation of microfibrils. The central empty portion is called the lumen.

2.1 Cellulose

The cellulose in each layer of cell wall occurs as long slender bundles composed of long chains of β-D-glucopyranose residues linked by (1→4)-glycosidic bonds (cellulose molecules) called elementary fibrils with a diameter of 35 Å. The length of the cellulose molecules in an elementary fibril varies from less than 15 β-D-glucopyranose residues in gamma-cellulose to as many as 10 000–14 000 per molecule in alpha-cellulose. The length of the cellulose molecule is measured as the degree of polymerization (DP) i.e. the number of β-D-glucopyranoside residues. Within each elementary fibril the cellulose molecules are bound laterally, with adjacent molecules running in opposite directions, by hydrogen bonds. They are associated in various degrees of parallelism—regions that contain highly oriented molecules are called 'crystalline cellulose' and those with less oriented molecules are called 'paracrystalline' or 'amorphous cellulose' (Hess *et al.*, 1954). A number of elementary fibrils form a microfibril 100 Å wide and 50 Å thick; the highly oriented central core of the microfibril is surrounded by a paracrystalline sheath of cellulose molecules and chains of non-cellulosic molecules (Preston & Cronshaw, 1958).

Several microfibrils when joined laterally form a macrofibril. The microfibrils or macrofibrils are differently oriented in each layer of the secondary wall to give structural strength to the cell wall. Microfibrils are covered with hemicelluloses and are also encrusted with lignin. The interspaces in the cell wall are also filled with hemicelluloses and lignin. Hemicelluloses are linked by covalent bonds with lignin (Fengel & Wegener, 1983). Therefore, the cellulose in nature is very well protected from biological degradation.

2.2 Hemicelluloses

Hemicelluloses are low-molecular-weight polysaccharides, most of which are more readily acid-hydrolyzed than cellulose. They are also easily hydrolyzed by alkali. Although the hemicelluloses are usually considered to be structural polysaccharides, it is convenient to include among them a few other plant polymers, such as the arabinogalactans which obviously have other functions. Hemicelluloses are built up from relatively few sugar residues, the most common of which are D-xylose, D-mannose, D-galactose, D-glucose, L-arabinose, 4-O-methyl-D-glucuronic acid, D-galacturonic acid, and D-glucuronic acid. Other rarer constituents of hemicelluloses are L-rhamnose, L-fucose, and various methylated neutral sugars. Xylan and mannan are the most prominent polysaccharides of hemicelluloses of hard and soft wood, respectively.

Xylan (O-acetyl-4-D-methylglucurono-xylan) consists of approximately 200 β-D-xylopyranose residues, linked together in (1→4)-glycosidic bonds. Some of the xylose units carry a single, terminal side chain consisting of a 4-O-methyl-α-D-glucuronic acid residue, attached directly to C-2 of the xylose. Seven out of ten xylose residues contain an O-acetyl group at C-2 or more frequently at C-3 (Timell, 1967).

Mannan (galactoglucomannan) consists of (1→4)-linked β-D-glucopyranose and β-D-mannopyranose residues distributed at random. Some of the hexose units carry a terminal residue of a D-galactopyranose attached to C-6. It is probable that all the galactoglucomannans are acetylated in their native state. The acetyl groups are attached to the mannose residues (Timell, 1967).

2.3 Lignin

Lignin is concentrated mainly in the spaces between the cell walls of adjacent cells (middle lamella) and in the S_2-layer of the cell wall where it is deposited during lignification of the plant (wood) tissues, although some lignin is also deposited in other layers.

Lignin is formed by dehydrogenative polymerization of *p*-hydroxycinnamyl alcohols. Guaiacyl lignin, which occurs in conifers, is mainly a dehydrogenation polymer of coniferyl alcohol. Guaiacyl–syringyl lignin, which occurs in angiosperms, is composed of a mixed dehydrogenation polymer of coniferyl and synapyl alcohols. Guaiacyl–syringyl–*p*–hydroxyphenyl lignin, which is found in grasses, is composed of a mixed dehydrogenation polymer of coniferyl, sinapyl, and *p*-coumaryl alcohols (Higuchi, 1980).

Lignin in the cell wall not only encrusts the cellulose microfibrils in a sheath-like manner, but is also bonded physically and chemically to hemicelluloses (Higuchi, 1971; Fengel & Wegener, 1983). Physically, lignin forms a barrier against the penetration of cellulases and hemicellulases (Kirk & Haskin, 1973).

3 PRETREATMENT OF LIGNOCELLULOSES

Crystallinity of cellulose and encrustation of cellulose fibrils with lignin make the lignocelluloses recalcitrant. Unless lignin is depolymerized, solubilized, or removed, cellulose and hemicelluloses cannot be easily hydrolyzed by cellulases and hemicellulases, respectively, for their bioconversion into various products. The crystallinity of cellulose is another hindrance in the bioconversion of cellulose. Thus, pretreatment of lignocelluloses will be necessary for their efficient conversion into various products.

To discuss the various pretreatments of lignocellulose is beyond the scope of the present chapter. However, some knowledge of them is necessary for a better understanding of the bioconversion processes. The reader should refer to Tarkow and Feist (1969), Millett *et al.* (1976) and Chahal (1991) for detailed information on pretreatment of lignocelluloses.

4 BIOCONVERSION OF LIGNOCELLULOSES INTO SINGLE-CELL PROTEIN

One of the earliest processes for single-cell protein (SCP) production from lignocelluloses was the production of yeast from wood hydrolyzate. During

World Wars I and II, Germany developed a process to produce food yeast from wood sugars obtained by acid hydrolysis and sulfite waste liquor. Later, commercial plants were set up at various places in the USA and other parts of the world (Harris & Belinger, 1946; Gilbert *et al.*, 1952; Underkofler & Hickey, 1954). Similarly, in the USSR the 'Hydrolysis Industry' to produce sugars from wood was created in the mid-1930s and continues to the present time. The sugar obtained from wood hydrolysis is used for SCP production (Laskin, 1977).

Keeping in view that lignocelluloses are constantly being replenished through photosynthesis, the attention of researchers all over the world has now been diverted to the utilization of this carbon source for SCP production. A number of processes are now available for the bioconversion of lignocelluloses into SCP. These may be grouped together into the following major processes.

4.1 Indirect Conversion of Lignocelluloses

The lignocelluloses are pretreated with various physicochemical methods to obtain cellulose, to be used in the following ways.

4.1.1 *Chemical Hydrolysis of Cellulose to Glucose and Growth of Microorganisms on the Hydrolyzate*

In this case wood is hydrolyzed into glucose with acid by the Bergius or Scholler process, and *Candida utilis,* a yeast, is grown on the glucose thus obtained to produce SCP. The Scholler process was operated on a large scale in Germany during World Wars I and II and production reached a level of 15 000 tons per year (Laskin, 1977). Microbial protein production from wood hydrolyzate could not survive because of its high cost of production. The Scholler process for hydrolyzing cellulose was upgraded at the US Forest Products Laboratories and a few wood sugar plants were built in the United States (Callihan & Clemmer, 1979). Meller (1969) did an in-depth economic analysis on hydrolysis and growth of yeast on the resultant sugar. He determined a production cost of the order of 30 US cents/kg of yeast produced.

Han *et al.* (1976) developed a process where *Aurobasidium pullulans,* a yeast, is grown on the acid hydrolyzate of ryegrass. A yield of 1·5 g cells/liter is obtained from a medium containing 6 g sugar/liter supplemented with 1·25 g yeast extract. The product contained 42·6% protein. Thus about 0·64 g protein/liter was obtained by adding double that quantity (1·25 g) of yeast extract, which is much more expensive than the product (SCP) itself.

4.1.2 *Enzymatic Hydrolysis of Cellulose to Glucose and Growth of Microorganisms on Glucose Obtained*

INDIAN INSTITUTE OF TECHNOLOGY PROCESS

This process consists of pulverizing the cellulosic material to a size of less than 25 μm and heating it to 200°C in an oxidizing atmosphere followed by enzymatic saccharification at pH 4·8 and 50°C (Das & Ghose, 1973). Residual enzyme and cellulose are recycled. About 7·3% sugar syrup is obtained by this method for SCP production with yeasts. Das and Ghose (1973) have projected production costs of 34·4, 17·1 and 14·1 US cents/kg of glucose from plant sizes of 10, 100 and 250 tons, respectively, per day. If the minimum cost of glucose production is taken as 14·1 US cents/kg, even then the cost of SCP production from such glucose syrup would still be higher than that of soybean meal.

UNIVERSITY OF CALIFORNIA PROCESS

In this process (Wilke *et al.*, 1976) cellulase enzyme is produced with *Trichoderma reesei* QM 9414 on hammered mill newspaper to hydrolyze the same substrate to produce sugar syrup. About 50% hydrolysis of the substrate has been reported by that cellulase enzyme. The sugar syrup obtained is concentrated before it is fermented into SCP because the hydrolyzate is too low in sugar content for economic fermentation. In this process, 59·4 tons/day of torula yeast and 81·5 tons/day of ethanol from 885 tons of pretreated newspaper were reported by Wilke *et al.* (1976).

The high cost of enzyme production, low yields (50%) of hydrolysis, and low concentration of sugar syrup which is to be concentrated before fermentation for SCP or ethanol, are the major drawbacks to this process becoming an economic enterprise.

KYOTO UNIVERSITY PROCESS

This process, developed by Tanaka and Matsuno (1985), uses hydrolyzate obtained by enzymatic hydrolysis of pretreated lignocelluloses. The hydrolyzate is a mixture of glucose, xylose and disaccharides (cellobiose and xylobiose). The hydrolyzate is first fermented in tank D_1 with *Saccharomyces* species to utilize glucose from the hydrolyzate for production of SCP (Fig. 2). The SCP thus produced is separated through a membrane and the residual hydrolyzate containing xylose and disaccharides is transferred into another tank, D_2. There it is fermented with *Candida* species for conversion of xylose into SCP. The SCP biomass is separated through a membrane and the residual hydrolyzate containing disaccharides (cellobiose and xylobiose) is supplied back to hydrolyzing tank B for further hydrolysis to monomer sugars. Part of these disaccharides is supplied to fermentor A for production of cellulases and xylanases. The immobilized β-glucosidase and β-xylosidase, column C, may be attached between the hydrolysis tank B and the fermentor D_1 to hydrolyze cellobiose and xylobiose to glucose and xylose, respectively.

The lignin after enzymatic hydrolysis must be removed by filtration as shown in Fig. 2. This process suffers from all the drawbacks pointed out earlier. However, it would also be very difficult to synchronize the production of enzymes, the rate of hydrolysis of lignocelluloses, and the assimilation of glucose and xylose by the respective yeasts in their respective fermentation tanks.

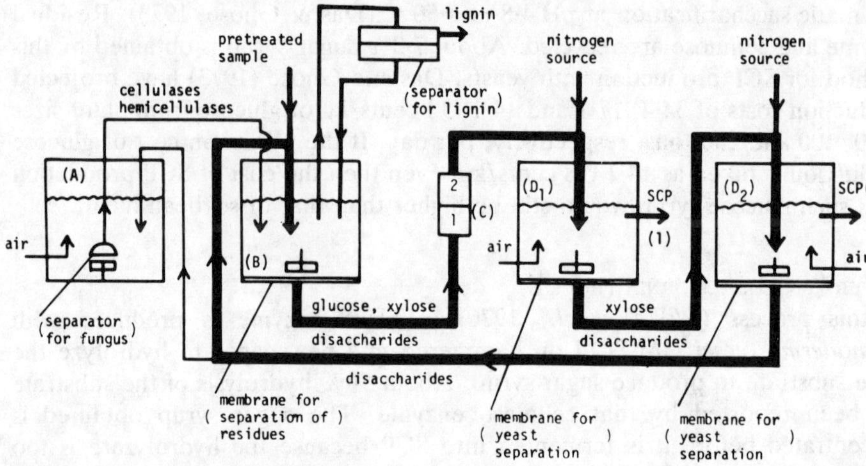

Fig. 2. Ideal scheme of SCP production process: (A) fermentor for cellulase and hemicellulase production; (B) hydrolysis tank; (C) columns (immobilized enzymes: 1, β-glucosidase; 2, β-xylosidase); (D_1) and (D_2) fermentors for SCP production. The bolder line of the liquid flow represents the larger flow rate. (Reproduced by permission of Butterworth Scientific Ltd from Tanaka & Matsuno, 1985.)

4.1.3 Mixed Culture

Cellulose hydrolysis by one intact organism and conversion of hydrolysis product into SCP by another intact organism is a very similar idea to that of the Symba process where *Endomycopsis fibuligera* produces amylases to hydrolyze starch and *Candida utilis* is grown for SCP production on the hydrolysis product, glucose (Wiken, 1972). However, no such process, where hydrolysis of cellulose or lignocellulose is achieved by one intact microorganism and production of SCP on their hydrolyzate by another intact microorganism, is available in the literature. Nevertheless, Peitersen (1975) tried to grow mixed cultures of *Trichoderma viride* and *Saccharomyces cerevisiae* to convert cellulose into SCP without any improvement in the yield as compared to the use of the cellulolytic microorganism *T. viride* alone.

4.2 Direct Conversion of Lignocelluloses

In this case lignocelluloses are also pretreated with various physicochemical methods, and hydrolysis of the substrate into utilizable compounds and their conversion into SCP is performed by the same organism.

The first attempt at conversion of lignocelluloses into animal feed with *Aspergillus fumigatus* was by Pringsheim and Lichtenstein in 1920 (Litchfield, 1968). During the last two decades a number of reports have appeared on the conversion of lignocelluloses into animal feed rich in protein (SCP) with various cellulolytic microorganisms (see Chahal & Moo-Young, 1981; Rolz, 1984). The Louisiana State University Process is discussed in detail here, although in this process isolated cellulose is used as the substrate instead of lignocelluloses.

LOUISIANA STATE UNIVERSITY PROCESS

In this process a bacterium, *Cellulomonas* sp., is used (Srinivasan & Han, 1969; Dunlap & Callihan, 1973). This organism has low β-glucosidase activity, thus there is a need for another organism, *Alcaligenes faecalis,* to remove the accumulated cellobiose which inhibits cellulase activity. Since it is endo-gluconase positive and lacks exo-glucocellobiose hydrolase, it needs drastic pretreatment with high concentration of alkali (30 g NaOH/100 g substrate at 30 psig for 4 h) to remove lignin and to swell the cellulose. About 68 g of cellulose is obtained by treating 100 g of bagasse with the above treatment. On fermentation with both the bacteria, a bacterial cell mass of 18–20 g was obtained and only 40 g of cellulose was utilized. Moreover, about 32 g of hemicelluloses and lignin were obtained as the effluents from the pretreatment of 100 g of bagasse. This effluent creates a big disposal problem as it has a very high biochemical oxygen demand (BOD).

Callihan and Clemmer (1979) have described the latest version of the above process where a yield of 0·12 g protein/g bagasse was reported. Because of the disposal problem of the effluents and the low yield of the product (SCP), the process seems to be unattractive as a commercial enterprise.

In another report, Han and Callihan (1974) obtained high yields of protein (1·76 g protein/liter) by growing a mixed culture of *Cellulomonas* sp. and *A. faecalis* on 1–5% slurry of alkali-treated rice straw. Here again the disposal problem of the effluent (hemicelluloses and lignin) has also not been discussed by them.

4.3 Search for New Cellulolytic Fungi

All the above processes require separation of cellulose from lignocelluloses and use of acid or enzymes for hydrolysis of cellulose into glucose. Hydrolysis of cellulose into glucose is an additional cost for SCP production from lignocelluloses. Moreover, hemicelluloses, the second major fraction of polysaccharides of lignocelluloses, were not being utilized in these processes. Thus they created disposal and pollution problems. Therefore the search continued in our laboratory to find cellulolytic fungi which could use cellulose as well as hemicelluloses for their conversion into SCP. This work, carried out by Chahal and various coworkers, was successful in finding such a fungus, which was named *Chaetomium cellulolyticum* Chahal et Hawks. (Chahal & Hawksworth, 1976). This fungus has proved to be the best for bioconversion of pure cellulose, hemicelluloses, mixtures of cellulose and hemicelluloses, and lignocelluloses without fractionation, both alone and also with admixtures of manures (Chahal *et al.*, 1977, 1987; Moo-Young *et al.*, 1977, 1980; Chahal & Wang, 1978; Chahal & Moo-Young, 1981; Chahal & Ishaque, 1986, 1988).

Our continuous research in this field led us to the development of a process which became popular as 'The Waterloo Process for SCP Production from Waste Biomass' (Moo-Young *et al.*, 1979). Unfortunately if was found that the fungus, *Chaetomium cellulolyticum,* would produce some toxins under certain cultural conditions (Sekita *et al.*, 1981), therefore commercialization of the Waterloo process became doubtful. Nevertheless, Chahal at the Institut Armand-Frappier continued to work on this project and developed a new mutant of *C. cellulolyticum* free of the toxin problem; he also developed a process free of the pollution problems associated with the other processes. This is described in the following sections.

4.4 The Institut Armand-Frappier Process

In the Institut Armand-Frappier (IAF) process *Chaetomium cellulolyticum* asporogenous mutant, nontoxin producer, and *Pleurotus sajor-caju,* an edible mushroom, are used to convert lignocelluloses into SCP. The other fungi used in this process are species of *Aspergillus* and *Penicillium.* These fungi have the ability to produce sufficient amounts of cellulases and hemicellulases to convert cellulose and hemicelluloses into SCP. Therefore, there is no need for prehydrolysis of lignocelluloses with acid or enzymes as is required in all the other processes described before.

In the IAF process lignocelluloses are fractionated into cellulose, hemicelluloses and lignin. It is recommended that the cellulose fraction should be used to produce paper or other valuable products rather than for conversion into SCP. However, the cellulose fraction thus obtained is very suitable for bioconversion into SCP of high quality to be used as food for human consumption. It could also be used to produce cellulase enzyme for hydrolysis of cellulose into pure glucose syrup to be used for fermentation into ethanol fuel or other pharmaceuticals. Lignocelluloses without fractionation and the hemicellulose fraction are used to produce SCP either as food for humans or as feed for animals depending on the purification of the SCP product after harvesting. Mixtures of hemicellulose and lignin can be fermented together, the hemicellulose fraction being converted into SCP and the lignin being exposed to the action of various enzymes produced by the fungi and degraded into lignin compounds of low molecular weight. The lignin obtained in this process could be used for production of adhesives and various chemicals, originally synthesized from hydrocarbons. In this process all the components are utilized for production of various products; therefore, the IAF process is also called 'An Integrated Process for Production of Food, Feed and Fuel and Biomass'. As nothing is left unutilized from the substrate, therefore, there are no pollution and disposal problems from the effluents from this integrated process.

The IAF Process is shown in Fig. 3. The biomass (lignocellulose) (1) is pretreated (2) with various methods as mentioned earlier and is fractionated into cellulose (3), lignin (4) and hemicelluloses (5). The pretreated biomass (2), without fractionation, is fermented with fungi (7) for production of SCP

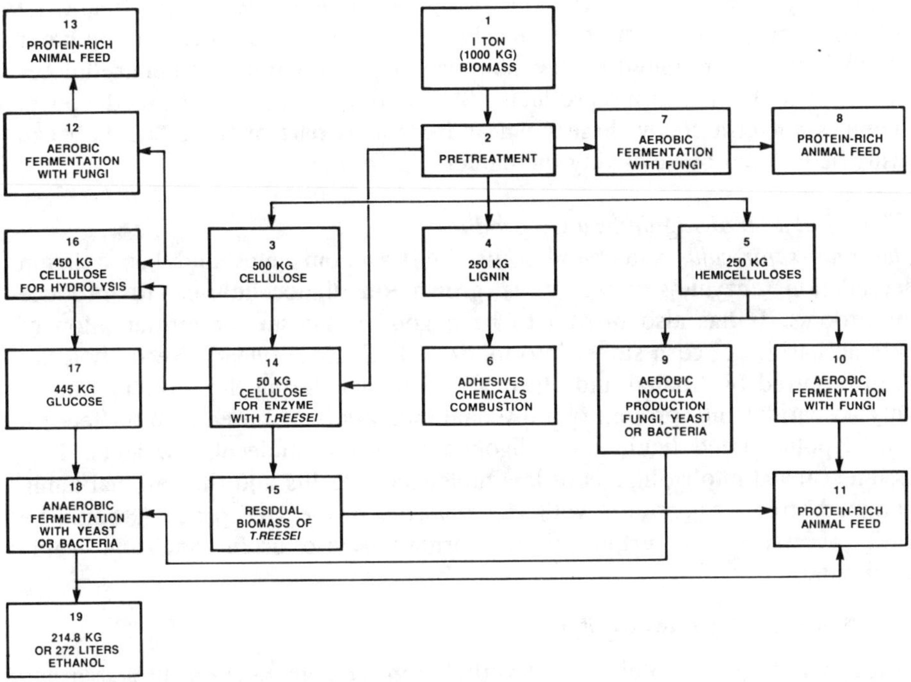

Fig. 3. The IAF Process: an integrated process for production of food, feed and fuel (ethanol) from biomass.

as animal feed (8). During the pretreatment some of the hemicelluloses and lignin are solubilized. The soluble portion is not removed as is done in other processes but is kept along with the cellulose in the fermentation medium. Cellulose and hemicelluloses are utilized by the organisms and lignin is left in the fermentation medium which can easily be isolated by precipitation after acidifying the spent fermentation liquor.

The fractionated cellulose (3) is used for production of paper or other cellulosic materials but it can also be fermented with fungi (12) for production of high quality SCP to be used for human consumption or as animal feed (13). The fractionated hemicelluloses (5) are utilized for production of inocula (9) of the fungi used for fermentation at steps 7 and 12, and for fermentation of surplus hemicelluloses at step 10 for production of SCP (11). The fractionated hemicelluloses (5) are also used for production of inocula of yeasts (*Saccharomyces cerevisiae, Kluyveromyces* spp.) at step 9, for the fermentation of glucose (17) at step 18 into ethanol (19) or other pharmaceuticals. About a tenth of the cellulose (3) is used for enzyme production (14) with *Trichoderma reesei* QMY-1 for hydrolysis of cellulose (16) received from step 3. The inoculum of *T. reesei* QMY-1 is also produced on hemicelluloses at step 9. The residual fungal biomass obtained at steps 15 and 18 is also used as animal feed (SCP).

It is easy to obtain two fractions of lignocelluloses after pretreatment: (i) cellulose, and (ii) a mixture of hemicelluloses and lignin. In this case the latter fraction can be fermented in the same way as the fractionated hemicelluloses are fermented for various products (SCP and inocula) as explained above. Lignin is not utilized by these fungi and becomes part of the effluent. It can easily be isolated by acidifying the effluent.

4.4.1 Isolation of Lignin as a Co-product

Pleurotus sajor-caju is a member of the Basidiomycetes and has a lignin degradation enzyme-system. It is grown on lignocelluloses to produce mushrooms. It has also proved to be a good organism for fermentation of polysaccharides of corn stover into food/feed (Chahal, 1989*a*). Recently it has been reported by Chahal and Hachey (1990) that during the fermentation of polysaccharides into protein-rich mycelial biomass (to be used as food/feed) it also depolymerized lignin into oligolignols of low molecular weights. It is assumed that the oligolignols of low molecular weights and the residual lignin obtained after fermentation with *P. sajor-caju* in the IAF process, would be easily accessible for further biotransformation into useful and high value compounds.

4.5 Nutritional Values of SCP

Production of mycoprotein (SCP) with *Fusarium graminearum* on starch and sugars, a process developed by Rank Hovis McDougall of the UK, has recently been commercialized for human consumption of SCP (Newark, 1980). Similarly Tate and Lyle, also of the UK, have developed a process to convert carob sugars, spoiled papaya, cassava, sulfite waste liquor and wastes from processing olives, palm oil, potatoes, dates and citrus into SCP (Righelato *et al.*, 1976). These two processes have been mentioned here to indicate that this type of SCP can also be produced from lignocelluloses as already explained.

The microbial food produced on pure carbohydrates (starches and sugars) contains more protein and is free from impurities than that produced on agricultural residues or other lignocelluloses. The low protein content in microbial biomass produced from lignocelluloses is mainly due to the presence of unutilized (residual) cellulose and lignin in the final product. The microbial biomass thus produced on such substrates would be suitable as animal feed. However, the microbial biomass produced from fractionated cellulose and from fractionated hemicelluloses (Fig. 3) in the IAF process would be as good as that produced on pure carbohydrates (starches and sugars). The fungal biomass produced on lignocelluloses without fractionation contains 22–45% protein and that produced on fractionated cellulose or hemicelluloses contains 35–50% protein (Chahal & Moo-Young, 1981; Chahal, 1989*a*).

The amino acid composition of various microorganisms used to produce food/feed from agricultural wastes and other carbohydrates is compared with that of alfalfa, a common animal fodder, soybean, a common protein source, and the FAO reference in Table 2. The comparison of all these organisms

Table 2
Essential Amino Acid Composition (% Total True Protein) of Proteins in *Chaetomium cellulolyticum*, Other Cellulolytic Organisms of SCP Interest, Alfalfa, Soya, and FAO Reference Protein

Amino acid	*C. celluloyticum*[a]	*C. cellulolyticum asporogenous mutant*[b]	*Fusarium graminearum*	*T. viride*[c]	*Cellulomonas*[d]	*Alfalfa*[e]	*Soybean*[f]	*FAO reference*
Threonine	6·14	4·78	5·1[g]	4·9	4·7	5·12	4·0	2·8
Valine	5·76	5·88	7·2[g]	4·4	6·79	6·70	5·0	4·2
Cystine	0·31	0·85	0·8[a]	1·45(?)	0·41	1·40	1·4	2·0
Methionine	2·33	2·27	2·1[a]	1·35	1·69	1·96	1·4	2·2
Isoleucine	4·70	5·07	4·3[g]	3·5	4·12	5·54	5·4	4·2
Leucine	7·54	7·93	6·5[g]	5·8	8·66	8·43	7·7	4·8
Tyrosine	3·26	3·68	4·1[g]	3·3	2·41	3·72	2·7	2·8
Phenylalanine	3·77	4·80	4·4[g]	3·7	3·69	5·75	5·1	2·8
Lysine	6·80	7·12	7·5[g]	4·4	8·0	6·70	6·5	4·2
Tryptophan	NA[i]	NA	NA	NA	NA	NA	NA	NA

[a] Moo-Young *et al.*, 1977. [b] Chahal, 1990. [c] Peiterson, 1975. [d] Han & Anderson, 1975; Han & Callihan, 1974. [e] Livingston *et al.*, 1971. [f] Shacklady, 1975. [g] Anderson *et al.*, 1975. Duthie, 1975. [i] NA = Not available.

meets the FAO reference requirement except for sulfur-containing amino acids. Anderson *et al.* (1975) have also reported similar findings by comparing the amino acid composition of various filamentous fungi grown on soluble sugars or starch. The deficiency of sulfur amino acids can be easily met by supplementing the microbial protein with methionine. The amino acid composition of *C. cellulolyticum* asporogenous mutant is as good as that of the parent strain, *T. viride, Cellulomonas* sp. and *F. graminearum*.

Feeding trials with various cellulolytic organisms (including *Chaetomium cellulolyticum*) used to produce microbial food/feed are still in their infancy. However, most of the feeding trials have indicated that up to 20–50% of the total protein requirement can be replaced with the protein from these organisms without any pathological problems (Duthie, 1975; Han & Anderson, 1975; Peitersen, 1975; Srivastava *et al.*, 1980; Chavez *et al.*, 1988). Some feeding trials arranged by Chavez *et al.* (1988) have indicated that the diet protein for rats, chicken and piglets can be replaced with that of *C. cellulolyticum* (parent strain) supplemented with methionine.

5 SOLID-STATE FERMENTATION

Solid-state fermentation (SSF) is simply defined as a process whereby an insoluble substrate is fermented with sufficient moisture but without free water. In liquid-state fermentation (LSF), on the other hand, the substrate is solubilized or suspended as fine particles in a large volume of water (Chahal, 1983, 1985*a*). Mudgett (1986) defined SSF as being distinguished from submerged cultures by the fact that microbial growth and product formation occur at or near the surfaces of solid materials with low moisture contents. However, this is not so because it was observed earlier (Chahal *et al.*, 1983) under an optical microscope that in SSF the organism (*Chaetomium cellulolyticum*) not only grows on the surface of the substrate (corn stover) but actually penetrates deep into the substrate and even into the cell lumen. Recently the hyphae of *Trichoderma reesei* QMY-1 were seen to have penetrated into intercellular spaces (middle lamella) as well as intracellular spaces (cell lumens) of wheat straw particles during SSF (Chahal, 1989*b*). This means that the organism in SSF not only grows at or near the surface of the substrate but actually penetrates deep into the intercellular and intracellular spaces of the substrate, showing a really close contact or association with the substrate.

Although SSF has many advantages over LSF, it has its own inherent problems (Hesseltine, 1972; Chahal, 1983; Mudgett, 1986). However, because of its many advantages over LSF it has been used for upgrading the protein values of lignocelluloses, i.e. agricultural residues (Han & Anderson, 1975; Yu *et al.*, 1976; Zadrazil, 1977; Chahal *et al.*, 1981), and for the production of aflatoxins (Hesseltine, 1972), gibberellic acid (Kumar & Lonsane, 1987), rennet (Brisk & Zuckermann, 1971; Thakur *et al.*, 1990), flavors (Seitz, 1989), antibiotics (Jermini & Demain, 1989), spores of mycoherbicides (Silman *et al.*, 1989), etc.

Mushroom production in SSF is one of the oldest microbiological processes known to man. It is not possible to discuss this aspect here as it is beyond the scope of this chapter. A few SSF processes where the lignocelluloses are upgraded for their protein and digestibility values and for the production of other useful industrial compounds, are discussed in the following sections.

5.1 Single-cell Protein Production in SSF

The use of SSF for the bioconversion of lignocelluloses into SCP as animal feed, and for increasing the in-vitro digestibility of agricultural residues for animal feed, is becoming popular especially in developing countries because of its low technology, low cost of dewatering of the final product and abundance of agricultrual residues.

Zadrazil (1977) reported that during SSF of wheat straw for 120 days with *Stropharia rugosannulata* or *Pleurotus cornucopiae,* in-vitro digestibility increased by up to 60–70%. Similarly Detroy *et al.* (1980) reported that growing *P. ostreatus* on wheat straw in SSF for 50 days increased its hydrolysis to 72% with cellulases. However, Tsang *et al.* (1987) showed that it was not practical simultaneously to produce *Pleurotus* mushrooms and a highly delignified residual straw by SSF.

There are many reports on increasing the protein values of lignocelluloses by using SSF. Protein content of newsprint was increased to 6·5% with *Sporotrichum thermophile* in 6 days (Barnes *et al.*, 1972) and of sawdust to 7·7% with *Chaetomium cellulolyticum* in 9 days (Pamment *et al.*, 1978). Three such processes to enhance the protein values of lignocelluloses are described briefly as follows.

5.1.1 Han's Process

In this process straw is chopped, mixed with three parts of 0·5N H_2SO_4 solution, and hydrolyzed in a pressure cooker under 15 lb steam for 30 min. The hydrolyzed straw is neutralized with ammonia or ammonium hydroxide to raise the pH to 4·5. The treated straw contains about 20% fermentable sugars. It is inoculated with *Aurobasidium pullulans, Candida utilis* or *Trichoderma viride.* The fermentation chamber is revolved to provide a constant tumbling motion to the straw and to permit free exchange of air.

The straw can also be treated with NaOH (4% dry wt). It is neutralized with acid, and ammonium sulfate is added as a nitrogen source. The treated straw is inoculated with a mixed culture of *Cellulomonas* sp. and *Alcaligenes faecalis* for production of SCP in SSF.

Protein contents of 14, 12.4 and 10.9% dry wt in the final products were reported when acid-treated straw was fermented with *Aurobasidium pullulans, Candida utilis* and *Trichioderma viride,* respectively. In the case of alkali-treated straw, a protein content of only 6·8% in the final product was reported when inoculated with a mixed culture of *Cellulomonas* sp. and *Alcaligenes faecalis* for 2–3 days (Han & Anderson, 1975; Han *et al.*, 1976; Yu *et al.*, 1976).

5.1.2 INRA–DIJON Process

In this process sugar beet pulp is fermented with a special strain of *T. viride* T. S. for production of SCP in SSF (Durand *et al.*, 1991). A new type of pilot reactor (1 ton capacity) and a new technology of microorganism cultivation in a substrate layer 1 meter deep has been developed at the INRA–DIJON research center in France. The process developed can be easily adapted to a continuous process and can be incorporated into a sugar factory.

During 48 h of cultivation, the protein content of pulp was increased up to 20–21% on the basis of dry matter. The fermentation process was carried out in nonsterile conditions, using simple and cheap mineral nutrients.

The amino acid content of the final product increased by about 50% during the fermentation. For essential amino acids, the increase was about 60%. The nitrogen value of the enriched pulp, which takes all essential amino acids into account, was slightly lower than that of soybeans (72–76%) with limiting factors mainly in sulfur amino acids and marginally in lysine and isoleucine.

No toxic effects were observed during the 11-week feeding trial of protein-enriched pulp on lambs and rabbits. The protein-enriched pulp can replace all of the soybean supplementation in lambs and rabbits.

The estimated cost price for production of protein-enriched pulp in SSF was about 2 FFr/kg dry wt in hypothetical production at 50 t/h in the new developed reactor.

5.1.3 SCP Production in SSF with *Chaetomium cellulolyticum*

With *C. cellulolyticum*, protein contents of 9·8, 10·7 and 19·7% in the final products were obtained from untreated, cold-ammonia pretreated, and washed alkali-pretreated wheat straw, respectively, in SSF in 10–12 days (Chahal *et al.*, 1981).

About 17·5% (dry wt) protein was obtained in the final product when cold- or hot-ammonia pretreated corn stover was fermented with *C. cellulolyticum* in SSF for 3–4 days. The highest protein content of 20–24% dry wt was obtained when alkali-pretreated corn stover containing ammonium sulfate was fermented with this fungus in SSF for 5 days (Chahal *et al.*, 1983).

Chaetomium cellulolyticum seems to be the best organism for production of SCP in SSF (Table 3). The high percentage of protein in the final product obtained in SSF with *C. cellulolyticum* is attributed to the fact that this fungus can penetrate deep into the solid substrate as well as in intercellular (middle lamella) and intracellular (cell lumen) spaces of the substrate for its bioconversion into SCP (Chahal *et al.*, 1983).

5.2 Production of Cellulase System in SSF

The 'Koji' process (SSF) is being extenstively used for production of amylases in Japan (Toyama, 1976). The production of amylases in SSF is discussed

Table 3
Comparison of SCP Production on Various Lignocelluloses with Various Cellulolytic Microorganisms in Solid-State Fermentation

Reference/organism	*Substrate/treatment*	*Crude protein in product (% DW)*	*Incubation time (days)*
1. Barnes *et al.*, 1972:			
Sporotrichum thermophile	Newsprint (untreated)	6·5	6
2. Han & Anderson, 1975:			
(a) *Aurobasidium pullulans*	Ryegrass (0·5 NH_2SO_4, 121°C, 30 min)	14·0	2–3
(b) *Candida utilis*	Ryegrass (0·5 NH_2SO_4, 121°C, 30 min)	12·4	2–3
(c) *Trichoderma viride*	Ryegrass (0·5 NH_2SO_4, 121°C, 30 min)	10·9	2–3
3. Yu *et al.*, 1976:			
(a) *Cellulomonas* sp. + *Alcaligenes faecalis*	Ryegrass (4% NaOH, *c.* 25°C)	6·8	2–3
(b) *Cellulomonas* sp. + *Alcaligenes faecalis*	Ryegrass (NH_3)	9·5	2–3
4. Pamment *et al.*, 1978:			
(a) *Chaetomium cellulolyticum*	Sawdust	7·7	9
(b) *Chaetomium cellulolyticum*	Sawdust (20% NaOH sashed)		
5. Chahal *et al.*, 1981:			
(a) *Chaetomium cellulolyticum*	Wheat straw (4% NaOH)	19·0	5
(b) *Chaetomium cellulolyticum*	Corn stover (4% NaOH)	20–24	4–5
(c) *Chaetomium cellulolyticum*	Corn stover (3% NH_3)	17	4
6. Durand *et al.*, 1991:			
Trichoderma viride T.S.	Sugar beet pulp	20–21	2

elsewhere in this book. However, the application of this process was extended to the production of cellulase on wheat bran and lignocelluloses in SSF by Toyama (1976), Chahal (1985*a*, 1986), Deschamps *et al.* (1985) and Shamala and Sreekantiah (1986, 1987). The production of a complete cellulase system in SSF as devised by Chahal (1985*a*, 1986, 1989*b*) is discussed here in detail.

For complete hydrolysis of cellulose into glucose the cellulase system must contain (i) endo-gluconase, (ii) exo-gluconase and (iii) β-glucosidase in the right proportions; it is due to their synergistic effect that cellulose is hydrolyzed to glucose. But the cellulase system of *Trichoderma reesei* produced under most cultural conditions is deficient in β-glucosidase. Due to this deficiency of β-glucosidase cellobiose, an intermediate product, accumulates in the hydrolyzate, which inhibits further saccharification of cellulose.

During the last 12 years of research, the cellulase activity per unit volume has been increased to a great extent by producing hypercellulase mutants of *T.*

reesei, e.g. MCG 77 by Gallo *et al.* (1978), Rut-C30 by Montenecourt and Eveleigh (1979), L27 by Shoemaker *et al.* (1981), C1847 by Warzywoda *et al.* (1983), D1-6 by Panda *et al.* (1983), and QMY-1 by Chahal (1985*a*). Similarly, the cellulase activity per unit volume has been increased by fed-batch fermentation (Hendy *et al.*, 1982; Watson & Nelligan, 1983; McLean & Podruzny, 1985), and by solid-state fermentation (SSF) (Chahal, 1985*a*). By these methods the filter paper cellulase activity has been increased up to 17–30 IU/ml.

On the other hand, to reduce the production cost of cellulases, the pure cellulose substrate has been successfully replaced with that of crude cellulose from pretreated lignocelluloses by Chahal *et al.* (1982), Chahal (1985*a*) and Mes-Hartree *et al.* (1988).

This survey of the literature indicates that very little is known about the production of a complete cellulase system containing all the enzymes as described above. To achieve this objective, solid-state fermentation (SSF) seems to be the most promising approach.

5.2.1 *Comparison of Cellulase Production in LSF and SSF*

Filter paper cellulase (FP cellulase) activity of 1·65 IU/ml and productivity of 165 IU/g substrate were obtained by growing *T. reesei* QMY-1 on 1% wheat straw in LSF. The enzyme activity increased to 6·0 IU/ml but the productivity decreased to 120 IU/g substrate when the concentration of the substrate was increased to the 5% level in LSF. On the other hand a high enzyme activity per unit volume (8·6 IU/ml) and a high productivity per unit of substrate (172 IU/g substrate) were obtained under SSF conditions (Table 4). The FP cellulase activity obtained under SSF is higher than that produced under the best conditions of LSF at the 1% substrate level. Complete analysis of other components of the cellulase system produced under SSF revealed that the ratio of FP cellulase to β-glucosidase was about 1:1·2. This ratio is comparable to the 1:1 recommended by Mandels *et al.* (1981) and Chahal *et al.* (1982) to

Table 4
Cellulase Production with *Trichoderma reesei* QMY-1 on Wheat Straw in Liquid-State Fermentation (LSF) and Solid-State Fermentation (SSF)[a]

Fermentation condition	*Time of fermentation (days)*	*FP cellulase*	
		IU/ml	*IU/g substrate*
LSF (1%)	7	1·65	165
LSF (5%)	11	6·0	120
SSF	22	8·6[b]	172

[a] Source: partly from Chahal, 1985*a*. [b] Five grams of original wheat straw of SSF was mixed in 100 ml of H_2O to extract enzyme to make it comparable to 5% wheat straw in LSF.

Table 5
Composition of Cellulase Systems of *Trichoderma reesei* QMY-1 Produced on Different Substrates under SSF

Substrate	*FP cellulase (IU/ml)*[a]	*β-glucosidase (IU/ml)*	*Ratio of FP cellulase: β-glucosidase*	*Xylanases (IU/ml)*	*Reference*
Wheat straw	8·6	10·6	1:1·2	270	Chahal, 1985*a*
Pro-cell(TM)	5·3	4·1	1:0·77	404	Chahal, 1986

[a] Five grams of original substrate of SSF mixed with 100 ml H_2O to extract enzymes.

obtain high hydrolysis of cellulose. The compositions of cellulase systems produced on wheat straw and Pro-cell(TM) (wood) with *T. reesei* QMY-1 are presented in Table 5. This shows clearly that a ratio of FP cellulase to β-glucosidase close to the desired ratio, i.e. 1:1, can be obtained in SSF.

The xylanase activity is also very high in the cellulase system obtained on these substrates under SSF.

5.2.2 Composition of Cellulase Systems of Various Fungi Produced in Solid-state Fermentation

Comparison of the composition of cellulase systems of various fungi produced in SSF indicated that *T. reesei* QMY-1 proved to be the best fungus for the production of a cellulase system containing the highest activities of all the enzymes, i.e. FP cellulase, β-glucosidase and xylanases, compared with all the other fungi tried by other researchers (Table 6). The higher productivity of *T. reesei* QMY-1 could be due to the effects of species differences and the methodology used, as well as to the ability of the hyphae of this fungus to penetrate deep into the substrate as well as into intercellular spaces (middle lamella) and intracellular spaces (cell lumens) (Chahal, 1989*b*).

Although xylanases are normally not considered as part of cellulase systems for hydrolysis of cellulose, these enzymes are needed when hydrolysis of cellulose and hemicelluloses, as found together in lignocelluloses, is required. For that purpose, the cellulase system of *T. reesei* QMY-1, which is also very rich in xylanases compared to those of other fungi (Table 6), will be the most suitable enzyme system to hydrolyze the polysaccharides of lignocelluloses into their monomer sugars.

5.2.3 Hydrolytic Potential of Cellulase System

A comparison of the hydrolysis of delignified wheat straw with the cellulase systems produced in SSF and LSF is given in Table 7. About 90% of delignified wheat straw was hydrolyzed into simple sugars within 96 and 72 h of incubation with the cellulase systems produced in SSF on wheat straw and Pro-cell(TM), respectively. It is interesting to note that the quantity of cellobiose in the hydrolyzate obtained with the cellulase system produced on Pro-cell(TM) was higher than that obtained with the cellulase system produced on wheat

Table 6
Composition of Cellulase Systems Produced in Solid-State Fermentation

Substrate	Organism	*Enzyme activity IU/g dry wt substrate*			
		FP cellulase[a]	*β-glucosidase*	*Xylanases*	*Reference*
1. Wheat straw	*T. reesei* QMY-1	172	212	5 400	Chahal, 1985*a*
2. Pro-cell(TM) (wood)	*T. reesei* QMY-1	106	82	8 080	Chahal, 1986
3. Wheat bran	*Aspergillus ustus*	3·7	60	615	Shamala & Sreekantiah, 1986
4. Rice straw	*Aspergillus ustus*	5·8	15·8	740	Shamala & Sreekantiah, 1986
5. Wheat bran and rice straw (1:1)	*Aspergillus ustus*	4·6	40·7	398	Shamala & Sreekantiah, 1986
6. Rice straw (recycled)	Various microorganisms	14	48	1 431	Shamala & Sreekantiah, 1986
7. Wheat bran (recycled)	Various microorganisms	8	81	788	Shamala & Sreekantiah, 1987
8. Rice straw (recycled 5 times)	*Sporotrichum pulverulentum*	9	15·8	385	Shamala & Sreekantiah, 1987
9. Wheat straw and wheat bran (80:20)	*Trichoderma harzianum*	18	—	—	Deschamps *et al.*, 1985
10. Straw and bran (1:1)	*Aspergillus niger*	—	—	2 500	Deschamps & Huet, 1985

[a] FP cellulase = Filter paper cellulase determined according to the method of Mandels *et al.* (1976). It is due to the synergistic effect of endo- and exogluconase on filter paper.
Note: The results of cellulase production in SSF of Toyama (1976) were not included in this table as these were not in comparable units of enzymes.

Table 7
Comparison of Hydrolysis of Delignified Wheat Straw with Cellulase Systems of *Trichoderma reesei* QMY-1 Produced Under Solid-State Fermentation[a]

Substrate for production of cellulase system	*Ratio of FP cellulase: β-glucosidase*	*Glucose (g/liter)*	*Cellobiose (g/liter)*	*Xylose (g/liter)*	*Arabinose (g/liter)*	*Total sugars*	*Hydrolysis (%)*[b]
Alkali-treated wheat straw (SSF) (96 h)[c]	1:1·2	68·18	3·19	26·71	1·67	99·75	89·77
Alkali-treated Pro-cell(TM) (SSF) (72 h)[d]	1:0·77	61·3	8·3	31·8	—	101·4	91·26
Alkali-treated wheat straw (LSF) (24 h)[e]	1:0·8	13·9	6·5	10·0	0·86	31·26[f]	56·2
(60 h)[e]		19·8	6·5	11·8	0·86	38·96[f]	70·1
(144 h)[e]		27·3	3·4	13·2	0·86	44·76[f]	80·5

[a] Delignified wheat straw (100 g/liter) was hydrolyzed with 20 IU FP cellulase per g substrate for 96 h with a cellulase system produced on alkali-treated wheat straw and for 72 h with a cellulase system produced on alkali-treated Pro-cell. Wheat straw was delignified by the method described by Toyama (1972).

[b] Percentage of hydrolysis $= \frac{\text{Total wt of sugar produced (g/liter)} \times 0{\cdot}9}{\text{wt of substrate (g/liter)}} \times 100.$

[c] Data from Chahal, 1985*a*. [d] Data from Chahal, 1986. [e] Data from Chahal, 1989*b*. [f] Total sugars from 50 g/liter delignified wheat straw.

straw. This could be due to the fact that the former has a lower ratio of FP cellulase to β-glucosidase (1:0·77) than that produced on wheat straw (1:1·2). However, the Pro-cell(TM) system had a faster rate of hydrolysis as it took 72 h to obtain 90% hydrolysis compared with 96 h for wheat straw.

The hydrolysis of delignified wheat straw with the cellulase system produced in LSF with 5% concentration of wheat straw was slow, taking 144 h to reach 80% hydrolysis. Although the FP cellulase to β-glucosidase ratio of the cellulase system produced in LSF was very close to that produced in SSF on Pro-cell(TM), it still gave a lower rate of hydrolysis (80%) even after 144 h, double the time the Pro-cell(TM) system took to reach 90%. This indicates that there may be some additional factor or enzyme produced under SSF conditions which could be responsible for the fast rate and high yields of hydrolysis.

5.3 Production of Other Compounds in SSF

5.3.1 Gibberellic Acid

Gibberellic acid (GA_3) is a potent plant growth regulator and is extensively used in agricultrue, nurseries, greenhouses, viticulture, tea gardens, etc., for elimination of dormancy in seeds, acceleration of seed germination, induction of flowering, improvement of crop yields, overcoming of dwarfism, etc. Currently it is produced in LSF but the cost of production is very high because of extremely low yields, extensive downstream processing and consequently high capital and operating expenses. In fact, constraints on its extensive use due to high cost are noticeable and hence any substantial lowering in its cost of production may trigger its use on a large scale (Kumar & Lonsane, 1987).

Kumar and Lonsane (1987) have developed a SSF process for production of gibberellic acid (GA_3) by using *Gibberella fujikuroi* P-3 on wheat bran. The accumulation of GA_3 in SSF was 1·63 times higher than that in LSF. On the basis of available carbohydrates in the medium, the conversion rates were 0·096% and 0·156% in LSF and SSF, respectively. The use of coarse wheat bran (0·3–0·4 cm) resulted in an increase of 2·5 times in the yield of GA_3. A yield of 1·05–1·20 g per kg of dry moldy bran establishes the potential and feasibility of SSF for its production on wheat bran. Kumer and Lonsane's preliminary cost analysis shows net savings of about 60% and 50% on fermentation medium cost and expenditure on downstream processing, respectively, compared with the costs of the presently employed LSF techniques.

5.4 Use of Lignocellulosic Wastes as Inert Supporting Substrates for SSF

5.4.1 Citric Acid Production

Citric acid, a tricarboxylic acid, was originally isolated from lime juice for various uses. Wehmer (1893) first reported that citric acid can be produced by fermenting sucrose with species of the genera *Penicillium* and *Mucor*. On an industrial scale citric acid is produced with *Aspergillus niger* under stationary or surface cultural conditions. These fermentation conditions have now been replaced by submerged fermentation (Röhr *et al.*, 1983).

The SSF for production of citric acid was introduced by Cahn as early as 1935 (Cahn, 1935). In this system the solid substrates, such as sugar cane bagasse, potato or beet pulp, pineapple pulp, etc., were impregnated with fermentation medium and were inoculated with the fungal spores. These substrates acted as inert meterials to hold the fermentation medium. Later, this method of SSF for citric acid production was patented by Hisanaga and Nishimura (1968) and Yo (1975). Lakshminarayana *et al.* (1975) reported 80% yield of citric acid by growing *A. niger* 3/1 on sugar cane bagasse impregnated with sucrose under SSF. The bagasse served just as an inert substrate to provide surface area for fermentation of impregnated sucrose solution in it. These yields were much higher than those obtained with surface or submerged fermentation. The production of citric acid by the Koji process (SSF) on wheat bran with *A. niger* is also being practiced in Japan (as reported by Röhr *et al.*, 1983).

5.4.2 Production of Spores

The use of spores as bioinsecticides and bioherbicides (mycoherbicides) is becoming very important in modern agriculture. The SSF is also becoming the chosen method for production of spores of a bacterium (*Bacillus thuringiensis*) and of many fungi (*Metarhizium anisopliae, Hirsutella thompsonii, Beauvaria bassiana, Verticillium lencanii,* etc.) to be used as bioinsecticides (Quinlan & Lisansky, 1983). The spores of *Colletotrichum gloeosporioides* are being sold under the brand name of Collego® to be used as a bioherbicide (Silman *et al.*, 1989). Wheat bran, ground corn, cottonseed meal, barley grains and wheat grains are preferred because these solid substrates also serve as nutrients in the SSF. On the other hand volcanic glass, diatomaceous earth, vermiculite, etc., are used as solid inert materials which are impregnated with soluble nutrients for SSF. Sugar cane bagasse or straw could also be used as an inert substrate for SSF after impregnation with suitable media for production of spores or other products.

TeBeest (1985) and Thomas *et al.* (1987) are of the opinion that production of spores in SSF is time consuming, labor intensive, prone to contamination, and uneconomic. Churchill (1982) claimed that the submerged production technique (LSF) is preferable because of the availability of the technology, and the relatively easy scale-up process, but this may not be true in the case of those microorganisms which cannot produce spores in LSF. However, it is worth mentioning here that recently Morin *et al.* (1990) succeeded in producing spores of such a microorganism, *Phomopsis convolvulus,* in LSF in modified Richard's medium with V-8. Nevertheless, SSF is becoming a subject of intensive study these days for production of various products, including spore production; thus it seems that real competition might arise between LSF and SSF for production of certain products in the near future.

6 PROBLEMS IN BIOCONVERSION AND FUTURE TRENDS

Collection of lignocellulosic wastes and their transportation to the processing sites are one of the major problems in their bioconversion into various products. Crop residues scattered in the fields and forestry wastes left after felling may be far away from the city where bioconversion processes are to be set up. However, some wastes, such as sawdust, bark and other wood wastes, are found in large quantities near the pulp and paper mills, and lumber mills. Similarly sugar cane bagasse is piled up near the sugar factories. In such cases bioconversion plants can be set up close to such sites.

The other major problem is that the polysaccharides of lignocelluloses are found in recalcitrant form and are well protected with lignin. Now a number of effective pretreatments, especially steam explosion and alkali treatment, are available to make the lignocelluloses readily available for bioconversion into various products. However, the cost of pretreatment of lignocelluloses is still a debatable question in making the bioconversion process economic, especially for low value products.

The lignocelluloses are composed of cellulose, hemicelluloses and lignin. On hydrolysis with chemicals or enzymes a mixture of various sugars and lignin (in various forms) is produced. There is no bioconversion process available which can use the mixture of all these components to synthesize a product or products. To overcome such problems, the 'Integrated Process for Production of Food, Feed and Fuel (Ethanol) from Lignocelluloses' was developed at the Institut Armand-Frappier (IAF). In this process various microorganisms are used to ferment various fractions of lignocellulose into food, feed and fuel (ethanol), and modified lignin is obtained as another co-product. Thus there is no fraction of the lignocellulose left unutilized. Such integrated processes for bioconversion of lignocelluloses into other useful products are needed in the future.

Lignin in terms of its weight is probably second only to cellulose in the lignocellulosic materials (Kirk *et al.*, 1980). It has been estimated that for every ton of lignocelluloses used in the above IAF integrated process approximately 250 kg of lignin will be released (Chahal, 1985*b*; Chahal *et al.*, 1987). It is also envisaged that enormous quantities of lignin will start accumulating as soon as the industries based on utilization of polysaccharides from lignocelluloses for production of food, feed, fuel (ethanol) and other products are set up. Moreover, sulfite mills are already producing lignin in the form of lignosulfonate in millions of tons every year. At present only a fraction of such lignin is being used in industry. Although a voluminous literature on the degradation of lignin is available, there is no process whereby such an enormous quantity of surplus lignin could be used to synthesize new products. Therefore, there is a dire need of new processes for chemical or biological conversion of lignin into new products before its disposal becomes a colossal problem.

REFERENCES

Anderson, C. *et al.* (1975). In *Single-cell Protein,* II, ed. S. R. Tannenbaum & D. I. C. Wang. MIT Press, Cambridge, Massachusetts, p. 314.

Barnes, T. G., Eggins, H. O. W. & Smith, E. L. (1972). *Int. Biodeterioration Bull.,* **8**(3), 112.

Brisk, J. L. & Zuckermann, S. S. (1971). *Schweizerische Milchzeitung,* **97,** 544.

Cahn, F. J. (1935). *Ind. Eng. Chem.,* **27,** 201.

Callihan, C. D. & Clemmer, J. E. (1979). In *Microbiol Biomass,* ed. A. H. Rose (*Economic Microbiology,* Vol. 4). Academic Press, New York, p. 271.

Chahal, D. S. (1983). In *Foundations of Biochemical Engineering Kinetics and Thermodynamics in Biological Systems.* ACS Symp. Ser., No. 207, p. 421.

Chahal, D. S. (1985*a*). *Appl. Environ. Microbiol.,* **49,** 205.

Chahal, D. S. (1985*b*). *Optimization of an improved process for single-cell protein (SCP) production from forest biomass.* Report submitted to Yves Levesque, Director General of Forest Industry, Government of Québec, Ministry of Energy and Resources, Ste-Foy, Québec, Canada.

Chahal, D. S. (1986). In *VII Int. Symp. on Alcohol Fuels.* Edition Technip, Paris, p. 48.

Chahal, D. S. (1989*a*). *J. Ferment. Bioeng.,* **68,** 334.

Chahal, D. S. (1989*b*). Paper presented at the 1989 International Chemical Congress of Pacific Basin Societies, Honolulu, Hawaii, 17–22 December 1989. Abstract 431.

Chahal, D. S. (1991). In *Food, Feed and Fuel from Biomass,* ed. D. S. Chahal. Oxford & IBH Publishing Co., New Delhi, p. 59.

Chahal, D. S. & Hachey, J.-M. (1990). In *Agricultural and Synthetic Polymers: Utilization and Biodegradability,* ed. J. E. Glass. ACS Symposium Series No. 433, p. 304.

Chahal, D. S. & Hawksworth, D. L. (1976). *Mycologia,* **68,** 600.

Chahal, D. S. & Ishaque, M. (1986). *Science et Techniques de l'Eau,* **19,** 339.

Chahal, D. S. & Ishaque, M. (1988). *Science et Techniques de l'Eau,* **21,** 27.

Chahal, D. S. & Moo-Young, M. (1981). *Dev. Ind. Microbiol.,* **22,** 143.

Chahal, D. S. & Wang, D. I. C. (1978). *Mycologia,* **70,** 160.

Chahal, D. S., Swan, J. E. & Moo-Young, M. (1977). *Dev. Ind. Microbiol.,* **18,** 433.

Chahal, D. S., Vlach, D. & Moo-Young, M. (1981). In *Advances in Biotechnology,* III, ed. M. Moo-Young. Pergamon Press, Toronto, p. 327.

Chahal, D. S., McGuire, S., Pikor, H. & Noble, G. (1982). *Biomass,* **2,** 127.

Chahal, D. S., Moo-Young, M. & Vlach, D. (1983). *Mycologia,* **75,** 597.

Chahal, D. S. *et al.* (1987). *J. Ind. Microbiol.,* **1,** 355.

Chavez, E. R., Touchburn, S. P. & Moo-Young, M. (1988). *Anim. Feed Sci. Technol.,* **22,** 23.

Churchill, B. W. (1982). In *Biological Control of Weeds with Plant Pathogens,* eds R. Charudattan & H. L. Walker. John Wiley, New York, p. 139.

Das, K. & Ghose, T. K. (1973). *J. Appl. Chem. Biotechnol.,* **23,** 829.

Deschamps, F. & Huet, M. C. (1985). *Appl. Microbiol. Biotechnol.,* **22,** 177.

Deschamps, F., Giuliano, C., Asther, M., Huet, M. C. & Roussos, S. (1985). *Biotechnol. Bioeng.,* **27,** 1385.

Detroy, R. W., Lindenfelser, L. A., St Julian Jr, G. & Orton, W. L. (1980). *Biotechnol. Bioeng. Symp.,* **10,** 135.

Dunlap, C. E. & Callihan, C. D. (1973). *Single-cell protein from waste cellulose.* Final report on grant EP00328-4 to the Federal Solid Waste Management Program, US Environmental Protection Agency.

Durand, A., Grajek, W. & Gervais, P. (1991). In *Food, Feed and Fuel from Biomass,* ed. D. S. Chahal. Oxford & IBH Publishing Co., New Delhi, p. 123.

Duthie, I. F. (1975). In *Single-cell Protein,* II, eds S. R. Tannenbaum & D. I. C. Wang. MIT Press, Cambridge, Massachusetts, p. 505.
Fengel, D. & Wegener, G. (1983). *Wood Chemistry, Ultrastructure Reactions.* Walter de Gruyter, Berlin & New York.
Gallo, B. J., Andreotti, R., Roche, C., Ryu, C. & Mandels, M. (1978). *Biotechnol. Bioeng. Symp.,* **8,** 89.
Gilbert, N., Hobbs, I. A. & Levine, J. D. (1952). *Ind. Eng. Chem.,* **44,** 1712.
Han, Y. W. & Anderson, A. W. (1975). *Appl. Microbiol.,* **30,** 930.
Han, Y. W. & Callihan, C. D. (1974). *Appl. Microbiol.,* **27,** 159.
Han, Y. W., Cheeke, P. R., Anderson, A. W. & Lekprayoon, C. (1976). *Appl. Environ. Microbiol.,* **32,** 799.
Harris, E. E. & Belinger, E. (1946). *Ind. Eng. Chem.,* **38,** 890.
Hendy, N., Wilke, C. & Blanch, H. (1982). *Biotechnol. Lett.,* **4,** 785.
Hess, K., Mahl, H. & Gutter, E. (1954). *Kolloid-Z.,* **155,** 1.
Hesseltine, C. W. (1972). *Biotechnol. Bioeng.,* **14,** 517.
Higuchi, T. (1971). *Adv. Enzymol.,* **34,** 207.
Higuchi, T. (1980). In *Lignin Biodegradation: Microbiology, Chemistry and Potential Applications,* eds T. K. Kirk, T. Higuchi & Hou-Min Chang. CRC Press, Boca Raton, Florida, p. 1.
Hisanaga, S. & Nishimura, Y. (1968). Japanese Patent 6820708, 5 September.
Jermini, M. F. G. & Demain, A. L. (1989). *Experientia,* in press.
Kirk, T. K. & Haskin, I. M. (1973). *Am. Soc. Chem. Eng. Symp. Series,* **69,** 124.
Kirk, T. K., Higuchi, T. & Chang, H.-M. (1980). In *Lignin Biodegradation: Microbiology, Chemistry and Potential Application,* eds T. K. Kirk, T. Higuchi & Hou-Min Chang. CRC Press, Boca Raton, Florida.
Kumar, P. K. R. & Lonsane, B. K. (1987). *Process Biochem.,* **22,** 139.
Lakshminarayana, K., Chaudhary, K., Ethiraj, S. & Tauro, P. (1975). *Biotechnol. Bioeng.,* **17,** 291.
Laskin, A. I. (1977). *Ann. Rep. Fermentation Processes,* **1,** 151.
Litchfield, J. (1968). In *Single-cell Protein,* eds R. I. Mateles & S. R. Tannanbaum. MIT Press, Cambridge, Massachusetts, p. 304.
Livingston, A. L., Allis, M. E. & Kohler, G. O. (1971). *J. Agric. Food Chem.,* **19,** 947.
Mandels, M., Andreotti, R. & Roche, C. (1976). *Biotechnol. Bioeng. Symp.,* **6,** 21.
Mandels, M., Medeiros, J. E., Andreotti, R. E. & Bisset, F. H. (1981). *Biotechnol. Bioeng.,* **23,** 2009.
McLean, D. & Podruzny, M. F. (1985). *Biotechnol. Lett.,* **7,** 683.
Meller, F. H. (1969). *Conversion of Organic Solid Wastes into Yeast.* Bureau of Solid Waste Management, Washington, DC.
Mes-Hartree, M., Hogan, C. M. & Saddler, J. N. (1988). *Biotechnol. Bioeng.,* **31,** 725.
Millett, M. A., Baker, A. J. & Sattar, L. D. (1976). *Biotechnol. Bioeng. Symp.,* **6,** 125.
Montenecourt, B. S. & Eveleigh, D. E. (1979). In *Fuel from Biomass Symposium,* eds W. W. Schuster & H. R. Bungay. Department of Energy Publication, Rensselaer Polytechnic Institute, Troy, New York, p. 613.
Moo-Young, M., Chahal, D. S., Swan, J. E. & Robinson, C. W. (1977). *Biotechnol. Bioeng.,* **19,** 527.
Moo-Young, M., Dougulis, A. J., Chahal, D. S. & Macdonald, D. G. (1979). *Process Biochem.,* October, 38.
Moo-Young, M., Chahal, D. S., Vlach, D. & Stickeny, B. (1980). *Biotechnol. Bioeng.,* **23,** 2407.
Morin, L., Watson, A. K. & Reeleder, R. D. (1990). *Can. J. Microbiol.,* **36,** 86.
Mudgett, R. E. (1986). In *Manual of Industrial Microbiology and Biotechnology,* eds A. L. Demain & N. A. Soloman. American Society for Microbiology, Washington, DC, p. 66.

Newark, P. (1980). *Nature,* **287,** 6.
Pamment, N., Robinson, C. W., Hilton, J. & Moo-Young, M. (1978). *Biotechnol. Bioeng.,* **20,** 1735.
Panda, J., Bisaria, V. S. & Ghose, T. K. (1983). *Biotechnol. Lett.,* **5,** 767.
Peitersen, N. (1975). *Biotechnol. Bioeng.,* **17,** 1291.
Preston, R. D. & Cronshaw, J. (1958). *Nature,* **181,** 248.
Quinlan, R. J. & Lisansky, S. G. (1983). In *Biotechnology,* Vol. 3, ed. H. Dellweg. Verlag Chemie, Deerfield Beach, Florida, p. 233.
Righelato, R. C., Imrie, F. K. E. & Vlitos, A. J. (1976). *Resour. Recovery Conserv.,* **1,** 257.
Röhr, M., Kubicek, C. P. & Kominek, J. (1983). In *Biotechnology,* Vol. 3, ed. H. Dellweg. Verlag Chemie, Deerfield Beach, Florida, p. 419.
Rolz, C. (1984). In *Fermentation Processes,* Vol. 7, ed. G. T. Tsao. Academic Press, New York, p. 213.
Seitz, E. W. (1989). *J. Dairy Sci.,* in press.
Sekita, S., Yoshihira, K. & Natori, S. (1981). *Can. J. Microbiol.,* **27,** 766.
Shacklady, C. A. (1975). In *Single-cell Protein,* II, eds S. R. Tannenbaum & D. I. C. Wang, MIT Press, Cambridge, Massachusetts, p. 489.
Shamala, T. R. & Sreekantiah, K. R. (1986). *Enzyme Microbiol. Technol.,* **8,** 178.
Shamala, T. R. & Sreekantiah, K. R. (1987). *Enzyme Microbiol. Technol.,* **9,** 97.
Shoemaker, S. P., Raymond, J. C. & Bruner, R. (1981). In *Trends in the Biology of Fermentation for Fuels and Chemicals,* eds A. Hollander *et al.* Plenum, New York, p. 89.
Silman, R. W., Nelsen, T. C. & Bothast, R. J. (1989). Paper presented at the 1989 International Chemical Congress of Pacific Basic Societies, Honolulu, Hawaii, 17–22 December 1989. Abstract 404.
Sloneker, J. H. (1976). *Biotechnol. Bioeng. Symp.,* **6,** 235.
Srinivasan, V. R. & Han, Y. W. (1969). *Adv. Chem. Ser.,* **95,** 447.
Srivastava, V. K., Mowat, D. N., Moo-Young, M., Daugulis, A. J. & Chahal, D. S. (1980). Paper presented at the VIth International Fermentation Symposium, London, Ontario, 20–25 July 1980.
Tanaka, M. & Matsuno, R. (1985). Enzyme Microbiol. Technol., **7,** 197.
Tarkow, H. & Feist, W. D. (1969). *Adv. Chem. Ser.,* **95,** 197.
TeBeest, D. O. (1985). *J. Agric. Entomol.,* **2,** 123.
Thakur, M. S., Karanth, N. G. & Nanad, K. (1990). *Appl. Microbiol. Biotechnol.,* **32,** 409.
Thomas, K. C., Khachatourians, G. G. & Ingledew, W. M. (1987). *Can. J. Microbiol.,* **33,** 12.
Timell, T. E. (1967). *Wood Sci. Technol.* **1,** 45.
Toyama, N. (1972). In *Fermentation Technology Today,* ed. G. Terui. Soc. Fermentation Technology Japan, Osaka, Japan, p. 743.
Toyama, N. (1976). *Biotechnol. Bioeng. Symp.,* **6,** 207.
Tsang, L. J., Reid, I. D. & Coxworth, E. C. (1987). *Appl. Environ. Microbiol.,* **53,** 1304.
Underkofler, L. A. & Hickey, R. J. (1954). *Industrial Fermentation,* Vol. 1. Chemical Publishing Co., New York.
Warzywoda, M., Ferré, V. & Pourquie, J. (1983). *Biotechnol. Bioeng.,* **25,** 3005.
Watson, T. G. & Nelligan, I. (1983). *Biotechnol. Lett.,* **5,** 25.
Wehmer, C. (1893). As quoted on p. 533 in Prescott, S. C. & Dunn, C. G. (1959). *Industrial Microbiology,* McGraw-Hill, New York. (French Patent 288554, 11 March 1893.)
Wenzel, H. F. J. (1970). *The Chemical Technology of Wood.* Academic Press, New York.

Wiken, J. O. (1972). In *Fermentation Technology Today*, ed. G. Terui. Soc. Fermentation Technology Japan, Osaka, Japan, p. 569.
Wilke, C. R., Cysewski, G. R., Yang, R. D. & von Stockar, U. (1976). *Biotechnol. Bioeng.*, **8,** 1315.
Yo, K. (1975). Japanese Patent 75154487, 12 December.
Yu, P. L., Han, Y. W. & Anderson, A. W. (1976). *Proc. Western Sec. Amer. Soc. Animal Sci.*, **27,** 189.
Zadrazil, F. (1977). *Eur. J. Appl. Microbiol.*, **4,** 273.

Chapter 13

CONVERSION OF LOW GRADE FATS BY BIOLOGICAL MEANS

EWA LIE & GÖRAN MOLIN

Applied Microbiology, Chemical Center, University of Lund, PO Box 124, S-221 00 Lund, Sweden

CONTENTS

1 INTRODUCTION

The non-enzymatic industrial process for the hydrolysis of fats and oils common today involves pressures of approximately 50 atm and temperatures of 250°C or higher for several hours to achieve 96–99% hydrolysis (Potts & Mukerheide, 1968). The result is an extremely dark fatty acid fraction and a discoloured aqueous solution of about 10% glycerol. The fatty acids are unusable as obtained, and must be redistilled to remove the coloured by-products. The glycerol fraction is usually also distilled after concentration for the same reason.

There are numerous patents on the use of lipases for hydrolytic cleavage. However, there seem to be few industrial processes in function. Miyoshi Oil and Fat Co. (Japan) has reported that a yeast lipase (*Candida cylindracea*) is used for hydrolysing oils for soap production (Anon., 1981). It is claimed that the process improved the odour and colour, and is cheaper overall.

From an industrial viewpoint the primary goal of glyceride synthesis by enzymes is the production of monoacylglycerols (Lazar *et al.*, 1986). These compounds are important because of their use in the food, cosmetic, and pharmaceutical industries. Esterification is also easier to control than hydrolysis. To produce triacylglycerols in high yields Morita *et al.* (1984) used diacylglycerols and free fatty acids as the start material in order to avoid the high yield of mono- and di-acylglycerols obtained when starting from free fatty acids and glycerol. Another application has been demonstrated by Seino *et al.* (1984), who tried to produce surface active agents by esterification between different sorts of sugars with free fatty acids. The highest conversion they could achieve was 68% when oleic acid was mixed with fructose.

Interesterification might be the most promising way to utilize lipases for bioorganic synthesis. The most well-studied application is the alteration of the melting point of fats. One well-known, patented application is the interesterification of palm oil (mainly dipalmitoyl-2-oleoyl glycerol) with stearic acid or tristearyl glycerol with a 1,3-specific lipase as catalyst (Macrae & How, 1983). The result is a product enriched in 1(3)-palmitoyl-3(1)-stearoyl-2-oleoyl glycerol and 1,3-distearoyl-2-oleoyl glycerol, which are the main components of the expensive cocoa butter. The interesterification yielded a product similar to cocoa butter and the price of cocoa butter greatly exceeds that of palm oil.

Pilot-scale experiments have been done in order to interesterify batches of shea oleine and shea oil with myristic acid or stearic acid (Wisdom *et al.*, 1987). These have been performed by the use of immobilized lipase from *Rhizopus arrhizus* which is a 1,3-position specific lipase.

Interest in lipases has steadily increased in recent years, both for oleochemical industry applications and because lipases show stereoselectivity with certain compounds. An example of the latter is the preparation of enantiomerically pure secondary alcohols (Laumen & Schneider, 1988) or resolution of D,L-menthol by interesterification with triacetin using *Candida cylindracea* lipase (Lokotsch *et al.*, 1989).

Table 1
Some Commercially Available Lipases (Catalogue Prices in Sweden, 1989)

Biological origin	*Amount*	*Price*	*Manufacturer*
Wheat germ	1 000 U	$33.65	Sigma
Porcine pancreas	100 000 U	$42.85	Sigma
Candida cylindracea	250 000 U	$8.60	Sigma
Pseudomonas sp.	1 000 U	$85.25	Sigma
Chromobacterium viscosum	10 000 U	$45.75	Sigma
Rhizopus arrhizus	400 000 U	$22.00	Sigma
Geotrichum candidum	465 000 U	£180.00	Biocatalysts
Aspergillus niger	100 g	£30.00	Biocatalysts

Many chemical companies sell lipase preparations (Table 1). Most commercial lipases are offered as free enzyme preparations, but Novo has an immobilized lipase, Lipozyme(TM). Most lipases are available in small quantities for research purposes but some are also available in bulk quantities.

2 HYDROLYSIS

2.1 General

The triacylglycerols of long-chain fatty acids are insoluble in water and the lipases are characterized by their ability to rapidly catalyse the hydrolysis of ester bonds. The reaction takes place at the interface between the insoluble substrate phase and the aqueous phase wherein the enzyme is soluble. The lipases are either positionally non-specific or specific: The non-specific enzymes release fatty acids from all three positions of the glycerol backbone and the final result is a complete breakdown of the molecule into free fatty acids and glycerol, even if 1,2-(2,3)-diacylglycerols, 1,3-diacylglycerols and monoacylglycerols appear as intermediates (Fig. 1).

2.2 Position Specificity

The 1,3-specific lipases release fatty acids only from the outer 1 and 3 positions of the glycerol molecule, and triacylglycerols are hydrolysed to give free fatty acids, 1,2-(2,3)-diacylglycerols and 2-monoacylglycerols. However, 1,2-(2,3)-diacylglycerols and 2-monoacylglycerols are chemically labile and easily

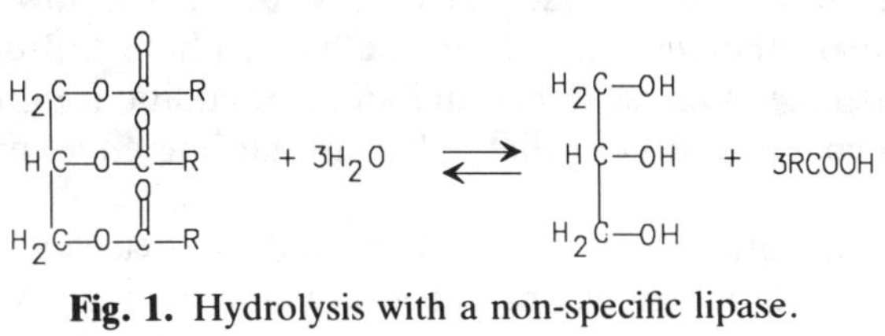

Fig. 1. Hydrolysis with a non-specific lipase.

undergo acyl migration which results in 1,3-diacylglycerols and 1-monoacylglycerols, respectively. Thus, prolonged incubation with a 1,3-specific lipase will result in a complete breakdown of some of the fats into glycerol and free fatty acids.

No data have been found which prove the existence of a lipase with specific activity towards the 2-position of the glycerol moiety. Furthermore, lipases appear to lack stereospecificity with respect to the triacylglycerol molecule, i.e. lipases remove fatty acids at a similar rate from the 1- and 3-positions of the glyceride.

2.3 Fatty Acid Specificity

A few examples of lipase fatty acid specificity are found in the literature. Thus, some microbiological lipases have low activity towards ester bonds with butyric acid, as demonstrated in milk fat (Lui *et al.*, 1973; Sugiura & Isobe, 1975; Aisaka & Terada, 1979), while pancreatic lipase catalyses the release of long chain ($>C_{18}$) polyunsaturated fatty acids in marine oils very slowly (Brockerhoff 1965). The only lipase which is known to possess an active, selective effect is lipase from *Geotrichum candidum*. This lipase preferentially releases long chain fatty acids containing a *cis* double bond in the 9th position, i.e. it splits the esters of oleic, palmitoleic, linoleic and linolenic acids (Jensen, 1974; Jensen & Pitas, 1976).

Finally, the fungus *Penicillium cyclopium* has been shown to produce a lipase which hydrolyses the partial acylglycerols, diolein and monoolein, more rapidly than the triacylglycerol triolein (Okumura *et al.*, 1980). This lipase, when combined with a conventional unspecific lipase, gives a more rapid total hydrolysis of triacylglycerols than is obtained by either of the lipases working alone.

3 ESTERIFICATION

The hydrolytic lipase reaction is reversible. In a system of low water activity, lipases can catalyse ester formation from free fatty acids and alcohols. However, there seem to be three major obstacles to this:

(1) In order to maintain a low water content of the reaction system throughout the reaction period, water must be withdrawn from the system at the same rate as it is released by the esterification reaction.
(2) The concentration of free fatty acid must be kept at low concentration (1mM–10 mM) otherwise the pH in the water phase will drop to a level where most lipases are severely inhibited. In addition, operating at such a low acid concentration probably has a retarding effect on the reaction rate.
(3) During esterification high yields of di- and monoacylglycerols are also obtained. This is a disadvantage because they interact in the interface

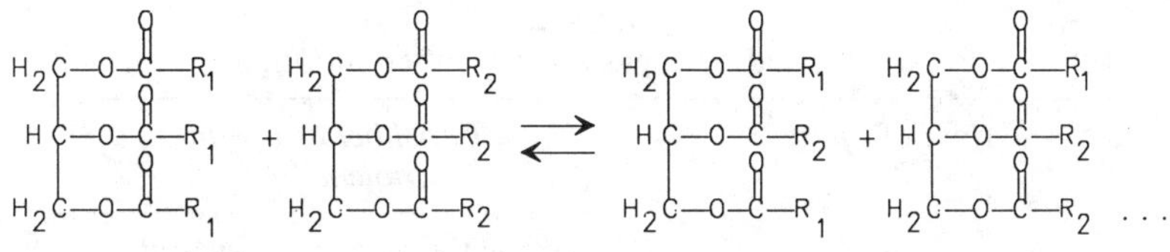

Fig. 2. Interesterification of two different triacylglycerols.

between the two phases and decrease the specific surface area (Brockerhoff & Jensen, 1974).

4 INTERESTERIFICATION

Interesterification can proceed along two lines: (1) Two types of triacylglycerols are mixed and the exchanges of acyl groups between the glycerol molecules are catalysed (Fig. 2); or (2) a triacylglycerol is mixed with free fatty acid or a fatty acid ester and the triacylglycerol becomes enriched in the added fatty acid (Fig. 3). If hydrolysis is supressed by operating under conditions of low water activity, the amount of free fatty acids in the reaction mixture will remain nearly constant, but the composition of free fatty acids after the reaction will be different.

Both reactions can be catalysed with lipases in systems of low water content (preferably less than 1% of the total volume). Furthermore, if a 1,3-position specific lipase is used, the acyl exchange is confined to the 1- and 3-positions of the acylglycerol; likewise if a fatty acid-specific lipase is used, a particular fatty acid from a mixture of fatty acids can be selectively introduced into the triacylglycerol. One advantage of interesterification between two different triacylglycerols is that the pH does not decrease, and thus the enzyme does not lose its stability and, hence, remains active longer.

5 PROCESS TECHNOLOGY

5.1 Organic Solvents

5.1.1 Lipase Sensitivity

Organic solvents can be used to make viscous oils and solid fats fluid at room temperature. Thus, in certain applications it might be necessary to perform the

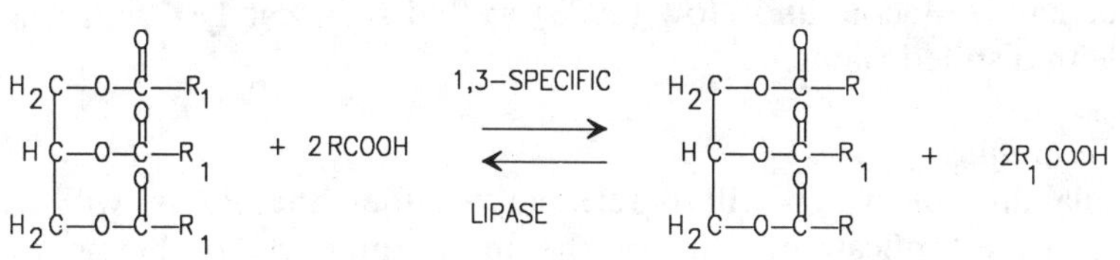

Fig. 3. Interesterification of a triacylglycerol and free fatty acid.

Table 2
FFA Liberated by Lipases from Different Origins[a]

Origin of lipase	*FFA liberated from triolein (%)*	
	with isooctane	*without isooctane*
Chromobacterium viscosum	42	60
Candida cylindracea	14	40
Pseudomonas sp.	10	50
Rhizopus arrhizus	2·5	7
Wheat germ	0	0
Porcine pancreas	30	30

[a] Source: E. Lie & G. Molin, unpublished results.

lipase-catalysed reaction in the presence of an organic solvent; heptane, hexane, isooctane or petroleum ether are most commonly used. Branched hydrocarbons were found to maintain the enzyme activity better than straight ones, isooctane being the best of those tested (Kim *et al.*, 1984).

The purpose of the organic solvent is not solely to decrease viscosity. It also solubilizes the product and makes it possible for the reactants to come into closer contact with the lipase. Furthermore, the organic solvent can enable reactions that would be thermodynamically impossible in a water system and sometimes the product recovery is easier from organic solvents than from water. Thus, lipase sensitivity to organic solvents is an important parameter when designing a process.

Organic solvents exert different influences on lipases from different origins which is demonstrated in Table 2. The hydrolytic activity of six different lipases was tested in the absence and presence of isooctane. Lipase from porcine pancreas and *Chromobacterium viscosum* in this trial were most resistant while *Pseudomonas* lipase was most sensitive to isooctane (E. Lie & G. Molin, unpublished results).

When operating in an organic solvent a small amount of water is needed in order to make the enzyme active. There are different opinions on how much water is needed for this activation. Zaks and Klibanov (1985) proposed that a few monolayers should be enough. They found that porcine pancreatic lipase with a water content of 0·5% acts as a catalyst in 99·98% organic medium. On the other hand, Macrae and How (1983) wetted their catalyst with 8% of its weight with distilled water.

5.1.2 Techniques

The availability of water will determine whether the lipase will catalyse hydrolysis of esterification. One of the major practical problems with esterification is maintenance of a sufficiently low water level. Released water must be continuously removed from the system if the reaction is to proceed. This can be done by continuous removal of water-saturated organic solvent and separation of the solvent from the water by distillation or drying with a molecular sieve, and recirculating the dry solvent into the reactor. Drying with zeolites has been practically demonstrated by Knox and Cliffe (1984). To let the water evaporate freely when using a temperature of 60°C has been proved to be efficient: 86% triolein was formed from oleic acid and glycerol (Ergan *et al.*, 1990).

In order to increase the interfacial area, a system of reversed micelles can be utilized. The major disadvantage of this type of system on an industrial scale is that it requires the presence of an efficient surfactant, e.g. bis(2-ethylhexyl)sodiumsulphosuccinate (AOT). Undesired side reactions between the surfactant and the reactant can be catalysed, and an initial vigorous stirring is also needed to provide the micellar solution.

One way to avoid emulsions and organic solvents is to use a membrane reactor; the lipase is immobilized on a porous hydrophobic membrane with the hydrophobic phase on one side of the membrane and the aqueous one on the other side. Continuous hydrolysis of tallow, olive oil and sunflower oil has been accomplished in this way (Hoq *et al.*, 1986; Taylor *et al.*, 1986; Rucka *et al.*, 1989).

5.2 Immobilization

5.2.1 General

Immobilization is one of the more promising approaches for enzyme stabilization; it has been used in hydrolysis, esterification and interesterification. The conformational structure of an enzyme immobilized on, or in, a rigid support becomes more resistant to distortion by, for example, organic solvents. Furthermore, it is well known that many enzymes *in vivo,* especially lipases, function in a membrane-bound state and that immobilization mimics this membrane-bound state. The enzyme can be immobilized on the carrier by (1) covalent attachment to the carrier, (2) adsorption of the enzyme on the carrier, (3) entrapment of the enzyme in the carrier.

5.2.2 Carrier Characteristics

Several studies have been done on the influence of the carrier on the catalytic effect. Lipase adsorbed on porous glass beads (Spherosil™), alumina, titania and celite were used in ester synthesis by Marlot *et al.* (1985). All enzyme preparations were active; the highest activity was observed with the preparations with celite and porous glass beads of high specific area. Even powdered pig bone has been used as support for lipase (Negishi *et al.*, 1989). Furthermore, the hydrolytic activity of lipase immobilized on pig bone was higher than on other supports, e.g. celite and amberlite. Working with celite, Wisdom *et al.* (1984) demonstrated that both the lipolytic activity and the

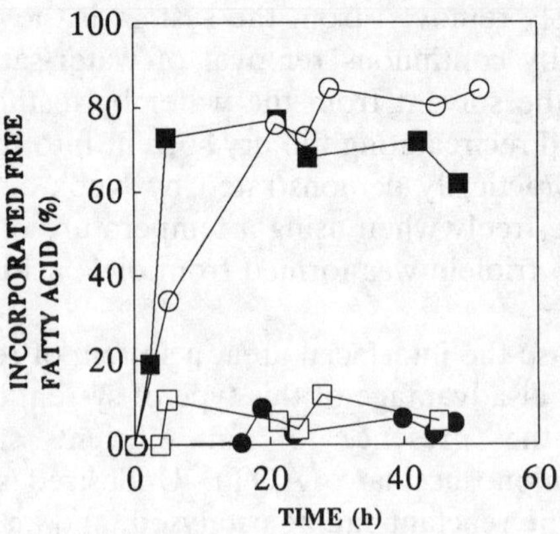

Fig. 4. Esterification with free *Candida cylindracea* lipase (○) and attached to celite (●), hydrophobic zeolite (□) and hydrophilic zeolite (■).

interesterification activity are dependent on the size of the pores and the size of the carrier particles. Large pores and small carrier particles were optimal.

Yokozeki *et al.* (1982) showed that the interesterification activity of lipases immobilized on celite and entrapped in photo-crosslinkable resin prepolymers was dependent on the hydrophobicity which determined the diffusion of the hydrophobic substrates.

The hydrophobicity of the carrier clearly influences the activity. Esterification between oleic acid and glycerol was carried out using zeolites of different hydrophobicity as carriers for lipase. The hydrophilic support gave results equal to the use of free enzyme, while lipase immobilized on celite and the hydrophobic support yielded preparations with very low activity (Fig. 4) (E. Lie & G. Molin, unpublished results). This is in contrast to the results obtained when chymotrypsin was adsorbed on different carriers and used for esterification between ethanol and *N*-acetyl-1-phenylalanine. Carriers that had a high ability to absorb water yielded an immobilized preparation with low activity while the ones with low ability, like celite, gave the highest activity. There was obviously a competition between the carrier and the enzyme for the water (Reslow *et al.*, 1988).

5.3 Temperature

The temperature optima of different lipases ranges between 15°C and 60°C. The temperature optima of seven different lipases was determined; see Table 3 (E. Lie & G. Molin, unpublished results). *Rhizopus arrhizus* had the lowest temperature optimum at 15°C and lipase from *Pseudomonas aureofaciens* had the highest temperature optimum of 60°C.

Temperature can also play an important role in the specificity of the hydrolysis. Alford and Pierce (1961) showed that below 0°C, microbial lipases released the unsaturated fatty acids from lard rather than the saturated ones.

Table 3
Temperature Optima of Different Lipases[a]

Lipase source	*Temperature optimum (°C)*
Chromobacterium viscosum	50
Candida cylindracea	45
Pseudomonas sp.	25
Rhizopus arrhizus	15
Wheat germ	25
Porcine pancreas	37
Pseudomonas aureofaciens	60

[a] Source: E. Lie & G. Molin, unpublished results.

In interesterification of a mixture of fatty acids with different melting points, the temperature could theoretically be used to determine which fatty acids participate in the reaction.

Lipase from *Pseudomonas* species has been shown to be a problem in the dairy industry since it is not completely deactivated at high temperatures. Milk is spoiled by the shorter free fatty acids that are liberated. *Pseudomonas* lipase retained 85% of its original activity at a UHT treatment of 150°C for 3·42 s (Swaisgood & Bozoglu, 1984).

5.4 pH

Lipases from different origins show different optimum values of pH ('pH-optima'). The hydrolytic activity of 14 different lipases on olive oil has been investigated and the pH-optima fall in a range between 5·6 to 8·0 (Lazar *et al.*, 1986). A lipase from *Pseudomonas fragi* was found to have an alkaline pH-optimum at pH 9 (Watanabe *et al.*, 1977). The hydrolytic activity of lipase from *Candida rugosa* on olive oil has been studied by several authors. The activity is almost constant between pH 4·8 and 7·2; above 7·2 the activity drops rapidly (Linfield *et al.*, 1984). This is in contrast to the findings of Wang *et al.* (1988) who found a sharp pH-optimum for hydrolysis of olive oil at pH 5·0. In an isooctane-bis(2-ethylhexyl)sodium sulphosuccinate (AOT) system a sharp pH-optimum at 7·1 was found (Han & Rhee, 1986). There are difficulties in measuring the actual pH in the suspensions described above because of the presence of fat or organic solvent.

5.5 Extraction of Fats with Supercritical Carbon Dioxide

A new mild way to separate triacylglycerols from free fatty acids is extraction with supercritical carbon dioxide ($SCCO_2$). Supercritical fluids offer a number of advantages over conventional liquid solvents. They have low viscosity and high diffusivity. The extraction of the solute proceeds more rapidly than with organic solvents.

The critical temperature of $SCCO_2$ is 31·3°C so thermal denaturation and decomposition are avoided. Carbon dioxide is also environmentally acceptable because it is non-toxic and non-flammable. The solubility of the solute can be controlled by regulation of temperature or pressure, and to increase the selectivity of the solubilities of different compounds an entrainer such as ethanol can also be added. An entrainer acts as a bridge between a polar compound and the carbon dioxide.

Brunetti *et al.* (1989) used $SCCO_2$ for deacidification of olive oil. When using pressures of 20–30 MPa and temperatures of 35–60°C they showed that CO_2 extracts fatty acids more selectively than triacylglycerols; the selectivity factor for fatty acids was at least 5.

Supercritical CO_2 has also been used as a solvent in interesterification reactions with lipases. Chi *et al.* (1988) interesterified triolein and stearic acid. Both compounds are soluble in $SCCO_2$, although stearic acid is less soluble than triolein. When using a lipase from *Mucor miehei* the initial interesterification rate was much higher in $SCCO_2$ than in *n*-hexane, especially at high water concentrations (20%).

Enzymatic oxidation of cholesterol aggregates by cholesterol oxidase has been done by Randolph *et al.* (1988). Variations in solvent power caused by pressure changes affected the rate of enzymatic oxidation of cholesterol. This was due to alteration of the structure of cholesterol aggregates. To study aggregate formation they used electron paramagnetic resonance spectroscopy.

6 CURRENT LIPASE-MEDIATED APPLICATIONS

6.1 Enzymatic Transesterification of Gluestock Fat

6.1.1 Process

Gluestock fat is a waste product from leather processing. The fat is mechanically removed from the backs of the hides together with tissue. This is done after unhairing of the hides and is called 'fleshing'. About 20–30% of the hide weight is removed in this way.

The composition of the fleshing is 80% water, 10% protein and 10% fat; the water is chemically bound to the protein because of the high pH. It is difficult to de-water the fleshing and to separate the fat, but the separation of fat can be done by heating the gluestock with steam by the 'Lamatic'(TM) process. There is also an improved version of the process called 'Secco-Lamatic'(TM) (Rydin 1989).

Later in the process the hides must be relubricated with fat, which is called fat liquoring, in order to obtain a soft leather with high mechanical strength. It would be an advantage if the gluestock fat could be used for fat liquoring. However, the gluestock fat is unusable due to its high melting point. This makes the leather stiff and increases the risk of spew formation. However, it has been possible to mix up to 20% of gluestock fat in the fat liquor (Pauckner, 1981).

Table 4
Fatty Acid Composition of Gluestock Fat, Olive Oil, Soybean Oil and Rapeseed Oil[a]

Fatty acid (no. of C atoms: no. of double bonds)	*Melting point (°C)*	*Glue-stock fat (%)*	*Olive oil (%)*	*Soy-bean oil (%)*	*Rape-seed oil (%)*
Myristic acid (14:0)	54	3	—	—	—
Palmitic acid (16:0)	63	25	17	10	5
Palmitoleic acid (16:1)	−1	8	2	—	—
Stearic acid (18:0)	70	9	3	4	2
Oleic acid (18:1)	16	43	62	23	63
Linoleic acid (18:2)	−5	3	15	55	20
Linolenic acid (18:3)	−11	3	1	7	9
Behenic acid (22:0)	80	—	—	1	—

[a] Source: Rydin, 1989.

6.1.2 Lowering of the Melting Point

The high melting point of gluestock fat is due to the high content of long chain saturated fatty acids, e.g. palmitic and stearic acid. These fatty acids are located mainly in the 1- and 3-positions of the triacylglycerol. Enzymatic transesterification of gluestock fat with a triacylglycerol containing unsaturated fatty acids in the 1- and 3-positions is one way to lower the melting point. Transesterification of gluestock fat by immobilized lipase from Novo has been done with three different oils, namely olive oil, soybean oil and rapeseed oil (Rydin, 1989). The composition of the different oils is shown in Table 4. When 50% gluestock fat and 50% rapeseed oil was transesterified at 60°C, the melting point of the product changed from 35°C to 24°C.

The converted gluestock fat could be mixed with commercial fat liquor in proportions up to 80% and used for lubrication of hides without any negative effect on hide quality. The converted gluestock fat was refined before use with sodium hydroxide in order to remove the free fatty acids. The refining also improved the evenness of colour and the uptake of fat.

6.2 Fat-containing Wastewaters

6.2.1 General

Wastewater containing fats, oils and grease can be a problem in many industries as well as in domestic wastewaters. Problems that arise include blocked sewers, plugged lines, blocked anaerobic biogas reactors and scum formation. Examples of industries with fat-containing wastewater are slaughterhouses, dairy industries, meat packaging industries and edible oil process industries such as palm or olive oil mills. Several strategies are currently applied in effluent handling; mechanical fat separation or flotation are common. However, the former method is insufficient if the fat is in a dispersed form.

Biological methods can also be applied. For example, slaughterhouse effluent treatment with thermophilic aerobic microorganisms reduced the soluble chemical oxygen demand (COD) by more than 90% at 52°C (Couillard *et al.*, 1989). The thermophilic process might also reduce the number of pathogens. The COD and the biochemical oxygen demand (BOD) increase with increasing fat-content.

6.2.2 *Palm Oil Mills*

The effluent from palm oil mills has a high BOD (about 20 000 mg oxygen/litre, measured over 5 days), but is essentially free of toxic constituents (Davis & Reilly, 1980). In order to reduce the BOD value different treatment methods have been exploited. For small palm oil mills with oil palm plantations near the mill, the effluent can be used as an irrigant/fertilizer if it is supplied in the right doses to prevent the soil from clogging (Davis & Reilly, 1980).

Palm oil effluent can also be degraded by anaerobic digestion. Biogas is obtained as a valuable product and can be used for electricity generation using a gas engine (Ma & Ong, 1986). Drawbacks are the generation of an unpleasant odour and the long detention time of the process. Anaerobic decomposition of fat is generally a slow process where the liberation of free fatty acids is the rate limiting step. However, addition of lipases to the wastewater might increase the efficiency of the process.

6.2.3 *Lipase Addition*

Lipases and cellulase have been added to the sludge of a municipal wastewater in order to enhance the degradation of the cellulosic and the oil- and grease-rich fractions (Higgins & Swartzbaugh, 1985). Cellulase was used prior to the anaerobic digestion while lipase was added directly to the anaerobic digester vessel. Pretreatment with cellulase gave a reduction in total solids of 64% and an increase in methane production per weight of primary sludge input of 166% for the test system. Lipase addition enhanced both the hydrolysis of oils and greases as well as their subsequent conversion to methane. An increase of 5% of the volume of total methane per unit weight volatile solids input was noted in the test system.

6.2.4 *Upgrading*

Single cell protein (SCP) production from fat-containing wastewaters has also been studied because of its potential for upgrading the value of wastewaters rich in fats. For example, the yeast *Candida tropicalis* was able to grow on the lipid fraction of a wastewater from a rendering plant (Rydin *et al.*, 1990). The lipid content of the wastewater decreased from 6 g/litre to 0·5 g/litre in 24 h. The biomass produced might be used as a feed supplement and the remaining water is suitable for methane production. Competing bacterial growth could be kept out of the process by running it at low pH; the yeast grew well but bacterial growth was inhibited. To enable microorganisms to grow near their maximum growth rate, the fat must be supplied in a dispersed form.

When using a solid fat as substrate, strong agitation is required to disperse the fat. When *Saccharomycopsis lipolytica* was cultivated on different animal fats, a stirring speed of 1200 rpm was required to obtain optimal growth (Tan & Gill, 1985). Lipase production may help the attacking microorganisms to maintain the dispersion by releasing hydrolysed acylglycerols into the growth medium.

6.3 Polyunsaturated Fatty Acids (PUFA)

6.3.1 General

Fats from fish contain much more polyunsaturated fatty acids (PUFA) than mammalian fats. The fatty acids from fish have been a focus of interest in recent years because of their nutritional value. Marine oils account for about 2% of the world's production of fats and oils. Fish body oils make up 97% of the supply. The fish are processed by cooking, pressing and separation for recovery of the oil. The price of marine oils on the world market is competitive with other fats and oils (Bimbo, 1987).

There are two kinds of polyunsaturated fatty acids which are essential for humans, namely the *n*-6 fatty acids, like linoleic acid, and the *n*-3 fatty acids (the '*n*' refers to the position of the first double bond). The *n*-6 fatty acids are major constituents of vegetable oils such as corn, sunflower and rapeseed oil (Hilditch & Williams, 1964) while the major constituents in marine fish oils and fats are the *n*-3 fatty acids; eicosapentaenoic acid (EPA, 20:5) and docosahexaenoic acid (22:6) are major constituents (Stansby, 1979). Arachidonic acid is an *n*-6 fatty acid and an important component of membranes and prostaglandins. The *n*-3 fatty acids affect the lipoproteins and lower the cholesterol content in blood (Becker, 1989).

6.3.2 Enzyme-mediated Reactions

Fish oils are usually manufactured from whole fish or prepared along with fish meal. The percentage of C_{20} and C_{22} in menhaden oil, for example, is around 30% from fish caught in the Atlantic (Stansby, 1979). The PUFA is generally situated in the 2 position in the triacylglycerol (Brockerhoff, 1967), with saturated and monosaturated acids in the 1 and 3 positions.

Triacylglycerols are easier for humans to digest than free fatty acids. Therefore, it would be an advantage to be able to increase the content of PUFA in the triacylglycerols. Because of the high degree of unsaturation of PUFA, traditional methods cannot be used since the PUFA is oxidized. Concentrating PUFA with urea complexing is one method (Ackman *et al.*, 1988). Lipase-catalysed reactions are also an alternative. Lipase from *Rhizopus* sp. and *Rhizopus delemar* were used to catalyse the transesterification of unsaturated lipids in a microemulsion (Osterberg *et al.*, 1989). The lipase activity was low when triacylglycerols with fatty acids containing an *n*-6 double bond, such as γ-linolenic acid, was involved. But free γ-linolenic acid could be incorporated into an unhindered triacylglycerol, so the active site could accommodate a single free fatty acid but not a triacylglycerol with an *n*-6 double bond. The esterification capacity of six different lipases towards PUFA has been tested in a mixture of glycerol and a 70% n-3 fatty acid. Lipase from *Chromobacterium viscosum* was most active; 75% of the free fatty acid was incorporated (E. Lie & G. Molin, unpublished results).

6.3.3 Single Cell Oil

Fat-accumulating microorganisms may convert a waste fat or a low value fat to a fat of higher value which can then be extracted from the microorganisms. Single cell oil (SCO) production is then an alternative to single cell protein, and a suitable SCO could command a higher price then SCP. Several species of soil fungi can utilize fatty acids as well as their sodium salts as their sole carbon and energy source (Radwan & Soliman, 1988).

Fat-accumulating microorganisms are most common among yeasts and moulds. The triacylglycerols they produce are similar to many plant oils with high contents of oleic acid, linoleic acid and palmitic acid (Ratledge, 1989). Some of the fungi species also produce arachidonic acid (C 20:4) and γ-linolenic acid (GLA; C 18:3). The latter can also be found in certain plants and is an important prostaglandin precursor. Production of GLA has been carried out on strains of the order Mucorales and especially *Mortierella* spp. They had a high lipid content of about 20–25% and a GLA content in the lipid of 15–20% (Hansson & Dostálek, 1988). Even higher concentrations of GLA have been achieved when cultivating *Mucor ruoxii* on glucose in a fed-batch. The GLA concentration of the fatty acids was then 40% (Hansson *et al.*, 1989).

Arachidonic acid, which is never found in plants and only in very small concentrations in animal tissue, is a precursor of prostaglandins of type 2. Arachidonic acid can be produced in high concentrations by several species of the fungus *Mortierella*. Under certain culture conditions nearly 60% of the lipid fraction of *Mortierella alpina* cells consists of arachidonic acid (A. M. Lindberg & G. Molin, unpublished results).

Fungal mycelia can also be a source of eicosapentaenoic acid (EPA). When fungi were grown at low temperatures (12°C), 14% of the lipid produced was found to be EPA (Shimizu *et al.*, 1988). When *Cryptococcus albidus* was grown on glycerol the lipid fraction of the biomass was 44%, which was higher than the value when grown on glucose (Hansson, 1986). Thus lipase addition to a fat-containing wastewater would yield glycerol and perhaps enhance the yield of SCO.

REFERENCES

Ackman, R. G., Ratnayake, W. M. N. & Olsson, B. (1988). *J. Am. Oil Chem. Soc.*, **65,** 136.
Aisaka, K. & Terada, O. (1979). *Agric. Biol. Chem.*, **43,** 2125.
Alford, J. A. & Pierce, D. A. (1961). *J. Food Sci.*, **26,** 518.
Anon. (1981). *Jap. Chem. Week*, 14 May, 2.

Becker, W. (1989). *Vår föda,* **41,** 94.
Bimbo, A. P. (1987). *J. Am. Oil Chem. Soc.,* **64,** 706.
Brockerhoff, H. (1965). *Arch. Biochem. Biophys.,* **110,** 586.
Brockerhoff, H. (1967). *J. Lipid Res.,* **6,** 10.
Brockerhoff, H. & Jensen, R. G. (1974). *Lipolytic Enzymes,* Academic Press, New York.
Brunetti, L., Daghetta, A., Kikic, I. & Zanderighi, L. (1989). *J. Am. Oil Chem. Soc.,* **66,** 209.
Chi, Y. M., Nakamura, K. & Yano, T. (1988). *Agric. Biol. Chem.,* **52,** 1541.
Couillard, D., Gariépy, S. & Tran, T. F. (1989). *Water Res.,* **23,** 573.
Davis, J. B. & Reilly, P. J. A. (1980). *Oléagineux,* **35,** 323.
Ergan, F., Trani, M. & André, G. (1990). *Biotechnol. Bioeng.,* **35,** 195.
Han, D. & Rhee, J. S. (1986). *Biotechnol. Bioeng.,* **28,** 1250.
Hansson, L. (1986). Influence of culture conditions on formation of intracellular products from microorganisms. Thesis, Department of Applied Microbiology, University of Lund, Sweden.
Hansson, L. & Dostálek, M. (1988). *Appl. Microbiol. Biotechnol.,* **28,** 240.
Hansson, L., Dostálek, M. & Sörenby, B. (1989). *Appl. Microbiol. Biotechnol.,* **31,** 223.
Higgins, G. & Swartzbaugh, J. T. (1985). Paper presented at the Symposium on Energy from Biomass Wastes IX, Lake, Buena Vista, Florida, p. 791.
Hilditch, T. P. & Williams, P. N. (1964). *Chemical Constitution of Natural Fats.* Chapman & Hall, London, p. 172.
Hoq, M. M., Yamane, T. & Shimizu, S. (1986). *Enzyme Microbiol. Technol.,* **8,** 236.
Jensen, R. G. (1974). *Lipids,* **9,** 149.
Jensen, R. G. & Pitas, R. E. (1976). In *Lipids,* Vol. 1, ed. R. Paolette, G. Porcellati & G. Jacini. Raven Press, New York, p. 141.
Kim, K. H., Kwon, D. Y. & Rhee, J. S. (1984). *Lipids,* **19,** 975.
Knox, T. & Cliffe, K. R. (1984). *Process Biochem.,* **19,** 188.
Laumen, K. & Schneider, M. P. (1988). *J. Chem. Soc., Chem. Commun.,* **9,** 598.
Lazar, G., Wiess, A. & Schmid, R. D. (1986). *Henkel-referate,* **22,** 5.
Linfield, W. M., O'Brian, D. J., Serota, S. & Barauskas, R. A. (1984). *J. Am. Oil Chem. Soc.,* **61,** 1067.
Lokotsch, W., Fritsche, K. & Syldatk, C. (1989). *Appl. Microbiol. Biotechnol.,* **31,** 467.
Lui, W. H., Beppu, T. & Arima, K. (1973). *Agric. Biol. Chem.,* **37,** 1349.
Ma, A. N. & Ong, A. S. H. (1986). *Water Sci. Tech.,* **18,** 35.
Macrae, A. R. & How, P. (1983). British Patent Application GB2119397A1, 16 November.
Marlot, C., Langrand, G., Triantaphylides, C. & Baratti, J. (1985). *Biotechnol. Lett.,* **7,** 647.
Morita, S., Narita, H., Matoba, T. & Kito, M. (1984). *J. Am. Oil Chem. Soc.,* **61,** 1571.
Negishi, S., Sato, S., Mukataka, S. & Takahashi, J. (1989). *J. Ferment. Bioeng.,* **67,** 350.
Okumura, S., Iwai, M. & Tsujisaka, Y. (1980). *J. Biochem.,* **87,** 205.
Osterberg, E., Blomstrom, A.-C. & Holmberg, K. (1989). *J. Am. Oil Chem. Soc.,* **66,** 1330.
Pauckner, W. (1981). *Leder und Häutemarkt,* **33,** 182.
Potts, R. H. & Mukerheide, V. J. (1968). In *Fatty Acids and Their Industrial Application,* ed. E. S. Pattison. Marcel Dekker, New York, p. 28.
Radwan, S. S. & Soliman, A. H. (1988). *J. Gen. Microbiol.,* **134,** 387.
Randolph, T. W., Clark, D. S., Blanch, H. W. & Prausnitz, J. M. (1988). *Science,* **239,** 387.

Ratledge, C. (1989). In *Microbial Lipids*, Vol. 2, eds C. Ratledge & S. G. Wilkinson. Academic Press, London, p. 567.
Reslow, M., Adlercreutz, P. & Mattiasson, B. (1988). *Eur. J. Biochem.*, **172,** 573.
Rucka, M., Turkiewicz, B., Tomaszewska, M. & Chlubek, N. (1989). *Biotechnol. Lett.*, **11,** 167.
Rydin, S. (1989). Biological conversion of low grade fats. Licentiate thesis, Department of Applied Microbiology, University of Lund, Sweden.
Rydin, S., Molin, G. & Nilsson, I. (1990). *Appl. Microbiol. Biotechnol.*, **33,** p. 473.
Seino, H., Uchibori, T., Nishitani, T. & Inamasu, S. (1984). *J. Am. Oil Chem. Soc.*, **61,** 1761.
Shimizu, S., Shinmen, Y., Kawashima, H., Akimoto, K. & Yamada, H. (1988). *Biochem. Biophys. Res. Commun.*, **150,** 335.
Stansby, M. E. (1979). *J. Am. Oil Chem. Soc.*, **56,** 793.
Sugiura, M. & Isobe, M. (1975). *Chem. Pharm. Bull.*, **23,** 1226.
Swaisgood, H. E. & Bozoglu, F. (1984). *J. Agric. Food Chem.*, **32,** 7.
Tan, K. H. & Gill, C. O. (1985). *Appl. Microbiol. Biotechnol.*, **21,** 292.
Taylor, F., Panzer, C. C., Craig, J. C. & O'Brian, D. J. (1986). *Biotechnol. Bioeng.*, **28,** 1318.
Wang, Y. J., Sheu, J. Y., Wang, F. F. & Shaw, J. F. (1988). *Biotechnol. Bioeng.*, **31,** 628.
Watanabe, N., Ota, Y., Minoda, Y. & Yamada, K. (1977). *Agric. Biol. Chem.*, **41,** 1353.
Wisdom, R. A., Dunnill, P., Lilly, M. D. & Macrae, A. (1984). *Enzyme Microbiol. Technol.*, **6,** 443.
Wisdom, R. A., Dunnill, P. & Lilly, M. D. (1987). *Biotechnol. Bioeng.*, **29,** 1081.
Yokozeki, K. *et al.* (1982). *Eur. J. Appl. Microbiol. Biotechnol.*, **14,** 1.
Zaks, A. & Klibanov, A. M. (1985). *Proc. Natl Acad. Sci. USA*, **82,** 3192.

Chapter 14

BIOCONVERSION OF WASTES FROM MARINE ORGANISMS

A. M. Martin & T. R. Patel

Department of Biochemistry, Memorial University of Newfoundland, St John's, Newfoundland, Canada A1B 3X9

CONTENTS

1 INTRODUCTION

Marine biomass constitutes an abundant and relatively inexpensive source of food and feed, and of raw materials for industry. Functional biopolymers, enzymes and pharmaceuticals are among the valuable products which can be extracted from marine organisms (Nestaas, 1986). The main activity relating to the exploitation of marine resources is fishing. Traditionally, fisheries wastes have been disposed of by returning them to the sea. This practice has contributed to the lag in the seafood waste processing industry, as compared to its terrestrial counterparts, concerning appropriate ways to utilize seafood processing wastes (Green & Mattick, 1979).

Besides the general lack of technological development which has characterized the fish processing industry, waste recovery and utilization has been hampered by several factors. These include the lack of suitable processes for the recovery and purification of individual components in the fish waste, the poor predictability of the quality of the marine catch, and the fact that most of the products that can be recovered from this kind of waste can, generally, be produced from other sources (Nestaas, 1986).

Lately, technological developments in the fisheries sector, which has concentrated larger amounts of fisheries wastes in large, centralized processing plants (in contrast to smaller plants disseminated over wide coastal areas), has produced the need for stricter regulations concerning waste disposal. In addition, the depletion of many fishing areas and the need for increased economic returns from the processing operations will require a more complete utilization of the total marine biomass catch. The conversion of much of the presently discharged fisheries wastes and by-products to value-added products is a solution to those problems. The success of this course of action should not be difficult to realize, considering the high biological value of many components of those wastes, such as high-quality protein and oils, enzymes, and complex polysaccharides.

Since fisheries constitute at present the main form of exploitation of marine resources, they are also the main source of marine wastes. This chapter will concentrate on those wastes. The development of aquaculture processes will contribute to the weight of this economic sector in the marine area. Another important activity related to the use of marine resources is seaweed cultivation and utilization, which has been studied by other researchers (Cheney, 1984; Renn, 1984).

Marine waste resource utilization will be presented here only from the point of view of biological (enzymatic and microbially-induced) processes, in accordance with the aims of this book. For marine-related wastes, most of those processes still are in the experimental stages, and most of them are, once again, related to fisheries wastes. In that sense, this chapter could also have been titled 'Bioconversion of Fisheries Wastes', or 'Bioconversion of Seafood Processing Wastes'. However, the authors prefer the more general title which it bears, as it emphasizes the whole of the marine sector as a source of potentially bioconvertible wastes.

1.1 Characteristics of the Main Sources of Marine Wastes

A potential large source of waste materials is the fisheries industry. Apparently, the fish frames and guts are the most important source of wastes. Approximately 40% of a head-on, gutted fish is the frame, that is, the portion left after the fillets have been removed. In 1988, almost 351 000 tonnes of groundfish were landed in the Canadian Province of Newfoundland, alone

Table 1
Fish Landed and Values: Newfoundland, 1988

Landings	*Quantities (×1000 tonnes)*	*Values ($ CAN millions)*
Groundfish[a]	351	167
Pelagic and other fin fish[b]	106	35
Molluscs and crustaceans[c]	28	58
Total	485	260

[a] Cod, flatfish, haddock, halibut, pollock, turbot, etc. [b] Herring, mackerel, salmon, capelin, char, trout, etc. [c] Clams, lobsters, shrimps, crabs, mussels, whelks, etc.
Source: Fisheries Statistics and Systems Branch, Department of Fisheries and Oceans, St John's, Canada.

(Table 1). This translates into about 140 000 tonnes of fish frames available as a source of raw material. In addition, about 32 000 tonnes of fish guts were generated, which could have been further processed for recovering useful enzymes as will be discussed below. The amounts of waste mentioned above increase greatly when one considers the total fisheries waste of Canada, which is one of the main fishing countries of the world.

Table 2 indicates the groundfish catches of major North Atlantic fishing countries. It can be observed that Canada catches over a third of the total, highlighting the importance for fish operators to develop new technologies focusing on the bioconversion of the resulting waste. Solid wastes generated by seafood plants range from about 30 to 85% of the weight of landed fish, depending upon the type of fishery. Processing of fin fish, crab, and shrimp can result in 30–60%, 75–85% and 40–80% waste, respectively (Green & Mattick, 1979). Table 2 reveals that a large potential source of raw material exists globally from the fisheries industry. The basic recoverable components of fish frames and guts are proteins and oils. Two-thirds of the crustacean shell waste

Table 2
Annual Groundfish Landings of Major North Atlantic Fishing Countries

Country	*Quantities (×1000 tonnes)*	*%*
Canada	757	34·1
Iceland	605	27·2
Norway	559	25·2
Denmark	180	8·1
USA	119	5·4
Total, listed countries	2 220	100·0

Source: Joyce, 1988.

is calcium carbonate and chitin, a polymer of *N*-acetylglucosamine that has a molecular chain structure similar to cellulose.

Conversion of some of this waste into fish meal occurs. However, fish meal production requires high capital investment, technical labor, and fuel, and is therefore not economical for small operators. Important considerations limiting investment in fish-processing waste elimination are the small size and the seasonal nature of most seafood processing plants. The other problem is that the disposal of fisheries wastes also involves high capital, labor and other operational costs which could be prohibitive for small plants. The review of fishery by-product utilization by Brody (1965) discusses many similar problems. Others have surveyed solid fisheries wastes, their utilization and methods of disposal (Martin, 1972; Kreag & Smith, 1973; Green & Mattick, 1979). The following sections will discuss processes involving biotransformation reactions in the utilization of fisheries wastes.

2 ENZYMATIC PROCESSES IN THE BIOCONVERSION OF FISHERIES WASTES

The proteolytic enzymes responsible for the bioconversion of fish proteins come from fisheries wastes and in some cases from microorganisms. A great variety of proteases have been found in fisheries wastes. The fish guts contain digestive enzymes such as pepsins from gastric mucosal glands, and trypsins and chymotrypsins secreted by the pancreatic tissue or pyloric caeca. Numerous exopeptidases are also secreted in the gut (Gildberg, 1982). Fish muscle tissue also harbors a battery of enzymes such as peptidases, cathepsins, transaminases, amidases, amino acid decarboxylases and glutamic dehydrogenases (Siebert & Schmitt, 1965). Cathepsins are intracellular enzymes which occur in muscle tissues and can bring about protein hydrolysis. These enzymes are mostly located in the lysosomes within the cells and in this respect differ from digestive enzymes which are secreted by cells (Mycek, 1970). Cathepsins are classified on the basis of their specificity towards substrates. Fish cathepsins which are active at acid and neutral pH have been described (Huang & Tappel, 1971; Makinonda & Ikeda, 1971).

In the degradation of fish waste, enzymes of microbial origin are often used. Usually eviscerated fish are fermented when exogenous enzymes from such sources are employed. *Aspergillus oryzae* and *Monascus purpureus* have been used as starter cultures in the fermentation of fish sauce (Amano, 1962; Miyazawa *et al.*, 1979).

Enzymes from the above-mentioned sources bring about biochemical changes during fermentation. Whitaker (1978) has discussed transformations occurring during fermentation of protein-rich food. Hydrolysis of proteins and decomposition of amino acids may result from biochemical reactions taking place during fermentation. The breakdown of proteins into peptides is brought about primarily by endoproteases, while exoproteases, including carboxypeptidases, amino peptidases, and di- and tri-peptidase, reduce the peptides into amino acids. The actions of decarboxylases, deaminases and transaminases result in amines, aliphatic acids and ammonia (Whitaker, 1978). Depending upon the type and quantity of hydrolases, the proportions of non-protein and protein nitrogen will vary. Factors such as temperature, pH and the presence of metal ions such as Ca^{2+} will also contribute to the levels of peptides, ammonia and aliphatic acids in the end product.

Saisithi (1967) showed that in Thai fish sauce the total soluble nitrogen increased during the fermentation process, as did total acid, which includes such acids as lactic acid and volatile acids. Ammoniacal nitrogen was found to increase during the first 6 months and then remain constant until the end of the process. The fish sauce fermentation occurred in three stages during which different enzymes played a role. Exopeptidases predominated during the early stage, in which soluble nitrogen and free amino acids increased rapidly (Tongthai & Okada, 1982). During the middle stage endopeptidases played a major role and the increase in free amino acids was very gradual. In the final stage halophilic bacteria appeared in the fermentation substrate and the release of free amino acid stopped.

In the case of fish silage, liquefaction is due almost entirely to autolytic proteolysis. The biochemistry of the process is not well understood in detail. Tatterson and Windsor (1974) have studied the formation of free amino acids and short chain peptides, and the corresponding reduction in TCA-precipitable protein during ensiling. The application of enzymes to food processing waste has been reviewed by Shoemaker (1986).

The idea of enzymatic reduction of food processing wastes has been discussed before by several authors (Cooper, 1975; Reed, 1975). Hydrolases such as amylase, pectinase, mannanase, cellulase, hemicellulase and lactase find applications in the fruit, vegetable, grain, coffee and dairy industries. For the present discussion, chitinases and proteases have been selected.

Chitinases have been used to depolymerize the shells of crustaceans (Cosio *et al.*, 1982). These authors attempted to utilize *N*-acetylglycosamine, released by chitinase reaction, to grow yeast to be used as single-cell protein. Chitosan, a product of deacetylation of chitin, as discussed above, can be hydrolyzed by chitosanases.

Proteases of fish origin play an important role in many products such as silage, fish sauce and fish solubles. New emerging technologies will employ fish enzymes for the deskinning of squid and fish, for caviar production and the cleaning of scallops (Raa, 1989). Such innovations in fish processing are likely to result in the reduction of fisheries wastes. Stefansson (1989) recently presented the main research areas of the Icelandic Fisheries Laboratories (IFL), and discussed the use of proteolytic enzymes from cod viscera for deskinning, membrane removal and scale removal.

2.1 Fish Silage

Fish silage is a liquid product, made from whole fish or parts of fish, to which no material has been added other than an acid, and in which liquefaction is carried out by enzymes already present in the fish. Acid aids in accelerating the process by creating the right conditions for the enzymes to work and by helping to break down bone. Acid also limits the growth of spoilage bacteria. After liquefaction, it is convenient to remove the oil which came from the raw material. The resulting product is a high quality animal feed with a long storage life (Tatterson & Windsor, 1974). Liquefaction can also take place under alkaline conditions. Hydrolyzates have been made in Iceland from fisheries wastes using the enzymes of the digestive tract in the presence of ammonia (Anon., 1975).

The application of the principle of acid preservation to fisheries wastes originated from earlier attempts to preserve animal feed materials. The first attempt to preserve and liquefy fish and fisheries wastes took place in the 1930s in Sweden. This appears to have been the origin of the product we know today as fish silage. In these initial trials a mixture of sulfuric and hydrochloric acids, also called AIV acid after Professor A. I. Virtenen, was used. Later, formic acid was used, and it remains the acid of choice today. Formic acid, beside aiding liquefaction, has a bactericidal action and preserves the silage at a pH of between 4·0 and 4·5, whereas a pH of 2·0 must be maintained if mineral acids are used alone. This is important because at pH 4·0 it is not necessary to neutralize the silage before use, and there are also fewer corrosion problems.

Production of fish silage on a commercial scale in Denmark started in 1948 using formic acid. Others also found formic acid to be a superior agent for silage (Hanson & Lovern, 1951). In Canada, Freeman and Hoogland (1956) used cod and haddock offal together with sulfuric acid to produce silage which was dried using vacuum drying, spray drying and drum drying. Good end products resulted with all three methods. Several reviews on silage have been reported (Petersen, 1953; Hansen, 1959). McBride *et al.* (1961) examined the hydrolysis of herring by different methods including proteolysis under acid conditions, high pressure steam liquefaction and proteolytic enzyme solubilization. The preservative effects of a variety of acids and their mixtures on various species of fish was studied by Lagunov *et al.* (1956).

The manufacture of silage is very simple, involving three steps, as illustrated in Fig. 1: (i) chopping or mincing; (ii) blending with the acid; and (iii) digestion and storage. When very oily raw fish is the starting material it is beneficial to remove the oil. Several processes have been worked out employing a variety of fish offals from many types of fish. Immediately after mincing, 3·5% (by weight) of 85% formic acid is added. The mixture must be at a pH level of 4 or lower to prevent bacterial action. The production tank can be of any size or shape, provided it is acid resistant.

The composition of fish silage is very similar to that of the material from which it is made. A typical analysis may yield 80% water, 15% protein and 4·5% ash. The fat composition will vary depending upon the nature of the fish offal (less or more oily). Fish silage of the correct acidity keeps at room temperature for at least 2 years without putrefaction.

Fish silage is used in the same way as fish meal in animal feed. Because of

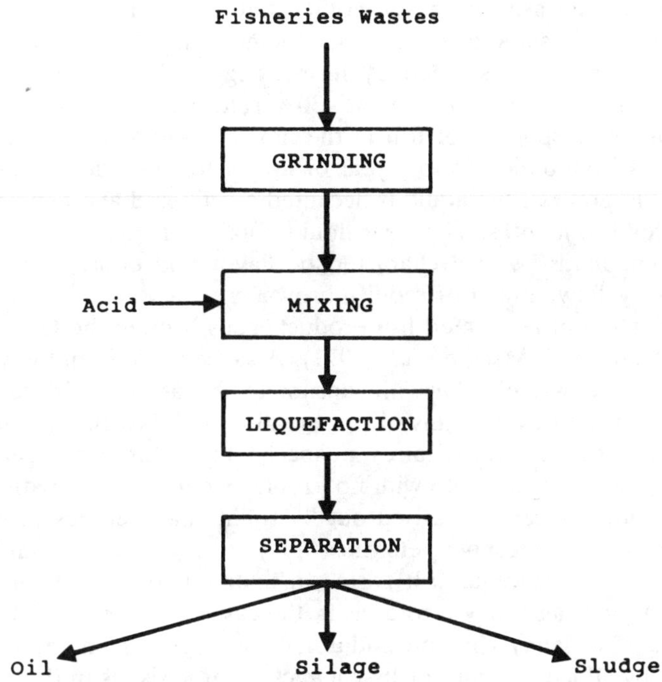

Fig. 1. Schematic diagram of the fish silage process.

the lower protein content of silage compared to fish meal, more silage is required for the same amount of protein intake. Fish silage has been used in feeding pigs, cows and hens with good results. Cow's milk and butter are untainted and egg production from hens is increased. The nutritional value of silage in animal feeds has been examined by many workers (Hillyer *et al.*, 1976; Smith & Adamson, 1976; Whittenmore & Taylor, 1976*a,b*).

The main advantages of fish silage are the recovery of fish offal and waste fish, low capital cost, no need for highly skilled workers, no offensive odors and good nutritional value for animals. Although, compared to fish meal, transport of silage is more expensive, the utilization of fisheries wastes for this purpose is an inexpensive way to eliminate the pollution problems associated with them.

2.2 Fish Sauce

Fish sauce is a well-known product used as a condiment in Southeast Asia, in countries such as the Philippines, Thailand and Vietnam, among others. It has been reported that fish sauce is part of the daily diet of hundreds of millions of people (Saisithi *et al.*, 1966). In addition, the demand for fish sauce is growing in other parts of the world, including North America, due to the increasing number of immigrants of Southeast Asian origin (Raksakulthai *et al.*, 1986).

The production of fish sauce is another alternative for the utilization of unwanted fish species and of refuse from commercial fisheries operations.

Generally, fish sauce is made by fermenting small fish such as sardines, anchovies, roundscale or capelin in 15–30% salt. The high salt concentration prevents microbial spoilage of fish in the hot and damp climate of Southeast Asia. It takes from 6 months to 1 year or longer for the fish fermentation. At the end of the process, the liquid is decanted off, filtered and allowed to ripen in the sun for 1–4 months. The clear liquid obtained is rich in salt and soluble nitrogen compounds, with a characteristic flavor and odor. The liquid may appear straw-yellow, amber or reddish brown.

The production of fermented fish products goes back to the Graeco-Roman times (Badham, 1954, Mackie *et al.*, 1971). A sauce made from the viscera and blood of mackerel was popular with consumers of that time. Fish fermentation has often been used as a means of preservation and flavoring (Liston, 1980). The process is inexpensive, requires no special apparatus or refrigeration, and the product has a long shelflife with no problems pertaining to pathogens.

Fish fermentation can be carried out by traditional methods in which only added salt and endogenous enzymes play a role, for example in the preparation of fish sauce or fish paste. In addition to endogenous enzymes, microbial enzymes also play a role, as in the cases of *mam-chau* (Cambodia), and *nampla* (Thailand). In non-traditional methods, the natural process is accelerated by added enzymes or by chemical hydrolysis, as in the preparation of fish solubles. Fish sauce, besides being used as a condiment, is also a good source of protein and calcium (Beauchat, 1983).

Different names have been coined for fish sauces originating from various geographic regions (Table 3). The source of raw materials also varies according to the region. For example, in Greece, a fish sauce called *garos* is made from the liver of a scomboid fish. In the south of France, *pissala* is made from larvae or small fish. A sauce called *shotturu* is produced in Japan from sand fish, sardines, anchovies, and molluscs, especially squid (Subba Rao, 1967).

Table 3
Fish Sauces from Various Countries

Country	*Product name*
Burma	Ngapi
Japan	Shotturu
Malaysia	Budu
Philippines	Patis
Thailand	Nampla
Vietnam	Nuoc-mam
France	Pissala
Greece	Garos
Cambodia	Mam-chau

Sources: Raksakulthai, 1986; Raksakulthai *et al.*, 1986.

The proteolytic enzymes responsible for the hydrolysis and ripening of fish sauce can be of endogenous, exogenous or microbial origin. Endogenous enzymes in fish come from the digestive tract, internal organs (kidney, liver, etc.) and muscle tissues. A battery of endogenous proteases originate from digestive organs; for example, pepsins are secreted by the pancreatic tissue and pyloric caeca. In recent years gastric proteases of fish have received attention from many workers. Proteases from Greenland cod (Squires *et al.*, 1986*a*), sardines (Noda *et al.*, 1982; Van *et al.*, 1983), capelin (Gildberg, 1982; Raksakulthai *et al.*, 1986), Arctic cod and American smelt (Haard *et al.*, 1982), and salmon and tuna (Norris & Mathies, 1953) have been investigated. Trypsins and chymotrypsins from capelin and herring (Kalac, 1978*a*,*b*), Greenland cod (Simpson & Haard, 1984), and capelin (Hjelmeland & Raa, 1982) have been reported. Cathepsins, amidases, peptidases and amino acid decarboxylases found in muscle tissue were described by Siebert and Schmitt (1965). Intracellular cathepsins from fish muscle tissue (Huang & Tappel, 1971; Makinonda & Ikeda, 1971), with activity under neutral and acidic conditions, have been examined. Del Rosario and Maddo (1984) studied the activity of cathepsins during the fermentation of fish sauce. Exogenous proteases used in traditional fermented fish come from fungi (Amano, 1962), and yeast and bacteria (Crisan & Sands, 1975). Proteolytic enzymes such as papain, bromelain, ficin, pronase, trypsin, etc., have also been tested in the preparation of fish sauce (Murayama *et al.*, 1962; Guevara *et al.*, 1972; Ooshiro *et al.*, 1981). However, none of these enzymes has been used in the production of fish sauce on a commercial scale. The exact role of microbial enzymes in fish sauce fermentation is still under investigation.

In fish sauce, the degradation of fish proteins into soluble proteins, peptides and amino acids is apparently due primarily to endogenous fish enzymes (Orejana, 1978; McIver *et al.*, 1982). Bacteria are believed to play an important role in the development of flavor (Saisithi, 1967; Beauchat, 1983). Ammonia may arise from two types of reactions (Amano, 1962): from the activity of enzymes in the fish, and from the activity of bacterial enzymes.

2.3 Chitin and Chitosan

Waste materials from shellfish processing present a more sophisticated technological problem, that of the disposal of relatively insoluble, environmentally persistent crustacean shells. These shells consist of approximately 33% protein and 33% chitin. Chitin and its deacetylated derivative, chitosan, are interesting polysaccharides with unique properties. Although chitin is insoluble in most solvents and resists most chemical reactions, it can be deacetylated to form chitosan by reacting with hot concentrated NaOH. Chitosan is soluble in organic acid solutions and has a variety of uses which make it valuable.

Chitosan is a biopolymer carrying positively charged amine groups. It is an attractive resource because of its abundance, its unique material properties and its various functional properties which can be exploited commercially. Potential commercial sources of chitin are the shells of crabs, shrimps, lobsters, clams, and oysters. Chitin is the most widely distributed polymer on earth, and is one of the most abundant (Austin *et al.*, 1981).

There are many potential industrial applications for chitosan and other derivatives of chitin. In industry, chitosan is used for reinforcing mechanical strength, enhancing dye-binding properties, reducing the consumption of sizing agents and preventing shrinkage of paper, pulp and textile products (Muzzarelli, 1973; Johnson & Carlson, 1978). Casey (1978) has discussed the application of chitosan and chitin in biomedical areas. They are believed to possess properties which promote and accelerate wound healing. Chitosan may find applications in improving the tactile properties of surgical gloves, and in making suture thread, bandages and sponges (Capozza, 1976; Austin, 1979). Chitosan membranes may be used in skin grafting procedures in the treatment of burns or in plastic surgery. Other applications of chitosan and its derivatives include dialysis, the deacidification of fruit, vegetables and coffee extracts, and in the future, possibly in some separation processes (Kesting, 1979; Brambilla & Horman, 1980). Chitosan is also used as a coagulant to recover proteins from waste streams. Separation chromatography and enzyme immobilization are other areas where chitosan has been very useful (Muzzarelli, 1973).

Chitin bioconversion has been proposed as a waste treatment alternative to the disposal of shellfish waste (Carroad & Tom, 1978; Revah-Moiseev & Carroad, 1981; Cosio *et al.*, 1982). Several microbes including bacteria (Tominaga & Stujisake, 1975; Davis & Eveleigh, 1984), fungi (Fenton & Eveleigh, 1981), myxobacteria (Hedges & Wolfe, 1974) and actinomycetes (Price & Stork, 1975; Ohtakara *et al.*, 1984) have been shown to produce chitosanases.

2.4 Extraction of Enzymes from Fisheries Wastes

By-product recovery methods employing old and new concepts are being reinvestigated for the fisheries. The possibility of recovering enzymes and other biological compounds from fish waste was suggested by Green and Mattick (1978, 1979). It should be pointed out that much of the work is in the exploratory stage and some is in the limited production stage. Selected examples that illustrate the potential and the utility of enzymes from fish waste will be presented.

According to Godfrey and Reichelt (1983), over 80% of all industrial enzymes are hydrolases of which approximately 60% are proteases, also known as proteinases or proteolytic enzymes. Of the enzymes used as food processing aids, the proteases are used most extensively to improve the quality, stability or solubility of foods, as in baking, brewing, cheese making and meat processing. Although all living organisms are potential sources of enzymes of industrial importance, the greatest variety of industrial enzymes are presently derived from microbial sources, with only a limited number coming from plant and animal sources. It is somewhat predictable that the use of proteases from plant and animal sources will increase substantially in the foreseeable future. This is because only a few microorganisms are recognized as safe and hence the problems in obtaining clearance for their use in foods will remain a major obstacle (Denner, 1983). Another factor in favor of animals and plants as sources of enzymes is that the use of those parts of animals and plants generally discarded as waste is more economical.

Fisheries waste is a large potential source of a variety of proteolytic enzymes with unique properties ready to be exploited. Proteolytic enzymes have been reported in various marine species including sardines (Noda & Murakami, 1981), salmon (Norris & Elam, 1940), tuna (Norris & Mathies, 1953), dogfish (Merrett *et al.*, 1969), bonito (Kubota & Ohnuma, 1970), trout (Owen & Wiggs, 1971), Atlantic cod (Brewer *et al.*, 1984), Greenland cod (Simpson & Haard, 1984), smelt (Haard *et al.*, 1982), Arctic cod (Arunachalam & Haard, 1985), and harp seals (Shamsuzzaman, 1983; Shamsuzzaman & Haard, 1984). Leading fish-producing and fish-exporting countries are putting more and more emphasis on new product development using underutilized fish. One outcome of expanding the fishing industry is the increased production of fish offal.

The utilization of fish offal as a source of industrial enzymes is very attractive since it would result in minimizing the need for waste disposal and would provide valuable biochemicals, such as proteases. Proteases from fish differ from those originating in other animals and in plants. Simpson and Haard (1987) have discussed the properties of some of the fish enzymes. Like some other poikilothermic organisms, fish are known to adapt to extremely low temperatures. This adaptation affects such enzyme properties as binding affinity, thermodynamic properties, molecular activity at low temperatures, specificity and cold stability (Simpson & Haard, 1987). Some of these properties may be exploited in certain food processing operations.

Trypsin was isolated from the pyloric caeca and the intestines of the Greenland cod (*Gadus ogac*), and characterized by Simpson (1983). The enzyme was distinct in its properties when compared to bovine trypsin. The author suggested its greater stability and activity at alkaline pH could be most suitable for industrial processes such as the hydrolysis of fish protein. Application of such proteases in the leather, detergent and dairy industries needs to be explored.

In another study, Squires *et al.* (1986*a,b*) examined proteases from subarctic specimens of Greenland cod (*G. ogac*). The zymogens of three gastric proteases were separated, purified and characterized. All three of the cod proteases had more alkaline pH optima with protein substrates than porcine pepsin. The cod enzymes were also found to differ from porcine pepsin in such other properties as heat stability, substrate specificity, milk clotting ability and amino acid composition (Squires, 1984).

Four zymogens of acidic proteases were isolated from the gastric mucosa of a harp seal (*Pagophilus groenlandicus*). Certain physical and catalytic properties of the proteases were compared with those of calf chymosin and porcine pepsin (Shamsuzzaman, 1983). Protease A was shown to be similar to calf chymosin in many respects. It has a higher ratio of milk clotting to proteolytic activity than seal protease C and porcine pepsin, clots milk up to pH 7·0, and has a pH optimum of 2·2–3·5 for hemoglobin hydrolysis. Experimental cheddar cheese made with protease A received higher sensory scores than those made with calf rennet, Formase (*Mucor miehei* protease) or crude seal gastric protease (SGP). The author concluded that protease A and SGP are promising as rennet substitutes. Other studies have shown that it is possible to prepare satisfactory cheddar cheese using proteases from marine organisms (Brewer *et al.*, 1984).

3 MICROBIAL PROCESSES IN THE CONVERSION OF FISHERIES WASTES

3.1 Lipolytic Fermentation

Any attempt to recover fish protein from solid fishery wastes must deal with the presence of lipids in the product. In addition, the lipid content of liquid fishery effluents constitutes a major source of environmental damage. Few fish plants have effluent treatment systems, and in those having some kind of treatment, the disposal of fish oils is one of the main problems. To most fishery operators, the removal and eventual recovery of fish oils is a high-capital, prohibitively expensive process.

Stickwater, an aqueous by-product of the fish meal industry, is usually evaporated to a form called 'fish solubles', which is used as an animal feed supplement. However, the presence of an average 11% of lipids (which are prone to developing oxidative rancidity) in 'fish solubles' limits the use of stickwater in feeds (Green *et al.*, 1976).

Good quality fish oil can be a marketable by-product, but its removal involves expensive equipment and skilled labor (Wignall & Tatterson, 1976), both prohibitive for most fishery operators. Therefore, inexpensive methods to reduce fish oil contents have always been pursued. Lipolytic fish fermentation tests have been conducted and reported to have been successful in decreasing the lipid content of fish and fish wastes (Burkholder *et al.*, 1968; Li *et al.*, 1970; Hottinger *et al.*, 1974*a*; Green *et al.*, 1976).

The biological removal of fish oils by lipolytic fermentation could be a practical way of decreasing the lipid content in fishery wastes and effluents, and an economical alternative to existing methods of oil removal. Moreover, it is expected that the growth of lipolytic microbial populations, by contributing microbial biomass protein, will improve the nutritional value of the recovered fish waste. This biological process could be more economical than other existing operations for lipid removal. Moreover, the nutritional value of the final product, enriched with the microbial biomass, could be higher.

Previous studies (Burkholder *et al.*, 1968) have shown that the yeasts

Candida lipolytica and *Geotrichum candidum* are promising microorganisms, which may be used for pure culture fermentation to produce high-quality food products. Those experiments produced a reduction of the fat content of young menhaden (*Brevortia tyrannus*) by 30–50% and increased the protein content by up to 20%. *Candida lipolytica* has been extensively tested for the production of microbial biomass protein (MBP) from hydrocarbons (Whitworth *et al.*, 1973; Gutierrez & Erickson, 1978; Kosaric *et al.*, 1979). Fermentation of fish lipid *in situ* (Burkholder *et al.*, 1968), of stickwater (Green *et al.*, 1976) and of fish oils (Li *et al.*, 1970; Hottinger *et al.*, 1974*a*) have been conducted with *C. lipolytica* and *G. candidum* on an experimental basis. In all cases, it has been reported that the lipid content was reduced and microbial cell growth was obtained. Nevertheless, little attempt was made to optimize the fermentation process itself or to design a process with possible commercial applications.

Difficulties in the utilization of fish oils as substrate in fermentation processes have been reported. Pan *et al.* (1959) observed that the high content of oleic and linoleic acids in vegetable oils increased penicillin production, but fish oils, which have a high content of polyenolic acids (polyunsaturated), were not as effective when used as substrate. Li *et al.* (1970) observed that growth in a fish lipid fermentation in shake flasks stopped before the expected level of consumption of the substrate was reached.

Fish oil contains a high concentration of polyunsaturated fatty acids which, in the presence of oxygen (as is the case in aerobic fermentations), produce high concentrations of peroxides. These are oxidative deterioration products, which could inhibit the growth of microorganisms. Higgs (1974) reported culture selection from three thermophilic enrichment systems and one mesophilic enrichment system. Selected microorganisms were tested for their ability to metabolize fish oil, and it was found that growth inhibition produced by high peroxide concentrations was eliminated by the addition of food grade antioxidants to the fish oil prior to fermentation, at concentrations of approximately 0·5–1·0%. Zajic *et al.* (1974) discussed the additional advantages that reduced oxidative deterioration could produce during fish oil fermentation, such as better emulsification of the oil during the process, and the prevention of the formation of polymers and soaps in the fermentation equipment, which decrease the fermentation yields.

3.2 Fisheries By-products as Energy Sources in Fermentation Processes

It has long been noted that the annual harvest of marine biomass represents an abundant and relatively inexpensive source of materials from which food and feed additives, enzymes, pharmaceuticals and other products could be extracted (Nestaas, 1986). One possibility is to obtain marine chemicals which could be employed as sources of nutrients and energy to support microbial growth. With the presently increasing development of biotechnological productions and the future prospects for this industrial sector, there is a need to find new, plentiful and inexpensive substrate sources for fermentation processes.

Although there has been much reference to the potential of chitin-derivatives from shellfish waste as an abundant source of marine polysaccharides, it has still not been fully developed. There remains a need to develop new enzymatic processes to improve the processing and quality of the chitin-based products. Because of the associated costs, it is possible that chitin will remain a source of materials for specialized productions, but not of large amounts of inexpensive raw materials such as are required in fermentations. However, it has been predicted that marine oils may become an increasingly important substrate source for fermentation processes. It has been stated that fish oil is a very attractive alternative to sugars for employment as fermentation substrate (Nestaas, 1986), considering that fish oil is, after crude oil and methanol, one of the cheapest available sources of carbon and energy for fermentation processes.

Nestaas (1986) has also presented tables showing that, on the basis of biomass yield (grams of biomass produced per gram of substrate consumed), it could be expected that the yield obtained when using fish oil as substrate for the production of microbial cells and products would be in the order of 0·70, which is considerably higher than the yields obtained with dextrose, methanol and fish meal (0·50, 0·45 and 0·25, respectively). The same author also presented information, based on costs assumed at the time of his article, showing that fish oil has the lowest substrate costs per unit of biomass produced when compared with the same other substrates mentioned above, and also per unit of product when compared with dextrose.

3.3 Composting

Around the world, many thousands of tonnes of fish scrap are disposed of as waste each year. Fisheries wastes can be used as a valuable organic fertilizer, but their foul smell prevents widespread use of this option. Mathur *et al.* (1986) have presented a new method of fish-waste composting, incorporating peat moss into the process. One of the main accomplishments of this procedure is that the NH_3 produced during the composting is absorbed as NH_4^+ by the moist, acidic peat fibres. It is claimed that the final product obtained is of high quality with an earthy odor and good concentrations of organic and inorganic nutrients (Mathur *et al.*, 1986). Mathur *et al.* (1988) presented information on the composition of raw materials used in the preparation of fish offal–peat compost. The authors also reported the major components and the mineral compositions, respectively, of various fish offal–peat composts prepared at Lameque, New Brunswick, Canada. This process is presented in more detail elsewhere in this book.

It has been reported that only two ingredients are required for producing fish compost: fisheries wastes and a bulking agent. The latter should allow aeration for the microbial population and provide an adequate carbon to

Table 4
Composition of Peat and of Fish Offal–Peat Compost

Components	*Concentration (% dry weight)*	
	Peat	*Compost*
Moisture (% wet weight)	$66{\cdot}6 \pm 1{\cdot}2^a$	$27{\cdot}07 \pm 0{\cdot}77^b$
Total lipids	$4{\cdot}8 \pm 1{\cdot}8^a$	$0{\cdot}90 \pm 0{\cdot}04^b$
Total nitrogen	$0{\cdot}66 \pm 0{\cdot}05^a$	$2{\cdot}5 \pm 0{\cdot}1^b$
Ash	$4{\cdot}3 \pm 0{\cdot}2^a$	$19{\cdot}5 \pm 2{\cdot}8^b$

[a,b] Mean values of three determinations ± standard deviations. Values in the same row with the same superscript are not statistically different ($P > 0{\cdot}05$).
Source: Martin & Chintalapati, 1989.

nitrogen ratio for an efficient composting without the production of offensive odors (Frederick *et al.*, 1989). These authors point out that, in general, wood wastes such as shredded brush, shredded bark and wood chips satisfy those needs.

Martin and Chintalapati (1989) produced liquid extracts, by chemical hydrolysis, of the fish offal–peat compost made by the process of Mathur *et al.* (1986), and employed those extracts as fermentation media for the growth of the acid-tolerant fungus *Scytalidium acidophilum*. The compost hydrolyzate contained higher concentrations of nitrogen than a liquid extract produced by a similar process of acid hydrolysis of peat, although the total carbohydrate concentration was higher in the latter (Tables 4 and 5). The authors concluded that fish offal–peat compost hydrolyzates were a good source of nutrients for the growth of the fungus, and did not require additional nutrient supplementation (with the possible exception of more fermentable carbohydrates) as do the

Table 5
Composition of Peat Hydrolyzate and of Fish Offal–Peat Compost Hydrolyzate

Components	*Concentration (g/liter)*	
	Peat hydrolyzate	*Compost hydrolyzate*
TCH (Total carbohydrate conc.)	$30{\cdot}25 \pm 1{\cdot}10^a$	$14{\cdot}45 \pm 0{\cdot}89^b$
Total lipids	$1{\cdot}70 \pm 0{\cdot}15^a$	$1{\cdot}93 \pm 0{\cdot}21^a$
Total nitrogen	$0{\cdot}65 \pm 0{\cdot}02^a$	$2{\cdot}70 \pm 0{\cdot}50^b$
Ash	$2{\cdot}75 \pm 0{\cdot}32^a$	$13{\cdot}85 \pm 0{\cdot}56^b$
Total solids	$60{\cdot}53 \pm 2{\cdot}78^a$	$43{\cdot}57 \pm 3{\cdot}33^b$
Dissolved solids	$45{\cdot}30 \pm 2{\cdot}57^a$	$40{\cdot}37 \pm 3{\cdot}51^b$
Total reducing sugars	$12{\cdot}10 \pm 0{\cdot}83^a$	$11{\cdot}64 \pm 0{\cdot}22^a$

[a,b] Mean values of three determinations ± standard deviations. Values in the same row with the same superscript are not statistically different ($P > 0{\cdot}05$).
Source: Martin & Chintalapati, 1989.

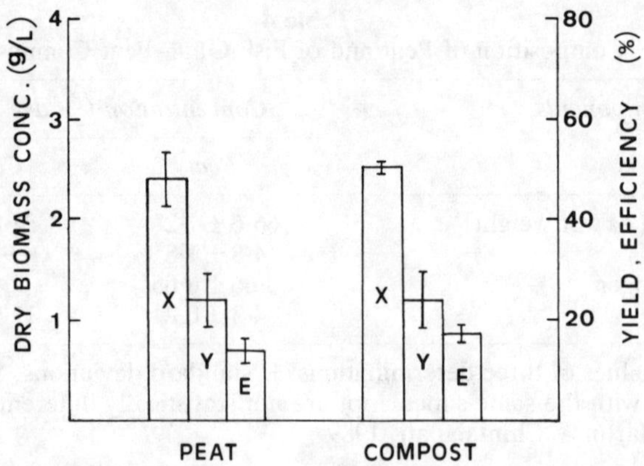

Fig. 2. Comparison of maximum growth parameters of *Scytalidium acidophilum* cultivated in peat hydrolyzate and fish offal–peat compost hydrolyzate. X = biomass concentration; Y = yield (g dry biomass produced per g substrate consumed); E = efficiency (g dry biomass produced per g initial substrate present in the medium). Source: Martin & Chintalapati, 1989.

peat extract and other carbohydrate-based culture media, which usually require nitrogen and other supplementary nutrients (Fig. 2).

3.4 Use of Fisheries Wastes as a Nutrient Source in Fermentation Processes

It has already been mentioned that industrial fermentation processes require the availability of abundant and inexpensive substrate sources. These characteristics are mostly associated with by-products of carbohydrate origin such as molasses, cellulosic hydrolyzates and other materials containing fermentable sugars. Carbohydrates, and lipids such as fish oils, when used as subsrates for the growth of microbial populations and the production of microbial metabolites, are basically sources of energy and carbon for the fermentation process. In those processes, it is usually necessary to add other ingredients as a source of required basic nutrients for the growing biomass. The expenses involved in incorporating those nutrient sources into the culture media could constitute an important factor in the total economy of the process.

The development of technologies for the utilization of nutrient-rich fisheries wastes as sources of nutrients for fermentation processes, such as the above-mentioned method reported by Martin and Chintalapati (1989), could open a new potential for their recovery. An example of the many existing possibilities in this regard is the use of culture media prepared from fish waste juice, reported by Hossain *et al.* (1988). Other notable cases are the production of microbial biomass protein employing fisheries wastes as substrate in submerged fermentation processes, and the utilization of fisheries by-products in the growth of mushrooms.

3.4.1 Production of Microbial Biomass Protein

Microbial biomass protein (MBP), also known as single-cell protein (SCP), is potentially useful as a high quality protein supplement for animal feeds and human foods.

Single-cell biomass production by the lipolytic yeasts *C. lipolytica* and *G. candidum* have been used for this process (Hottinger, 1972), utilizing a concentration of up to 5% fish oil in the culture media. The author reported an average crude protein content of 40% for both microorganism species. Hottinger *et al.* (1974*a*) reported alewife (*Alosa pseudoharingus*) oil in a basal medium was as effective a nutrient supplement as yeast extract or cornsteep liquor for the above-mentioned yeasts. The increase in cell yield stopped at the 5% oil level. The operating conditions for this process were optimized (Hottinger *et al.*, 1974*b*), and those studies suggested the possibility of obtaining 800 g of dry yeast biomass with a crude protein yield of 320 g from 1 kg of fish oil using both batch fermentation and continuous fermentation processes.

The fungus *Scytalidium acidophilum*, which was utilized by Martin and Chintalapati (1989) in their experiments using fish offal–peat compost extract as a substrate source (Fig. 3), is also considered to be a promising source of microbial protein, being able to attain a 47% crude protein content, relatively low levels of nucleic acids, and a good amino acid composition (Ivarson & Morita, 1982; Martin & White, 1985).

As was mentioned above, Carroad and Tom (1978) proposed chitin bioconversion to yeast single-cell protein as a waste treatment alternative to the disposal of shellfish waste. *Pichia kudriavzevii* grew well on chitin hydrolyzates and yielded a microbial protein with acceptable amino acid composition (Revah-Moiseev & Carroad, 1981). Cosio *et al.* (1982) studied the

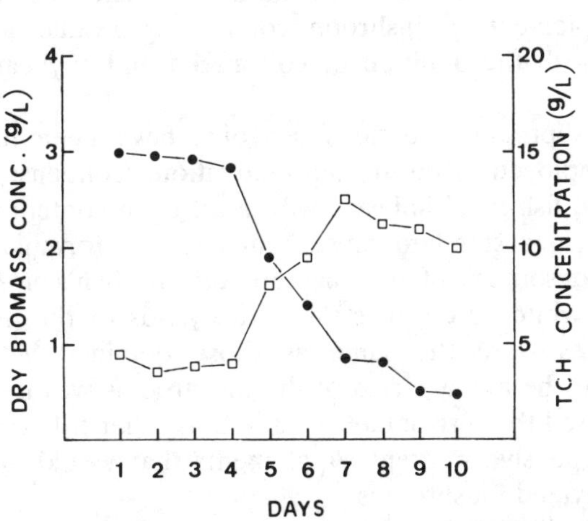

Fig. 3. Growth curve of *Scytalidium acidophilum* cultivated in fish offal–peat compost hydrolyzate. □, Dry biomass concentration; ●, total carbohydrate concentration. Source: Martin & Chintalapati, 1989.

conditions for chitin waste pretreatment (size reduction, deproteination and demineralization) and the chitinase production by *Serratia marcescens*. These authors also presented information on process design and economic analysis for an integrated process for the conversion of shrimp shell chitin waste to microbial biomass.

Welsh and Zall (1984) described a method to use spent food processing brines in the production of *Candida utilis* single-cell protein. Salt brines are used as contact refrigerants aboard fishing boats and, in the process, become contaminated with organic matter from the fish. Before being discarded, the brines need to be treated, and the fermentation process could make it possible to recycle the brines while producing SCP.

The treatment of wastewater from a dried-bonito processing factory using yeast was reported by Kakuta *et al.* (1985). One strain of *Hansenula anomala* showed the highest potential to utilize organic substances, reducing the wastewater chemical oxygen demand (COD) from 24 600 ppm to 6450 ppm, and the nitrogen content from 3610 ppm to 1050 ppm. The growth of five microfungi species in mussel steam-opening effluents was also studied, with the aim of developing a combined process for waste depuration and microbial protein production (Siso *et al.*, 1987).

These methods illustrate the great number and variety of potentials existing in the fisheries industry for employing waste materials in the production of microbial biomass.

3.4.2 Fisheries By-products in the Growth of Mushrooms

Evidence for a relationship between lipid metabolism and the initiation of fruiting in the cultivated mushroom *Agaricus bisporus* (Lange) Sing. was reported by Schisler and Sinden (1966), who found that supplementation with refined and crude seed oils increased mushroom yield. The use of menhaden fish oil as a supplement to mushroom composting produced results similar to those obtained with the addition of cottonseed and soybean oil (Schisler & Patton, 1974).

Other fisheries products, besides fish oils, have been found to enhance growth when employed in mushroom cultivation. Condensed fish solubles, a by-product of the fish meal industry with a nitrogen content of approximately 5·1%, was used in a standard mushroom compost formula in lieu of more traditionally used sources of organic supplement (Schisler & Patton, 1974). With either the white or cream strains, the yields of the mushrooms grown with fish solubles were the same as those obtained with other organic supplements, and the average size of the mushrooms was increased. Green *et al.* (1973) suggested that fish solubles, as well as other fisheries wastes, may be regarded as inexpensive nutrient supplements that would increase the yields and sizes of cultivated mushrooms.

Martin and Bassler (1989) cultivated *Pleurotus ostreatus*, commonly known as the oyster mushroom, with fisheries by-products, including fish meal, fish oil and fish offal–peat compost, as nutrient supplements. The mushrooms grew well in all the substrates employed, with the exception of those containing fish meal, in which no growth occurred. Fish oil was considered to produce the best results.

3.5 Production of Fish Protein Concentrates by Biological Methods

The production of fish protein concentrates (FPC) has been attempted with the objective of obtaining a new food source by mechanical, chemical or biochemical methods. It was expected that this process could help to reduce some of the fishing industry's problems, such as incomplete utilization of the caught fish and perishability of the product, among others (Green & Mattick, 1979). Early work concentrated on solvent extraction for the production of FPC (Knobl, 1967; Finch, 1970). However, industrial acceptance was poor due to a lack of functional properties in the product. More recent work has also included the use of enzymatic hydrolysis (Hale, 1974), and Spinelli *et al.* (1972) reported the production of FPC with functional properties. A process for the enzymatic conversion of underutilized fish to a liquefied product was patented by Keys and Meinke (1966). Venugopal *et al.* (1989) immobilized cells of *Bacillus megaterium, Aeromonas hydrophila* and *Pseudomonas marinoglutinosa* and applied them to fish meat suspension in water. The cells secreted protease which solubilized the fish meat. *Bacillus megaterium* was the most efficient hydrolyzing agent, producing the solubilization of 30% of the fish protein.

It has been stated that, although the technology for inexpensive fish liquefaction processes by either chemical or enzymatic methods has been developed and is ready to be applied, the overall process has become expensive due to the cost of the drying process involved (Green & Mattick, 1979).

4 CONCLUSIONS

4.1 Present Situation

The fishing industry is still far from the accelerated technological development which characterizes the avant-garde group of processing industries. Therefore, the utilization of fisheries wastes, although conducted in many cases at a primary level (the use of wastes as fertilizer and for feeding some animals), still awaits the widespread application of new technologies. Generally, fish processing operations are conducted in the vicinity of the sea, and most of the fisheries wastes end up being returned to it. However, it is expected that pollution control measures and a better understanding of the valuable composition of the biomass extracted from the sea will promote its recovery and the development of new products from it.

4.2 Future Trends

Physical and chemical processes have been applied with limited success for the recovery of fisheries wastes and the production of quality products from them. Because of this, it is expected that biological processes will be prominent among those which will need to be developed for fisheries waste processing. The general development of biotechnology should accelerate the application of biotechnological processes to fisheries, which has been almost completely lacking, and thus be conducive to closing the present technological gap between fisheries and the other biochemical and food industries. From the materials mentioned above in this chapter, the following lists presents possible future developments in the bioconversion of fisheries wastes to industrial products:

(a) Processes will be developed for the extraction of specific high-value-added products such as enzymes, hormones and other chemicals. The potential for the application of biological methods to those processes needs to be developed.
(b) Fish protein concentrate will probably find new uses and markets. Enzymatic and fermentation methods, aided by new biotechnological development, could contribute significantly to this.
(c) Low-technology processes such as the production of fish silage and fish compost will be applied more extensively, mostly in developing countries.
(d) Fish oils and, on a smaller scale, other nutrients from fish wastes will become increasingly important substrates for fermentation processes.
(e) Enzymatic processes will be applied to improve the quality of chitin-based products obtained from shellfish waste.

REFERENCES

Amano, K. (1962). In *Fish in Nutrition,* eds E. Heen & R. Kreuzer. Fishing News (Books) Ltd, London, p. 180.
Anon. (1975). Icelandic Fisheries Laboratory Annual Report. Reykjavik, Iceland.
Arunachalam, K. & Haard, N. F. (1985). *Comp. Biochem. Physiol.*, **80B,** 467.
Austin, P. R. (1979). US Patent 4165433.
Austin, P. R., Brine, C. J., Castel, J. E. & Zikakis, J. P. (1981). *Science,* **212,** 749.
Badham, C. D. (1954). *Prase Halienties or Ancient and Modern Fish Tattle.* Parker & Son, London, p. 552.
Beauchat, L. R. (1983). In *Biotechnology,* Vol. 5, ed. G. Reed. Verlag Chemie, Deerfield Beach, Florida, p. 509.
Brambilla, A. & Horman, J. (1980). German Patent 3014408.
Brewer, P., Helbig, N. & Haard, N. F. (1984). *Can. Inst. Food Sci. Technol. J.*, **17,** 38.
Brody, J. (1965). *Fishery By-products Technology.* AVI, Westport, Connecticut.
Burkholder, L., Burkholder, P. R., Chu, A., Kostyk, N. & Roels, O. A. (1968). *Food Technol.*, **22,** 1278.
Capozza, R. C. (1976). US Patent 3989535.

Carroad, P. A. & Tom, R. A. (1978). *J. Food Sci.*, **43,** 1158.

Casey, D. J. (1978). US Patent 4069757.

Cheney, D. P. (1984). In *Biotechnology in the Marine Sciences,* eds R. R. Colwell, A. J. Sinskey & E. R. Pariser. John Wiley, New York, p. 161.

Cooper, J. L. (1975). Paper presented at Symposium on Enzymatic Conversion of Cellulosic Materials: Technology and Applications, Newton, Massachusetts, 8–10 September.

Cosio, I. G., Fisher, R. A. & Carroad, P. A. (1982). *J. Food Sci.*, **47,** 901.

Crisan, E. V. & Sands, A. (1975). *Appl. Microbiol.*, **29,** 106.

Davis, B. & Eveleigh, D. E. (1984). In *Chitin, Chitosan and Related Enzymes,* ed. J. P. Zikakis. Academic Press, New York, p. 161.

del Rosario, R. R. & Maddo, S. M. (1984). *Philipp. Agric.*, **67,** 167.

Denner, W. H. B. (1983). In *Industrial Enzymology—The Application of Enzymes in Industry,* eds T. Godfrey & J. Reichelt. The Nature Press, New York, p. 111.

Fenton, D. M. & Eveleigh, D. E. (1981). *J. Gen. Microbiol.*, **126,** 151.

Finch, R. (1970). *CRC Crit. Rev. Food Technol.*, **1,** 519.

Frederick, L., Harris, R., Peterson, L. & Kehrmeyer, S. (1989). *The compost solution to dockside fish wastes.* University of Wisconsin Sea Grant Advisory Report No. WIS-SG-89-434.

Freeman, H. E. & Hoogland, P. L. (1956). *Fisheries Research Board of Canada*: *Progress Report of the Atlantic Coast Stations,* **65,** 24.

Gildberg, A. (1982). Autolysis of fish tissue—general aspects. PhD thesis, Institute of Fisheries, University of Tromsø, Tromsø, Norway.

Godfrey, T. & Reichelt, J. (1983). In *Industrial Enzymology—The Application of Enzymes in Industry,* eds T. Godfrey & J. Reichelt. The Nature Press, New York, p. 1.

Green, J. H. & Mattick, J. F. (1978). *J. Food Quality,* **1,** 239.

Green, J. H. & Mattick, J. F. (1979). In *Food Processing Wastes,* eds J. H. Green & A. Kramer. AVI, Westport, Connecticut, p. 202.

Green, J. H., Paskell, S. L., Goldmintz, D. & Schisler, L. C. (1973). In *Food Processing Waste Management*: *Proc. 1973 Cornell Agric. Waste Mgmt Conf.*, Syracuse, NY, p. 51.

Green, J. H., Paskell, S. L. & Goldmintz, D. (1976). *Appl. Environ. Microbiol.*, **31,** 569.

Guevara, G., Matias, V. C., & Delta-Pena, P. O. (1972). *Philipp. J. Fish.*, **10,** 30.

Gutierrez, J. R. & Erickson, L. E. (1978). *Biotechnol. Bioeng.*, **20,** 1833.

Haard, N. F., Feltham, L. A. W., Helbig, N. & Squires, E. J. (1982). In *Advances in Chemistry Series No.* 198: *Modification of Protein. Food. Nutrition and Pharmacological Aspects,* eds R. E. Feeney & J. R. Whitaker, American Chemical Society, Washington, DC, p. 223.

Hale, M. B. (1974). *Mar. Fisheries Rev.*, **36**(2), 15.

Hansen, P. (1959). *Ensilering of fik og fiskeaffald.* Fiskoministeries Forgslaboratorium, Copenhagen, Denmark.

Hanson, S. W. F. & Lovern, J. H. (1951). *The Fishing News.* Fine Technical Publications, London, p. 11.

Hedges, A. & Wolfe, R. S. (1974). *J. Bacteriol.*, **120,** 844.

Higgs, T. W. (1974). Enhanced Fish Oil Fermentation. MEng Thesis, The University of Western Ontario, London, Ontario, Canada.

Hillyer, G. M., Peers, D. G., Morrison, R., Parry, D. A. & Woods, M. P. (1976). *Proc. Torry Res. Sta. Symp. Fish Silage,* Paper No. 4. Aberdeen, Scotland.

Hjelmeland, K. & Raa, J. (1982). *Comp. Biochem. Physiol.*, **71B,** 557.

Hossain, M. A., Furuichi, M. & Yone, Y. (1988). *Nippon Suisan Gakkaishi,* **54,** 469.

Hottinger, H. H. (1972). Single cell biomass production from fish oil by *Candida lipolytica* and *Geotrichum candidum.* MSc Thesis, University of Wisconsin, Madison, Wisconsin.

Hottinger, H. H., Richardson, T., Amundson, C. H. & Stuiber, D. A. (1974*a*). *J. Milk Food Technol.*, **37,** 463.
Hottinger, H. H., Richardson, T., Amundson, C. H. & Stuiber, D. A. (1974*b*). *J. Milk Food Technol.*, **37,** 522.
Huang, F. L. & Tappel, A. L. (1971). *Biochim. Biophys. Acta,* **236,** 739.
Ivarson, K. C. & Morita, H. (1982). *Appl. Environ. Microbiol.*, **43,** 643.
Johnson, M. A., & Carlson, J. A. (1978). *Biotechnol. Bioeng.*, **20,** 1063.
Joyce, A. (1988). *The Canadian east coast fish and seafood industry.* Report to Burns Fry Limited Investor Services, Toronto, Ontario, Canada.
Kakuta, T., Koizumi, T., Yoshizawa, K., Kodama, K. & Nojiro, K. (1985). *Nippon Shokuhin Kogyo Gakkaishi,* **32,** 260.
Kalac, J. (1978*a*). *Biologia* (Bratislava), **33,** 704.
Kalac, J. (1978*b*). *Biologia* (Bratislava), **33,** 939.
Kesting, R. E. (1979). German Patent 2849863.
Keys, C. W. & Meinke, W. W. (1966). US Patent 3249442.
Knobl Jr, G. M. (1967). *Food Technol.*, **21**(8), 56.
Kosaric, N., Zajic, J. E., Aboue, G., Jack, T. & Gerson, D. (1979). *Biotechnol. Bioeng.*, **21,** 1133.
Kreag, R. & Smith, F. J. (1973). *Seafood solid waste in Oregon: Disposal or recovery?* Oreg. State Univ. Ext. Spec. Rep. 395.
Kubota, M. & Ohnuma, A. (1970). *Bull. Jap. Soc. Sci. Fisheries,* **36,** 1147.
Lagunov, L. L., Egorova, L. N., Rekhina, N. I. & Eremeev, M. N. (1956). *Chem. Abstr.*, **52,** 12263.
Li, C. F., Stuiber, D., Richardson, T. & Amundson, C. H. (1970). *Food Prod. Devel.*, **4,** 28, 102.
Liston, J. (1980). In *Advances in Fish Science and Technology,* ed. J. J. Connell. Fishing News (Books) Ltd, Farnham, Surrey, UK, p. 138.
Mackie, I. M., Hardy, R. & Hobbs, M. (1971). *FAO Fisheries Rep.*, **100,** 54.
Makinonda, V. & Ikeda, S. (1971). *Bull. Jap. Soc. Sci. Fisheries,* **37,** 1002.
Martin, A. M. & Bassler, E. C. (1989). Paper presented at the 1989 Annual Meeting of the Institute of Food Technologists, Chicago, Illinois, 25–29 June.
Martin, A. M. & Chintalapati, S. P. (1989). *Biol. Wastes,* **27,** 281.
Martin, A. M. & White, M. D. (1985). *J. Food Sci.*, **50,** 197.
Martin, R. E. (1972). *Oak Brook Seminar, Mechanical Recovery and Utilization of Fish Flesh.* National Fisheries Institute, Washington, DC.
Mathur, S. P., Daigle, J.-Y., Levesque, M. & Dinal, H. (1986). *Biol. Agric. Hortic.*, **4,** 27.
Mathur, S. P., Daigle, J.-Y., Brooks, J. L., Levesque, M. & Arsenault, J. (1988). *BioCycle,* September, 44.
McBride, J. R., Idler, D. R., & McLeod, R. A. (1961). *J. Fish. Res. Board Can.*, **18,** 93.
McIver, R. C., Brooks, R. I. & Reineccius, G. A. (1982). *J. Agric. Food Chem.*, **30,** 1017.
Merrett, T. G., Bar-Eli, E. & Vunakis, H. V. (1969). *Biochemistry,* **8,** 3696.
Miyazawa, K., VanLe, C., Ito, K. & Matsumoto, F. (1979). *J. Fac. Appl. Biol. Sci. Hiroshima Univ.*, **18,** 55.
Murayama, S., Calvez, D. L. & Nitayachin, P. (1962). *Bull. Tokai Reg. Lab.*, **32,** 155.
Muzzarelli, R. A. A. (1973). In *Natural Chelating Polymers: Alginic Acid, Chitin and Chitosan.* Pergamon Press, New York, p. 177.
Mycek, M. J. (1970). In *Methods in Enzymology,* Vol. 19, *Proteolytic Enzymes,* eds G. E. Perlmann & L. Larand. Academic Press, New York, p. 285.
Nestaas, E. (1986). Paper presented at Nor-Fishing '86, 11th International Fisheries Fair, Trondheim, Norway.
Noda, M. & Murakami, K. (1981). *Biochim. Biophys. Acta,* **658,** 27.

Noda, M., Van, T. V., Kusukabe, I. & Murakami, K. (1982). *Agric. Biol. Chem.*, **46,** 1565.
Norris, E. R. & Elam, D. W. (1940). *J. Biol. Chem.*, **134,** 443.
Norris, E. R. & Mathies, J. C. (1953). *J. Biol. Chem.*, **204,** 673.
Ohtakara, A., Ogata, H., Taketomi, Y. & Mitsutomi, M. (1984). In *Chitin, Chitosan and Related Enzymes,* ed. J. P. Zikakis. Academic Press, New York, p. 147.
Ooshiro, Z., Ok, T., Une, H., Hayashi, S. & Itakura, T. (1981). *Mem. Fac. Fish., Kagoshima Univ.*, **30,** 383.
Orejana, F. M. (1978). Proteolysis and control mechanisms in fish sauce fermentation. PhD Thesis, University of Washington, Seattle, Washington.
Owen, T. G. & Wiggs, A. J. (1971). *Comp. Biochem. Physiol.*, **40B,** 465.
Pan, S. C., Bonnanno, S. & Wagman, G. H. (1959). *Appl. Microbiol.*, **7,** 176.
Petersen, H. (1953). *FAO Fisheries Bulletin,* **6**(1–2), 18.
Price, J. S. & Stork, R. (1975). *J. Bacteriol.*, **124,** 1574.
Raa, J. (1989). Paper presented at the Atlantic Fisheries Technological Conference, St John's, Newfoundland, Canada, August 1989.
Raksakulthai, N. (1986). Role of protein degradation in fermentation of fish sauce. PhD Thesis, Memorial University of Newfoundland, St John's, Newfoundland, Canada.
Raksakulthai, N., Lee, Y. Z. & Haard, N. F. (1986). *Can. Inst. Food Sci. Technol. J.*, **19,** 28.
Reed, G. (1975). In *Enzymes in Food Processing,* ed. G. Reed. Academic Press, New York, pp. 7, 494.
Renn, D. W. (1984). In *Biotechnology in the Marine Sciences,* eds R. R. Colwell, A. J. Sinskey & E. R. Pariser. John Wiley, New York, p. 191.
Revah-Moiseev, S. & Carroad, P. A. (1981). *Biotechnol. Bioeng.*, **23,** 1067.
Saisithi, P. (1967). Studies on the origins and development of typical flavor and aroma of Thai fish sauce. PhD Thesis, University of Washington, Seattle, Washington.
Saisithi, P., Kasemsarn, B.-O., Liston, J. & Dollar, A. M. (1966). *J. Food Sci.*, **31,** 105.
Schisler, L. C. & Patton, T. G. (1974). In *Mushroom Science IX. Part I. Proc. Ninth Int. Sci. Congr. on the Cultivation of Edible Fungi.* International Society of Mushroom Science, Tokyo, p. 175.
Schisler, L. C. & Sinden, J. W. (1966). *Can. J. Botany,* **44,** 1063.
Shamsuzzaman, K. (1983). Isolation, properties and the use of a chymosin-like enzyme from harp seal (*Pagophilus groenlandicus*). PhD Thesis, Memorial University of Newfoundland, St John's, Newfoundland, Canada.
Shamsuzzaman, K. & Haard, N. F. (1984). *Can. J. Biochem. Cell Biol.*, **22,** 699.
Shoemaker, S. (1986). In *Biotechnology in Food Processing,* eds S. K. Harlander & T. P. Labuza. Noyes Publications, Park Ridge, New Jersey, p. 259.
Siebert, G. & Schmitt, A. (1965). In *Technology of Fish Utilization,* ed. R. Kreuzer. Fishing News (Books) Ltd, London, p. 47.
Simpson, B. K. (1983). Isolation, characterization and some application of trypsin from Greenland cod (*Gadus ogac*). PhD Thesis, Memorial University of Newfoundland, St John's, Newfoundland, Canada.
Simpson, B. K. & Haard, N. F. (1984). *Can. J. Biochem. Cell Biol.*, **62,** 894.
Simpson, B. K. & Haard, N. F. (1987). In *Food Biotechnology,* ed. D. Knorr. Marcel Dekker, New York, p. 495.
Siso, M. I. G., Murrado, M. A. & Franco, J. M. (1987). In *Proc. II World Congr. Food Sci. Technol.*, Barcelona, Spain, 3–7 March. Inter-Congrès, Barcelona, Spain, p. 677.
Smith, P. & Adamson, A. H. (1976). *Proc. Torry Res. Sta. Symp. Fish Silage,* Paper No. 3. Aberdeen, Scotland.
Spinelli, J., Koury, B. & Miller, R. (1972). *J. Food Sci.*, **37,** 599.

Squires, E. J. (1984). Isolation and characterization of gastric proteases from the Greenland cod (*Gadus ogac*). PhD Thesis, Memorial University of Newfoundland, St John's, Newfoundland, Canada.
Squires, E. J., Haard, N. F. & Feltham, L. A. W. (1986*a*). *Can. J. Biochem. Cell Biol.*, **64,** 205.
Squires, E. J., Haard, N. F. & Feltham, L. A. W. (1986*b*). *Can. J. Biochem. Cell Biol.*, **64,** 215.
Stefansson, G. (1989). Paper presented at the Atlantic Fisheries Technological Conference, St John's, Newfoundland, Canada, August 1989.
Subba Rao, G. N. (1967). *Fish processing in the Indo-Pacific area.* Regional studies No. 4, FAO Regional Office for Asia and the Far East, Bangkok, Thailand, p. 149.
Tatterson, I. N. & Windsor, M. L. (1974). *J. Sci. Food Agric.*, **25,** 369.
Tominaga, Y. & Stujisake, Y. (1975). *Biochim. Biophys. Acta,* **410,** 145.
Tongthai, C. & Okada, H. (1982). In *Microbial Utilization of Renewable Resources,* Vol. 2, ed. H. Taguchi. ICME Faculty of Engineering, Osaka University, Osaka, Japan, p. 304.
Van, T. V., Kusakabe, I. & Murakami, K. (1983). *Agr. Biol. Chem.*, **47,** 2453.
Venugopal, V., Alur, M. D. & Nerkar, D. P. (1989). *Biotechnol. Bioeng.*, **33,** 1098.
Welsh, F. W. & Zall, R. R. (1984). *Process Biochem.*, **19**(3), 122.
Whitaker, J. R. (1978). *Food Technol.*, **32,** 175.
Whittenmore, C. T. & Taylor, A. G. (1976*a*). *Proc. Torry Res. Sta. Symp. Fish Silage,* Paper No. 2. Aberdeen, Scotland.
Whittenmore, C. T. & Taylor, A. G. (1976*b*). *J. Sci. Food Agric.*, **27,** 239.
Whitworth, D. A., Moo-Young, M. & Viswanatha, T. (1973). *Biotechnol. Bioeng.*, **15,** 649.
Wignall, J. & Tatterson, I. (1976). *Process Biochem.*, **11,** 17.
Zajic, J. E., Higgs, T. W. & Kosaric, N. H. (1974). Paper presented at the GVC/AIChE-Joint Meeting and Jahrestreffen 1974 der Verfahrens-Ingenieure, Munich, Germany, 17–20 September.

Chapter 15

BIORECOVERY OF METALS FROM MINING, INDUSTRIAL AND URBAN WASTES

DAVID S. HOLMES

Department of Biology, Clarkson University, Potsdam, New York, USA

CONTENTS

1 SCOPE OF THE CHAPTER

Two different approaches for the biological recovery of metals from wastes are discussed in this chapter. In the first approach, termed bioleaching, metals are solubilized into dilute acid solution from waste rocks using living microorganisms. The solubilized metals are then recovered from the dilute acid solution by conventional technologies such as solvent extraction or electroplating. In the USA 25% of the copper recovered is thought to result from bioleaching and it has been estimated that over 1000 million dollars' worth of copper and uranium are recovered annually worldwide using bioleaching (Nicolaides, 1987; Holmes *et al.*, 1988*a*).

In the second approach, living or dead microorganisms or biological components (e.g. chitosan) are used to trap or precipitate metals from industrial and urban wastes. A major objective of this approach is to develop technology for the mitigation of environmental pollution problems. However, the potential also exists to recover and recycle the waste metals, especially the more valuable ones, and so a description of this approach comes under the purview of this chapter.

Two related applications of biologically assisted metal-recovery will not be described in this chapter because they do not strictly involve the treatment of waste deposits. The first of these is the biological beneficiation of gold ore. This is a very recently commercialized technology involving biological removal of pyrite from gold-bearing pyritic ores, permitting the gold to be economically recovered from the residue by a chemical cyanidation procedure. The other application is the in-situ bioleaching of underground ore bodies. Although this application permits the utilization of material that could not be exploited in the absence of biotechnology, the in-situ ore deposits can not be strictly considered as waste. Interested readers are directed to several recent papers describing gold biotechnology (Lawrence & Bruynesteyn, 1983; Marchant & Lawrence, 1989) and in-situ bioleaching (Ismay *et al.*, 1986; Rossi *et al.*, 1986). Also, a recently published book has chapters describing biotreatment of gold ores and in-situ bioleaching (Ehrlich & Brierley, 1990).

2 BIORECOVERY OF METALS FROM MINING WASTES

2.1 Brief History

Although it is thought that the Romans made use of bioleaching to recover copper from the Rio Tinto area of Spain nearly 2000 years ago, one of the first written accounts of biologically assisted metal recovery is in *The Book Concerning the Tincture of the Philosophers,* written in about 1510 by Aureolus Bombast. In this account, the practice is described whereby copper can be deposited from acid solution onto iron followed by the recovery of copper by smelting: 'For truly, when the rustics in Hungary cast iron at the proper season into a certain fountain, commonly called Zifferbrünnen, it is consumed into rust and when this is liquified with a blast-fire, it soon exists as pure Venus (copper), and nevermore returns to iron'. Almost certainly the Zifferbrünnen fountain included the acid drainage from local copper mines which contained copper solubilized by bacterial action.

By 1750, 200 tons of copper were obtained annually from the Zifferbrünnen region of Hungary. Biological leaching of copper was also practiced in Germany in the 16th century and at the Rio Tinto mine in Spain by 1725 (Karavaiko & Groudev, 1985). More recently, biological leaching of copper was begun in the USSR at the end of the last century and in the USA by the 1920s. Now almost every copper mine in the world has some form of solution technology for copper recovery in which, in most cases, microorganisms play a major role.

Pioneering work by Rudolfs and Helbronner (1922), Waksman and Joffe (1922) and Carpenter and Herndon (1933) on bacteria capable of oxidizing sulfur compounds to sulfuric acid and by Colmer and Hinkle (1947) on the iron and sulfur oxidizing bacterium *Thiobacillus ferrooxidans* formed the foundation for future research into the role of microorganisms in metal solubilization. The concept of using microorganisms for the recovery of metals dates to the early 1950s when Bryner *et al.* (1954) described how iron pyrites and copper sulfides could be oxidized by acidic thiobacilli from the mine drainage of Kennecott's Bingham Canyon open pit mine. These observations led to the issue of the first patent (Zimmerley *et al.*, 1958) assigned to the Kennecott Copper Corporation for a cyclic leaching process employing acidophilic iron-oxidizing bacteria. This initial patent led to the issue of a series of additional patents.

Despite these efforts the belief that copper solubilization was purely a chemical process was commonly held by many mining operations even up to the 1970s. A curious event in 1967 has been said to have contributed to the acceptance of a biological role in copper solubilization. At that time the Anaconda Company was recovering 100 tons of copper per day from their waste dumps in Butte, Montana. Part of the copper solubilization process called for dilute sulfuric acid to be sprinkled onto the dumps. It was this acid that mining engineers believed to be responsible for the resulting copper solubilization. It was brought home sharply that microorganisms were involved in the process when the company, during a prolonged strike, purchased sulfuric acid that was subsequently shown to be contaminated with 50 ppm mercury. Using the contaminated acid the solubilization of copper dropped to nearly zero within 3 months. The mercury was clearly inhibiting microbial activity. The Anaconda leaching operation never reopened after this unfortunate but enlightening incident.

2.2 Bioleaching of Copper from Waste Dumps

2.2.1 Typical Dump Construction and Operation

Nearly all copper mining companies recover copper from dumps of waste rock. These dumps are built from the overburden of strip mines or from rock that

Fig. 1. Terraced dump at the Kennecott Bingham Canyon Mine in Utah. As much as 200 tons of copper per day have been recovered from this dump (Boseker & Kursten, 1978), accounting for about 25% of the total copper production of the mine, averaged over a year.

comes from the mines but is deemed to be too low in copper content, usually 0·1–0·5%, to warrant recovery by conventional processes. In the majority of mining operations the rock waste is hauled by trunks to form terraced dumps often 100 or more meters high and wide (see Fig. 1). To recover the copper, dilute sulfuric acid is sprinkled or sprayed onto the surface of the dump (see Fig. 2). As the acid percolates down through the dump it lowers the pH to around 2–3, promoting the growth of acidophilic microorganisms such as *Thiobacillus ferrooxidans*. These microorganisms then oxidize the copper in the ore both directly by enzymatic attack and indirectly by producing ferric iron, a chemical oxidant, that oxidizes the copper. The microorganisms also oxidize reduced sulfur compounds, resulting in the production of additional sulfuric acid. The oxidized copper goes readily into solution in the acid. The acid solution, sometimes blue with dissolved copper, is collected at the bottom of the dump and is pumped to a nearby metal-recovery plant. Until a few years ago copper was recovered from the acid solution by cementation onto scrap iron. However, recent developments in solvent extraction and electrowinning have revolutionized the economics of copper recovery and these techniques have now been adopted almost universally.

A common factor in dump operations is the recycling of the spent leach solution from the cementation or solvent extraction process. This is because many dump operations are in arid areas of the world such as the western United States, and water needs to be preserved. There is also a need to avoid stream pollution. Unfortunately, the recycled solution is frequently rich in ferric iron, derived both from the microbial activity in the dump and from the cementation process. The dissolved ferric iron is precipitated from solution at pH values above about 2·0. Precipitated iron salts tend to coat mineral surfaces and seal areas in the dump, rapidly decreasing the availability of the copper minerals for leaching. For this reason mine operators usually add additional sulfuric acid to the recycled leach solution, maintaining a pH of about 2.

Microorganisms are used to recover copper commercially from a wide range of minerals: pyrite and marcasite (FeS_2), pyrrhotite (FeS), chalcopyrite ($CuFeS_2$), bornite (Cu_5FeS_4), covellite (CuS), chalcocite (Cu_2S), tetrahedrite ($Cu_8Sb_2S_7$), and enargite ($3Cu_2S\ As_2S_5$) (Torma, 1978; Lundgren, 1980).

Examples of commercial dump leaching operations are given in Table 1.

2.2.2 Problems Encountered in Dump Bioleaching

The construction of dumps has historically proceeded with little regard for the optimization of biological activity and often with little regard to engineering optimization. This can lead to inefficient recovery of copper. Some problems frequently encountered are as follows.

(1) Material is dumped in sizes ranging from large boulders to sand. The former present very small surface to volume ratios and cannot be considered efficient biological substrates. The latter can clog the flow of the leach solution.

(2) No attempt is usually made to separate barren rock from copper-bearing ore.

(3) Compaction of the waste material by the haulage trucks impedes the flow of acid solution and oxygen into the dump.

(4) Dumps grow by the deposition of waste rock over the edge of a terrace. This causes significant segregation of material according to size as it rolls down the terrace slope. Frequently, lenses of fine material build up on the upper slopes of a terrace. When these are later covered by newer waste they prevent the percolation of the solution and cause serious channeling problems.

(5) Oxygen is reported to diffuse very poorly in dumps. The actual extent of diffusion depends a lot on the individual dump, but concentrations as low as 1% only 1 m from the surface have been reported (Harries *et al.*, 1988). Thus the interior of dumps are frequently anaerobic or anoxic, inhibiting the biological oxidation of metals. Such conditions also promote the activity of sulfate-reducing microorganisms which can lead to the reprecipitation of already solubilized metals.

(6) An ore may contain too much of an acid consuming component like $CaCO_3$. Also, in some instances the biooxidation of an ore component such as chalcocite (Cu_2S) may consume enough acid to bring the process to a stop. Both these problems can be overcome by the addition of sulfuric acid.

(7) An ore may contain a toxic compound. For example *T. ferrooxidans* is inhibited by molybdate ion. Molybdate toxicity, however, can be overcome if the bioleaching process generates ferric iron which forms ferric molybdate, lowering its toxicity towards *T. ferrooxidans*. Other ore constituents that are potentially troublesome include mercury and silver containing minerals.

(a)

(b)

Fig. 2. Distribution of acid leaching solution to a copper dump at the Cypress Johnson Mine, Arizona. (a) Pumping Station; (b) Main distribution to heap.

(8) The non-mineral part of the ore (gangue) may have a significant absorption capacity for metal ions, such as cupric ions. This problem can generally be alleviated by proper pH control.

(9) Galena (PbS) can present a special problem because the product of bioleaching ($PbSO_4$) is a relatively insoluble compound and tends to accumulate on the mineral surface, preventing further oxidation by microorganisms.

(c)

(d)

Fig. 2—*contd.* (c) Distribution of leaching solutions on heap surface; (d) collection of pregnant solution at base of heap.

Technically, most of these problems can be solved. However, their solution is not always practical or economically justified. For example, it is not practical to separate gangue from low grade ore during dump construction because the work force involved in dump construction does not have the technical knowledge to distinguish gangue from ore. In many mining operations problems associated with poor oxygen diffusion in large dumps have been overcome by going to an altered dump construction design. In this new design,

Table 1
Examples of Commercial Dump Leaching Operations

Mine location and/or company	*Mineralization*			*Mean total weight of rock being leached (kg)*	*Waste (ore) body height/width*	*Irrigation rate and mode (influent flow rate liter/min)*	*Area irrigated (m^2)*
	Primary sulfides or oxides	*Total Cu (%)*	*Host rock*				
Bluebird Mine, Miami, Arizona, USA (Ranchers Exploration and Development)	100% chrysocholla	0·5	Schist	4×10^{10}	61 m/183 m	8 505 (2 300 gal/min) intermittent	92 000
Phelps Dodge Corp., Tyrone, New Mexico, USA	Pyrite, covellite, chrysocholla, tenorite, chalcocite, chalcanthite	0·35	Quartz, monzonite	$3{\cdot}9 \times 10^{10}$	71 m/792 m	12 490 (330 gal/min) intermittent (2 cycles/year)	—
Baghdad, Arizona, USA	Mixed oxides	0·30	—	—	—	12 600 (3 328 gal/min) continuous	—
Duval Esparanza Mine, Arizona, USA	Chalcocite, malachite, azurite, chalcopyrite	0·15–0·20	Quartz, monzonite and diorite	$2{\cdot}7 \times 10^{10}$	91 m/366 m	8 516 (2 250 gal/min) continuous	—
Duval, Mineral Park, Arizona, USA	Chalcocite	0·14	Monzonite	64×10^{10}	152 m/91 m	7 570 (2 000 gal/min) intermittent, periodic rest cycles	457 640
Duval, Copper Basin, Arizona, USA	85% sulfide	0·31	Sandstone, siltstone	$2{\cdot}4 \times 10^{10}$	46 m/366 m	5 678 (1 500 gal/min) intermittent, periodic rotation	74 000
Duval, Copper Canyon, Arizona, USA	Oxide	0·25	Sandstone, porphyry	$3{\cdot}9 \times 10^{10}$	122 m/349 m	4 542 (1 200 gal/min) intermittent, periodic rotation	74 000
Deglyansky, Urals, USSR	Chalcopyrite, chalcocite, pyrite	—	—	—	—	—	—
Rio Tinto, Spain	Chalcocite, chalcopyrite, pyrite	variable	—	—	—	—	—
Cananea, Mexico	Chalcocite, pyrite	variable	—	—	—	—	—

waste material is arranged into finger dumps several hundred feet long and 30–50 ft high (Fig. 3). Also, surface compaction due to haulage trucks is frequently relieved by ripping the surface of the dump with machines or by the use of explosive charges.

2.2.3 Economics of Dump Bioleaching

Biological copper recovery worldwide is thought to exceed US $1000 million annually and to account for about 25% of world copper production. In 1988, US production of copper by biological leaching exceeded 250 000 tons, valued in excess of US $650 million.

Very few data exist on the cost of production of individual dump leaching operations. Estimates within the industry for total copper production costs can be as low as 30¢/pound (LB) for large operations in the southwest USA where solvent extraction is widely used. This is about 50% less than conventional copper recovery techniques.

It should be noted that one of the reasons that biological recovery of copper

Starting solution (g/liter)				*Effluent flow rate (liter/min)*	*Finishing solution (effluent) (g/liter)*				*Recovery method*	*Average copper production (kg)*	*Years of operation*	*Estimated percent Cu recovered*
Cu	Fe^{2+}	Fe^{3+}	*pH*		*Cu*	Fe^{2+}	Fe^{3+}	*pH*				
0·20	0·03	0·63	6 g g/liter H_2SO_4	8 705 (2 300 gal/min)	1·80	0·03	0·63	3 g/liter H_2SO_4	Solvent extraction	$6{\cdot}8\times10^6$	16	40
0·30	1·52	1·20	3·8	12 112 (3 200 gal/min)	0·78	4·10	0·10	2·5	Cementation on iron scrap	$4{\cdot}5\times10^6$	9	25
0·2	4·5	0·2	2·0	12 100 (3 200 gal/min)	1·0	0·1	3·0	2·5	—	$13{\cdot}6\times10^6$	—	—
0·12	3·3	0·10	1·3	6 056 (1 600 gal/min)	0·90	0·30	1·60	2·0	Cementation on iron scrap	$2{\cdot}5\times10^6$	15	60
0·05	2·85	<0·01	3·8	6 813 (1 800 gal/min)	0·55	0·25	0·95	2·5	Cementation on iron scrap	$1{\cdot}5\times10^6$	14	40
0·08	2·50	2·50	1·7	4 164 (1 100 gal/min)	1·1	2·5	2·5	2·2	Solvent extraction in 1980, cementation piror to 1980	$2{\cdot}3\times10^6$	13	40
0·02	1·2	0·10	1·9	3 785 (1 000 gal/min)	0·90	0·02	0·20	2·9	Cementation on iron scrap	$1{\cdot}7\times10^6$	13	25
0·1	0·2	2·0	2·5–2·9	—	0·3	1·5	0·7	2·9	Cementation on iron scrap	$\sim9\times10^5$	~20	—
0·2	4·0	1·0	1 g/liter H_2SO_4	—	2·2	18	2	11 g/liter H_2SO_4	Cementation on iron scrap	8×10^6	hundreds	unknown
0·35	7·1	1·8	2·7	—	3·3	3·0	7·4	2·3	Cementation on iron scrap	$8{\cdot}7\times10^6$	—	—

from dumps is economically attractive is that the waste rock had to be moved in any case to provide access to the higher grade ore. Therefore, most dump leaching operations do not bear the costs of rock removal, rock transport or comminution (rock breakage).

A breakdown of individual costs for dump leaching is largely unavailable and is considered to be proprietary information. Schlitt (1980) estimated the

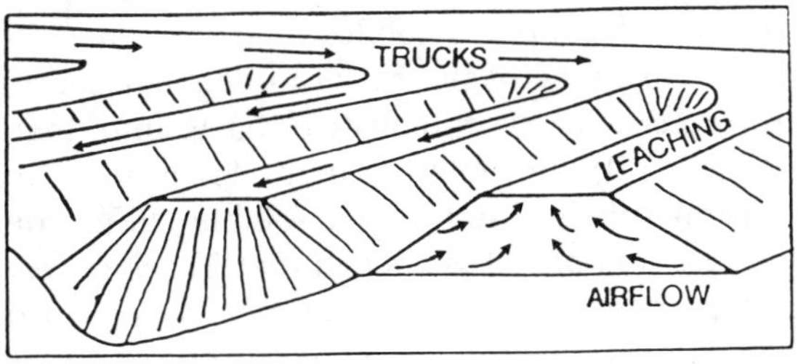

Fig. 3. Diagramatic representation of finger dumps.

costs of dump leaching of copper for Kennecott's Bingham Canyon operation, where the bacterially assisted leaching of existing waste dumps containing 0·15–0·25% Cu (predominantly as $CuFeS_2$) is credited as having produced 200 tons per day (TPD) of saleable cement copper at a 1980 operating cost of 35–38¢/lb. Of this, about 40% (15¢/lb) is attributed to the leaching operation proper, where the major cost is for power to pump the leach solutions in three stages through the 2700 ft head between the base and the top of the dumps. The cost of copper recovery accounts for another 40% (15¢/lb), most being spent on detinned scrap iron for cementation. The remaining cost of 5–8¢/lb is for labor and maintenance.

There is a limited market for cement copper in the manufacture of clutch plates and in antifouling marine paints. The low purity (55–80%) cement copper is not generally considered an acceptable final product, and further upgrading by smelting and refining is required. The cost of producing cathode copper this way at Bingham Canyon is 62–67¢/lb, leaving a comfortable operating margin of between 20 and 45¢/lb (based on copper prices of 85¢/lb to $1.10/lb).

Copper extracted pyrometallurgically from high grade ore from the mine costs $1.05–$1.10/lb. Thus conventional operation is economically viable only because of substantial credits for gold, silver, and molybdenite byproducts, which reduce the net copper production cost to about 58¢/lb.

These figures were calculated in 1980. Since that time most major copper leaching operations have made the transition from cementation to solvent extraction with reported cost savings exceeding 50% in some operations. As noted, about 50% of the cost of dump leaching are manifested in scrap iron costs for the cementation processes. However, it should be noted that extraction is a capital intensive process relative to cementation and, therefore, may not be suitable to scaling down to relatively small operations.

2.2.4 Biochemistry of Copper Bioleaching

The biological solubilization of copper from sulfidic ores (e.g. pyrite) is not completely understood. It is known to involve both a direct enzymatic oxidation of the mineral and associated copper and also an indirect oxidation of the mineral and associated copper by ferric iron which is produced biologically by oxidation of ferrous iron in the ore.

Several different types of microorganisms and perhaps several different biological mechanisms of copper solubilization are found in dump leaching operations (Norris, 1990). One of the better understood mechanisms is carried out by *T. Ferrooxidans*, a Gram-negative bacterium that lives in acidic environments (pH 1·2–4·0) and obtains energy for growth from the oxidation of ferrous iron, metal sulfides, sulfur and soluble sulfur compounds. It was clearly demonstrated a number of years ago that *T. ferrooxidans* can oxidize a number of insoluble copper bearing sulfide minerals, oxidizing continuously both the iron and sulfur moieties of the minerals (Torma *et al.*, 1970; LeRoux & Mehta, 1978).

For example, *T. ferrooxidans* is capable of direct enzymatic attack on the crystal lattice of pyrite yielding energy for growth:

$$4FeS_2 + 15O_2 + 2H_2O \xrightarrow[\text{energy}]{\text{bacteria}} 2Fe_2(SO_4)_3 + 2H_2SO_4 \quad (1)$$

producing ferric sulfate and sulfuric acid. The resulting ferric iron is a powerful chemical oxidant and can oxidize pyrite:

$$FeS_2 + 7Fe_2(SO_4)_3 + 8H_2O \rightarrow 15FeSO_4 + 8H_2SO_4 \quad (2)$$

$$FeS_2 + Fe_2(SO_4)_3 \rightarrow 3FeSO_4 + 2S \quad (3)$$

The resulting ferrous sulfate and sulfur can be oxidized by microorganisms such as *T. ferrooxidans* to ferric iron and sulfuric acid respectively, yielding biologically usable energy in the process:

$$4FeSO_4 + O_2 + 2H_2SO_4 \rightarrow 2Fe_2(SO_4)_3 + 2H_2O \quad (4)$$

$$2S + 3O_2 + 2H_2O \rightarrow 2H_2SO_4 \quad (5)$$

The ferric iron, in turn, can chemically oxidize more pyrite. Thus an important role for microorganisms in copper solubilization is to create the chemical oxidant, ferric iron, from ferrous iron, completing a biological–chemical cycle as shown in Fig. 4. This is known as the indirect biological attack on pyrite, because the microorganisms are creating an oxidant, ferric iron, which can chemically oxidize pyrite at a distance. The direct, enzymatic attack on pyrite described in eqn (1) is thought to require the attachment of the microorganisms to the pyrite.

Direct and indirect oxidation result in the breakdown of the crystal lattice of the pyrite, leading to the eventual disintegration of the mineral and the

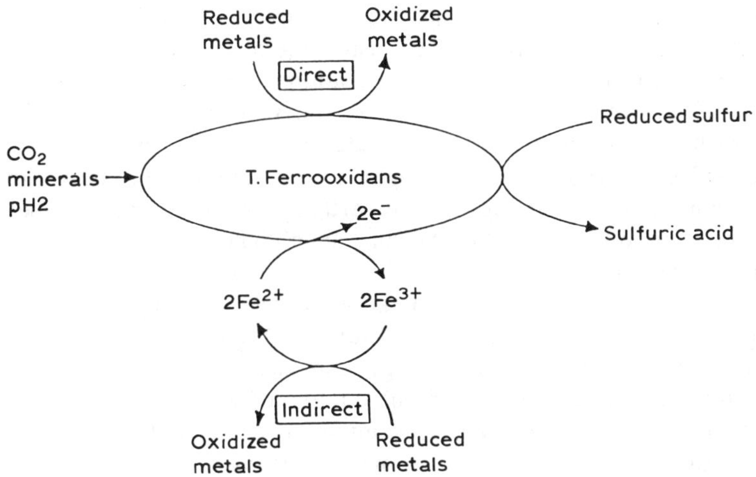

Fig. 4. Illustration of the role of *Thiobacillus ferrooxidans* in both the direct and indirect oxidation of copper from a pyritic copper bearing ore.

formation of sulfuric acid which can assist in the oxidation of copper and promote the solubility of the oxidized copper. However, the relative importance of the direct versus indirect biological attack on pyrite for the solubilization of copper in dump leaching operations is still being debated.

The copper associated with the pyrite is oxidized by three mechanisms. First, as mentioned above, bacterially and chemically produced sulfuric acid can act on some copper minerals, e.g.

$$Cu_2S + H_2SO_4 + \tfrac{1}{2}O_2 \rightarrow CuS + CuSO_4 + H_2O \quad (6)$$

Second, by bacterially (or chemically) produced ferric iron:

$$CuFeS_2 + 2Fe_2(SO_4)_3 + 2H_2O + 3O_2 \rightarrow CuSO_4 + 5FeSO_4 + 2H_2SO_4 \quad (7)$$

Note that in this reaction ferrous iron is produced which becomes available for bacterial oxidation.

Third, by direct bacterial oxidation:

$$CuFeS_2 + 4O_2 \rightarrow CuSO_4 + FeSO_4 \quad (8)$$

Again, the relative importance of the direct versus the indirect oxidation of copper is not completely understood.

2.3 Microorganisms Involved in Bioleaching

The full range of microorganisms involved in metal leaching from dumps is not completely known. Early work by Ehrlich showed that yeasts, flagellates and amoebae were present in acid mine drainage in addition to *Thiobacillus*-type microorganisms, (Ehrlich, 1963). A survey of microorganisms in a copper dump in Bulgaria (Groudev *et al.*, 1978) and an underground mine in Italy (Rossi *et al.*, 1986) indicated the presence of many other types of microorganisms in addition to the above mentioned, including heterotrophic and sulfate reducing bacteria. Table 2 lists some of the microorganisms thought to play an important role in the solubilization of metals in dumps. This list is not comprehensive, but is provided to give the reader a feel for the diversity of microorganisms and their activities in a dump leaching operation.

Thiobacillus ferrooxidans is thought to be one of the most important microorganisms in bioleaching and it is certainly one of the most studied. It is a Gram-negative rod, a strict aerobe and an obligate chemolithoautotroph. It obtains energy by the oxidation of iron and reduced sulfur compounds. It derives carbon by carbon dioxide fixation. It has a pH optimum for growth around 2–2·5 and a temperature optimum for growth around 30°C. Being a strict autotroph it cannot obtain energy from the oxidation of reduced carbon compounds and, in fact, some carbon compounds are inhibitory to its growth. In dump leaching it oxidizes pyrite, forming ferric iron and sulfuric acid. It can directly oxidize copper and it can reoxidize ferrous iron, formed when ferric iron chemically oxidizes copper, back to ferric iron. *Leptospirillum ferrooxidans* is Gram-negative, motile and spiral shaped. It is a strict aerobe and

Table 2
Examples of Microorganisms Involved in Bioleaching of Metals

Name of organism	*Optimum growth temperature (°C)*	*Special physiological traits*	*References*
Thiobacillus ferrooxidans	30	Oxidizes Fe^{2+}, S°, S_2O^{2-}, U(IV), metal sulfides, CuS	Buchanan & Gibbons, 1974
Leptospirillum ferrooxidans	30	Oxidizes Fe^{2+} but not S°	Balashova *et al.*, 1974
Thiobacillus thiooxidans	30	Oxidizes S°, $S_2O_3^{2-}$, but not Fe^{2+} or metal sulfides	Buchanan & Gibbons, 1974
Sulfolobus brierleyi	70	Oxidizes Fe^{2+}, S°, MoS_2, $CuFeS_2$	Brierley & Brierley, 1973
Thiobacillus TH-1	55	Oxidizes Fe^{2+}, S°, metal sulfides, and needs organic carbon (e.g. yeast extract)	Brierley *et al.*, 1978
Sulfobacillus thermosulfidooxidans	50	Oxidizes Fe^{2+}, S°, forms endospores	Karavaiko *et al.*, 1988*a*
Desulfovibrio spp.	30	Reduce SO_4^{2-}, S° and some partially oxidized S compounds; form metal sulfides	Postgate, 1979
Acidophilium cryptum	30	Heterotrophic, can scavenge carbon from organic excretions of *T. ferrooxidans* in the laboratory	Harrison, 1984
Protozoa	30	Prey on bioleaching microorgansims	Ehrlich, 1963
Thiobacillus neopolitanus	30	Can oxidize S° at pH 6·0	

obligate chemolithautotroph, using iron or pyrite as an energy source. It cannot obtain energy from reduced sulfur compounds. It has a temperature and pH optimum similar to those of *T. ferrooxidans*. An important difference, however, is that it grows and can oxidize iron at much lower values of pH than *T. ferrooxidans,* for example between pH 1 and 1·5. *Thiobacillus thiooxidans* is a close relative to *T. ferrooxidans* with similar properties except that it can not oxidize iron.

Deep inside dumps temperatures as high as 60–80°C have been recorded, due to heat liberated by biological and chemical oxidation processes. In such areas many types of thermophiles have been recorded, for example the sulfur, iron and pyrite oxidizing archaebacterium *Sulfolobus brierleyi* (Brierley & Brierley, 1973). This microorganism has a temperature optimum of around 70°C. Other thermophiles include iron and sulfur oxidizing Gram-negative *Thiobacillus* Th-1 with a temperature optimum of 50°C, and the Gram-positive, spore forming *Sulfobacillus thermosulfidooxidans* which can oxidize iron, reduced sulfur and pyrite. The latter has a wide temperature optimum between

50 and 60°C. All the thermophiles recorded so far are strict aerobes, and presumably contribute significantly to the solubilization of copper in hot regions of dump leaching operations.

An example of an anaerobe found in anoxic regions of dumps is the sulfate reducing bacterium *Desulfovibrio* spp. These and other Gram-negative sulfate reducers such as *Desulfotomaculum* spp. and *Desulfobacter* spp. affect mineral recovery by forming metal sulfides that precipitate out of solution, making the metals unavailable for recovery and also coating the surfaces of minerals, preventing further metal solubilization. Such activity highlights the need to build dumps in which oxygen can freely diffuse to all areas in order to optimize bioleaching.

Also found in dump leaching operations are an enormous variety of acidophilic, heterotrophic and mesophilic bacteria (e.g. *T. acidophilus*), fungi and protozoa. Such organisms are thought to affect leaching processes in at least two ways. Bacteria such as *Acidophilium cryptum* have been shown in the laboratory to promote the growth of *T. ferrooxidans* by metabolizing organic excretions that are otherwise inhibitory to the growth of *T. ferrooxidans*. A similar process is presumed to occur in dumps. Second, protozoa, such as amoebae, prey on the microorganisms found in dumps and presumably affect population levels of microorganisms involved in metal solubilization.

There are no reports of bacteriophages of *T. ferrooxidans,* although it is almost certain that they must exist. There is a preliminary report of cryptic bacteriophages of the acidophilic heterotroph *Acidophilium* sp. (Ward *et al.*, 1990).

Finally, there are a group of mesophilic, sulfur oxidizing microorganisms exemplified by *T. neapolitanus* that oxidize reduced sulfur compounds at values of pH between 4 and 7. These microorganisms play an important role in the natural development of a dump in which exogenous acid is not applied by the mining company. When a freshly prepared dump is first exposed to rain and the pH is around neutral or higher, these microorganisms oxidize available reduced sulfur compounds to sulfuric acid. Ultimately, they lower the pH to a level where microorganisms such as *T. ferrooxidans* can take over.

A picture has emerged of a complex collection of interacting microorganisms changing dynamically in time as the dump matures from neutrality to low pH and varying in space from the surface to the inside of the dump. The surface of the dump is characterized by mesophilic aerobes while the interior contains mesophilic anaerobes and thermophilic aerobes. There are probably thermophilic anaerobes inside dumps also but none have been described, most likely due to difficulties in their isolation and culture.

Microbial diversity and dynamics are widely recognized as being very important for metal bioleaching. What is not clear is the complete extent of this diversity. Nor is there a thorough understanding of all the ways in which the microorganisms interact with each other to promote or retard metal solubilization. For example, biofilms have been observed on rocks in dump leaching operations and are presumed to play a key role in bioleaching.

However, there is only a rudimentary knowledge about the microbial composition and activity of the biofilms, and, unfortunately, this important area of research is still largely neglected (Geesey & Jang, 1990).

One problem that has impeded progress in the general area of microbial ecology of dump leaching is the inadequacy of techniques for microbial strain identification and enumeration. For example, conventional techniques of strain identification, relying on biochemical and morphological characteristics, have placed *Thiobacillus ferrooxidans, T. ferrooxidans* M1, *T. acidophilus, T. novellus, T. versutus* and *T. neapolitanus* in the same genus. This designation was based largely on a common ability to oxidize sulfur. However, preliminary work using whole genome DNA hybridization (Harrison, 1982) and more recently the use of cloned DNA probes (Yates *et al.*, 1986) and ribosomal RNA sequencing (Lane *et al.*, 1985) have convincingly demonstrated that these different species of *Thiobacillus* are well separated phylogenetically and should not be classified in the same genus. The recent introduction of DNA probe techniques including the Polymerase Chain Reaction (PCR) technique and rRNA sequencing promises to revolutionize our understanding of the diversity and phylogenetic relationship of bioleaching microorganisms.

A second problem concerns the strong potential that exists for the biased sampling of microbial species from bioleaching operations (Holmes, 1990). Current techniques for the collection of microorganisms from bioleaching dumps strongly favor the recovery of microbes that are more readily accessible, for example those in solution rather than those that are firmly bound on the ore particles. Moreover, existing methods of cultivation of microorganisms in the laboratory probably do not permit the growth, or allow only reduced growth, of some of the microorganisms in a bioleaching operation. It is quite possible, in fact almost certain, that some strains of bioleaching microorganisms are being underestimated or missed altogether due to these problems. It is hoped that DNA probe technology, coupled with PCR amplification techniques with the ability to detect single microbial cells in environmental samples (Steffan & Atlas, 1988), will solve some of the above-mentioned problems and present us with a clearer picture of the microbial diversity of a bioleaching dump. This type of information is essential for the advancement of our understanding of microbial metal solubilization.

2.4 Uranium Bioleaching

Microorganisms are being used commercially to recover uranium from uranium oxide ores. For example, uranium is produced by the action of *T. ferrooxidans* on uranium minerals when these are found in association with iron sulfides. Under these circumstances the *T. ferrooxidans* produces acidic ferric sulfate by oxidation of the iron sulfide and this, in turn, oxidizes tetravalent uranium into the acid soluble hexavalent form:

$$UO_2 + Fe_2(SO_4)_3 + 2H_2SO_4 \rightarrow UO_2(SO_4)_3^{4-} + 2FeSO_4 + 4H^+ \quad (9)$$

Microorganisms, such as *T. ferrooxidans*, are also able to oxidize uranium directly (Dispirito & Tuovinen, 1982):

$$2U^{4+} + O_2 + 2H_2O \rightarrow 2UO_2^{2+} + 4H^+ \quad (10)$$

However, in uranium bioleaching operations the indirect oxidation of uranium by the biologically produced ferric ion is considered more important.

Bioleaching has been used to obtain uranium from waste gold ore at Buffelofontein in South Africa (Anon., 1979). Uranium has also been obtained by bioleaching from the walls and debris from several underground uranium mines in the Elliot Lake region of Canada including the Stanrock, Denison and Milliken mines. In the late 1960s about 4500 kg of U_3O_8 per month were obtained by bioleaching from the Stanrock mines and 4800 kg/month from the Milliken mines. Uranium production from the Denison mines was reported to be around 32 000 kg/month in 1987 (McCready, 1988). Uranium has also been successfully bioleached from Agnew Lakes Mines in Canada (Roswell, 1977).

2.5 Bioleaching of Other Metals

In the laboratory it has been shown that microorganisms can extract arsenic, cobalt, nickel, cadmium, antimony, zinc, lead, gallium, indium, manganese and tin from a range of sulfured ores (Karavaiko, 1988). The process of metal solubilization from the sulfide ore is thought to involve primarily oxidation of the metal by biologically produced ferric iron.

Not only can metals be recovered from solid ores but waste products of metal refining can also be bioleached. For instance, lead blast furnace matter (Corrick & Sutton, 1968), oxidic fly ash from a pyrite roasting plant and basic slag from a lead-processing process (Ebner, 1978) have been reported to be capable of being bioleached. Also, vanadium and nickel have been recovered from fly ash produced by a petroleum-burning power station (Szolnoki, 1977). Copper slag is the residue from pyrometallurgical processes. It is a vitreous, nearly homogeneous mass with a copper content of 0·4–0·6%. This is about twice the copper content of waste ores used for dump bioleaching. Genchev (1985) has shown that after reoxidation of the slag with combustion with air 70–80% of the copper could be recovered by bioleaching. Genchev (1985) also showed that copper could be successfully bioleached from waste mine tailings which result from flotation techniques for copper recovery. However, none of these processes have been commercialized.

Microorganisms are being used in field studies to alter manganese ores beneficially. For example, microorganisms can convert hausmanite (Mn_3O_4) to the more valuable pyrolusite (MnO_2) which is used in dry cell batteries. Microorganisms can also be used to remove impurities such as phosphorus and iron from manganese ores for use by the ferro-manganese alloys industry (Agate & Deshpande, 1977).

2.6 Development of New Strains of Microorganisms

2.6.1 Introduction

At the present time only indigenous microorganisms that are found occurring naturally in dumps or mine run-off are used in bioleaching operations. This is equivalent, in a sense, to using wild wheat for farming or an unselected strain of *Penicillium* mold for producing penicillin. In both these cases genetic selection and breeding have been used to develop strains that outperform their wild-type ancestors by a factor of hundreds or thousands respectively. There is no reason to think that genetic manipulation, including genetic engineering, could not also significantly improve bioleaching microorganisms such as *T. ferrooxidans* for commercial applications. This would make existing applications of bioleaching microorganisms more attractive and could extend the use of microorganisms to totally new opportunities for metal recovery and ore beneficiation.

New strains of microorganisms can be developed for commercial applications by four procedures:

—the isolation of new strains of microorganisms from the environment;
—the selection and adaptation of naturally occurring strains;
—the modification of strains by classical genetic techniques;
—genetic engineering.

2.6.2 Isolation of New Microorganisms from the Environment

We probably know only a few percent of the microorganisms that could be of value for biological metal recovery applications and, of those, only a handful have been studied in any detail. Also, of the handful that are being studied in laboratories around the world it is only very recently that an effort has been made to evaluate the commercial potential of any of them except those of the *Thiobacillus* genus. For example, there appear to be potential advantages to *Sulfolobus* spp. for high temperature applications (Norris & Barr, 1988; Norris, 1990) and for *Leptospirillum* spp. for very low pH (~1·0) applications (Karavaiko *et al.*, 1988), and for manganese oxidizing and reducing microorganisms from deep sea vents (Ehrlich, 1987).

It is almost certain that there exists a huge repository of microorganisms in natural habitats such as mine dumps, mine drainages, coal refuse piles, hot springs, toxic waste sites, deep sea vents, psychrophilic habitats, etc., some of which could exhibit interesting new enzymes and properties for metal recovery applications.

Another neglected area of research concerns the evaluation of mixed cultures of microorganisms for metal recovery. Although symbiosis and commensalism are recognized as being important for bioleaching in a copper ore dump, little effort is being made to develop mixed cultures for commercial metal recovery applications. The issue here is that, frequently, mixed cultures can do things that the individual strains of microorganisms that compose the mixed culture cannot do alone. It is suggested that this area of research should be more thoroughly explored.

2.6.3 Selection and Adaptation of Naturally Occurring Strains

In the previous section it was indicated that mixed cultures frequently exhibit synergism. Mixed cultures frequently exhibit another important property, namely the ability to give rise to new strains of microorganisms. These strains are thought to arise by two mechanisms: (1) by the transfer of genes between microorganisms by conjugation, resulting in a new strain with properties from both parents; and (2) by mutation.

Even though an appropriate new strain may arise infrequently in a mixed culture, its appearance and propagation can often be selected by chemostat culturing. This technique has been widely used for the development of new strains for toxic waste degradation (Holmes, 1989).

Although chemostat selection has been widely used for microbial strain development in other areas, it has been neglected as a tool for the development of microbial strains for metal recovery. It was recently shown that *T. ferrooxidans* undergoes mutation at a rate that far exceeds normal point mutation rates (Holmes *et al.*, 1988*b*). This high mutation rate increases the probability that chemostat selection could be used successfully to select for interesting strains. The high mutation rate has been shown to be due to the presence of different classes of repetitive DNA sequences in the chromosomes and plasmids of *T. ferrooxidans*. One of the classes of these repetitive DNA sequences has been shown to be an insertion sequence (Yates *et al.*, 1988). Also, recent evidence indicates that the repetitive DNA sequences are mobile, moving around the genome, and that this movement can cause mutations to arise (Holmes *et al.*, 1988*a*), some of which may create useful commercial strains (Holmes & Haq, 1990).

Much work remains to be done on evaluating the mechanism of action and the importance of these repetitive DNA sequences for the generation of commercial strains of *T. ferrooxidans*. There are also interesting questions to be addressed concerning the frequency and specificity of the genetic changes. For example, do the repetitive DNA sequences move at random around the genome of *T. ferrooxidans* or do they integrate into and excise from only particular regions of the genome, perhaps in a genetically programmed fashion? Can the rate of movement be influenced by external conditions, including conditions known to induce physiological stress such as heat shock? Are there other classes of repetitive DNA elements in addition to family 1 and family 2 that remain to be discovered?

A key advantage of using such naturally adapted strains for commercial applications is that they do not come under the guidelines and regulations governing the use of genetically engineered strains.

2.6.4 Classical Genetic Manipulation

Classical genetic techniques for the development of new microbial strains include mutation, conjugation, transformation and transduction. Very little work has been done in these areas with bioleaching microorganisms. Pioneering studies were carried out by Groudeva and Groudev (1980) using nitrosoguanidine to mutate and select for mutants of *T. ferrooxidans*. However, none of these studies has been followed up.

Conjugation systems for some thiobacilli have been developed (e.g. Davidson & Summers, 1983; Kulpa *et al.*, 1983; Roberto *et al.*, 1990) but none are available for acidophilic iron-oxidizing thiobacilli. Bacteriophages for transduction have not been reported for thiobacilli, although there is a recent report of bacteriophage-like particles in acidophilic heterotrophs (Ward *et al.*, 1990) which may prove valuable for developing transduction systems.

2.6.5 Genetic Engineering

Advances in the genetic engineering of microorganisms involved in metal recovery, especially *T. ferrooxidans,* have been reviewed recently by Holmes and Osgood (1990), Holmes and Yates (1990) and Rawlings (1990). Also there is an excellent report on earlier work in the area of genetic engineering (Bell *et al.*, 1987).

3 BIORECOVERY OF METALS FROM INDUSTRIAL AND URBAN WASTES

3.1 Introduction to the Problem

Efforts to reduce future environmental pollution and to remedy existing pollution problems continue to have a very high priority in many countries around the world. Problems due to the release of metals from industrial and urban wastes constitute a significant proportion of total environmental pollution. For example, in the USA, the Environmental Protection Agency estimated that over US $58 million of metals were disposed of in 1985 from the aqueous solutions of the electroplating industry alone. Other industries that yield significant metal wastes include wire manufacturers (nickel, copper), light bulb manufacturers (mercury), plastics manufacturers (copper), metal refining industries (zinc, nickel, copper, iron), circuit board manufacturers (platinum), power companies (fly ash with many metals), film manufacturers and processors (silver), and leather tanners (chromium). In New York alone in 1987 the EPA estimates that 15 000 tons of chromium, 19 000 tons of lead and 29 000 tons of other heavy metals were disposed of. In addition, solid waste from urban sewage treatment is frequently contaminated with unacceptable levels of heavy metals. These wastes must presently be treated as hazardous or toxic, and disposed of at great expense. The recovery of metals could significantly subsidize waste stream detoxification costs, provide a valuable source of certain metals and make the wastes considerably safer.

There are several existing technologies for the treatment of metal wastes including solvent extraction, hydroxide precipitation, evaporative concentration, electrowinning and membrane concentrations. One of the most successful of these is the hydroxide precipitation of metals. Advantages of this approach are that it is capable of dealing with very large aqueous volumes and with relatively high concentrates of metals. Two disadvantages are that certain metals are not efficiently precipitated and the technique often does not reduce the metal content of the solution to acceptable levels for discharge.

Biological materials, usually live or dead microorganisms, or biologically derived molecules such as chitosan or proteins, have long been considered a resource for trapping or precipitating unwanted metals. Biological materials frequently have very high capacities for metal binding, some concentrating metals by orders of magnitude (Geesey & Jang, 1990; McLean & Beveridge, 1990). They can also exhibit metal binding specificity (Muzzarelli, 1977). This latter attribute can be very significant in those cases where the required metal is associated with very high concentrations of competing ions. In fact, it may be possible to utilize this biological specificity to concentrate one metal in the presence of other metals.

To be commercially attractive, a bioaccumulator, whether intact cell or an isolated cell component, must (1) have a high selective affinity for metals, (2) be readily and cheaply producible, (3) be reusable after elution of the concentrated metal, (4) be chemically stable, and (5) be cheaply disposable without causing a pollution problem after its usefulness is exhausted.

Biological recovery of metals can be divided into the following categories for discussion:

(1) Processes requiring metabolically active cells
 —Intracellular metal accumulation
 —Extracellular metal accumulation
 —Precipitation
(2) Processes that do not require metabolically active cells
 —Cellular metal accumulation
(3) Processes that use isolated biological components
 —Cellular metal accumulation

3.2 Biological Processes for Metal Recovery

3.2.1 *Processes Requiring Metabolically Active Cells*

INTRACELLULAR METAL ACCUMULATION

In general, metals which are required nutritionally (e.g. Co, Cu, Fe, Mg, Mn, Mo, Ni, Se) by microbial cells are taken up by special mechanisms which make possible their passage through the cell membrane. Since at high concentration these metals are toxic to the cell, their uptake must be carefully regulated by these mechanisms. Hardly ever do metal ions enter a healthy, living cell by simple diffusion.

Metal-binding proteins can accumulate metals intracellularly. For example, yeast metallothionein (Winge *et al.*, 1986) and *Neurospora* metallothionein

(Munger *et al.*, 1985) accumulate copper intracellularly, although *in vitro* they can also bind Zn, Ag and Hg ions.

There are no commercial examples of the use of metabolically active intracellular accumulation for metal recovery.

EXTRACELLULAR METAL ACCUMULATION

Gallionella ferruginea can grow autotrophically on Fe(II) in a partially reduced environment, depositing the ferric iron as oxide in its lateral stalk (Hanert, 1981). *Leptothrix ochracea* and other species of *Leptothrix* have been reported to oxidize Fe(II) or Mn(II) and deposit the corresponding oxides in their sheaths (see; Lundgren & Dean, 1979; Ehrlich, 1981). It is not certain at this time how *Leptothrix* spp. oxidize Fe(II) or whether they can obtain energy from the oxidation of either Fe(II) or Mn(II). Encrustation of the sheath by either iron or manganese oxide may be so heavy as to completely obscure it and the cell it encloses. Members of the family Siderocapasaceae may be able to oxidize Fe(II) or Mn(II). Even if incapable of oxidizing these substances, they accumulate the Fe(II) or Mn(IV) oxides in their capsules (Hanert, 1981).

Accumulation of metals in extracellular secretions from living organisms probably occurs in activated sludge processes (Romanenko & Koren'kov, 1977; Sterritt *et al.*, 1981).

PRECIPITATION

Heavy metals can also be precipitated by H_2S generated by sulfate-reducing bacteria (see Ehrlich, 1981). Besides the classical *Desulfovibrio* and *Desulfotomaculum* strains (Postgate, 1979), a variety of new sulfate-reducing bacteria have been discovered in the last few years. They differ in part with respect to the energy sources they utilize (Pfennig *et al.*, 1981). The organisms include autotrophic, hydrogen-utilizing sulfate-reducers which use acetate as reductant.

According to Cameron and Koch (1980), anaerobic microbial digestion of trace-metal containing leachates from landfills has been shown to lead to the precipitation of >85% of Al, Ba, Cd, Hg, Ni, and Zn, 80% of the Fe, 40–70% of the Cr, Cu, Pb, and Mn, 30% of the Ca, and <10% of the Mg, K and Na as insoluble compounds (sulfides) or organic complexes. The metals in the leachate did not affect the anaerobic digestion adversely (Cameron & Koch, 1980). The metals precipitated from the leachate could be extracted from the precipitate by the method of Schonborn and Hartmann (1978). This method involves recovery of the metals, present as sulfides (and probably also as organic complexes), in sewage sludge by enriching the sludge with 1% native sulfuric acid from the sulfur and inoculating with *T. thiooxidans,* to generate sulfuric acid from the sulfur for lowering the pH, and with *T. ferrooxidans,* to oxidize the metal sulfides. The metals they solubilized in this way included Fe, Co, Zn, Ni and Cd.

3.2.2 *Processes That Do Not Require Metabolically Active Cells*

CELLULAR METAL ACCUMULATION

Many microorganisms accumulate metals, generally on their cell walls. Usually this does not require actively metabolizing cells but works with dead cells also. Table 3 is a partial list of such microorganisms.

Some of the cell-surface components that bind metals include cell-wall polymers like peptidoglycans, teichoic and teichuronic acids in Gram-positive bacteria, all membranes and peptidoglycan in Gram-negative eubacteria, and slime-layer polysaccharides in various prokaryotes; chitin and chitosan and probably some other glycans of fungal cell walls (Gadd, 1990), and unidentified polymers in algal cell walls (Greene & Darnall, 1990). The affinity for metal cations of these polymers is due to negatively charged groups (chiefly carboxyl groups and phenolic hydroxyl groups) and also sulfhydryl groups (—SH). The affinity for metal anions is mainly due to protonated amino groups.

COMMERCIAL APPLICATIONS

Three procedures for using dead biomass to trap metals from wastes have been recently commercialized. These are the AMT-Bioclaim process and two processes using immobililized algae on an inorganic matrix (Bio-Recovery Systems, Inc., Las Cruces, New Mexico, and B. B. Sorbex, c/o McGill University, Department of Chemical Engineering, Montreal, Canada). The AMT-Bioclaim process, developed and marketed by AMT, Inc., is now owned by Visatech Partnership Ltd USA, who are licensing the technology through InterPro.

The core of the AMT-Bioclaim process is based on organic granules which the company call MRA (metal recovery agent). These granules are essentially a hard, insoluble, solid biomass product derived chemically from the microorganism *Bacillus subtilis*. The metal uptake capacity of this microorganism is improved by treating the microorganism with a caustic solution which converts it into a 'solid' biomass. The MRA exists as a small (approximately 0·1 mm) granular material with a density slightly greater than water. The technology employed in forming the MRA granules distributes metal-binding sites not only on the surface of the MRA granule but throughout the granule itself. This dramatically increases the area of metal uptake. The AMT-Bioclaim MRA granules can be used for recovery of precious metals from jewelry and plating manufacturers' wastewaters. The MRA can accumulate these metals even when they are complexed with cyanide.

The MRA granules accumulate metals directly from wastewater streams, eliminating costly and cumbersome chemical pretreatment. Sludge, the unavoidable byproduct of chemical pretreatment, is not produced by the AMT-Bioclaim process.

AMT has produced two grades of MRA: (1) cationic MRA, and (2) oxy-anionic MRA. Cationic MRA accumulates toxic cationic metal ions such

Table 3
Examples of Microbial Metal Accumulation

Microbial type	*Name of organism*	*Site of metal accumulation*	*Types of metal accumulated (and extent)*	*References*
Bacteria	*Pseudomonas aeruginosa*	Cytoplasm cell envelope	Cd^{2+}	Horitsu *et al.*, 1979
	Pseudomonas aeruginosa	Intracellular	U (10–15% of cell, dry wt)	Strandberg *et al.*, 1981
	Escherichia coli K-12	Membrane-polar head groups or peptidoglycan layer in wall	Hf, Os, Pb, Zn, Zr, Fe(III), Mn, Mo, Mg, Co, Ce(IV), Na, K, Rb, Ca, Cr, Cu, Sc, La, Pr, Sm, U, Fe(II), Ru, Ni, Hg, Pb, Pt, Pd, Au, In	Beveridge & Koval, 1981; Brierley, 1990
Fungi	*Saccharomyces cerevisiae*	Intact cells	Co, Cd, Zn, Ni, Cu, Mn	Norris & Kelly, 1979; Gadd, 1990
Algae	*Navicula pelliculosa, Chlamydomonas, C. pyrenoidosa, Mongiotia* sp., *S. obliquus, Ulothrix fimbrinata, Zygenema* sp., *Nostoc* 586, *N. muscorum, Oscillatoria, S. calcicola*	?	Cd, Pb, Hg (concentration factor of 1 310–46 700 at pH 7 depending on alga)	Hassett *et al.*, 1981; Greene & Darnall, 1990
	Chlorella vulgaris	Cell surface	U	Horikoshi *et al.*, 1979; Nakajima *et al.*, 1979
	Hormidium fluitans	Cell wall	Ag, Al, Cu, Fe, Mo, Ti, V	Madgwick & Ralph, 1977

as chromium(III), lead, silver, cadmium, zinc, copper, mercury, and nickel. Oxy-anionic MRA accumulates toxic oxy-anions of metal and metalloid complexes such as chromate, arsenates, and selenates.

A significant characteristic of MRA is its ability to be repeatedly regenerated. Regeneration of loaded MRA is accomplished through a two-step process. In the first step, the loaded metals are stripped from the MRA by the application of one of three chemical solutions: an acid, a caustic, or a proprietary neutral solution. In the second step, granules are washed and reactivated. Regeneration of MRA takes place in a container separate from the one in which active metal uptake is occurring.

Metals which have been stripped from the loaded MRA are concentrated in solution and may be: (1) recycled and reused by the producer, (2) recovered from the solution by an electrowinning process to yield a cathodic metal, or (3) added back into the process for pretreatment by a precipitation technology that may be in place.

The process economics of the AMT-Bioclaim system, as with any wastewater clean-up technology, is site specific. Costs will vary as certain parameters such as metal concentration and flow rates vary, not to speak of variations in the toxic elements present in the solution. (AMT claims that their system is effective in removing any metal from solution with an atomic weight greater than that of aluminum.) For this reason, it is particularly difficult to compare the economics of differing systems without holding some variable constant. However, a careful analysis of the AMT-Bioclaim process (Holmes *et al.*, 1988*b*) indicates that the process has significantly reduced metal recovery costs to the point that in some process-waters, containing high levels of relatively valuable metals, the costs of removing the metals may be exceeded by the value of the recovered metals.

Furthermore, the AMT-Bioclaim process has proved to be a low cost metal removal system even when the metals are not reclaimed in both secondary and primary wastewater treatment. It appears that for high volume, dilute solutions which are typically encountered in mining, milling and smelting operations, the process economics are further enhanced. On the other hand, due to the nature of the process and the regeneration of MRA, for very high concentrations of metals the process loses its cost advantage over competitive processes when the metals are not recovered.

3.2.3 Processes That Use Isolated Biological Components

Exopolysaccharides of Zoogloea ramigera: The bacterium *Zoogloea ramigera* produces substantial amounts of extracellular polysaccharides. The polymer comprises glucose, galactose and pyruvate, and is active in binding metals (Friedman & Dugan, 1968; Norberg & Person, 1984).

Biomass of *Z. ramigera,* when added to solutions containing heavy metal cations and stirred at 800 rpm, is saturated with metals and flocculates in 15 min. The activity of exopolysaccharides of *Z. ramigera* in the sorption of

Table 4
Characteristics of Sorption Properties of Exopolysaccharides of *Zoogloea ramigera*

Metal cation	*Optimum pH values for sorption*	*Biosorbent loading capacity, mg/g biomass*
U	3·5	0·37
Cd	5·5	0·233
Cu	6·5	0·323

uranium, cadmium and copper cations is characterized by the data presented in Table 4. Optimum pH values for the sorption of these cations are different. This allows for selective metal recovery from solutions.

For regeneration of the biosorbent, flocs consisting of exopolysaccharide and biomass with adsorbed metals are diluted with water, acidified with HCl to pH 3·0. The selectively adsorbed cadmium, copper and uranium are solubilized and separated from the flocs. The flocs are again dissolved in water under intensive stirring and are further used for metal sorption.

Emulsan and Emulsanosol: Emulsan, produced by *Acinetobacter calcoaceticus*, is a polyanion heteropolysaccharide with main components of *N*-acetyl-D-galactosamine (25%) and *N*-acetyl-aminuronic acid (33%). Emulsan possesses a metal-sorbing capacity and can be used as a biosorbent either as homogeneous aqueous solution or as hydrocarbon emulsion (emulsanosol), the uranium-binding capacity of the latter exceeding that of emulsan (Zosim *et al.*, 1983). At initial concentrations of UO_2^{2+} of 100 μg/ml over 90% of uranium is bound by 1 mg of emulsan. Maximum binding capacity of 1 mg of emulsan is 0·9 μg of UO_2^{2+}.

Emulsanosol is an aqueous emulsion of hydrocarbons (in particular, hexadecane) stabilized with emulsan. For preparation of emulsanosol, a mixture of hexadecane and aqueous solution of emulsan is treated with ultrasound. In solutions containing 67·5 μg UO_2^{2+}/ml, emulsanosol binds 42 μg UO_2^{2+}/ml.

Capacity of emulsan and emulsanosol to bind uranium decreases below pH 4·0. In the case of an emulsan solution, even at pH 1·0 over 50% of UO_2^{2+} remains bound, while with emulsanosol more than 70% of the bound uranium is solubilized at pH 3·0. Thus, emulsanosol possesses other advantages in addition to the higher uranium-binding capacity compared to emulsan. Emulsanosol is easier to regenerate. For this, emulsanosol with bound uranium is separated from the purified solution and pH is adjusted to 2·0. More than 90% of the bound uranium and less than 10% of emulsan is released into the solution. Regenerated emulsan is recycled into the process. Emulsanosol exceeds other biosorbents of microbiological origin in uranium-binding capacity (see Table 5). The binding by emulsanosol of UO_2^{2+} and hydrolysis products, $(UO_2)_2(OH)_2^{2+}$ and $(UO_2)_3(OH)_5^{+}$, mainly present at pH > 3·0, is

Table 5
Comparative Uranium-binding Capacity of Emulsanosol and Other Biosorbents

Biosorbent	*Bound uranium (mg/g biosorbent)*
Chitin from cell walls of *Rhizopus arrhizus*	6
R. arrhizus biomass	180
Emulsan (1 g) in emulsanosol	800

believed to occur not only through interaction with the carboxyl groups of the polysaccharide but also through hydrogen bonds with free hydroxyl and *N*-acyl groups.

Emulsan is ecologically safe as it does not contain phosphorus, is non-toxic and is biodegradable. Zosim *et al.* (1983) believe that emulsanosol is a promising biosorbent for uranium removal from uranium industry wastewaters.

Bisorbent M: Biosorbent M is obtained using biomass *Penicillin chrysogenum,* a waste product of antibiotics production cycle (Jctek, 1978). The biomass is treated with urea-formaldehyde polycondensate yielding a solid product which comprises mycelium and its components, polycondensate, and the reaction products of urea and formaldehyde with mycelium components. The product is crushed to grains of 0·3–0·8 mm. The resulting substance is hydrophilic with grains swelling in water and in electrolyte solutions, mechanically strong and stable to the effect of physico-chemical parameters. Biosorbent M has been used in test studies to remove radioactive elements from wastewaters in a semi-industrial plant. The process uses two columns in series, each filled with Biosorbent M. Wastewaters are fed to the columns at a specific rate. Under similar conditions, commercial ion exchange resins were unable to reduce uranium concentration to the desired level of 0·05 mg/liter and Ra concentration to 3–30 pCi/liter while Biosorbent M achieved this level of metal removal.

Biosorbent M can be regenerated by elution of bound uranium with a solution containing 10% NaCl and 0·5% Na_2CO_3. Radium is eluted with 0·1N HCl.

Biomass of Rhizopus arrhizus: Comparative investigations of several types of biomass on wastes of biotechnological processes revealed a high capacity of *R. arrhizus* biomass for uranium and thorium sorption (Khraptsova, 1988). The loading capacities of dry biomass of *R. arrhizus* for uranium and thorium were found to be 180 and 185 mg/g of biosorbent, respectively, at an optimum pH of 4·0–5·0.

At equilibrium *R. arrhizus* biomass absorbs, respectively, 2·5 and 3·3 times more uranium than ion exchange resins or activated carbon, procedures that are used in industry for removal of uranium from wastewaters.

Uranium-loaded biomass can be regenerated with a 1N solution of $NaHCO_3$ at a solid-to-liquid ratio of 100:1.

A technique for immobilizing *R. arrhizus* biomass has been developed, using a polymer material not exceeding 15% of the total weight of the preparation. Immobilization increases the life of biosorbent and yields biosorbent particles of standard size with controlled strength and porosity.

The disadvantage of this type of biosorbent is the inhibition of uranium biosorption by divalent and trivalent iron cations and by sulfate ions. Therefore, the possibility of using this type of biosorbent for uranium removal from acid mine waters in the process of underground leaching appears to be problematic at this stage.

Chitin and Chitosan: Chitin or chitosan (deacetylated chitin), from the walls of fungi and the exoskeleton of arthropods, has an affinity for U, Th, and Ra, and several heavy metals (Yoshinara & Subramanian, 1976; Muzzarelli, 1977). Purified chitin from lobster shells exhibited some selectivity toward heavy metal ions (Zn > Ni > Co > Cu > Pb) according to Yoshinara and Subramanian (1976). Muzzarelli (1977) has observed that transition metals of the first series (Ti, V, Cr, Fe, Co, Ni and Cu) are bound by purified chitin and chitosan from fungi, forming variously colored complexes.

Fungal chitosan exhibits differential affinity for metal cations. Sulfate anions enhance metal uptake while chloride anions suppress it. Treatment of chitosan can enhance its adsorption capacity. Elution of sorbed metals is said to be easy. However, Fe(III) can only be eluted after reduction to Fe(II) (Muzzarelli, 1977). In general, the mechanism of binding of metal cations is thought to be through formation of coordination complexes with $—NH_2$ groups of conditioned chitosan. Oxyanions of metals react by anion exchange on protonated amino groups of conditioned chitosan. Untreated chitosan probably binds metal ions by ion exchange sorption and chelation. A chitosan–glucan complex (200 mg of 100–200 mesh powder) from the fungus *Aspergillus niger* bound up to 98% of Cr^{3+}, 82% of Mn^{2+}, 96% of Co^{2+}, 98% of Ni^{2+}, 100% of Cu^{2+}, 90% of Zn^{2+}, 90% of Cd^{2+}, and 100% of Pb^{2+} from 50 ml of 0·5mM metal sulfate solutions (except for lead nitrate) at pH 5·6 (final pH 7·5) after 1 h shaking at room temperature (Muzzarelli *et al.*, 1980*a*,*b*).

Isolated biopolymers from cell surfaces, like chitin (especially after conversion to chitosan by deacetylation in concentrated NaOH solution), can be made specific by further pretreatment and by carefully controlling the conditions of metal sorption. The basis for this is the different affinities for different metal cations of the chitosan preparations, and the fact that chitosan can be transformed into an anion exchange resin which has an affinity for metal oxyanions but not cations (Muzzarelli, 1977).

For example, although peptidoglycan of bacterial cell walls and chitosan seems to bind equivalent amounts of Cu^{2+} (in the order of 50 kg of Cu^{2+} per ton), chitosan is so much easier to prepare on a large scale and its source material, fungal mycelium, so readily available in large quantities as a byproduct of the fermentation industry, that by the above criteria it should be preferable commercially as a bioaccumulator. Muzzarelli (1977) estimated the cost of production of chitosan in 1977 to be between \$1·98 and \$4·84 per kg, which is still very expensive unless chitosan can be reused a great many times.

3.3 Comparison of Living versus Non-living Biosorbents

In utilizing living microbial cultures as biosorbents two major problems are to be solved:

(1) how to maintain the bacterial activity, given constant changes in the chemical composition of industrial wastewaters and wide variations in the concentration of heavy metal cations; and
(2) how to separate the microbial biomass from purified solutions.

Utilization of living microbial cultures in the biosorption processes is not always possible because of the high toxicity of some wastewater components. In addition, there is often variability of the chemical composition of wastewaters and the necessity of adding nutrient substrate and maintaining physico-chemical parameters optimal for microbial growth and activity. Also, regulatory problems are sometimes encountered when using live biomass. These problems can be avoided if preparations based on non-living biomass are used as biosorbents. It is significant that the only three processes of biological metal recovery successfully commercialized are based on dead biomass derived itself from cheap sources.

3.4 Genetic Engineering of Biological Metal-binding Processes

3.4.1 Introduction

Genetic engineering, and also classical genetic manipulation, could be used to generate new strains of microorganisms with superior metal-binding properties. To date, there has been no published work in this area. Part of the reason for this is that the mechanism(s) of metal-binding are often complicated, involving several components including, but not limited to, proteins and polysaccharides, and also the mechanism(s) are frequently poorly understood. The most likely near-term impact of genetic engineering will be as a research tool to unravel the physiology and biochemistry of biological metal binding and detoxification.

Another future role for genetic engineering might be to produce specific metal-binding proteins that could be used directly for industrial applications. Naturally occurring metal-binding proteins, such as metallothionein, frequently exhibit a degree of selectivity for specific metal ions that exceed by orders of magnitude the selectivity exhibited by commercially available chelating agents, thus reducing the problem of competition with common ions. Their binding constants can also be extremely high, making them effective in recovering metals from very dilute solutions.

On the other hand, proteins are expensive and there is insufficient documentation on their properties and stability in industrial situations. Compared to conventional chelating agents, proteins might be less robust and relatively ineffective at non-biological temperatures and pH, or in non-aqueous solvents. Even in aqueous solution around neutral pH and ambient temperature, proteins might be unstable due to denaturation, oxidation of proteolysis. Due to the high cost involved in the production or genetically engineered proteins, it is important that they are robust enough to be used in many cycles of metal binding and metal removal. They must also offer significant advantages compared to currently available metal-chelating agents.

3.4.2 Metallothionein

An example of an investigation into the feasibility of producing a genetically engineered protein for metal chelation is metallothionein. Metallothionein has very high selectivity for zinc, copper, mercury and silver (Kagi, 1970; Beltramini *et al.*, 1984). There is also a considerable body of information on its properties (Vasak *et al.*, 1981), including an X-ray structure of a mammalian metallothionein (Furey *et al.*, 1985). These properties make it an ideal candidate for genetic engineering. Leib and Holmes (1987) have investigated the properties of several metal-binding protein analogs of *Neuropora* metallothionein artificially synthesized on an automatic peptide synthesizer. They conclude that:

(1) Chemical synthesis is a feasible route for the preparation of small, metal-binding proteins, despite the potential problems of cysteine oxidation during synthesis and purification.
(2) Oxidation of cysteines, subsequent to the synthesis and preparation of the synthetic *Neurospora* metallothionein, significantly impairs its metal-complexing capabilities. Therefore, metallothioneins could only be of value as industrial metal-chelators when synthesized and utilized in an anaerobic environment such as inside an organism.
(3) When histidine was substituted for cysteine, problems associated with oxidation were alleviated. However, the resulting peptides had diminished metal-binding strengths for zinc and cadmium, but undiminished binding of mercury. Therefore, these substituted peptides could have potential applications for mercury removal.
(4) The synthetic peptides cost about $100–1000 per gram, depending on the quantity being synthesized. At this cost it would be essential to be able to recycle the metallothionein many times for metal detoxification or metal recovery in order to make these processes economically competitive. The feasibility of multiple recycling remains to be investigated.

3.4.3 Conclusions

Despite several successful attempts at demonstrating the utility of chelating agents as metal processing agents on a laboratory scale, there are no commercial plants using these reagents (other than the commonly used xanthates, dithiophosphates, and fatty acids) for metal detoxification or metal recovery. One of the major disadvantages of using purified proteins as reagents for flotation and flocculation applications or metal detoxification appears to be their prohibitively high cost. It is to be recognized, however, that there are a number of long-standing toxic metal problems not amenable to treatment with conventional reagent combinations and where the superior performance of new reagents could easily justify the higher cost. It is thus projected that in coming decades, the commercialization of a number of reagents already tested in the laboratory and at pilot plant scale is going to increase.

The principal objective of research programs in this area must be to establish a proper theoretical basis for predicting selectivity in chelating reagent systems and preferably develop reagents that are metal specific rather than cation specific. As demonstrated by the work carried out in the recent past, this requires a collaborative effort among genetic engineers, microbiologists, biochemists and chemists.

4 CONCLUSIONS AND FUTURE DIRECTIONS

1. Bioleaching has won wide acceptance for the production of copper, especially from dumps of waste ore. Although the copper industry sometimes overlooks the microbiological nature of their leaching operations, bioleaching accounts for approximately 25% of the world's copper production. It should be noted that despite their unsung role, microorganisms are probably responsible for the financial solvency of several major US copper producers during the 1982–86 period of overproduction and low prices, as several companies shut down their high-cost conventional production systems but continued to bioleach copper. Most notable of these was the Kennecott Bingham Canyon operation in Utah which closed its mining but continued its dump leaching.

   Microorganisms can also oxidize waste gold ores from current or abandoned refractory gold mining operations. Because these are waste piles none of the mining or comminution costs are added to the gold recovery process and, for this reason, returns on investment could be spectacular. It is expected that bioxidation of refractory gold waste products will be a rapid growth area.
2. Near-term improvements in metal-recovery from waste dumps are more likely to come from engineering advances, such as improved dump design, rather than from new biology. However, it should be pointed out that many of the new engineering advances will take into account biological principles of design.
3. The following areas of biological research will have an impact on the commercial biotechnology of metal recovery.

   Genetic engineering studies are necessary, initially to enhance the understanding of the biochemistry and physiology of the relevant microorganisms, and subsequently to use this knowledge to generate superior strains of microorganisms.

A second research area concerns the survivability of laboratory strains, including genetically engineered strains in commercial bioleaching operations. Since the microbial substrates are not sterile, competition with naturally occurring fauna and flora is expected.

Third, bioleaching of metals from waste dumps involves many different types of microorganisms and, of these, only a small percentage have been investigated in any detail. It would almost certainly repay efforts to discover and evaluate the metal recovery capabilities of novel microorganisms in habitats such as deep sea vents, thermal springs and toxic waste sites.

4. Although in the past waste metals have been ignored or largely viewed as a nuisance by industry, the possibility exists that these waste metals may prove to be a valuable resource, particularly if the value of these metals exceeds the costs of removing and recovering them. Biological metal concentration has often been mentioned as a process with the potential for reducing the costs of metal recovery from wastes sufficiently to be able to regard these waste metals as a resource. The general economic characteristics of biological processes (low capital costs and the potential of very low or negligible operating costs) are ideally suited to such an application. At the same time, the primary economic drawback to biological processes, namely slow reaction times, is not relevant to biosorption processes where the rate-limiting parameter is not the metabolism of the cell but rather the rate of binding to the biomass, or fractions thereof, and the metals in the solution.

ACKNOWLEDGEMENTS

My grateful thanks to Drs Marcy Osgood and Ramananda Rao for their critical review of the manuscript.

REFERENCES

Agate, A. & Deshpande, H. A. (1977). In *Conference Bacterial Leaching*, ed. W. Schate, GBF Monograph Series No. 4. Verlag Chemie, New York, p. 243.

Anon. (1979). *Quarry mine*, **18,** 17.

Balashova, V. V., Vedenina, I. Y., Markosyan, G. E. & Zavarzin, G. A. (1974). *Microbiologiya*, **43,** 581. (English Transl. pp. 491–4.)

Bell, P. R., McBride, B. C. & Spiegelman, G. B. (1987). Potential genetic engineering of *Thiobacillus ferrooxidans* to optimize the leaching of mineral values. Canmet Report Serial No. Osq85-0023.

Beltramini, M., Lerch, K. & Vasak, M. (1984). *Biochemistry*, **23,** 3422.

Beveridge, T. J. & Koval, S. F. (1981). *Appl. Environ. Microbiol.*, **42,** 325.

Boseker, K. & Kursten, M. (1978). *Process Biochem.*, October, 2.
Brierley, C. L. (1990). In *Microbial Mineral Recovery,* eds H. L. Ehrlich & C. L. Brierley. McGraw-Hill New York p. 303.
Brierley, C. L. & Brierley, J. A. (1973). *Can. J. Microbiol.* **19,** 183.
Brierley, J. A., Norris, P. R., Kelly, D. P. & LeRoux, N. (1978). *Eur. J. Appl. Microbiol. Biotechnol.* **5,** 291.
Bryner, L. C., Beck, J. F., Davis, D. B. & Wilson, D. G. (1954). *Ind. Eng. Chem.*, **46,** 2587.
Buchanan, R. E. & Gibbons, N. E. (eds). (1974). *Bergey's Manual of Determinative Bacteriology,* 8th edition. Williams & Wilkins, Baltimore, Maryland.
Cameron, R. D. & Koch, F. A. (1980). *J. Water Pollut. Control Fed.*, **52**:282.
Carpenter, L. & Herndon, J. (1933). *West Virginia University Eng. Exploit Stat. Res. Bull.*, **10.**
Colmer, A. R. & Hinkle, M. E. (1947). *Science,* **106,** 253.
Corrick, J. D. & Sutton, J. A. (1968). Oxidation of lead blast furnace matte by *Ferrobacillus ferrooxidans* or a dilute acid solution. US Bureau of Mines Report 7126. US Department of the Interior, Washington, DC.
Davidson, M. S. & Summers, A. D. (1983). *Appl. Environ. Microbiol.*, **46,** 565.
Dispirito, A. A. & Tuovinen, O. H. (1982). *Arch. Microbiol.*, **133,** 33.
Ebner, H. G. (1978). In *Metallurgical Applications of Bacterial Leaching and Related Microbiological Phenomena,* eds L. E. Murr, A. E. Torma & J. A. Brierley. Academic Press, New York, p. 195.
Ehrlich, H. (1963). *J. Bacteriol.*, **86,** 350.
Ehrlich, H. L. (1981). *Geomicrobiology*. Marcel Dekker, New York.
Ehrlich, H. L. (1987). *Geomicrobiol J.*, **5,** 423.
Ehrlich, H. L. & Brierley, C. L. (eds) (1990). *Microbial Mineral Recovery*. McGraw-Hill, New York.
Friedman, B. A. & Dugan, P. R. (1968). *Dev. Ind. Microbiol.*, **9,** 381.
Furey, W. F. *et al.* (1985). *Science,* **231,** 704.
Gadd, G. M. (1990). In *Microbial Mineral Recovery,* eds H. L. Ehrlich & C. L. Brierley, McGraw-Hill, New York, p. 249.
Geesey, G. & Jang, L. (1990). In *Microbial Mineral Recovery,* eds H. L. Ehrlich & C. L. Brierley. McGraw-Hill, New York, p. 223.
Genchev, F. N. (1985). In *Biotechnology of Metals,* eds G. I. Karavaiko & S. N. Groudev. UNEP, Centre Int. Projects GKNT, Moscow, p. 327.
Greene, B. & Darnall, D. W. (1990). In *Microbial Mineral Recovery,* eds H. L. Ehrlich & C. L. Brierley. McGraw-Hill, New York, p. 277.
Groudev, S. N., Genchev, E. N. & Gaidarjiev, S. S. (1978). In *Metallurgical Applications of Bacterial Leaching and Related Microbiological Phenomena,* eds L. E. Murr, A. E. Torma, & J. A. Brierley. Academic Press, New York, p. 253.
Groudeva, V. & Groudev, S. (1980). *Mikrobiologiya,* **17,** 33.
Hanert, H. H. (1981). In *The Prokaryotes. A Handbook on Habitats, Isolation, and Identification of Bacteria,* eds M. P. Starr, H. Stolp, H. G. Trueper, A. Balows & H. G. Schlegel. Springer-Verlag, Berlin, Heidelberg & New York, p. 509.
Harries, J. R., Hendy, N. & Ritchie, A. I. M. (1988). In *Biohydrometallurgy '87,* eds P. R Norris & D. P. Kelly, Science & Technology Letters, Kew, Surrey, UK, p. 233.
Harrison, A. P. (1982). *Arch. Microbiol.*, **131,** 68.
Harrison, A. P. (1984). *Ann. Rev. Microbiol.*, **38,** 265.
Hassett, J. M., Jennett, J. C. & Smith, J. E. (1981). *Appl. Environ. Microbiol.*, **41,** 1097.
Holmes, D. S. (1989). In *Hazardous Waste Treatment: Biosystems for Pollution Control, Proc. 1989 Int. Symp. A & WMA/EPA,* Cincinnati, Ohio, p. 29.
Holmes, D. S. (1990). In *Engineering of Organisms for Open Systems: Challenges and Opportunities,* ed. S. Silver, Academic Press, Orlando, USA in press.

Holmes, D. S. & Haq, R. U. (1990). In *Biohydrometallurgy 1989*, eds J. Salley, R. G. L. McCready & P. L. Wichlacz. Canmet SP89-10, Canada, p. 115.

Holmes, D. S. & Yates, J. R. (1990). In *Microbial Mineral Recovery*, eds H. L. Ehrlich & C. L. Brierley. McGraw-Hill, New York, p. 29.

Holmes, D. S., Debus, D. & Watson, T. (1988*a*). *Bioextraction. Biorecovery and Biodetoxification of Metals. Technology Assessment, Economic Evaluation, Market Potential and New Business Opportunities, 1988–1998.* Falmouth Assoc., Falmouth, Maine.

Holmes, D. S., Yates, J. R. & Schrader, J. A. (1988*b*). In *Biohydrometallurgy '87*, eds P. R. Norris & D. P. Kelly, Science & Technology Letters, Kew, Surrey, UK, p. 153.

Horikoshi, T., Nakajima, A. & Sakaguchi, T. (1979). *Agric. Biol. Chem.*, **43,** 617.

Horitsu, H. A., Kato, H. & Tomoyeda, M. (1979). *J. Ferment. Technol.*, **57,** 273.

Ismay, A., Rosato, L. & McKinnon, D. (1986). In *Fundamental and Applied Biohydrometallurgy,* eds R. W. Lawrence, R. M. R. Branion & H. G. Ebner, Elsevier, Amsterdam, p. 191.

Jctek, A. (1978). *Biosorbent M,* published by the Scientific and Technological Information Department of the Czechoslovak Atomic Energy Commission, Zbraslow-on-Vltava, p. 11.

Kagi, J. H. R. & Nordberg, M. (1989). In *Proc. Fifth Int. Congr. Biochem.*, eds Metallothionein & Birkhauser, Basel, Switzerland, p. 130.

Karavaiko, G. I. (1988). In *Biotechnology of Metals Manual,* eds G. I. Karavaiko, G. Rossi, A. D. Agate, S. N. Groudev & Z. A. Avakyan. Centre Int. Projects GKNT, Moscow, p. 10.

Karavaiko, G. I. & Groudev, S. N. (1985). In *Biotechnology of Metals,* eds G. I. Karavaiko & S. N. Groudev. Centre Int. Projects GKNT, Moscow, p. 6.

Karavaiko, G. I., Golovacheva, R. S., Pivovarova, T. A., Tzaplina, I. A. & Vartanjan, N. S. (1988*a*). In *Biohydrometallurgy,* eds P. R. Norris & D. P. Kelly. A. Rowe Ltd, Chippenham, UK, p. 29.

Khraptsova, A. I. (1988). *Methods for Evaluation of Bacterial Activity in Metal Biosorption and Sedimentation.* UNEP Program, Moscow.

Kulpa, C. F., Roskey, M. T. & Travis, M. T. (1983). *J. Bacteriol.*, **156,** 434.

Lane, D. J., Stahl, D. A., Olsen, G. I., Heller, D. J. & Pace, N. R. (1985). *J. Bacteriol.* **163,** 75.

Lawrence, R. W. & Bruynesteyn, A. (1983). In *Biohydrometallurgy '87,* eds P. R. Norris & D. P. Kelly, Science & Technology Letters, Kew, Surrey, UK, p. 359.

Leib, T. K. & Holmes, D. S. (1987). *Biotech. 87,* Online Publ. Inc., p. 208.

LeRoux, N. M. & Mehta, K. B. (1978). In *Metallurgical Applications of Bacterial Leaching and Related Microbiological Phenomena,* eds L. E. Murr, A. E. Torma & J. A. Brierley. Academic Press, New York, p. 463.

Lundgren, D. G. (1980). *Ann. Rev. Microbiol.*, **34,** 263.

Lundgren, D. G., & Dean, W. (1979). In *Biogeochemical Cycling of Mineral Forming Elements,* eds P. A. Trudinger & P. J. Swaine, Elsevier, Amsterdam, p. 211.

Madgwick, J. C. & Ralph, B. J. (1977). In *Bacterial Leaching,* ed. W. Schwartz. Verlag Chemie, Weinheim, Germany, p. 95.

Marchant, P. B. & Lawrence, R. W. (1989). In *Biohydrometallurgy 1989*, eds J. Salley, R. G. L. McCready & P. L. Wichlacz. Canmet SP89-10, Canada, p. 205.

McLean, R. J. C. & Beveridge, T. J. (1990). In *Microbial Mineral Recovery,* eds H. L. Ehrlich & C. L. Brierley, McGraw-Hill, New York, p. 185.

McCready, R. G. L. (1988). In *Biohydrometallurgy '87,* eds P. R. Norris & D. P. Kelly, Science & Technology Letters, Kew, Surrey, UK, p. 177.

Munger, K., Germann, V. A. & Lerch, K. (1985). *EMBO J.*, **4,** 2665.

Muzzarelli, R. A. A. (1977). *Chitin.* Pergamon Press, Oxford, UK, p. 309.

Muzzarelli, R. A. A., Tafini, T. & Scarpini, G. (1980*a*). *Biotechnol. Bioeng.*, **22,** 885.

Muzzarelli, R. A. A., Tafani, T., Scarpini, G. & Tucci, G. (1980*b*). *J. Appl. Biochem.*, **2,** 54.

Nakajima, A., Horikoshi, T. & Sakaguchi, T. (1979). *Agric. Biol. Chem.*, **43,** 625.
Nicolaides, A. A. (1987). *J. Chem. Tech. Biotechnol.*, **38,** 167.
Norberg, A. B. & Person, H. (1984). *Biotechnol. Bioeng.*, **26,** 265.
Norris, P. (1990). In *Microbial Mineral Recovery,* eds H. L. Ehrlich & C. L. Brierley. McGraw-Hill, New York, p. 3.
Norris, P. R. & Barr, D. W. (1988). In *Biohydrometallurgy '87,* eds D. P. Kelly & P. R. Norris, Science & Technology Letters, Kew, Surrey, UK, p. 532.
Norris, P. R. & Kelly, D. P. (1979). *J. Gen. Microbiol.*, **99,** 317.
Pfennig, N., Widdel, F. & Trueper, H. G. (1981). In *The Prokaryotes. A Handbook on Habitats, Isolation, and Identification of Bacteria,* eds M. P. Starr, H. Stolp, H. G. Trueper, A. Balows & H. G. Schlegel. Springer-Verlag, Berlin, Heidelberg & New York, p. 926.
Postgate, J. R. (1979). *The Sulfate Reducing Bacteria.* Cambridge University Press, Cambridge, UK.
Rawlings, D. (1990). Molecular biology of *Thiobacillus ferrooxidans.* In *Engineering of Organisms for Open Systems*: *Challenges and Opportunities,* ed. S. Silver, Academic Press, Orlando, USA, in press.
Roberto, F. F., Glenn, A. W. Rowland, M. L. & Ward, T. E. (1990). In *Biohydrometallurgy 1989*, eds J. Salley, R. G. L. McCready & P. L. Wichlacz. Canmet SP89-10, Canada, p. 137.
Romanenko, V. I. & Koren'kov, V. N. (1977). *Mikrobiologiya,* **46,** 414 (English transl., pp. 329–32.)
Rossi, G., Trois, P. & Visca, P. (1986). In *Fundamental and Applied Biohydrometallurgy,* eds R. W. Lawrence, R. M. R. Branion & H. G. Ebner. Elsevier, Amsterdam, p. 173.
Roswell, M. D. (1977). Presentation to the Canadian Nuclear Associaton, Montreal.
Rudolfs, W. & Helbronner, A. (1922). *Soil Sci.*, **14,** 459.
Schlitt, W. J. (1980). In *Leaching and Recovering Copper from As-mined Materials,* ed. W. J. Schlitt, SME-AIME, Report from Amer. Soc. Mech. Eng. & Amer. Inst. Mining Met. Petrol. Eng., United Eng. Center, New York, USA.
Schonborn, W. & Hartmann, H. (1978). *Eur. J. Appl. Microbiol. Technol.*, **5,** 305.
Steffan, R. J. & Atlas, R. M. (1988). *Appl. Environ. Microbiol.*, **54,** 2185.
Sterritt, R. M., Brown, M. J. & Lester, J. N. (1981). *Env. Poll.* (Ser. A), **24,** 313.
Strandberg, G. W., Shumate, S. E. & Parrott, J. R. (1981). *Appl. Environ. Miocrobiol.*, **41,** 237.
Szolnoki, J. (1977). In *Bacterial Leaching,* ed. W. Schwartz. Verlag Chemie, Weinheim, Germany, p. 223.
Torma, A. E. (1978). *Can. J. Microbiol.*, **24,** 888.
Torma, A. E., Walden, C. C. & Branion, R. M. R. (1970). *Biotechnol. Bioeng.*, **12,** 501.
Vasak, M., Kagi, J. H. R. & Hill, H. A. O. (1981). *Biochemistry,* **20,** 2852.
Waksman, S. A. & Joffe, I. S. (1922). *J. Bacteriol.*, **7,** 239.
Ward, T. E., Rowland, M. L., Watkins, C. S., Bruan, D. F. & Roberto, F. R. (1990). In *Biohyrometallurgy 1989*, eds J. Salley, R. G. L. McCready & P. L. Wichlacz. Canmet SP89-10, Canada, p. 159.
Winge, D. R., Nielson, R. B., Gray, W. R. & Hamer, D. H. (1986). *J. Biol. Chem.*, **260,** 14464.
Yates, J. R., Lobos, J. H. & Holmes, D. S. (1986). *J. Ind. Microbiol.*, **1,** 129.
Yates, J. R., Cunningham, R. P. & Holmes, D. S. (1988). *Proc. Natl. Acad. Sci. USA*, **85,** 7284.
Yoshinara, T. & Subramanian, V. (1976). In *Environmental Biogeochemistry,* Vol. 2, ed. J. O. Nriagu. Ann Arbor Science, Ann Arbor, Michigan, p. 541.
Zimmerley, S. R., Wilson, D. G. & Pratter, J. F. (1958). US Patent 2829964.
Zosim, Z., Iutnick, P. & Rosenberg, C. (1983). *Bioeng.*, **25,** 1725.

Chapter 16

MICROBIAL CONVERSION OF HYDROCARBONS TO PRODUCTS: COMMERCIAL PERSPECTIVES

W. R. FINNERTY

Finnerty Enterprises, Inc, 160 Chinquapin Place, Athens, Georgia, 30605, USA

CONTENTS

1 INTRODUCTION

Hydrocarbon microbiology has compiled comprehensive databases concerning the biotransformation of various paraffinic and aromatic hydrocarbons. Reviews concerning developments in hydrocarbon microbiology have appeared at regular intervals, with a general emphasis on the physiology and biochemistry of hydrocarbon-utilizing microorganisms (Ratledge, 1980; Buhler & Schindler, 1984; Singer & Finnerty, 1984*a*). Exploitation of various facets of hydrocarbon microbiology are few, with developments in the application of hydrocarbon-utilizing microorganisms and their physiological processes suggesting a potential for commercialization of useful products (Finnerty & Singer, 1983). Numerous products resulting from hydrocarbon oxidations appear of importance as value-added articles of commerce. This chapter discusses the use of hydrocarbon-utilizing microorganisms in the production of specialty chemicals, the microbial beneficiation of fossil fuels, the development of biosurfactants, single cell protein and single cell oil technologies and microbial treatment of petroleum wastes and hazardous chemicals. Emphasis is directed to those aspects of hydrocarbon microbiology amenable to technological development and, where possible, an examination of the economics and technical constraints associated with such applications.

2 SPECIALTY CHEMICALS

The 1970s witnessed intensive research into the development of chemicals from microorganisms. The production of unique products and biochemicals by hydrocarbon-utilizing microorganisms was recognized early with respect to their metabolic potential to effect special biotransformations of an unusual nature and specificity. The use of hydrocarbon-utilizing microorganisms for the production of bioproducts has certain attractive features meriting commercial interest. Advantages noted are the specificity of the primary oxidative attack on hydrocarbons yielding products difficult to obtain by other means, and the ability of such biological systems to operate at ambient temperatures and pressures with low activation energies and high turnover numbers per catalytic site (May, 1979; Drozd, 1980). Disadvantages associated with hydrocarbon fermentations are the poorly understood enzymological parameters of many of the reactions, release of the products into a dilute aqueous stream making recovery difficult and expensive, and the many products which are toxic to the producing microorganism. Nevertheless, the development of hydrocarbon fermentations has produced commercial scale production facilities capable of significant output. Numerous publications and patents emerged on the production of amino acids, fatty acids, vitamins, flavor enhancers such as 5′-nucleotides, polyhydroxy alcohols, pigments, antibiotics, wax esters, tricarboxylic acid cycle intermediates and enzymes, all of which are related to expanding the potential of single cell protein technology for added-value economic benefits. Today, few chemicals, if any, are manufactured commercially by hydrocarbon fermentations. The production of various bioproducts from specific hydrocarbons has generated an extensive diversity of products (Abbott & Gledhill, 1971; Bird & Molton, 1972; Miall, 1978; Ratledge, 1980; Buhler & Schindler, 1984; Singer & Finnerty, 1984*a*). The types of products developed from hydrocarbon fermentations are listed in Table 1.

2.1 General Aspects of Hydrocarbon Oxidations

The metabolic capacity of bacteria, yeasts and fungi to utilize hydrocarbons as sole sources of carbon and energy is of ubiquitous distribution in nature. The

Table 1
Substrate–Product Relationships in Microbial Hydrocarbon Oxidations

Substrate	*Products*
Alkanes–Alkenes	
Propane	Acetone
Butane	Butan-2-one
Pentane	Pentan-2-one
Hexane	Hexan-2-ol, hexan-2-one, hexanoic acid
Heptane	Propionic, pentanoic and heptanoic acids
Octane	Octanoic acid, 1,8-octanedioic acid, 1,6-hexanedioic acid
Decane	Decan-1-ol, decanoic acid, 2,3,4,5-decanols, 2,3,4,5-decanones, 10-hydroxydecanoic acid, 1,10-decanedioic acid
Undecane	Undecanoic acid, 1,11-undecanedioic acid
Dodecane	Dodecanoic acid, decylhexadecanoate, 1,12-dodecanedioic acid
Tridecane	1,13-Tridecanedioic acid
Tetradecane	Tetradecan-1-ol, tetradecanoic acid, tetradecylhexadecanoate, 1,14-tetradecanedioic acid
Pentadecane	Pentadecylhexadecanoate
Hexadecane	Mono-, di- and triacylglycerides, hexadecylhexadecanoate
Heptadecane	Heptadecylheptadecanoate, heptadecylhexadecanoate, heptadecyloctadecanoate
Octadecane	Octadecylhexadecanoate, octadecyloctadecanoate
Hepten-1-ene	Penten-4-oic acid
Oct-1-ene	1,2-Octane epoxide
Hexadec-1-ene	Hexadec-15-enylhexadecanoate, 1,2-dihydroxy hexadecane
Cycloalkanes and cycloalkenes	
Cyclopropane	Propionaldehyde
Cyclopentane	Cyclopentanone
Cyclohexane	Cyclohexanol, cyclohexane epoxide, adipic acid, glutaric acid
Cyclohexene	Cyclohexanone, cyclohexane epoxide, cyclohexene-1-ol, cyclohexanediol
Cycloheptane	Cycloheptanone
Cyclooctane	Cyclooctanone
Decalin	Adipic acid, pimelic acid
Tetralin	α-Tetralone, α-tetralol
Ethylcyclohexane	4-Ethylcyclohexanol
Bicyclohexyl	4,4-Dihydroxybicyclohexyl
Phenylcyclohexane	3- and 4-Phenylcyclohexanols
Monocyclic aromatics	
Benzene	Catechol, muconic acid, phenol, 2-ketoadipate, succinate
Toluene	Benzoic acid, benzalaldehyde, benzyl alcohol, 3-methylcatechol
o,m,p-Xylenes	α,α-Dimethyl muconic acid, *o,m,p*-toluic acids, 2,3-dihydroxy-*p*-toluic acid, 3,6-dimethylcatechol
Mesitylene	3,5-Dimethylbenzoic acid, 1,2,4,5-tetramethylbenzene, 2,4,5-trimethylbenzoic acid
Ditoyl	3-Carboxy-4-methyldiphenyl
Alkylbenzenes	Phenylpropionate, phenylacetate, *n*-tolylpropionate, cinnamic acid, 5-phenylpentanoic acid
Polycyclic aromatics	
Naphthalene	Salicyclic acid, catechol, coumarin, 1,2-naphthoquinone, *o*-hydroxycinnamic acid, melilotic acid, salicylaldehyde, *o*-hydroxybenzalpyruvic acid, gentisic acid

(*continued*)

Table 1—*contd.*

Substrate	*Products*
Anthracene	2-Hydroxy-3-naphthoic acid, salicylic acid
Phenanthrene	1-Hydroxy-2-naphthoic acid, salicyclic acid, 7,8-benzocoumarin, 3,4-dihydroxyphenanthrene
Terpene hydrocarbons	
α-Pinene	Borneol, myrtenol, D-verbenone, D-*cis*-verbenol, D-*trans*-sorbreol
α-Santalene	*tere*-Santalic acid, *tere*-santalol
Limonene	Carvone, carveol, perillic acid, 4,9-dihydroxy-1-*p*-menthen-7-oic acid, D-α-terpineol, 8-*p*-menthene-1,2-*cis*-diol
p-Cymene	Cumic acid, α-*p*-tolyl-isopropanol, cumyl alcohol, β-isopropylpyruvate
p-Menthene	Methyl isopropyl ketone, β-isopropylpimelic acid, *trans*-*p*-menthene-1,2-diol

pathway of alkane oxidation is:

$$RCH_2CH_3 + O_2 \rightarrow RCH_2CH_2OH \rightarrow RCH_2CHO \rightarrow RCH_2COOH$$

These oxidations are obligately O_2-dependent reactions, catalyzed by monooxygenases or dioxygenases. Variations on this reaction sequence are known, representing diterminal and subterminal oxidations (Singer & Finnerty, 1984*a*):

$$HOOCCH_2(CH_2)_nCH_2COOH \leftarrow CH_3CH_2(CH_2)_nCH_2CH_3 \rightarrow CH_3CH_2(CH_2)_nCOCH_3$$

In all instances, fatty acids or methyl ketones arise and either undergo further degradation, are assimilated into the cellular lipids or accumulate as extracellular products. Qualitative and quantitative differences occur in the products resulting from alkane metabolism as compared to more conventional substrates. A relationship between the chain length of the alkane substrate and the chain length of cellular and extracellular fatty acids has been demonstrated in certain bacteria (Makula & Finnerty, 1968; Dunlop & Perry, 1975). In general, the growth of the microorganisms on alkanes as a sole source of carbon and energy reflect the following characteristics:

(1) Even-carbon-numbered alkanes (C_{12}, C_{14}, C_{16}, C_{18}) yield even-carbon-numbered fatty acids, the cellular fatty acid composition reflecting the alkane carbon number.
(2) Odd-carbon-numbered alkanes (C_{11}, C_{13}, C_{15}, C_{17}) yield odd-carbon-numbered fatty acids, the cellular fatty acid composition reflecting the alkane carbon number.

Studies describing the quantitative and qualitative changes that occur in the cellular and extracellular lipids of *Acinetobacter* species HO1-N have been reported (Makula *et al.*, 1975; Singer *et al.*, 1985). These results are

Table 2
Cellular and Extracellular Lipids of *Acinetobacter*

Lipid class	*Cellular lipid µmol/g dry wt*			*Extracellular lipid µmol/liter spent broth*		
	NBYE[a]	C_{16}[b]	$C_{16}OH$[c]	*NBYE*	C_{16}	$C_{16}OH$
Phospholipid	46·00	129·00	100·00	0·00	0·00	0·00
Mono- and diglycerides	0·38	6·80	20·40	0·00	410·00	0·00
Triglyceride	1·78	2·50	30·00	2·4	25·6	30·00
Free fatty acid	7·50	8·20	0·30	4·0	60·0	3·5
Fatty alcohol	trace	2·60	3·50	0·00	0·5	ND[e]
Wax esters[d]	11·50	18·00	306·00	0·00	280·00	280·00
Hexadecane	0·00	360·00	trace	0·00	ND	0·00

[a] NBYE is 0·8% nutrient broth–0·5% yeast extract growth medium. [b] C_{16} is hexadecane grown cells. [c] $C_{16}OH$ is hexadecan-1-ol grown cells. [d] Esters derived from NBYE-grown cells were low molecular weight, water-soluble esters. Esters derived from hexadecane and hexadecanol-grown cells were 32 carbon esters. [e] Not determined.

summarized in Table 2 to demonstrate the effect of substrate on lipid production and composition. The cellular phospholipids, mono- and diglycerides increase 3-, 18- and 50-fold, respectively, in alkane- and fatty alcohol-grown cells. Extracellular lipid accumulation is significantly increased in alkane- and fatty alcohol-grown cells, the fatty acid carbon number representing >95% 16-carbon fatty acids in hexadecane- and hexadecan-1-ol grown cells. Alternate alkane substrates (C_{13}, C_{14}, C_{15}, C_{17}) reflect the same relationship between alkane carbon number and the carbon number of the resulting fatty acids. Such substrate directed physiological relationships offer the ability to tailor-make lipids with specific fatty acid compositions and high added-value potential through hydrocarbon fermentations.

The oxidation of aromatic hydrocarbons is an obligately O_2-dependent reaction found in bacteria, yeasts and fungi capable of utilizing substrates such as benzene, naphthalene, anthracene and phenanthrene as sole sources of carbon and energy. The generalized pathway of aromatic hydrocarbon oxidation is:

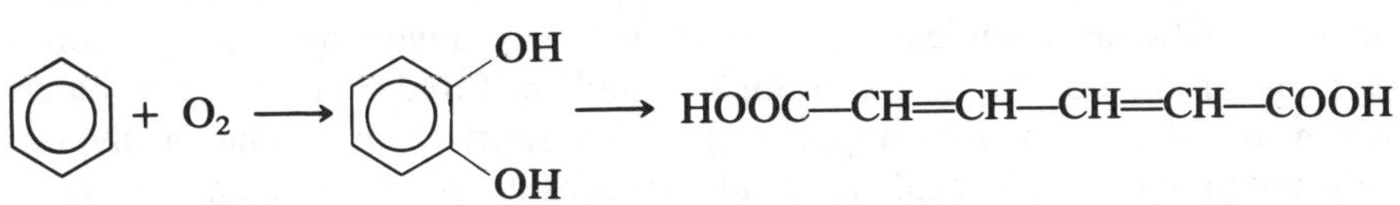

The bioconversion of simple aromatics, substituted aromatics and polycyclic aromatic hydrocarbons is well documented in the microbiological literature (Abbott & Gledhill, 1971; Cain, 1980; Cerniglia, 1984). The biotransformation of methylbenzenes to corresponding acids, dihydroxy derivatives, methyl-substituted muconic acid, dihydroxy-*p*-toluic acid and salicylate are a few examples of conversions effected to date. The genetics of aromatic hydrocarbon oxidation are particularly well studied in comparison to the genetics of alkane oxidations (Singer & Finnerty, 1984*b*). The involvement of plasmid-localized genes coding for aromatic hydrocarbon oxidations and the ability to manipulate these genes into expression vectors containing strong promoters offers exciting possibilities for the development of cost-effective processes in the use of genetically-engineered microorganisms for the production of value-added specialty chemicals. An example of such developments concerns the isolation of new soil isolates capable of converting naphthalene to salicylate and gentisate (Monticello *et al.*, 1985*a*). Transpositional mutagenesis of pDBT2 harbored in *Pseudomonas alcaligenes* yielded Tn5-mutants which accumulated gentisate in the culture broth.

2.2 Biotransformation of Alkane Analogs

The oxidation of a homologous series of symmetrical dialkyl ethers varying in carbon number from C_{14} to C_{20} was studied in *Acinetobacter* species HO1-N (Modrzakowski *et al.*, 1977; Modrazakowski & Finnerty, 1980). These substrates were selected on the basis of their structural similarity to alkanes of identical carbon number, differing only by the presence of an ether oxygen positioned in the middle of the molecule. This study described a new pathway in prokaryotic bacteria involving an internal carbon–carbon scission reaction, resulting in the formation of a homologous series of alkoxyacetic acids as nonmetabolizable extracellular endproducts. These products accumulate in the culture broth in direct stoichiometric relationship to the disappearance of the dialkyl ether substrate. Concomitantly, an alkanedioic acid is formed which serves as the sole source of carbon and energy for growth of the organism (see Table 3). The pathway of dialkyl ether metabolism is:

$$CH_3(CH_2)_n\text{—O—}(CH_2)_nCH_3 + O_2 \rightarrow CH_3(CH_2)_n\text{—O—}(CH_2)_nCOOH \rightarrow$$
$$CH_3(CH_2)_n\text{—O—}CH_2COOH + HOOC(CH_2)_nCOOH$$

These studies were extended further to examine the types of endproducts resulting from the oxidation of phenoxy-substituted fatty acids, *n*-fatty acids and alkanes (see Table 3). The nature of products arising from the oxidation of such substrates indicates a novel pathway involving the scission of carbon–carbon bonds. In all examples cited in Table 3, three facts emerge which are of fundamental importance: (1) a short-chain alkane, a phenoxy-substituted short-chain acid or an alkoxy-substituted short-chain acid results from the oxidation of alkanes, fatty acids, phenoxy-substituted fatty acids and alkoxyalkanes; (2) alkanedioic acids are produced which constitute the sole source of carbon and energy for cell growth; and (3) the ether oxygen of the dialkyl ether is metabolically refractory. Such studies illustrate the potential that exists for future basic research initiatives that offer biotechnological opportunities for the production of cost-effective, high value-added speciality chemicals.

Table 3
Products from Oxidation of Dialkyl Ethers, Alkanes, Fatty Acids and Substituted Fatty Acids

Substrate	*Products*
Diheptyl ether	Heptoxy-2-acetic acid; glutaric acid
Dioctyl ether	Octoxy-2-acetic acid; adipic acid
Dinonyl ether	Nonoxy-2-acetic acid; pimelic acid
Didecyl ether	Decoxy-2-acetic acid; suberic acid
Pentadecane	Pentane; 1,10-decanedioic acid
Hexadecane	Hexane; 1,10-decanedioic acid
Heptadecane	Heptane; 1.10-decanedioic acid
Pentadecanoic acid	Pentane; 1,10-decanedioic acid
Hexadecanoic acid	Hexane; 1,10-decanedioic acid
Heptadecanoic acid	Heptane; 1,10-decanedioic acid
Phenoxy-11-undecanoic acid	Phenoxy-3-propionic acid; 1,8-octanedioic acid
Phenoxy-12-dodecanoic acid	Phenoxy-4-butyric acid; 1,8-octanedioic acid

2.3 Specialty Chemicals from Hydrocarbon Fermentations

A summary discussion follows to document the various initiatives occurring in hydrocarbon microbiology and commercial development over the past several years and to indicate, where possible, the reasons for the absence today of viable full-scale production of specialty chemicals through hydrocarbon fermentations. Programs for the development of specialty chemicals through biological processes developed rapidly in Japan in the 1960s and 1970s and remain today an active focus of the Japanese industry. Hydrocarbon fermentations for the production of extracellular amino acids developed early in Japan, although efficient chemical syntheses exist for most of the important amino acids. Chemical syntheses, however, yield racemic mixtures which are usually difficult and expensive to separate; whereas biological systems offered advantages in producing almost exclusively the L-isomer of amino acids. Comprehensive reviews concerning the production of amino acids by hydrocarbon fermentations are published which document the details of amino acid production (Abbott & Glenhill, 1971; Kinoshita & Nakayama, 1978; Fukui & Tanaka, 1980).

Tricarboxylic acid (TCA) cycle intermediates have been produced through hydrocarbon fermentations employing paraffinic substrates. However, only citric acid production reached serious levels of consideration for commercial development. The increases in crude oil prices that developed in the late 1970s made further commercial development uneconomic and projects were cancelled. Other TCA cycle intermediates such as malate, fumarate and succinate are either of marginal commercial importance or are synthesized more economically by chemical means. Early patents concerning citric acid production from the fermentation of paraffin mixtures (C_9 to C_{20}) by *Yarrowia lipolytica* taught reduced fermentation residence time and increased product yields. However, the simultaneous production of isocitrate reduced citric acid yields. Later work focused on successfully reducing the conversion of citrate to isocitrate with mutants blocked at aconitate hydratase, competitive inhibitors and uncoupling agents. Currently, the production of citric acid through hydrocarbon fermentations is not practiced on a commercial scale, probably for reasons of the associated economics in using hydrocarbon feedstocks.

The production of specific fatty acids through directed hydrocarbon fermentations offers some possibilities for the development of added-value products. The major difficulty in the microbial production of fatty acids resides in the fact that few microorganisms accumulate significant amounts of free fatty acid as either cellular or extracellular products. In addition, if accumulation does occur, product buildup becomes toxic to the microorganism. However, the long-chain fatty acids and alkanedioic acids, ω and ω-1 hydroxy fatty acids and primary and secondary alcohols can be produced through directed hydrocarbon fermentations in low yields without degradation of the parent alkane carbon chain. Such products have commercial applications due to their surface activity and ability to stabilize emulsions as well as uses in the formulation of plasticizers, synthetic resins and polymers, lubricants and cosmetics.

Extracellular wax esters are produced by *Acinetobacter, Moraxella* and *Nocardia* species grown on alkanes. Alkane-grown *Acinetobacter* species HO1-N produces extracellular wax esters, the composition of which depends on the chain length of the alkane substrate (see Table 4). Interestingly, hexadecan-1-ol- grown cells produced a total of 273 mg cetylpalmitate/liter as compared to 138 mg cetylpalmitate/liter for hexadecane-grown cells (Makula *et al.*, 1975; Singer *et al.*, 1985). Similar studies were reported where wax ester yields were 275 mg/liter from a mutant of *Acinetobacter* unable to degrade the wax ester product (Geigert *et al.*, 1984). Interest in microbially-produced wax

Table 4
Extracellular Wax Esters Produced by *Acinetobacter*

Substrate	*Products*
Decane	Decylhexadecanoate
Dodecane	Dodecylhexadecanoate
Tetradecane	Tetradecylhexadecanoate
Pentadecane	Pentadecylhexadecanoate
Hexadecane	Hexadecylhexadecanoate
Heptadecane	Heptadecylhexadecanoate
	Heptadecylheptadecanoate
	Heptadecyloctadecanoate
Octadecane	Octadecylhexadecanoate
	Octadecyloctadecanoate
Hexadec-1-ene	Hexadec-15-enylhexadecanoate
Octadec-1-ene	Octadec-17-enylhexadecanoate
	Octadec-17-enylheptadecanoate
	Octadec-17-enyloctadecanoate

esters stems from the possibility of developing alternative sources of cetylpalmitate, which is currently obtained from the sperm whale, and for developing long-chain, di-unsaturated wax esters as jojoba oil substitutes.

The production of biologically-derived polyesters through polymerization of hydroxy fatty acids has attracted attention for years. These biological polyesters are biodegradable, unlike the synthetic polymers. Serious attention is currently focused on the commercial development of poly-β-hydroxybutyrate (PHB) as a biodegradable, thermoplastic polyester with properties similar to polyethylene and polypropylene. Commercial applications suggested for PHB are degradable sutures, controlled drug release and biodegradable wrappers. Full-scale commercial production of PHB is under evaluation, although a production fact of considerable economic magnitude is that 2·5 tons of glucose are required to produce 1 ton of PHB. Again, the economics of production becomes a contributing factor to success or failure, with success depending on improvements to the production process and the price the market is willing to pay for a biodegradable polyester product.

The modification of polyester synthesis has been explored by employing alternative secondary substrates for the formation of different poly-hydroxyalkanoates (PHA). Reports have appeared on the synthesis of poly-β-hydroxyoctanoate and poly-β-hydroxypentanoate following growth of PHB-producing microorganisms on decane (deSmet *et al.*, 1983) and valerate (Doi *et al.*, 1987), respectively. These polyesters are produced along with PHB, the ratio of PHB to PHA being dependent on the concentration of the secondary substrate. The PHB–PHA heteropolymers exhibit improved properties suitable for plastic molding and appear certain to secure a favorable market niche due to their biodegradability.

The formation of alkanedioic acids has been frequently reported as extracellular products resulting from the microbial oxidation of alkanes. A mutant strain of *Torulopsis candida* was reported as producing decanedioic acid when grown on decane with yields increasing to 33 g/liter of decanedioic acid (Kaneyuki *et al.*, 1980). Recently, a mutant strain of *Candida tropicalis* was reported as producing dodecanedioic acid when exposed to dodecane (Hill *et al.*, 1986). High yields of alkanedioic acids resulted from the oxidation of C_{12} to C_{16} alkanes. Another laboratory has reported on alkanedioic acid formation by *C. tropicalis* yielding tridecanedioic acid and dodecanedioic acid from the oxidation of tridecane and dodecane, respectively (Yi & Rehm, 1982*a,b,c*). A biotransformation of alkanes to alkanedioic acids was developed employing *C. tropicalis* immobilized in carrageenan (Yi & Rehm, 1982*d*).

Several process patents have appeared concerning the production of alkanedioic acid (Buhler & Shindler, 1984). An example of one patent concerns a process whereby a mutant strain of *C. tropicalis* is grown on a carbohydrate-containing medium and then transferred to a non-growth medium containing hexadecane or hexadecan-1-ol. These substrates are oxidized to hexadecanedioic acid with yields approaching 4 g/liter, representing a 40% conversion of substrate to product (Phillips Petroleum Co., 1976). Commercial production of tridecanedioic acid from tridecane is operated by the Nippon Mining Company, Japan, at a scale of 150 tons/year (Uemura *et al.*, 1989). The process employs a mutant of *C. tropicalis* with production yields of 70% alkanedioic acids resulting from the oxidation of alkanes. There are no other commercial operations known in this area, due mainly to the marginal cost-effectiveness in comparison to chemical syntheses.

3 MICROBIAL BENEFICIATION OF FOSSIL FUELS

The application of microbial processes to various aspects of energy production and technology has received renewed interest. The use of fossil fuels for power generation is expected to increase, despite diminishing world reserves of high quality crude oil and coal. The demand for low-sulfur and low-nitrogen fossil fuels has been intensified by increasingly stringent regulatory standards for reduced levels of sulfur and nitrogen oxides in released emissions. Amendments to the 1964 Clean Air Act require not only compliance with emission standards but also a percentage reduction in emissions that otherwise would occur from the burning of untreated fuel, requiring that low-quality fuel be subjected to pre- and postcombustion desulfurization, as mandated by EPA in 1989. The import of acid rain is of international significance and may prompt similar legislative action to further limit emissions in other industralized countries. Postcombustion, flue-gas desulfurization is estimated to cost over $2800 million by 1990 in the United States and may represent a burden of $200–300 \times 10^9$ dollars to consumers over the next 30 years.

The objection to high-sulfur and high-nitrogen fossil fuel combustion is the generation of sulfur and nitrogen oxides which appear to be involved in the formation of atmospheric acid deposition. Such acidic depositions produce detrimental effects on complex ecosystems and have become an issue of international significance. Sulfur in coal and oil is known to exert serious corrosive properties, its removal incurring considerable time and expense. Postcombustion desulfurization technology, which is used almost exclusively, is inconvenient and expensive, and fails to address the economic aspects of sulfur and/or nitrogen contamination of fossil fuels other than emission control. These considerations plus general concern over acid depositions have renewed interest in the development of microbially-assisted organic and inorganic precombustion desulfurization technology for fossil fuels (Monticello & Finnerty, 1985).

3.1 Types of Sulfur in Fossil Fuels

Crude oil: The sulfur content of crude oil ranges from 0·025% to 5% depending on geographical source. Over 200 sulfur-containing organic compounds have been identified in oil in addition to elemental sulfur, sulfate, sulfite, thiosulfate and sulfide. The type of organic compounds identified include sulfide, thiols, thiophenes, substituted benzo- and dibenzothiophenes plus sulfur-containing polycyclic aromatic molecules (Willey *et al.*, 1981). The sulfur content of oil generally increases in the following sequence: paraffin fraction < simple aromatic fraction < polycyclic aromatic fraction < resin acid fraction < asphaltene fraction. Sulfur-containing constituents in the high-boiling asphaltene or residue fractions are predominantly benzo- or dibenzothiophene linkages (Speight & Moschopedis, 1981). Some Texas crude oils have been characterized as containing 70% dibenzothiophene compounds, whereas several Middle East oils contain 40% substituted benzo- and dibenzothiophenes.

Coal: The sulfur content of coal varies widely, ranging from 0·5% to over 11%. Coals from the western United States are generally low in sulfur (less than 0·6%), whereas eastern US coals range between 0·2 and 7·0%. Sulfur in coal occurs in either inorganic or organic forms with elemental sulfur rarely present. Inorganic sulfur occurs mainly as iron pyrite. Lead, zinc and copper sulfides are observed as well as sulfate, usually as gypsum. Sulfate in unweathered coal is usually less than 0·1%, although levels of sulfate can increase during transport and storage. Pyrite is the predominant inorganic form of sulfur in coal, occurring as nodules, lenses or finely disseminated inclusions. Pyritic sulfur equals or exceeds organic sulfur, although exceptions are common. Some Yugoslavian coals exceed 11% total sulfur, with organic sulfur the major portion (Attar & Hendrickson, 1982).

Organic sulfur present in coal appears to be covalently bound in the complex coal matrix as thiols, mercaptans, sulfides, disulfides and complex thiophene moieties. Complete structural elucidation of any one type of coal remains to be accomplished, although a considerable database exists on the various chemical and physical characteristics of coal (Given, 1985). The covalent bonding of the sulfur heteroatom in the macromolecular matrix of coal represents a stable molecular configuration. Solubilization studies of low rank coals indicate ester bond functionality, whereas high rank coals do not show such reactions. Solvent extraction of high rank bituminous coal has, however, yielded mixtures of complex organic products. The application of microbial processes to fossil fuel upgrading is included in this chapter in that many of the microorganisms studied in coal microbiology exhibit reactions and biological characteristics similar to those studied in petroleum microbiology. Many of the same types of complex polycyclic complexes are present in coal as are found in crude oil, particularly the heavy ends of heavy oils. Therefore, in view of the interest in the development of microbial processes for fossil fuels beneficiation and the general similarities that exist between coal and oil, it would be remiss not to review the current state of developments in this area.

3.2 Microbial Desulfurization Technology

Soil microorganisms have been isolated which partially oxidize benzo- and dibenzothiophene as model substrates to water-soluble products (Yamada *et al.*, 1968; Kodama *et al.*, 1970; Sagardia *et al.*, 1975; Hou & Laskin, 1976;

Laborde & Gibson, 1977; Malik, 1978; Monticello *et al.*, 1985*b*). The reaction mechanism for these oxidations represents ring hydroxylation followed by ring cleavage yielding water-soluble products retaining the sulfur heteroatom. The pathway described for the partial oxidation of dibenzothiophene to water-soluble products is:

dibenzothiophene
↓
1,2-dihydro-1,2-dihydroxydibenzothiophene
↓
1,2-dihydroxydibenzothiophene
↓
cis-4[2-(3-hydroxy)-thionaphthenyl]-2-oxo-butenoic acid
↓
trans-4[2-(3-hydroxy)-thionaphthenyl]-2-oxo-butenoic acid
↓
3-hydroxy-2-formylbenzothiophene

Soil isolates have been characterized which efficiently oxidize dibenzothiophene to water-soluble products (Finnerty & Robinson, 1986). These isolates, classified as *Pseudomonas alcaligenes* (DBT-2), *P. stutzeri* (DBT-3) and *P. putida* (DBT-4), were unable to grow at the expense of dibenzothiophene, oxidized polycyclic aromatic hydrocarbons slowly and were metabolically inert to long-chain alkanes (Table 5). Accordingly, these microorganisms were developed as strict biocatalysts which carried out specialized biotransformations under nongrowing conditions. Additional parameters of importance in the development of a microbially based desulfurization process are (1) product inhibition, and (2) biocatalyst stability. Certain water-soluble products resulting from the oxidation of dibenzothiophene are inhibitory to the reaction. This inhibition was relieved by addition of ion exchange resins to the reaction mixture with a 10-fold stimulation of the rate of dibenzothiophene oxidation. Biocatalyst stability was achieved by lyophilization of active cultures under special conditions with biocatalysts remaining active in excess of 3 years. A biological system has been defined which exhibits specificity towards conversion of sulfur heterocycles to water-soluble products. Approaches to strain development were indicated in the context of physiological and biochemical improvements through genetic means.

Table 5
Characteristics of Dibenzothiophene-oxidizing Microorganisms

Characteristic	*DBT*-2	*DBT*-3	*DBT*-4
Plasmid	72 kB	none	72 kB
Rate of DBT oxidation (μmol/h)	325	675	368
Rate of PAH oxidation (μmol/h)	8	14	11
Rate of alkane oxidation	0·00	0·00	0·00
Oxygen requirement	O_2-dependent	O_2-dependent	O_2-dependent
Maximum temperature optimum	55°C	55°C	55°C
Biocatalyst stability	2 years	2 years	2 years

3.3 Approaches to Strain Improvement

The oxidation of dibenzothiophene by DBT-2 is encoded on a 72-kilobase plasmid (pDBT-2) with the loss of plasmid DNA resulting in the concomitant loss of dibenzothiophene oxidation (Monticello *et al.*, 1985*b*). The plasmid pDBT-2 was restricted and a 25-kilobase fragment was ligated into pIMS6026 (Finnerty & Robinson, 1986). The recombinant plasmid (pC1) was purified from *Escherchia coli* DH1 and introduced into a plasmidless strain of DBT-2 which regained the ability to oxidize dibenzothiophene. Studies to optimize the reaction kinetics for DBT-2 (pC1) yielded rates of dibenzothiophene oxidation in excess of 1360 μmol/h, an approximate fourfold increase over the wild-type strain.

Transpositional mutagenesis of pDBT-2 with Tn5 yielded over 48 independent pDBT-2::Tn5 strains. Several pDBT-2::Tn5 mutant phenotypes were obtained and classified into three distinct groups: Dbt-I mutants, Dbt-II mutants, and Dbt-III mutants. Type I mutants did not produce water-soluble products and accumulated dibenzothiophene dihydrodiol; Type II mutants accumulated the first ring cleavage product and did not produce hydroxyformylbenzothiophene; Type III mutants produced a black pigment tentatively identified as a condensation product of dibenzothiophene-1,2-diol.

These studies illustrate the applicability of genetic engineering to strain improvement. Progress in the cloning of a multi-gene complex directed to the oxidation of sulfur-containing heterocycles is possible as well as the inactivation of specific genes by transpositional mutagenesis to alter endproduct inhibition of a specific reaction sequence.

3.4 Biodesulfurization of Crude Oil

The application of this bacterial system was extended to crude oil. The oxidation of dibenzothiophene by the DBT-2 biocatalyst was optimized with respect to pH, substrate concentration, cell concentration and temperature. These optimized reaction parameters allow for a reduction of reactor residence time from greater than 24 h to less than 1 h with a 100% conversion of dibenzothiophene to water-soluble products. Assessment of these reaction parameters to crude oil containing heterocyclic sulfur was determined by the addition of dibenzothiophene to low-sulfur oil. The ratio of oil to water was 1:1 with dibenzothiophene at 0·1% concentration. The biocatalyst converted 100% of the dibenzothiophene to water-soluble products within 1 h. The treatment of high-sulfur crude oil with the biocatalyst resulted in the loss of greater than 70% organic sulfur from the pentane-soluble fraction of the crude oil (Finnerty & Robinson, 1986). Major deficiencies exist in this bioprocess technology. Foremost, these biocatalysts do not remove organic sulfur from the heavy fractions (asphaltenes and resin acids) of crude oil which contain the major concentrations of organic sulfur. Secondly, the biocatalysts cause a downgrading in oil quality by virtue of their carbon-targeted reactions, converting sulfur-containing organic molecules into sulfur-containing, water-soluble molecules with a significant loss of BTU content in the treated oil. Thirdly, a byproduct is produced of sulfur-containing, water-soluble products for which there is no apparent end use. It is estimated that for every 100 lb of 3% crude oil processed, 16 lb of water-soluble byproduct is produced. Conversion of byproducts to higher value chemicals, such as phenol and benzene, offers possible alternatives to waste material and becomes important to the overall process economics.

3.5 Biodesulfurization of Coal

The selection of microorganisms capable of modifying various forms of sulfur has been of recent interest with respect to their application to coal biodesulfurization. Considerable research has occurred in the development of microorganisms capable of removing inorganic forms of sulfur through a bioleaching process (Dugan, 1986; Uhl *et al.*, 1989). The basis for commercialization of bioleaching processes for removal of inorganic sulfur from coal has been established and is effective on bituminous coals and lignite. The development of specialized cultures for the removal of pyritic sulfur and inorganic sulfur represents technology that removes over 90% of the pyritic sulfur in addition to minerals equivalent to 5–10% of the raw coal weight without significant loss of BTUs. These microorganisms are able to utilize both the sulfur and iron of pyrite, the energy released through the oxidation being used to support their growth. The bacterial oxidation of FeS_2 in the presence of O_2 and water yields H_2SO_4 and iron hydroxide.

The selection of microorganisms capable of selectively abstracting the sulfur heteroatom from organic molecules has been of recent interest with respect to coal biodesulfurization. The isolation of an aerobic, heterotrophic, nonacidophilic microorganism (CB1) has been studied which released the sulfur heteroatom of dibenzothiophene as sulfate (Isbister & Kobylinski, 1985). It was further reported that up to 47% of the organic sulfur present in coal was removed in bench-scale, continuous reactions. A small scale microbial desulfurization pilot plant operation was reported capable of processing 2500 lb of coal per day. New soil isolates were reported which abstracted organic sulfur from four different coals, yielding a 35–45% reduction in the organic sulfur present in the coal (Finnerty & Robinson, 1986). The isolation of soil microorganisms as mixed cultures was reported as capable of using thiophenic sulfur as a sole source of sulfur for growth (Kilbane *et al.*, 1989).

The development of microbial processes effective in reducing the levels of organic sulfur present in coal requires that: (1) the microorganisms exhibit targeted specificity to carbon–sulfur bonds; (2) the abstraction of organic sulfur from organic moieties should be unrelated to the general physiology and growth requirements of the microorganism; (3) the microorganisms should be developed as specific biocatalysts exhibiting long-term stability; and (4) since coal is a solid and by necessity a surface-oriented catalytic reaction occurs on the surface of a coal particle and the surface of a bacterial cell, modification of the coal surface is required to allow for the effective interaction of the coal particle and the biocatalyst. Evidence to date indicates that microorganisms can be developed which reduce the organic sulfur content of coal and upgrade the quality of the product. Further research is warranted to develop such biological systems for applications in microbially-assisted coal desulfurization technologies.

3.6 Biodesulfurization Process Economics

Few studies are available concerning the process economics associated with different biologically-mediated desulfurization processes. Bioleaching processes are probably the best documented, demonstrating a highly favorable cost/benefit ratio for commercialization of the technology (Dugan, 1986). The case of organic sulfur biodesulfurization processes is more difficult, since most of this work is of a proprietary nature and the details of process design and cost factors associated with the technology are not available.

A process economic evaluation of biodesulfurization of high-sulfur crude oil was performed (Hartdegan *et al.*, 1984). This analysis was directed to developing economic insight into both biological and economic parameters relating to full-scale process technology and the potential for cost-effective technology. Such analyses additionally serve to identify priorities for guiding further research and development efforts.

The engineering parameters were based on specific assumptions concerning the conditions and output of the final process. These parameters were: conversion of 3·0% high-sulfur residual fuel oil to 0·7% low-sulfur fuel oil; reactor design is an immobilized whole cell packed reactor bed; reactor operating conditions of 45°C and pH 7·0; an oil to water ratio of 1:1; process capability of 3 400 000 barrels/year; and the water-soluble, sulfur-containing byproducts are processed to benzene and phenol. Specific cost estimate assumptions were made and projected to desired cost approximations (see Table 6).

A linear scale-up of existing operational costs for biodesulfurization was determined and compared to the costs associated with hydrodesulfurization. These analyses were conducted in 1983 and reflected pricings existing at that time. The analyses assumed a price of $21.28/bbl of high-sulfur residual oil with a desulfurized product competing with the direct purchase of low-sulfur oil priced at $24.95/bbl. Given these assumptions, hydrodesulfurization is uneconomic in practice since this product is more costly by $6.95 than the direct purchase of low-sulfur oil (see Table 6). The catalyst cost for hydrodesulfurization was estimated at $3.40/bbl while the biocatalyst cost was estimated at $16.18/bbl of oil. However, sensitivity analyses indicated that by reducing substrate costs for growth of microorganisms in the preparation of the biocatalyst, reductions in reactor residence time to less than 1 h and significant increases in the half-life of the biocatalyst, the cost of biocatalyst could be

Table 6
Preliminary Process Economics of Microbial Desulfurization

	$ per barrel produced		
	Hydrodesulfurization	*Microbial*	
		Current	*Goal*
Raw material (residual oil)	21·28	24·64	24·64
Utilities operating overhead	2·51	3·63	3·63
Subtotal	23·79	28·27	28·27
Byproduct credit	−1·21	−6·29	−6·29
Subtotal	22·58	21·98	21·98
Catalyst	3·47	16·18	<0·01
Subtotal	26·05	38·16	21·98
Depreciation + 10% discounted cash flow	5·85	10·24	8·02
Total	31·90	48·40	30·00
Difference	−6·95	−23·45	−5·05

reduced to less than 1 ¢/bbl of oil. Such projections result in an idealized biodesulfurization process slightly more advantageous than hydrodesulfurization but still more expensive than the direct purchase of low-sulfur oil.

This economic evaluation makes assumptions which could be considered simplistic in view of the complex issues determining the future availability and pricing of low-sulfur oil and technological improvements to the process. The value of biological desulfurization technologies is more dependent upon the differential that exists between low- and high-sulfur oil feedstocks. Although oil prices have declined over the past several years, the differential pricing of low- and high-sulfur oil may or may not change, depending on the continued availability of low-sulfur oil in future years.

It is considered that biological process improvements are possible and have been achieved, impacting on the overall process economics of biodesulfurization technology. Improvements to date concern developments in genetic engineering for construction of a production strain and improvements in the long-term stability of the biocatalyst. Further, reactor design parameters are conceptual at this time, with significant changes to be realized in the most effective and efficient design module for performance-effective biodesulfurization processes. The economic evaluation did, however, point realistically to the process improvements necessary for successful commercial application.

4 MICROBIAL BIOSURFACTANTS

Biosurfactants are surface-active agents derived from biological sources which, like synthetic surfactants, exhibit characteristic physical and chemical properties. Structurally, surface-active molecules contain hydrophilic and hydrophobic components, a property which enables such molecules to concentrate at interfaces of air/water or oil/water and thereby to reduce surface tension and interfacial tension, respectively. At the critical micelle concentration (CMC), surface-active molecules form micellar aggregates with surface activity reduced to minimum values. Surfactants and biosurfactants are classified into four groups based on charge-type: anionic, cationic, nonionic and amphoteric. A number of surface-active microbial products which exhibit surface activity have been identified. However, limited information exists on the definitive chemical and physical properties of these surface-active products or on the types and distribution of surfactant-producing microorganisms.

4.1 Microbial Sources

Biosurfactants are produced by a variety of microorganisms including bacteria, yeasts and fungi (Table 7). The topic of microbial biosurfactants and bioemulsifiers, their occurrence, chemistry and surface properties have been reviewed (Zajic & Panchal, 1976; Cooper & Zajic, 1980; Zajic & Mahomedy, 1984). The term bioemulsifier refers to biologically-produced emulsion stabilizers which may or may not exhibit surface activity. Biosurfactants have generally been associated with the growth of microorganisms on alkanes. Most of the biosurfactants listed in Table 7 were discovered as products of alkane metabolism with a physiological role related to the transport and uptake of alkane by the microbial cell.

Biosurfactants, in certain cases, are produced during the growth of alkane-utilizing microorganisms on nonhydrocarbon substrates. Sophorolipids

Table 7
Microbial Biosurfactants and Bioemulsifiers

Microorganism	*Biosurfactant/bioemulsifier*
Torulopsis species	Sophorolipids
Pseudomonas species	Rhamnolipids
Rhodococcus erythropolis	Trehalose lipids
	Trehalose mycolates
	Sucrose and fructose lipids
Rhodococcus species H13-A	Trehalose lipids
Candida species	Mannosyl erythritol lipid
Candida bogoriensis	Sophorolipid
Acinetobacter species	Fatty acid, glycerides, emulsan
Corynebacterium lepus	Corynemycolic acids
Candida petrophilum	Peptidolipid
Bacillus subtilis	Cyclic lipopeptide
Bacillus licheniformis	Cyclic lipopeptide
Candida tropicalis	Mannan–fatty acid complex
Corynebacterium hydrocarboclastus	Proteo–lipid–carbohydrate complex

are produced by *Torulopsis* species during growth on alkanes or on glucose supplemented with vegetable oil. Rhamnolipids are produced by *Pseudomonas* species on both alkane and nonalkane growth substrates. *Acinetobacter calcoaceticus* RAG-1 produces an emulsion stabilizer on either alkane or nonalkane growth substrates. Biosurfactants produced by nonhydrocarbon-utilizing microorganisms have been described for the production of surfactin by *Bacillus subtilis,* sophorolipid by *Candida bogoriensis* from a glucose-yeast extract medium, and a surface-active lipopeptide from *B. licheniformis* grown on sucrose.

Microbial biosurfactants are lipoidal molecules which include glycolipids, neutral lipids, lipopeptides, fatty acids, phospholipids and high molecular weight polymers such as lipoproteins, lipopolysaccharide–protein complexes and polysaccharide–protein–fatty acid complexes. Corynemycolic acids are α-branched, β-hydroxy long-chain fatty acids produced by *Corynebacterium* species as cellular or extracellular products. Sophorolipids are extracellular anionic glycolipids produced by *Torulopsis bombicola* and *Candida bogorinesis*. The *Torulopsis* glycolipid consists of sophorose and 17-hydroxyoctadecanoic acid plus esterified acetate, whereas sophorolipid produced from *C. bogoriensis* contains sophorose glycosidically linked to docosanoic acid with acetate esterified to the 6 and 6′ hydroxyls of sophorose. Rhamnolipods are anionic extracellular glycolipids produced by *Pseudomonas* species consisting of two molecules of L-rhamnose and two molecules of β-hydroxydecanoic acid. Trehalose dimycolate is a cell-bound glycolipid produced by *Rhodococcus erythropolis*. This glycolipid consists of α-branched, β-hydroxy mycolic acids esterified to the 6 and 6′ hydroxyl of trehalose. Surfactin is an extracellular lipopeptide produced by *Bacillus subtilis*. Surfactin is a cyclic lipopeptide containing seven amino acids which are amide-linked to β-hydroxyisopentadecanoic acid at the glutamate residue and ester-linked to the fatty acid at the leucine residue.

4.2 A Biosurfactant Case Study

A soil microorganism was isolated and identified as a *Rhodococcus* species which produced an extracellular biosurfactant when grown on long-chain alkanes or high-paraffin based crude oils (Finnerty & Singer, 1984). Characterization of this extracellular surface-active product established the chemical and physical properties of the molecule. The product is an anionic glycolipid consisting of trehalose and a complex mixture of fatty acids and mycolic acids. The biosurfactant exhibits a CMC of 1·5 mg/ml with minimum interfacial tension values of 0·02 mN/m and 0·000 06 mN/m in the presence of a coagent. The surface activity is not significantly affected by brine, extreme pH values, divalent cations or temperature. The biosurfactant forms stable oil-in-water emulsions and exhibits phase behavior with the formation of a middle phase.

The application of biosurfactant H13-A to heavy oils such as Venezuelan Monogas and Cerro Negro crudes, Westminster crude and California crudes having API gravity values between 8 and 15 results in the formation of stable oil-in-water emulsions with a greater than 95% reduction in their relative viscosities. Characteristics of the treated oil were: (1) formation of a stable emulsion containing approximately 30% water; (2) improved rheological properties with flow characteristics significantly improved from the original crude oil, including reduced adherence to surfaces; and (3) destabilization of the emulsion by increases in temperature to 60°C. This biosurfactant exhibits a potential for the pipeline transportation of heavy crude oils, a technology that depends on the use of gas distillates, a commodity of ever-increasing value.

Studies were conducted on the application of the H13-A biosurfactant as an effective chemical-enhanced oil recovery agent for displacement of oil from a rock matrix. Core flood studies employed a radial core configuration consisting of Berea sandstone. The biosurfactant was injected at a rate equivalent to 1 ft/day to displace the residual in-place oil following core saturation and water flooding. Tertiary oil recovery was approximately 15% of the in-place oil with the loss of pH control and biosurfactant fractionation within the core matrix. However, greater than 95% of the biosurfactant injected was recovered from the core, indicating insignificant losses due to adsorption to the rock matrix. Preliminary studies indicate a potential for the application of biosurfactants as chemical-enhanced oil recovery agents in the tertiary recovery of known oil deposits. A biosurfactant-producing *Bacillus licheniformis* has been described which mobilized crude oil with oil bank formation in sand packs (Jenneman *et al.*, 1983). The reader is referred to a comprehensive treatise concerning additional information and details on the subject of biosurfactants (Kosaric *et al.*, 1987).

5 SINGLE CELL PROTEIN AND SINGLE CELL OIL

The subject of single cell protein (SCP) production through fermentation of hydrocarbon feedstocks as the sole source of carbon and energy has attracted industrial interest for decades as a source of supplementary protein for various end uses. The technical feasibility for large scale, continuous or batch fermentations is a well-researched, proven fermentation technology which yields a high-quality protein product. Despite the existence of this exploitable commercial technology, the industrial use of hydrocarbon feedstocks for SCP or other products is largely nonexistent within the industrialized world. Historically, the successes and failures of SCP as a viable commodity on the world market appear to relate directly to the changes in oil pricings that have characterized the last few decades (Shennan, 1984). Other factors contributing to the demise of SCP are market acceptability, competition from soy and fish meal as protein supplements, and the economics associated with energy input and feedstock prices which became prohibitive by the late 1970s. There existed, however, intense activity in basic and applied research where achievements in process design and bioengineering accomplished the construction of 100 000 tons per year SCP production facilities. A voluminous literature exists on developments in SCP production. The reader is advised to consult the authoritative review of Shennan (1984). SCP production is a major scientific and technical success of the twentieth century, reflecting the achievements that are possible through concentrated efforts in bioengineering and biotechnology.

Alternatives to paraffin-based SCP technology have occurred through the use of gaseous alkanes as feedstocks. Numerous problems are encountered in the process engineering of gaseous alkanes, such as mass transfer and mixing of the heterogeneous mixture of methane, oxygen and liquid medium. Most pilot plant studies utilized relatively pure methane, whereas commercial scale plants would employ natural gas. Commercial scale SCP production from natural gas should be of interest to middle East countries such as Saudi Arabia due to large feedstock supplies.

A nontraditional feedstock for SCP production has been methanol. ICI, Hoechst and Phillips Petroleum Co. have developed a highly efficient methanol-based SCP production technology. These more progressive commercial projects have attempted to develop processes that yield products other than SCP, hopefully to improve on the overall process economics. The Phillips high-density yeast process technology illustrates developments in the current state-of-the-art in SCP production. This fermentation process develops extremely high cell mass yields per liter of fermentation broth (see Table 8). The growth medium is fed to the fermentor at a rate designed to maintain the cell density and productivity at the desired level, with the process adaptable to either batch or continuous fermentations. The high cell density simplifies downstream recovery processes, eliminating centrifugal recovery of biomass in many applications, making it possible to directly spray-dry the product with advantages occurring in production economics. This high density process technology has been successfully applied to the growth of *Bacillus subitlis*, *B. megaterium*, *Methylomonas* species and *Escherichia coli* K-12, yielding dry cell

Table 8
Biomass Yields from Various Feedstocks

Feedstock:	*Microorganisms*	*Dry cell yield (g/liter)*	*Cell productivity (g/liter/h)*
Methanol:	*Pichia pastoris*	126	12
Ethanol:	*Pichia pastoris*	108	11
	Candida utilis	122	33
Sucrose:	*Candida utilis*	150	25
	Saccharomyces cerevisiae	138	28
Beet molasses:	*Candida utilis*	160	27
Cane molasses:	*Candida utilis*	118	19
Whey permeate:	*Kluyveromyces fragilis*	132	22
Glucose:	*Candida utilis*	136	27

yields (in g/liter) of 57, 110, 114 and 110, respectively. These specially engineered fermentors are licensed to Daniel International Corporation to manufacture and feature Phillips proprietary high density design which results in high rates of oxygen transfer and heat transfer.

Improvements in the high cell density process have been achieved through molecular genetics. An expression system was developed with *Pichia pastoris* for production of recombinant DNA proteins. The first enzyme in the methanol oxidation pathway is an alcohol oxidase. Several genes in the pathway were cloned into a *Pichia* expression vector including the regulatory regions (Ellis *et al.*, 1985). These cloned DNA fragments have been sequenced and used to construct expression vectors which are tightly regulated in the expression of recombinant proteins. Two types of expression vectors have been constructed: (1) autonomously replicating sequences (ABS), and (2) integrative vectors. The integrative expression vectors offer long-term stability in continuous fermentation applications. Applications of these expression vectors have been successfully developed for the production of human necrosis factor, streptokinase, hepatitis surface B antigen and invertase in genetically-engineered *Pichia pastoris*. With the strong alcohol oxidase promoter controlling expression, high levels of specific protein production coupled to high cell density result in increased levels of recombinant protein per fermentor volume. This high yield of recombinant protein results in lower production cost by combining SCP production with other byproducts of commercial value. The application of high cell density process technology to other fermentation scenarios such as sucrose, ethanol, molasses and whey feedstocks has been successfully developed for the production of useful food products (Shay & Wegner, 1985, 1986).

Single cell oil: The production of single cell oil (SCO) was selected as a target for industrial development over 50 years ago. Renewed interest developed in the 1970s due to technological improvements in full-scale SCP production and the possibility of deriving new or added-value products from SCP production. Ratledge has championed and researched the subject of SCO for several years (Ratledge, 1984, 1989). Interest in SCO is related to economic forecasts of increased future demands for plant derived oils in the face of developing shortages on the world market. The bulk of edible oils and technical grade oils (over 80%) is derived from plant seeds, the remainder coming from animal and fish sources. In Europe, up to 10 million tons of oils are imported, with a predicted shortage of 18 million tons/year by the early 1990s. The pricing for such oils is highly variable, supply and demand fluctuating widely from year to year. However, worldwide demand is increasing on an annual basis and traditional sources of oils may well be unable to supply the volume demand. Accordingly, SCO coupled to SCP production may well have a future (Moreton, 1989).

SCO production based on current pricing and associated economics offers only a negative forecast with respect to economic viability. If one assumes, however, that significant technological achievements can be made through basic research to increase SCO yields, then perhaps economic viability would more closely approach parity. Precedences do exist for the improvement of SCP production, as well as of other microorganisms, where doubling and tripling of the cellular lipid content was achieved through genetic manipulations. We have seen how substrate choice significantly influenced the total lipid content of hydrocarbon-utilizing microorganisms (see section 2.1). Further examples of producing high value-added oils concern reports on the synthesis of γ-linolenic acid from decane-grown *Mortiella isabellina* and *M. vinacea* (Yokochi & Suzuki, 1984) and the commercialization of γ-linolenic acid production by fungi (Suzuki, 1989). Such high value specialty oils could well stimulate further commercial development in the production of SCO.

Microorganisms involved in considerations of SCO are the oleaginous yeasts and fungi (Ratledge, 1989). The oleaginous yeasts as a group are relatively small, while the oleaginous fungi represent a considerably larger number of organisms. Additionally, the fungi synthesize a considerably broader diversity of fatty acids as compared to the oleaginous yeasts, possibly impacting on their potentially greater economic value. SCO is generally considered to represent lipid accumulation in the form of triacylglycerols. Oleaginous microorganisms accumulate such lipids in ranges varying from 20% to over 85% of the total biomass. The attractiveness of SCO centers on the nature of the fatty acids in the triacylglycerols. Most yeast fatty acids are not particularly unusual and are generally similar to fatty acids present in seed oils. Yeast fat, with a high stearate content, is a potential target of interest due to its similarity to cocoa butter—the most expensive of all bulk commercial oils at $5500/ton.

The advantages of using fungi rather than yeasts for SCO production stem from the ability of the fungi both to utilize a wider diversity of waste materials than yeasts and to produce a wider spectrum of fatty acids, particularly polyunsaturated fatty acids. Generalizations that can be stated about the fatty acids of oleaginous microorganisms are as follows.

(1) Fungi exhibit more diversity of fatty acids than yeast.
(2) Higher amounts of the polyunsaturated fatty acids, 18:2 and 18:3, occur in the fungi than in yeasts.
(3) The genus *Mucorales* uniquely contains γ-linolenic acid[18:3(ω6)] as the sole 18:3 fatty acid rather than the usual α-linolenic acid[18:3(ω3)].
(4) The highly polyunsaturated fatty acids such as arachidonic acid[20:4(ω6)] and eicosapentaenoic acid[20:5(ω3)] are found in the fungi but not in the yeasts.
(5) A number of unusual fatty acids including branched-chain and β-hydroxy fatty acids occur in the fungi but not in the yeasts.

Biotechnological exploitation of certain oleaginous fungi is under development by J. & E. Sturge (UK) and Idemitsu Co. (Japan) for the production of SCO rich in γ-linolenic acid.

6 MICROBIAL TREATMENT OF PETROLEUM WASTES AND HAZARDOUS CHEMICALS

The fate and effect of oil in the marine, freshwater and soil environments has been extensively studied and reviewed in detail (Atlas, 1984). The general pattern that characterizes the interactions that exist between oil and the microflora has been recognized with the conclusion that microorganisms are the major biological agents responsible for oil degradation in marine, freshwater and soil ecosystems. Petroleum-degrading microorganisms are naturally occurring populations which increase significantly in population density in those areas receiving oil exposure. Ecological niches in need of further study are tropical ecosystems, particularly mangrove swamps. Estuarine environments are biologically, hydrographically and chemically complex systems with comparatively few data available on the effect of oil on the microbiology of the system. Also, the relationship that exists between microbial oil degradation and the protozoa, microfauna and zooplankton requires further investigations.

Oil pollution of ecosystems results from accidental spills and seeps, deliberate dumping of petroleum wastes, and natural gas and petroleum seeps. Soils are abundantly rich in microbial life with non-contaminated soils containing a heterogeneous microflora, including naturally-occurring hydrocarbon-degrading microorganisms which impart a large hydrocarbon degradation potential to the soil. The addition of petroleum to soil selectively enriches that segment of the total microbial population which is able to adapt to and utilize the new substrate. Generalizations concerning the microbial response of soils to oil contamination are difficult to formulate, primarily due to the diversity of soils and oils studied in different geographical and climatic regions, making comparisons difficult. However, certain trends emerge which appear representative of most conditions. Microbial numbers and activity increase in contaminated soils up to 5% oil contamination levels. The potential toxicity or inhibitory effect of oil does not necessarily occur in soils where there exist favorable conditions for biodegradation. Toxic components tend to volatilize rapidly or become immobilized by adsorption to soil particles. Bacteria and fungi appear to represent the principal agents of petroleum degradation, although the relative contribution made by each group is not clear.

It is important to recognize that natural ecosystems harbor an enormous biological potential to degrade waste and hazardous materials, providing proper management of such systems occurs. Continued research to identify and isolate this metabolic potential offers significant opportunities for the development of novel biotechnologies in the treatment of toxic and hazardous wastes. An estimated 40–45 million tons of hazardous waste are produced annually in the United States. Hazardous wastes, whether from oil refineries or other industrial or domestic sources, are treated similarly in that suspended solids are separated by settling and dissolved organics are partially oxidized and partially converted to settleable solids. Refinery wastes are often complex mixtures consisting of emulsified oil, oil sludges, residues from tank bottoms, accumulated sediments from oxidation ponds, oil-contaminated soils and spent lubricants. Most, if not all, of these waste materials have been classified as hazardous wastes and their disposal is subject to strict regulations.

Land treatment and incineration are widely used disposal procedures for wastes occurring from refinery operations. Alternative disposal practices are landfill and impoundment, deep well injection and solidification. Land treatment is a widely used means for disposal of petroleum wastes (Bartha, 1986). This technique relies on soil microorganisms to degrade petroleum wastes and represents a procedure that is cost-effective. The American Petroleum Institute summarized the results of extensive research into land treatment procedures in the form of an operational manual. The desirable features associated with land treatment for disposal of hazardous wastes are that it involves relatively inexpensive natural biological processes and there appears to be no horizontal or vertical movement of waste into the adjacent areas or associated groundwaters. The negative aspects surrounding disposal of hazardous wastes by land treatment are that the process is slow and incomplete. Typically, only 50–70% of the applied waste is degraded before the next application occurs. Additionally, an accumulation of polycyclic aromatic hydrocarbons occurs following repeated applications, with total biodegradation requiring years rather than months. Complete or partial biodegradation leaves an accumulation of heavy metals in the soil so that land treatment areas are not used for agronomic crop production or livestock grazing.

Other forms of biological treatment for refinery effluents are oxidation ponds, activated sludge treatments, trickling filters and rotating biological contactors. The biological contactors require less energy than the activated sludge process, do not produce aerosols, are free of the 'bulking' problems and yield sludge of excellent settling properties. Considerable attention has been drawn to genetic engineering for solutions to problems of bioremediation in the environment. It is cautioned that the principles of physiological ecology need to interface with and complement the genetic engineering efforts for the development of effective solutions to environmental pollution problems of all types.

REFERENCES

Abbott, B. J. & Gledhill, W. E. (1971). *Adv. Appl. Microbiol.*, **14,** 249.

Atlas, R. M. (ed.) (1984). *Petroleum Microbiology,* Chapters 8, 9, 10, 11, 12, 14 and 15. Macmillan, New York.

Attar, A. & Hendrickson, G. G. (1982). In *Coal Structure,* ed. R. A. Meyers. Academic Press, New York, p. 132.

Bartha, R. (1986). *Microbiol. Ecology,* **12,** 155.

Bird, C. W. & Molton, P. M. (1972). *Top. Lipid Chem.*, **6,** 125.

Buhler, M. & Schindler, J. (1984). In *Biotechnology,* Vol. 6a, eds H.-J. Rehm & G. Reed. Verlag Chemie, Weinheim, Germany, p. 329.
Cain, R. B. (1980). In *Hydrocarbons in Biotechnology,* eds D. E. F. Harrison, I. J. Higgins & R. Watkinson. Heyden & Sons, London, p. 99.
Cerniglia, C. E. (1984). In *Petroleum Microbiology,* ed. R. M. Atlas. Macmillan, New York, p. 99.
Cooper, D. G. & Zajic, J. E. (1980). *Adv. Appl. Microbiol.,* **26,** 229.
deSmet, M., Eggink, G., Witholt, B., Kingma, J. & Wynberg, H. (1983). *J. Bacteriol.,* **154,** 870.
Doi, Y., Tamaki, A., Kunioka, M. & Soga, M. (1987). *J. Chem. Soc., Chem. Commun.,* 1635.
Drozd, J. W. (1980). In *Hydrocarbons in Biotechnology,* eds D. E. F. Harrison, I. J. Higgins & R. Watkinson. Heyden & Sons, London, p. 75.
Dugan, P. (1986). In *Workshop on Biotechnology for the Mining, Metals Refining and Fossil Fuel Industries,* eds H. L. Erhlich & D. S. Holmes. John Wiley, New York, p. 185.
Dunlop, K. R. & Perry, J. J. (1975). *J. Bacteriol.,* **121,** 250.
Ellis, D. B. *et al.* (1985). *Mol. Cell Biol.,* **5,** 1111.
Finnerty, W. R. & Robinson, M. (1986). In *Workshop on Biotechnology for the Mining, Metals Refining and Fossil Fuel Industries,* eds H. L. Erhlich & D. S. Holmes, John Wiley, New York, p. 205.
Finnerty, W. R. & Singer, M. E. (1983). *Biotechnol.,* **1,** 47.
Finnerty, W. R. & Singer, M. E. (1984). *Dev. Ind. Microbiol.,* **45,** 31.
Fukui, S. & Tanaka, A. (1980). *Adv. Biochem. Eng.,* **17,** 1.
Geigert, J., Neidleman, S. L. & DeWitt, S. H. (1984). *J. Am. Oil Chem. Soc.,* **61,** 1747.
Given, P. (1985). In *Coal,* Vol. III, ed. P. Given. Academic Press, New York, p. 63.
Hartdegan, F. J., Coburn, J. M. & Roberts, R. L. (1984). *Chem. Eng. Progr.,* **80,** 63.
Hill, F. F., Venn, I. & Lukas, K. L. (1986). *Appl. Microbial Biotechnol.,* **24,** 168.
Hou, C. T. & Laskin, A. I. (1976). *Dev. Ind. Microbiol.,* **17,** 351.
Isbister, J. B. & Kobylinski, E. A. (1985). In *Coal Science and Technology: Processing and Utilization of High Sulfur Coals,* ed. Y. A. Attia. Elsevier, New York, p. 627.
Jenneman, G. E. *et al.* (1983). *Dev. Ind. Microbiol.,* **24,** 485.
Kaneyuki, H., Deno, H., Hiratsuka, J., Matsuyoshi, T. & Furukawa, T. (1980). *J. Ferment. Technol.,* **58,** 405.
Kilbane, J., Maka, A. & Akin, C. (1989). Paper presented at First Electric Power Research Institute Symposium, Palo Alto, California, May 1989.
Kinoshita, S. & Nakayama, K. (1978). In *Economic Microbiology,* Vol. 4, ed. A. H. Rose. Academic Press, New York, p. 209.
Kodama, K. *et al.* (1970). *Agr. Biol. Chem.,* **34,** 1320.
Kosaric, N., Cairns, W. L. & Gray, N. C. C. (eds) (1987). *Biosurfactants and Biotechnology.* Marcel Dekker, New York.
Laborde, A. L. & Gibson, D. T. (1977). *Appl. Environ. Microbiol.,* **34,** 783.
Makula, R. A. & Finnerty, W. R. (1968). *J. Bacteriol.,* **95,** 2102.
Makula, R. A., Lockwood, P. J. & Finnerty, W. R. (1975). *J. Bacteriol.,* **121,** 250.
Malik, K. A. (1978). *Process Biochem.,* **13,** 10.
May, S. W. (1979). *Enzyme Microbiol. Technol.,* **1,** 15.
Miall, L. M. (1978). In *Economic Microbiology,* Vol. 2, Ed. A. H. Rose. Academic Press, New York, p. 47.
Modrzakowski, M. C. & Finnerty, W. R. (1980). *Arch. Microbiol.,* **126,** 285.
Modrzakowski, M. C., Makula, R. A. & Finnerty, W. R. (1977). *J. Bacteriol.,* **131,** 92.
Monticello, D. & Finnerty, W. R. (1985). *Ann. Rev. Microbiol.,* **39,** 371.
Monticello, D., Bakker, D., Schell, M. & Finnerty, W. R. (1985*a*). *Appl. Environ. Microbiol.,* **49,** 761.

Monticello, D., Bakker, D. & Finnerty, W. R. (1985*b*). *Appl. Environ. Microbiol.*, **49,** 756.
Moreton, R. S. (1989). In *World Conference on Biotechnology for the Fats and Oils Industry,* ed. T. H. Applewhite. American Oil Chemists Society, Urbana, Illinois p. 148.
Phillips Petroleum Co. (1976). US Patent 3975234.
Ratledge, C. (1980). In *Hydrocarbons in Biotechnology,* eds D. E. F. Harrison, I. J. Higgins & R. Watkinson. Heyden & Sons, London, p. 133.
Ratledge, C. (1984). In *Biotechnology in the Oils and Fats Industry,* ed. C. Ratledge, P. S. S. Dawson & J. B. M. Rattray. American Oil Chemists Society, Urbana, Illinois, p. 119.
Ratledge, C. (1989). In *Microbial Lipids,* Vol. 2, eds C. Ratledge & J. Wilkinson. Academic Press, New York, p. 567.
Sagardia, F. *et al.* (1975). *Appl. Microbiol.*, **27,** 722.
Shay, L. K. & Wegner, G. (1985). *Food Technol.*, **39,** 61.
Shay, L. K. & Wegner, G. (1986). *J. Dairy Sci.*, **69,** 676.
Shennan, J. L. (1984). In *Petroleum Microbiology,* ed. R. M. Atlas. Macmillan, New York, p. 643.
Singer, M. E. & Finnerty, W. R. (1984*a*). In *Petroleum Microbiology,* ed. R. M. Atlas. Macmillan, New York, p. 1.
Singer, J. T. & Finnerty, W. R. (1984*b*). In *Petroleum Microbiology,* ed. R. M. Atlas. Macmillan, New York, p. 299.
Singer, M. E., Tyler, S. & Finnerty, W. R. (1985). *J. Bacteriol.*, **162,** 162.
Speight, J. G. & Moschopedis, S. E. (1981). In *Chemistry of Asphaltenes,* eds J. W. Bunger & N. C. Li. American Chemical Society, Washington, DC, p. 1.
Suzuki, O. (1989). In *World Conference on Biotechnology for the Fats and Oils Industry,* ed. T. H. Applewhite. American Oil Chemists Society, Urbana, Illinois, p. 110.
Uemura, N., Taoka, A. & Takagi, M. (1989). In *World Conference on Biotechnology for the Fats and Oils Industry,* ed. T. H. Applewhite. American Oil Chemists Society, Urbana, Illinois, p. 148.
Uhl, W., Hone, H.-J., Beyer, M. & Klein, J. (1989). *Biotechnol. Bioeng.*, **34,** 1341.
Willey, C., Iwao, M., Castle, R. N. & Lee, M. L. (1981). *Anal. Chem.*, **53,** 400.
Yamada, K., Minoda, K., Kodama, S., Nakatani, S. & Akasaki, T. (1968). *Agr. Biol. Chem.*, **7,** 840.
Yi, Z.-H. & Rehm. H.-J. (1982*a*). *Eur. J. Appl. Biotechnol.*, **14,** 144.
Yi, Z.-H. & Rehm, H.-J. (1982*b*). *Eur. J. Appl. Biotechnol.*, **14,** 254.
Yi, Z.-H. & Rehm, H.-J. (1982*c*). *Eur. J. Appl. Biotechnol.*, **15,** 175.
Yi, Z.-H., & Rehm, H.-J. (1982*d*). *Eur. J. Appl. Biotechnol.*, **16,** 1.
Yokochi, T. & Suzuki, O. (1984). Japanese Patent 59 130191-A.
Zajic, J. E. & Mahomedy, A. Y. (1984). In *Petroleum Microbiology,* ed. R. M. Atlas. Macmillan, New York, p. 221.
Zajic, J. E. & Panchal, C. J. (1976). *CRC Crit. Rev. Microbiol.*, **5,** 39.

INDEX